THE RADIATING ATMOSPHERE

ASTROPHYSICS AND SPACE SCIENCE LIBRARY

A SERIES OF BOOKS ON THE RECENT DEVELOPMENTS
OF SPACE SCIENCE AND OF GENERAL GEOPHYSICS AND ASTROPHYSICS
PUBLISHED IN CONNECTION WITH THE JOURNAL
SPACE SCIENCE REVIEWS

VOLUME 24

THE RADIATING ATMOSPHERE

PROCEEDINGS OF A SYMPOSIUM ORGANIZED BY THE SUMMER ADVANCED STUDY INSTITUTE, HELD AT QUEEN'S UNIVERSITY, KINGSTON, ONTARIO, AUGUST 3–14, 1970

Edited by

B. M. McCORMAC
Lockheed Palo Alto Research Laboratory,
Palo Alto, Calif., U.S.A.

D. REIDEL PUBLISHING COMPANY

DORDRECHT-HOLLAND

Library of Congress Catalog Card Number 70–154742
ISBN-13: 978-94-010-3092-2 e-ISBN-13: 978-94-010-3090-8
DOI: 10.1007/978-94-010-3090-8

Softcover reprint of the hardcover 1st edition 1971

PREFACE

This book contains the lectures presented at the Summer Advanced Study Institute, 'Aurora and Airglow, 1970' which was held at Queen's University, Kingston, Ontario, during the period August 3 through 14, 1970. One-hundred twenty-two persons from 15 different countries attended the Institute.

The authors and the publisher have made a special effort for rapid publication of an up-to-date status of upper atmospheric emissions, which is an ever-changing research area. Special thanks are due to the lecturers for their diligent preparation and excellent presentations. The individual lectures and the published papers were deliberately limited; the author's cooperation in conforming to these specifications is greatly appreciated. The contents of the book are organized by subject area rather than in the order in which the papers were presented during the Institute. Many thanks are due to Drs Jacques Blamont, B. W. Currie, Alv Egeland, Donald M. Hunten, Lawrence Kavanagh, D. C. Rose, Lance Thomas, Alister Vallance Jones, and Martin Walt who served as session chairmen during the Institute and contributed greatly to its success by skillfully directing the discussion period in a stimulating manner after each lecture. Dr Alister Vallance Jones and the Summary Panel worked hard to prepare an excellent summary of various aspects of aurora and airglow at the end of the Institute.

Many persons contributed to the success of the Institute. The Institute co-chairman, Dr Alister Vallance Jones, was most helpful during all phases of the Institute preparation and planning, and especially in selecting the site in Canada. Drs Donald M. Hunten, Günther Lange-Hesse, Anders Omholt, Albert G. Opp, Lance Thomas, Alister Vallance Jones, Martin Walt, and Gilbert Weill were especially helpful in preparing the technical program. Mr James Brownlee, Department of Conferences, Queen's University, was especially cooperative in providing very excellent living and conference facilities. The assistant editor, Mrs Diana R. McCormac, checked the manuscripts and proofs and worked hard to achieve a uniform style in this book.

Direct financial support was provided the Institute by: the Advanced Research Projects Agency, the Defense Atomic Support Agency, Lockheed Palo Alto Research Laboratory, and the Office of Naval Research. ESSA Research Laboratories and the National Research Council of Canada also sponsored the Institute. The European Space Research Organization provided a number of scholarships.

BILLY M. McCORMAC

Palo Alto
March 1971

TABLE OF CONTENTS

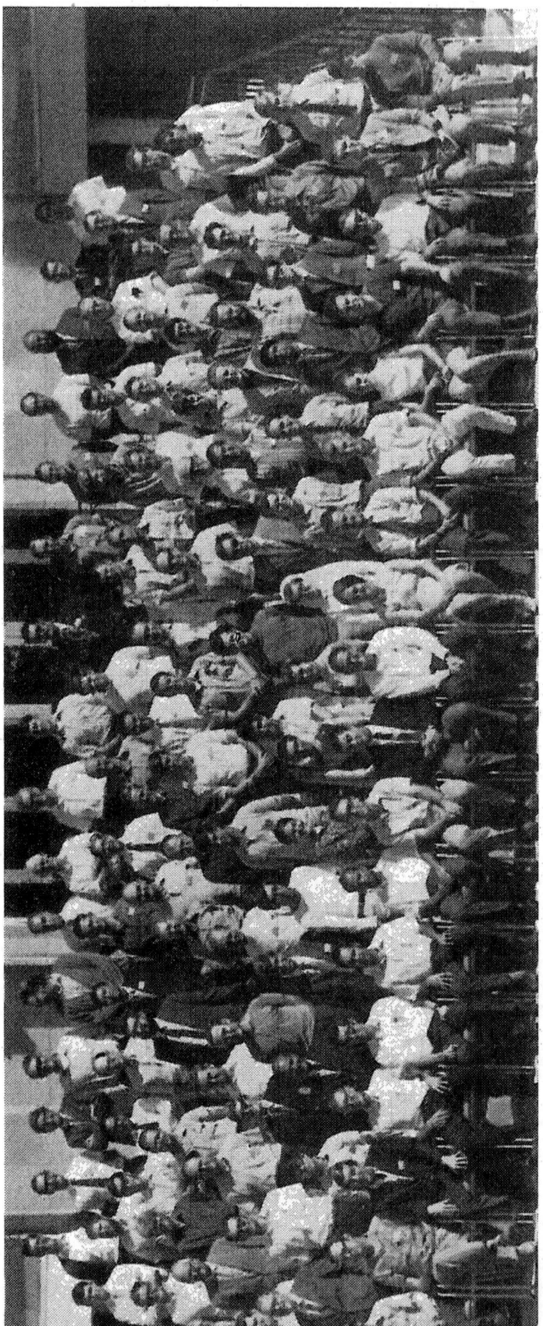

Participants

1. Sam LaValle
2. J. Ronald Burrows
3. John Miller
4. Gary Culp
5. Melvin Heaps
6. Kay Baker
7. Vasant Agashe
8. Helge Pettersen
9. Gerard Thuillier
10. Martin Walt
11. Catherine Vallance Jones
12. Alister Vallance Jones
13. Billy McCormac
14. Diana McCormac
15. Efstathios Kamaratos
16. L. Herman
17. Hans Lauche
18. Michel Hamelin
19. David Nicholls
20. Wayne Evans
21. Guy Moreels
22. Mukhtar Ahmed
23. Gordon G. Shepherd
24. Richard Johnson
25. Margaret Heeran
26. Doris Jelly
27. Jacques Vigneron
28. Jeannine Christophe
29. A. Levasseur
30. Dieter Beran
31. Axel Korth
32. G. M. Shah
33. Anthoney Lui
34. Harry Sullivan
35. James McElaney
36. Giovanni Gregori
37. William Sawchuk
38. Hugh Wood
39. Donald Hunten
40. Donald McEwen
41. Jean-Claude Gérard
42. Rainer Kist
43. R. Pottelette
44. Paul Simon
45. William Sheldon
46. Eric Carmen
47. D. R. Snelling
48. Edward Llewellyn
49. Walter Brown
50. A. Egeland
51. J. R. Sternberg
52. David Marsh
53. Brian Whalen
54. Lawrence Kavanagh
55. Steve Mende
56. David Winningham
57. Lawrence Choy
58. Richard Chappell
59. Walter Williams
60. Günther Lange-Hesse
61. David Bleil
62. Hans Mayr
63. Roger Smith
64. H. N. Rundle
65. Michael Isherwood
66. John Evans
67. Eric Pemberton
68. Guido Visconti
69. Johannes Steyn
70. Lawrence Spitz
71. L. E. Montbriand
72. S. R. Pal
73. John Noxon
74. Akira Hasegawa
75. George Stuart
76. E. C. Zipf
77. T. R. Hartz
78. Jeffrey Pearce
79. Robert Schaeffer
80. Dan Holland
81. Derek Cunnold
82. Rolf Boström
83. James Coleman
84. Lance Thomas
85. Adolph Johanson
86. Duncan Bryant
87. R. L. Gattinger
88. James Bradbury
89. Thomas Hallinan
90. Roland Johnson
91. Kenneth McWatters
92. Michael Schwitters
93. Frank Bunn
94. Robert Sears
95. David Cartwright
96. R. D. Rawcliffe
97. R. S. Unwin
98. A. G. McNamara
99. John Meriwether
100. Neil Davis
101. Charles Wilson
102. Donald McDiarmid
103. Michael Watson
104. Robert Eather
105. Luke Sennema
106. Fokke Creutzberg

PART I

ATMOSPHERIC AIRGLOW EMISSIONS

AIRGLOW – INTRODUCTION AND REVIEW

DONALD M. HUNTEN

Kitt Peak National Observatory, Tucson, Ariz., U.S.A.

1. Introduction

The processes leading to optical emissions in the airglow (nightglow, dayglow, twilight) are defined and discussed. Review papers and more recent literature are surveyed. An extensive table summarizes the known emissions, their intensities, locations, and production processes, for the Earth; the inner planets are discussed briefly.

The subject of airglow is considerably older than the space age, and it no longer occupies the high place it once did in the interest of atmospheric physicists (now known as aeronomers). Nevertheless, this interest is still present, though the reasons have changed somewhat; the new tools offered by rockets and satellites have enormously expanded the observational side of the subject. Airglow still offers the possibility of remotely monitoring certain processes, not only from the ground but now also from satellites and space probes. Local measurements from rockets are important too; they give direct information on processes not easily studied otherwise, especially the production, transformation, and loss of excited metastable atoms and molecules.

Curiously, the definition of airglow is negative, obtained by excluding aurora, Rayleigh and other nonselective scattering and thermal emission. This situation is not really surprising because airglow encompasses a variety of production processes, some of which are not fully understood. The original energy source is solar radiation; the prompt emission is dayglow, and emission with a sufficient delay is nightglow. Twilight is dayglow observed from the night side, but it has many problems and a history of its own, and is conveniently considered as a third subject.

2. Excitation Processes

Table I shows the known emissions in order of increasing wavelength. For band systems the wavelength given usually refers to the 0–0 band, but in a few cases the brightest has been chosen instead. The dominant excitation processes are given for each emission; where a blank is left, no observations exist, even if the presence of the emission can reasonably be inferrred. We next describe each of the excitation processes in turn; other summaries are given by Wallace and McElroy (1966), Hunten (1967), and Dalgarno *et al.* (1969).

Resonance Scattering by an atom or molecule refers strictly to an absorption into the lowest excited state for an allowed transition, followed by re-emission. Thus, in

B. M. McCormac (ed.), The Radiating Atmosphere, 3–16. *All Rights Reserved.*

TAB

Known airglow emissions for the earth. As explained in the text, production processes are R, resonance scat-
Production rate factors are g and

λ(Å)	Emitter State	Day			Twilight		Night	
		Intensity	Height	Process	Intensity	Height	Intensity	Height
304	He^+	Present		R			(4.8)	
584	He	Present		R			(12)	
834	O^+	Present		R				
1025	H Lyβ	Present	200–10^4	R			10 R	200
1200	N(^4P)	400 R	180	R?				
1216	H Lyα	6 kR	100–10^5	R			2 kR	100–10^5
1302-4-6	O(^3S)	7.5 kR	190	eFR				
1356	O(^5S)	350 R	140	e				
1300–1500	$N_2(a^1\Pi_g)$	Present		e				
1493, 1744	N	Present		e				
2000–4000	$N_2(A^3\Sigma_u^+)$	Present		e				
2160 etc.	NO($A^2\Sigma^+$)	1 kR	70–150	R				
3371 etc.	$N_2(C^3\Pi_u)$	900 R	130 up	e				
2600–3800	$O_2(A^3\Sigma_u^+)$						600 R	90
3466	N(^2P)	Present		e				
3889	He(^3P)			R	1 R	> 400?		
3914 etc.	$N_2^+(B^2\Sigma_u^+)$	2.0 kR	150	RF	200–500 R	300	< 1 R	
3933,68	Ca^+(^2P)				⩽ 100 R	80–200		
4368	O(4^3P)				1 R			
5200	N(^2D)	90 R	~ 200	I	10 R		1 R	~ 250
5000–6500	NO_2?						1 R/Å	~ 90
5577	O(^1S)	3.0 kR	90; 175	Ce	400 R	200?	250 R	90; 300
5893	Na(^2P)	30 kR	92	R	1–4 kR	92	20–150 R	~ 92
6300, 64	O(^1D)	2–20 kR	250	FIe	1-kR	300	10–500 R	300
6563	H(3^2P)						3 R	200
6708	Li(^2P)				10–1000 R	~ 90		
7619 etc.	$O_2(^1\Sigma)$	300 kR	40–120	RFT			6 kR	~ 80
7699	K(^2P)				40 R	~ 90		
7774, 8446	O	1.6, 1.1 kR	~ 150	e				
10510 etc.	$N_2(B^3\Pi_g)$	900 R	150	e				
10830	He(^3P)				3 kR	500		
11036 etc.	$N_2^+(A^2\Pi_u)$	4 kR	150	RF				
12700 etc.	$O_2(^1\Delta)$	20 MR	50	F	5 MR	80	80 kR	90?
28000 etc.	OH($v \leqslant 9$)	4.5 MR		C			4.5 MR	90

the familiar example of the Na D lines strict resonance scattering occurs only in D_1. In a more extended usage any scattering from the ground state without change of frequency qualifies as resonance (or 'resonance-like') scattering. The strict version does not have a useful application to molecules, with their multiplicity of rotational and vibrational levels.

The convenient concept of *emission-rate factor* or *g-factor* was introduced by Chamberlain (1961). It is the number of photons per second scattered by a single atom or molecule exposed to unattenuated sunlight (that is, at zero optical depth). If the solar photon flux is $\pi\mathscr{F}_\nu$ per unit frequency in the neighborhood of the line,

LE I

tering; *F*, fluorescence; *C*, chemical association; *I*, ionic reactions; *e*, photoelectrons; *T*, excitation transfer. d; h_q is the quenching height

Process	$g\,(\text{sec}^{-1})$	d [source]	h_q, km	Remarks
	1.1×10^{-4}			Nightglow radiation could be either or both
	1.7×10^{-5}			
R	2.6×10^{-6}			
R	2.1×10^{-3}	$(4.5 \times 10^{-8}\,[H_2])$		
	1.0×10^{-4}			
				Lyman-Birge-Hopfield
				Vegard-Kaplan
	4.0×10^{-6}			γ bands; g for 1–0 band
				2nd Positive
C				Herzberg
	0.1			Scatterer is $He(^3S)$
	0.050	$10^{-8}\,[N_2]$	46	1st Negative
	0.3, 0.15			
I	(6×10^{-11})		~ 200	Also quenched by electrons
C				Continuum
C, I	(1×10^{-11})	$3 \times 10^{-8}\,[O_2]$	94	2972 Å (5%)
C	0.80		40	
I	$(4\ 5 \times 10^{-10})$	$5.8 \times 10^{-6}\,[O_2]$	340	
F	2.6×10^{-6}			
	16			May be of artificial origin
C	6.3×10^{-9}	$\leqslant 5 \times 10^{-3}\,[O_3]$	90	Atmospheric
	1.67			
			37	1st Positive
	16.8		~ 400	Scatterer is $He(^3S)$; h_q for its destruction
	0.042	2.8×10^{-8}		Meinel; g, d for 1–0 band (9200 Å)
C?	(9.4×10^{-11})	$9.5 \times 10^{-3}\,[O_3]$	75	IR Atm; 0–1 band 1.58 μ; Noxon bands 1.9 μ
C				Meinel; 4.5 μ to 3816 Å

the g-factor for strict resonance scattering is

$$g = \pi\mathscr{F}_\nu \int \alpha_\nu \, d\nu = \pi\mathscr{F}_\nu \frac{\pi e^2}{mc} f = \pi\mathscr{F}_\nu \frac{\lambda^2}{8\pi} \frac{\varpi_2}{\varpi_1} A \,. \qquad (1)$$

Here α_ν is the absorption cross section, f is the oscillator strength, the ϖ's are statistical weights, and A is the transition probability; the integral is over the line of interest. Convenient tabulations of f and $\pi\mathscr{F}_\nu$ are given by Allen (1963). Many of the g's in Table I are from Chamberlain (1961) or Hunten (1967); the rest are explained in the text or have been newly calculated.

For resonance-like transitions it is often necessary to multiply Equation (1) by a branching ratio. A familiar example is Lyβ of H; the upper state radiates Hα 11.56% of the time (a fluorescent process) and Lyβ the remaining 88.44%. With molecular bands it is necessary to consider simultaneously several vibrational transitions, as well as the rotational structure. The procedure is tedious and not always straightforward, but can be carried through. Other modifications must be made if the solar radiation is attenuated or if the excited state is collisionally quenched as discussed below.

It must be remembered that the g-factors are tabulated for direct sunlight as the only source. At wavelengths above 3000 Å there is an additional radiation field present in the daytime due to earth albedo. Although the correction factor could in principle be as large as 3 for a perfectly diffusing surface, 1.5 is more usual. A detailed discussion is given by McElroy and Hunten (1966).

Fluorescence refers to processes in which the emitted wavelength differs from the exciting wavelength. It includes fluorescent scattering, as with Hα, but a more prominent case is the radiation accompanying photodissociation or photoionization. For example, dissociation of both O_2 and O_3 produces one atom in the ^1D state which emits the 'auroral' red lines, and ionization of N_2 can produce excited ions emitting the Meinel and First Negative band systems. A *production-rate factor*, d, can be defined in analogy to g, as the number of productions per molecule-second at zero optical depth:

$$d = \int \pi \mathscr{F}_\nu \alpha_\nu \, d\nu . \tag{2}$$

The integral is over the wavelength region relevant to the particular process. Because this region can be hundreds of Ångstroms wide, the total production in an atmospheric column can be much larger than for resonance. But the cross section α_ν is so much smaller that typical values of d and g are in the same range.

Albedo is not a factor for most fluorescent processes; for absorption continua, in particular, the atmosphere is almost always optically thick and therefore has zero albedo.

Chemical Association operates day and night. The important reactants are O, O_3, H, and probably NO: the latter two can be important, despite their small concentrations, as catalysts that are regenerated and used over and over. Thus, the basic store of energy is dissociated oxygen, produced during the day by the wavelength region 1000 to 1750 Å. (Wavelengths up to 2400 Å produce dissociation, but at heights lower than concern us here.) Averaged over the globe, O_2 molecules are dissociated at a rate of 5×10^{11} cm^{-2} sec^{-1}; the association of the atoms could provide photons at the same rate, which corresponds to 500 kR.

The Rayleigh (R) corresponds to an emission of 10^6 photons cm^{-2}, in simple cases (Chamberlain, 1961) (Appendix II). The threshold for practical detection in the visible is about 1 R. For a layer 10 km thick it corresponds to a volume emission rate of 1 photon cm^{-3} sec^{-1}.

The principal reactions are:

$$O + O + M \rightarrow O_2 + M + 5.17 \text{ eV} \quad (3)$$

$$H + O_3 \rightarrow OH + O_2 + 3.34 \text{ eV} \quad (4)$$

$$O + O_3 \rightarrow O_2 + O_2 + 4.06 \text{ eV}. \quad (5)$$

Ozone and H are produced in

$$O + O_2 + M \rightarrow O_3 + M + 1.10 \text{ eV} \quad (6)$$

$$OH + O \rightarrow O_2 + H + 0.72 \text{ eV}. \quad (7)$$

The net effect is always association of O atoms. Hydrogen is not used up; a very small supply from photolysis of H_2O is sufficient.

Reaction (4) gives enough energy to excite the 9th, but not the 10th, vibrational level of OH, as observed both in nightglow and laboratory. Reaction (3), the 'Chapman reaction', can excite all the nightglow bands of O_2 and lines of O (the latter if M=O or by subsequent excitation transfer). The relative roles of Reactions (3) and (5) are not established. All three reactions have their maximum rates in the 70 to 90 km region, and indeed most nightglow emissions are found near 90 km.

Ionic Reactions operate day and night also; but at night fewer of them are important, and they are confined to the region above 200 km. Dissociative recombination can leave one or both atoms excited:

$$O_2^+ + e \rightarrow O + O + 6.96 \text{ eV} \quad (8)$$

$$NO^+ + e \rightarrow N + O + 2.74 \text{ eV}. \quad (9)$$

The three states (emissions) expected are $O(^1S)$ (5577 Å), $O(^1D)$ (6300 Å), and $N(^2D)$ (5200 Å); all are observed, day and night, in the expected region. Ion-neutral reactions are also potentially important, though no case is firmly established as a light source; an example is

$$N_2^+ + O \rightarrow NO^+ + N + 3.08 \text{ eV}. \quad (10)$$

At night N_2^+ is essentially absent because there is no strong source; O^+ remains from the daytime and produces NO_2^+ and O_2^+ as secondaries.

Photoelectrons are produced in daytime with a mean energy of 10 eV; in slowing down they produce essentially a weak auroral spectrum whose intensity is similar to that of an aurora at the threshold of visibility (International Brightness Coefficient I). They also raise all the ambient ionospheric electrons to a temperature much greater than the gas temperature. At times these *hot electrons* can excite low-lying states, especially $O(^1D)$. States that can only be reached by a spin change are excited much more efficiently by electrons than by photons. Included are the low states of O and N, $O(^5S)$ (1356 Å), and N_2 $(A, B^3\Pi)$ (First and Second Positive bands). A special case is the lowest triplet state of He, (2^3S), the ground state of 'orthohelium'. When produced by electron impact, it can scatter lines at 10830 Å and 3889 Å, which have been observed in twilight.

Photoelectrons can arrive in a dark atmosphere when the magnetic conjugate point is sunlit. These effects are discussed by Noxon (1971).

Transfer of Excitation has been studied in the laboratory for decades under the name "sensitized fluorescence" (cf., Mitchell and Zemansky, 1934). But only recently have we had good evidence of such processes in the atmosphere. By far the best documented is

$$O(^1D) + O_2 \rightarrow O + O_2(^1\Sigma). \tag{11}$$

This reaction is a strong source of O_2 Atmospheric emission in the 100 km region. There have been many suggestions about transfer from vibrationally excited N_2 and O_2, but quantitative discussion is still impossible because the lifetimes of these levels in the atmosphere are unknown. Reactions (3) to (5) can in principle produce O_2 in several metastable electronic states, as well as vibrationally excited O_2 and OH. All these states are potential candidates as sources of transferred energy.

Excitation transfer is both a source and a sink of excitation, and is a special case of *quenching*, which is important for many metastable species. A survey of quenching coefficients has been given by Zipf (1969); an earlier review by Hunten and McElroy (1966) has been largely superseded, but does contain some additional material. Quenching rates by electrons have recently been obtained by Henry and Williams (1968). In Table I the quenching rates are represented indirectly by the *quenching height*, h_q, at which the rate of quenching is equal to the spontaneous decay rate. More generally, the ratio of emission to production is given by the *quenching factor*

$$Q = \frac{A}{A + k_q[Q]}, \tag{12}$$

where k_q is the quenching rate coefficient and $[Q]$ is the concentration of quenching species. If there is more than one significant quencher, the term $k_q[Q]$ is replaced by a summation. At the height h_q we have $k_q[Q]=A$, and $Q=\frac{1}{2}$.

Table I shows a very wide range of quenching heights, and only a loose correlation with the radiative lifetime. The most 'quenchable' radiation is the O red lines, with a lifetime of 110 sec and a quenching height of 340 km; the Infrared Atmospheric bands of O_2 have a lifetime of 1 hr and a quenching height of 75 km.

3. Recent Developments

In this section we attempt to bring up to date the reviews of Hunten (1967), Noxon (1968), Krassovsky and Shefov (1965), and Roach (1963). There are shorter reviews by Gadsden (1967), Noxon (1967), and several more in the present volume (Broadfoot, 1971; Llewellyn and Evans, 1971; Noxon, 1971; Pearce, 1971; Rundle, 1971). Broadfoot and Kendall (1968) have published an atlas of the nightglow spectrum. We discuss only those emissions for which there have been significant developments in the last few years. The bibliographical coverage is not complete: use is made of recent reviews and research papers that refer to earlier work.

The far UV lines (Lyβ and shorter) have been detected in dayglow by Donahue (1971) and some of them by Bowyer (1970). Young *et al.* (1968) have detected nightglow Lyβ and a shorter-wavelength radiation that could be from either He^+ or He. The Lyβ data have been analyzed, along with Hα, by Meier (1969). Lyβ is probably important in maintaining the nighttime E layer (Tohmatsu and Wakai, 1970).

The 1200 Å line of N was probably observed by Fastie *et al.* (1964), but the data have never been analyzed.

Lyα is really a twilight emission observable in the night sky because of its great height. Observations of the line profile have been reported by Winter and Chub (1967). Mariner 5, on its way to Venus, measured the dayglow out to 10^5 km (Wallace *et al.*, 1970). Mange and Meier (1970) report on data from OGO 3. All this work, as well as that of Kurt and Dostovalov (1968) makes it clear that there is a background of about 500 R, presumably galactic, or from the outer solar system, or both. Thus, many of the earlier attempts to understand geocoronal scattering need drastic revision. Many more satellite measurements have been made and a wealth of data can be expected to appear in the next few years.

The oxygen resonance triplet in dayglow has recently been treated in detail by Strickland and Donahue (1970). Multiple scattering makes the problem very difficult: the optical depth for the strongest line can exceed 10^4. Resonance scattering of solar radiation is found to contribute in the 500 km region, but it is unimportant in the 180 km range, where the observed intensity is greatest. Photoelectrons are probably able to explain this peak, but a contribution by dissociative fluorescence of O_2 is also possible.

Equatorial airglow of the oxygen lines is discussed by Noxon (1971). Pearce (1971) has reported the presence of N_2 bands of the Lyman-Birge-Hopfield and Vegard-Kaplan systems, allowed N lines at 1493 Å, and 1744 Å, and the forbidden 'trans-auroral' line of N, 3466 Å.

Not included in Table I is dayglow fluorescence of the O_2 Schumann-Runge bands, which has possibly been observed by Doherty (1968) with a UV camera onboard an X-15 aircraft. Not enough information is given to permit an assessment of the probability of stray light. Large intensities have been predicted by Tohmatsu (1963) and Barth (1966). But no allowance was made for predissociation of the upper state; Hudson *et al.* (1969) have pointed out that this phenomenon should suppress most of the fluorescence, in addition to providing an important source of O atoms.

The NO bands at low altitudes have been observed by Pearce (1969) and Meira (1970). The constant mixing ratio obtained by Pearce has been shown by Meira to be due to contamination by Rayleigh-scattered continuum. Meira finds a nearly constant NO concentration of (5 to 10) $\times 10^7$ cm^{-3} from 70 to 110 km. Norton and Barth (1970) and Strobel *et al.* (1970) explain the NO production by means of a reaction between $N(^2D)$ and O_2, combined with downward eddy mixing. There is thus a curious link with the 5200 Å dayglow.

The 'green continuum' in the nightglow has long been supposed due to the reaction

$$NO + O \rightarrow NO_2 + h\nu . \qquad (13)$$

From the data of Fontijn *et al.* (1964) the rate coefficient of Reaction (13) per Ångstrom at 5000 Å is 7.6×10^{-21} cm^3 sec^{-1} Å^{-1}. If the O concentration is 2×10^{11} cm^{-3} over an effective thickness of 30 km, Reaction (13) yields 0.25 to 0.5 R Å^{-1}, in reasonable agreement with approximately 1 R Å^{-1} observed.

Hennes (1966) obtained a low-resolution photographic spectrum in the UV by observing the nightglow edge-on from a rocket. He found the O_2 Herzberg bands and the 2972 Å line of O, as well as 5 unidentified bands. Chamberlain (1961) also notes the presence of unidentified features, some of which might be the $B\,^3\Delta_u - a\,^1\Delta_g$ system of O_2.

The excitation of N_2^+ bands in dayglow and twilight has been extensively discussed by Broadfoot (1967, 1968, 1971). The second paper shows the very large effect of tropospheric Rayleigh scattering on twilight measurements in the near UV, and finds the observations to be consistent with photoionization of N_2, followed by resonance-like scattering. Anomalies in the rotational and vibrational distributions, however, suggest an important contribution by charge transfer from metastable O^+:

$$O^+(^2D) + N_2 \rightarrow O + N_2^+ . \qquad (14)$$

The First Negative and Meinel systems are coupled by the absorption of sunlight and subsequent re-emission. The vibrational degree of freedom is uncoupled from the atmosphere, and the ions tend towards an equilibrium with solar radiation. At the lower altitudes this approach is terminated when the ion reacts or recombines. There is some difficulty fitting all the data into a consistent framework. Though the Meinel system has been detected in the dayglow (Wallace and Broadfoot, 1969), better observations could help to clarify the matter.

Broadfoot and Hunten (1966) have set an upper limit of 1 R to the brightness of the 3914 Å band in the nightglow. (Care is necessary to avoid the twilight, which can extend to solar depression angles of 26°.) This result was surprising, because cosmic-ray excitation would be expected to produce about 2.4 R – a suggestion made long ago by Meinel and revived by O'Brien. A resolution of this curious discrepancy is now possible, on the basis of the quenching coefficient reported by Stair and Gauvin (1967), and the corresponding quenching height given in Table I. The peak ionization (for 38° N, a slightly higher latitude than Kitt Peak) is 35 cm^{-3} sec^{-1} at 13 km, where the quenching factor is less than 10^{-2}. The unquenched 2.4 R is thus reduced to 0.04 R.

The impressive work of Weill and collaborators on nightglow 5200 Å has been thoroughly described by Weill (1969). The source has been established as Reaction (9), the quenching has been studied, and a plausible argument given that $N(^2D)$ transfers excitation to produce $O(^1D)$ and therefore produces red-line emission. Another study is due to Hernandez and Turtle (1969).

A paper by Gulledge *et al.* (1968) describes a rocket measurement of nightglow green and red lines at high altitudes. The existence of a second layer of green emission, produced by Reaction (8), was confirmed, and the quenching of the corresponding red emission was observed. Recently, Peterson (1970) has commented that the quenching rate could reasonably be adjusted to agree better with accepted values from dayglow and laboratory. Though Packer *et al.* (1970) objected to this modification, the writer is inclined to agree that it is reasonable.

Tinsley (1970) has observed a weak twilight and post-wilight decay of a permitted O line at 4368 Å.

Sporadic twilight emission of the green line has been reported by Deehr (1969). The review by Hunten (1967) overlooked the source suggested by Deehr, dissociative fluorescence from O_2. The sporadic nature of the phenomenon must presumably be explained by occasional enhancements of thermospheric O_2 concentrations. That the same process produces the red lines was established a decade ago by Barbier, and further observations are given by Deehr and by Noxon and Johanson (1970). Over the observable height range, excitation is produced from O_2 and quenched by N_2; the O_2/N_2 ratio in the upper thermosphere can therefore be obtained.

The tropical nightglow of the red lines has been discussed by Roach (1969), who emphasizes the drain on the ionospheric electron content implied by the observed number of photons in certain events.

The behavior of the Na dayglow is still causing surprises. Zeeman photometers are now giving reliable diurnal variations, and similar results have been obtained by Gadsden and Purdy (1970), Albano *et al.* (1970), and Donahue (1969b). There is a large maximum near noon, the values for low Sun being less by a factor between 2 and 5. Twilight intensities have not been obtained for comparison, but may be somewhat smaller still. Gadsden (1968, 1969b, 1970) has explored some of the properties following from a concentrated source (meteor ablation) at 92 km, and has been able to obtain the very steep topsides that are observed by rocket dayglow probes and in twilight. Unfortunately, his models of meteor ablation do not show the required concentration; nevertheless, the general approach is promising. The model fails to account for the large seasonal abundance variation at high latitudes. The same is true of the large noon maximum, and one is again led to wonder whether another excitation source is supplementing resonance scattering. The 90 km region is just optically thick to Schumann-Runge dissociation of O_2, and the production of $O(^1D)$, for example, varies in the required manner. But this state is too low to excite $Na(^2P)$. Perhaps $O(^1D)$ with kinetic energy can excite vibrational levels of N_2, which in turn excite Na? Against these speculations is the evidence given by Donahue and Meier (1967) that the optical thickness of their Na layer was appropriate to the abundance deduced on the assumption of pure resonance excitation.

The shortcomings of the meteor-ablation model have also been pointed out by Rundle (1971), and he has reminded us of the 'dust-bank' hypothesis of Wallace and Hunten (1968). Particles of 100 Å diameter can be supported by eddy mixing to heights of around 75 km; below this height the particles should be uniformly mixed with the atmosphere. Above 75 km the slope of the distribution becomes steeper and steeper, and at 90 km its gradient corresponds to a scale height of less than 1 km. Such a dust bank would be an admirable source of Na atoms if a reasonable mechanism for evaporation could be found. It is less satisfactory as a sink, but the problem is just as bad for a meteoric source. With the dust bank, the observed correlations with enhanced vertical mixing (especially at high latitudes) are naturally explained.

Gadsden (1969b) has found the twilight abundance of Na to be roughly the same at the South Pole as at middle latitudes. Donahue *et al.* (1970) report a huge enhancement, observed from OGO 6, of the dayglow over each pole for a short time around local midsummer; both 5893 and 5577 Å are found to have a slant emission rate of 1.5 MR (normal values are 300 and 80 kR).

Gault and Rundle (1969) have presented observations of Na, K, and Li twilight over a year or more. The identification of the K line was verified by resonance absorption in a vapor cell. The abundance shows no seasonal variation, in striking contrast to Na. The K g-factor of Table I is greater by a factor of 2.5 than the one used by Gault and Rundle, as well as most previous investigators; thus, K abundances must be reduced by the same factor. The change is the result of new measurements of the depth of the solar line, which turns out to be not as deep as had previously been guessed. Albano *et al.* (1970) give 19.35% residual intensity; a value of 20.0% has been obtained at Kitt Peak for the whole solar disc (Kvifte *et al.*, 1970).

Gault and Rundle (1969) have investigated the limitations of the twilight method for finding density distributions; they find that sharp variations may not be resolvable. Though this result was to be expected, it is valuable for its quantitative nature. Corrections are given to the screening heights quoted by Hunten (1967) for Li and K. The principal change is to raise the Li layer about 5 km, so that its height now agrees with that for Na. The K seems to be somewhat lower, but there are unavoidable uncertainties in the transmission function that could perhaps be responsible. (At 7700 Å rays penetrating the troposphere are transmitted by clear air; the amount of cloud or aerosol thus becomes important.) Sporadic enhancements of Li continue to be observed (Gault and Rundle, 1969; Sullivan, 1970).

Albano *et al.* (1970) have attempted to measure the K dayglow. A careful series of experiments showed convincingly that they were seeing a continuum, not a line. This 'Ring effect' is discussed in some detail by Noxon (1967), and partial explanations are offered by Brinkmann (1968) and Hunten (1970).

Gadsen (1969a) has made a careful search for lines of several elements that should be produced by meteor ablation, and some of which have been observed by positive-ion mass spectrometers. Though success might have been expected in some cases, no lines were observed.

A thorough treatment of the Atmospheric bands of O_2 in the dayglow has been given by Wallace and Hunten (1968). They find that three main processes are required: transfer from $O(^1D)$ (Reaction 11), resonance scattering, and fluorescent photolysis of O_3. The laboratory evidence for Reaction (11) is now unanimous (Clark,

1970). The nightglow is also treated; the Chapman Reaction (3) gives a good fit, with a yield of a few percent; but other processes are not excluded.

A recent laboratory experiment by Gauthier and Snelling (1970) finds no production of $O_2(^1\Sigma)$ in ozone photolysis at 2537 Å. If this result continues to hold at shorter wavelengths, another source will be necessary in the 50 km region. Rundle (1971) has pointed out that (11) is probably faster than Wallace and Hunten assumed; this modification may permit the data to be fitted. If not, there remains the possibility that ozone photolysis by wavelengths greater than 3000 Å may produce $O_2(^1\Sigma)+O(^3P)$.

The Infrared Atmospheric bands of O_2 have become among the best studied; observations are available for dayglow, twilight, nightglow, and solar eclipses. This work is reviewed by Evans and Llewellyn (1970). The daytime excitation by O_3 photolysis is well understood; the nightglow is doubtless produced during oxygen association, but details are lacking. Recently, interest has focused on the remarkably intense auroral emission, whose energy source is not understood (Noxon, 1970).

Wallace and Broadfoot (1969) report daytime observations of the O lines, 7774 Å and 8446 Å, and N_2 1P bands (as well as the Meinel N_2^+ mentioned above). Again, the resemblance between a weak aurora and the photoelectron-excited dayglow is noted.

Tinsley (1968) discusses his extensive twilight observations of the 10830 Å emission of orthohelium (triplet He). The most striking result is a large winter maximum, along with a distinct enhancement toward the north. It is suggested that these effects reflect the 'winter He bulge' observed by high altitude satellites. A few calculations of orthohelium distributions are given by Patterson (1967), and additional observations by Harrison (1969).

Curiously little has been published about the intense OH emission since Chamberlain's book (1961) appeared. Lytle and Hampson (1964) have established its presence in the dayglow at essentially the nighttime intensity. Ferguson and Parkinson (1963) have emphasized that the total intensity cannot be established without measurements of the $\Delta v=1$ sequence between 2.5 μ and 4 μ. MacDonald *et al.* (1968) have observed it in the nightglow, and Stair and Gauvin (1967) in the laboratory, but no intensities are given. We must continue, therefore, to use Chamberlain's estimate of 4.5 MR. We pointed out above that the available rate from O_2 photolysis is about 0.5 MR; however, the discrepancy is not as great as it seems, because cascading can give several photons for each event of excitation.

Some of the work reported in this volume bears on the problem of the total intensity. The balloon measurements of Gush (1970) suggested that only the 0–1, 1–2, and 1–3 bands are bright enough to be important. The laboratory work of R. Murphy, reported by Stair *et al.* (1971), is in agreement, and the transition probabilities obtained will permit a full solution of the problem. It seems clear that the important transitions from the higher vibrational levels involve quantum jumps of 2 or more, and that the system intensity is of the order of 1 MR. With allowance for cascading, the energy source is in reasonable balance. We have, moreover, a demonstration that the H reactions are an important sink of dissociated oxygen.

Shefov (1969a, b) has presented the results of a long term monitoring program. He suggests that magnetic storms may trigger the release of nightglow energy, and that waves of excitation move away from the auroral zone at 5 to 10 m/sec. In the absence of any plausible mechanism, such ideas are difficult to accept; nevertheless, the amount of observational material is impressive.

4. Other Planets

Lyα emission has been observed from Venus (Barth *et al.*, 1967; Moos *et al.*, 1969), Mars (Barth *et al.*, 1969), and Jupiter (Moos *et al.*, 1969). The Venus results seemed to require two components, tentatively identified with H and D by Donahue (1969a) and McElroy and Hunten (1969). The Jupiter data can be interpreted to give H abundances and infer the degree of dissociation of H_2 (Hunten, 1969; Carlson and Judge, 1971). Moos *et al.* (1969) observed a feature which may have been 1304 Å of O, but it was not seen by Mariner 5 (Barth *et al.*, 1967). This spacecraft did, however, detect a weak UV nightglow at the dark limb; some possible sources have been suggested by McElroy and Strobel (1969).

Mariners 6 and 7 observed a rich dayglow spectrum, including 1304 Å, 1356 Å, and 2972 Å of O, two band systems of CO_2^+, and the Cameron bands of CO, in addition to the Lyα (Barth *et al.*, 1969).

Visual observers have sporadically reported an 'ashen light' in the dark hemisphere of Venus. If real, it could be airglow or aurora; the latter interpretation is advocated by Levine (1969). Several spectroscopic searches have been made, some marginally successful; the latest, which gave a negative result, is that of Goody and McCord (1968).

Acknowledgments

Kitt Peak National Observatory is operated by the Association of Universities for research in Astronomy, Inc. under contract with the National Science Foundation. I am indebted to several speakers at the Aurora and Airglow Institute (referred to in the text) for valuable suggestions and discussions.

References

Albano, J., Blamont, J. E., Chanin, M. L., and Petitdidier, M.: 1970, *Ann. Geophys.* **26**, 151.
Allen, C. W.: 1963, *Astrophysical Quantities*, Athlone Press, London.
Barth, C. A.: 1966, in *The Middle Ultraviolet: Its Science and Technology* (ed. by A. E. S. Green), John Wiley and Sons, New York, p. 177.
Barth, C. A., Pearce, J. B., Kelly, K. K., Wallace, L., and Fastie, W. G.: 1967, *Science* **158**, 1675.
Barth, C. A., Fastie, W. G., Hord, C. W., Pearce, J. B., Kelly, K. K., Stewart, A. I., Thomas, G. E., Anderson, G. P , and Raper, O. F.: 1969, *Science* **165**, 1004.
Bowyer, C. S.: 1970, private communication.
Brinkmann, R. T.: 1968, *Astrophys. J.* **154**, 1087.
Broadfoot, A. L.: 1967, *Planetary Space Sci.* **15**, 1801.
Broadfoot, A. L.: 1968, *Planetary Space Sci.* **16**, 693.
Broadfoot, A. L.: 1971, this volume, p. 34.
Broadfoot, A. L. and Hunten, D. M.: 1966, *Planetary Space Sci.* **14**, 1303.

Broadfoot, A. L. and Kendall, K. R.: 1968, *J. Geophys. Res.* **73**, 426.
Carlson, R. W. and Judge, D. L.: 1971, *Planetary Space Sci.* to be published.
Chamberlain, J. W.: 1961, *Physics of the Aurora and Airglow*, Academic Press, New York.
Clark, I. D.: 1970, *Chem. Phys. Letters* **5**, 317.
Dalgarno, A., McElroy, M. B., and Stewart, A. I.: 1969, *J. Atmospheric Sci.* **26**, 753.
Deehr, C. S.: 1969, *Ann. Geophys.* **25**, 881.
Doherty, L. R.: 1968, *J. Geophys. Res.* **73**, 3597.
Donahue, T. M.: 1969a, *J. Geophys. Res.* **74**, 1128.
Donahue, T. M.: 1969b, Comment at the IAGA General Scientific Assembly, Madrid.
Donahue, T. M.: 1971, *J. Geophys. Res.* **76**, 145; and private communication.
Donahue, T. M., Guenther, B., and Blamont, J. E.: 1970, *Trans. Amer. Geophys. Union* **51**, 587.
Donahue, T. M. and Meier, R. R.: 1967, *J. Geophys. Res.* **72**, 2803.
Evans, W. F. J. and Llewellyn, E. J.: 1970, *Ann. Geophys.* **26**, 167.
Fastie, W. G., Crosswhite H. M., and Heath, D. F.: 1964, *J. Geophys. Res.* **69**, 4129.
Ferguson, A. F. and Parkinson, D.: 1963, *Planetary Space Sci.* **11**, 149.
Fontijn, A., Meyer, C. B., and Schiff, H. I.: 1964, *J. Chem. Phys.* **40**, 64.
Gadsden, M.: 1967, in *Aurora and Airglow* (ed. by B. M. McCormac), Reinhold Publishing Corporation, New York, p. 109.
Gadsden, M.: 1968, *J. Atmospheric Terrest. Phys.* **30**, 151.
Gadsden, M.: 1969a, *Ann. Geophys.* **25**, 667.
Gadsden, M.: 1969b, *Ann. Geophys.* **25**, 721.
Gadsden, M.: 1970, *Ann. Geophys.* **26**, 141.
Gadsden, M. and Purdy, C. M.: 1970, *Ann. Geophys.* **26**, 43.
Gault, W. A. and Rundle, H. N.: 1969, *Can. J. Phys.* **47**, 85.
Gauthier, M. and Snelling, D. R.: 1970, *Chem. Phys. Letters* **5**, 93.
Goody, R. and McCord, T.: 1968, *Planetary Space Sci.* **16**, 343.
Gulledge, I. S., Packer, D. M., Tilford, S. G., and Vanderslice, J. T.: 1968, *J. Geophys. Res.* **73**, 5535.
Gush, H. P.: 1970, unpublished work.
Harrison, A. W.: 1969, *Can. J. Phys.* **47**, 1377.
Hennes, J. P.: 1966, *J. Geophys. Res.* **71**, 763.
Henry, R. J. W. and Williams, R. E.: 1968, *Pub. Astron. Soc. Pacific* **80**, 669.
Hernandez, G. and Turtle, J. P.: 1969, *Planetary Space Sci.* **17**, 675.
Hudson, R. D., Carter, V. L., and Breig, E. L.: 1969, *J. Geophys. Res.* **74**, 4079.
Hunten, D. M.: 1967, *Space Sci. Rev.* **6**, 493.
Hunten, D. M.: 1969, *J. Atmospheric Sci.* **26**, 826.
Hunten, D. M.: 1970, *Astrophys. J.* **159**, 1107.
Hunten, D. M. and McElroy, M. B.: 1966, *Rev. Geophys.* **4**, 303.
Krassovsky, V. I. and Shefov, N. N.: 1965, *Space Sci. Rev.* **4**, 176.
Kurt, V. G. and Dostovalov, S. B.: 1968, *Nature* **218**, 258.
Kvifte, G., Hunten, D. M., and Slaughter, C. D.: 1970, unpublished work.
Levine, J. S.: 1969, *Planetary Space Sci.* **17**, 1081.
Llewellyn, E. J. and Evans, W. F. J.: 1971, this volume, p. 17.
Lytle, E. A. and Hampson, J.: 1964, *Nature* **202**, 76.
MacDonald, R., Buijs, H. L., and Gush, H. P.: 1968, *Can. J. Phys.* **46**, 2575.
Mange, P. and Meier, R. R.: 1970, *J. Geophys. Res.* **75**, 1837.
McElroy, M. B. and Hunten, D. M.: 1966, *J. Geophys. Res.* **71**, 3635.
McElroy, M. B. and Hunten, D. M.: 1969, *J. Geophys. Res.* **74**, 1720.
McElroy, M. B. and Strobel, D. F.: 1969, *J. Geophys. Res.* **74**, 1118.
Meier, R. R.: 1969, *J. Geophys. Res.* **74**, 3561.
Meira, L. G., Jr.: 1971, Ph.D. Thesis, Univ. of Colorado; *J. Geophys. Res.* **76**, 202.
Mitchell, A. C. G. and Zemansky, M. W.: 1934, *Resonance Radiation and Excited Atoms*, Cambridge University Press, Cambridge.
Moos, H. W., Fastie, W. G., and Bottema, M.: 1969, *Astrophys. J.* **155**, 887.
Norton, R. B. and Barth, C. A.: 1970, *J. Geophys. Res.* **75**, 3903.
Noxon, J. F.: 1967, in *Aurora and Airglow* (ed. by B. M. McCormac), Reinhold Publishing Corporation, New York, p. 123.
Noxon, J. F.: 1968, *Space Sci. Rev.* **7**, 92.

Noxon, J. F.: 1970, *J. Geophys. Res.* **75**, 1879.
Noxon, J. F.: 1971, this volume, p. 64.
Noxon, J. F. and Johanson, A. E.: 1970, *Planetary Space Sci.* **18**, 1367.
Packer, I. G., Packer, D. M., Tilford, S. G., and Vanderslice, J. T.: 1970, *J. Geophys. Res.* **75**, 2616.
Patterson, T. N. L.: 1967, *Planetary Space Sci.* **15**, 1219.
Pearce, J. B.: 1971, this volume, p. 45.
Pearce, J. B.: 1969, *J. Geophys. Res.* **74**, 853.
Peterson, V. L.: 1970, *J. Geophys. Res.* **75**, 2613.
Roach, F. E.: 1963, in *Advances in Electronics and Electron Physics* (ed. by L. Marton), Academic Press, New York, p. 1.
Roach, F. E.: 1969, in *Atmospheric Emissions* (ed. by B. M. McCormac and A. Omholt), Van Nostrand Reinhold Company, New York, p. 439.
Rundle, H. N.: 1971, this volume, p. 90.
Shefov, N. N.: 1969a, *Planetary Space Sci.* **17**, 797.
Shefov, N. N.: 1969b, *Planetary Space Sci.* **17**, 1629.
Stair, A. T., Jr., Huppi, E. R., Sandford, B. P., Murphy, R. E., O'Neil, R. R., Hart, A. M., Huppi, R. J., and Pendleton, W. R.: 1971, this volume, p. 185.
Stair, A. T. and Gauvin, H. P.: 1967, in *Aurora and Airglow* (ed. by B. M. McCormac), Reinhold Publishing Corporation, New York, p. 365.
Strickland, D. J. and Donahue, T. M.: 1970, *Planetary Space Sci.* **18**, 661.
Strobel, D. F., Hunten, D. M., and McElroy, M. B.: 1970, *J. Geophys. Res.* **75**, 4307.
Sullivan, H. M.: 1970, *Ann. Geophys.* **26**, 161.
Tinsley, B. A.: 1968, *Planetary Space Sci.* **16**, 91.
Tinsley, B. A.: 1970, *J. Geophys. Res.* **75**, 3932.
Tohmatsu, T.: 1963, *Rep. Ionosphere Space Res. Japan*, **17**, 187.
Tohmatsu, T. and Wakai, N.: 1970, *Ann. Geophys.* **26**, 201.
Wallace, L., Barth, C. A., Pearce, J. B., Kelly, K. K., Anderson, D. E., and Fastie, W. G.: 1970, *J. Geophys. Res.* **75**, 3769.
Wallace, L. and Broadfoot, A. L.: 1969, *Planetary Space Sci.* **17**, 975.
Wallace, L. and Hunten, D. M.: 1968, *J. Geophys. Res.* **73**, 4813.
Wallace, L. and McElroy, M. B.: 1966, *Planetary Space Sci.* **14**, 677.
Weill, G. M.: 1969, in *Atmospheric Emissions* (ed. by B. M. McCormac and A. Omholt), Van Nostrand Reinhold Company, New York, p. 449.
Winter, T. C. and Chubb, T. A.: 1967, *J. Geophys. Res.* **72**, 4405.
Young, J. M., Carruthers, G. R., Holmes, J. C., Johnson, C. Y., and Patterson, N. P.: 1968, *Science* **160**, 990.
Zipf, E. C.: 1969, *Can. J. Chem.* **47**, 1863.

THE DAYGLOW

E. J. LLEWELLYN and W. F. J. EVANS
Institute of Space and Atmospheric Studies, University of Saskatchewan, Saskatoon

1. Introduction

The continuum day sky brightness in the visible region of the spectrum is typically 5×10^6 RÅ^{-1} (Noxon, 1968) but is complicated by the presence of structure due to Fraunhofer lines and telluric absorption and is itself a complicated function of the direction of observation (Coulson *et al.*, 1960). Dayglow observations, therefore, require the subtraction of this high intensity complex background spectrum, or at least its reduction to a very low level, to leave the desired dayglow feature. It is obvious that the intensity of this scattered continuum is proportional to the number of scatterers, at least for single scattering, so that for small scattering optical depths the scattered intensity will be a linear function of the local atmospheric density (Hughes, 1964; Lloyd *et al.*, 1965). However, as expected for a Rayleigh scattering atmosphere the continuum sky-light has a significant degree of polarization which is also a function of the direction of observation and it is, therefore, necessary to either eliminate, or at least have a knowledge of the polarization characteristics of the instruments used to measure the dayglow (Evans, 1967; Evans *et al.*, 1968; Noxon, 1968). The direct reduction of the continuum brightness possible through observations from ballons, rockets and aircraft frequently involves difficult and expensive methods so that the need for ground based observations for synoptic studies is apparent; even for high altitude observations the effects of the scattered background must be considered. Dayglow instrumentation has been developed by many investigators and has been discussed in the extensive reviews of Hunten (1967) and Noxon (1968).

2. Observations

The dayglow observations may be divided into two classes:

(a) those which provide the integrated intensity of the emission and its temporal variation (e.g., Moreels *et al.*, 1970); and

(b) rocket observations which provide a height profile of the emission at some given time (e.g., Wood *et al.*, 1970).

Naturally these observations complement each other and it is of great advantage if both may be undertaken at the same time (e.g., Hunten and Wallace, 1967). The most extensive dayglow observations have been for the O red and green lines, the IR atmospheric system of O_2 and Na. However, there have been very recent attempts to measure the OH dayglow which extends throughout the visible and IR regions of the spectrum and constitutes one of the stronger emission features in the dayglow.

B. M. McCormac (ed.), The Radiating Atmosphere, 17–33.

It is probable that these latter observations will provide extremely valuable information on the atmospheric chemistry in the *D* region.

A. OH EMISSION

The first observations of the OH Meinel emission in the dayglow were made by Lytle and Hampson (1964) with a spectrometer which detected the $\Delta v = 1$ sequence of bands near 3 μ. These observations indicated that the day/night intensity ratio was approximately unity and it has been suggested by Shefov (1969) that this is in agreement with the excitation mechanism proposed by Krassovsky (1956). However, a subsequent observation by Lytle (Hunten, 1967) of the morning twilight emission has shown that the intensity decreases rapidly and then recovers to near the night-

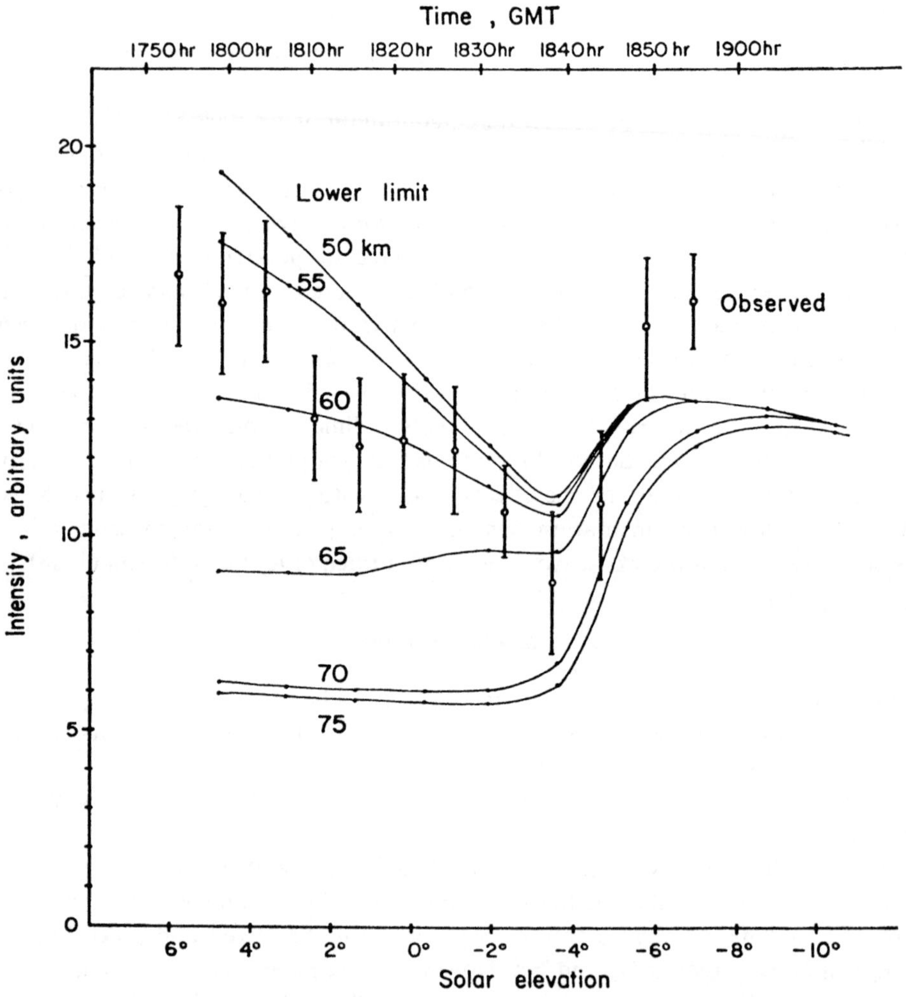

Fig. 1. A balloon observation of the variation of the 8–6 OH band during the evening twilight (Moreels *et al.*, 1970).

time value which is in agreement with the H–O_3 mechanism. Shefov (1969) has analyzed many years of observations and indicates other types of pre-dawn variation may occur although the observations are not directly comparable. To provide further information of the twilight variations there have been a number of recent balloon observations. It is extremely unlikely that ground based measurements can detect these variations as the maximum band intensity is approximately 2 kR in the photomultiplier region. A measurement of the OH emission in the evening twilight has been reported by Moreels *et al.* (1970) who used a balloon borne multichannel photometer which was pointed in azimuth. This type of observation has the distinct advantage that the solar-zenith distance along the line of sight is approximately constant and the degree of polarization of the scattered continuum also remains constant. The intensity variation obtained by these investigators is shown in Figure 1 together with the theoretical predictions of Gattinger (1969b). The minimum intensity at 3° solar depression is readily apparent. A similar experiment (Pick *et al.*, 1971) was flown at Fort Churchill on August 1, 1968 and the results are shown in Figure 2a; the theoretical curve is due to Isaksen *et al.* (1971). It is possible that a part of the emission decay recorded before sunset may have been due to scattered continuum. In this case the experiment continued into the morning twilight and the decrease in the morning intensity was again observed (Figure 2b). These observations confirm the identification of the $H+O_3$ reaction for the Meinel OH emission in the atmosphere and the predicted diurnal variation of the emission intensity is shown in Figure 3 (Isaksen *et al.*, 1970). It would appear from a comparison of this calculation with the observed variations that the dayglow originates from altitudes above 50 km while the nightglow originates from

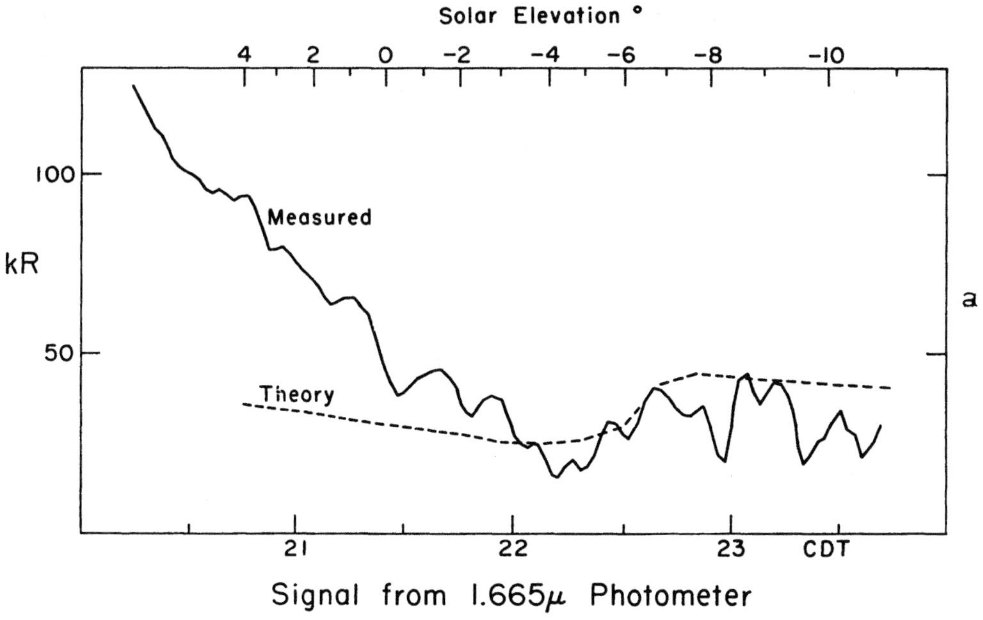

Fig. 2a.

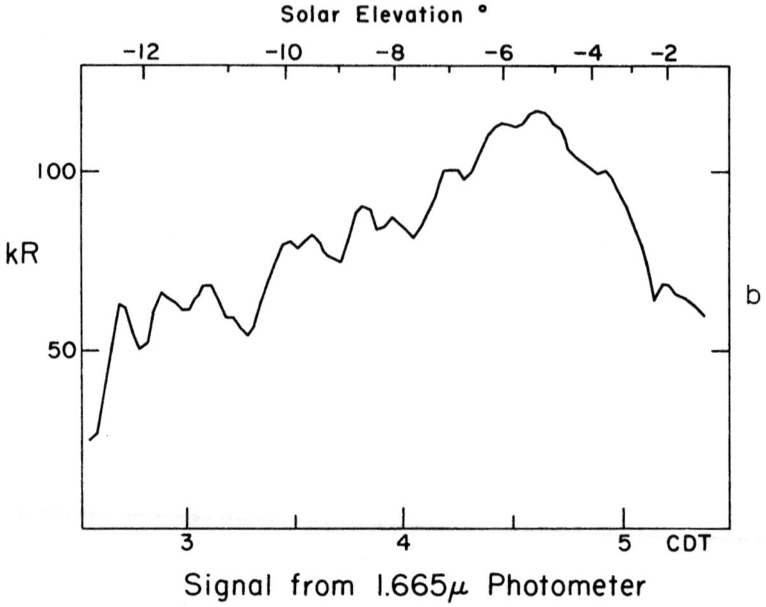

Fig. 2b.

Fig. 2. The temporal variation of the OH emission at 1.665 μ (a) during the evening twilight and (b) during the morning twilight (Pick *et al.*, 1971).

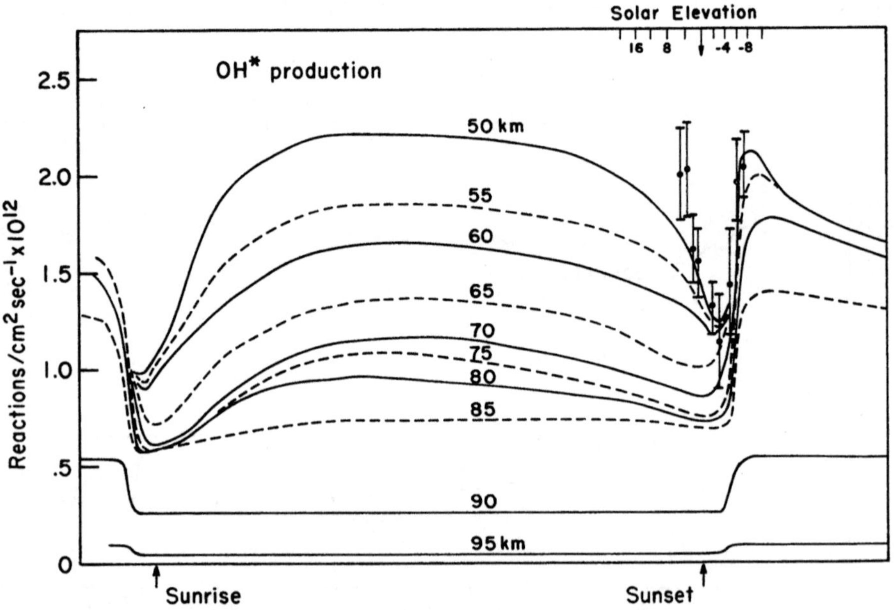

Fig. 3. The predicted diurnal variation of the OH* production rate (Isaksen *et al.*, 1970). The different curves represent the total production rate above the marked altitude and the error bars enclose the observations of Moreels *et al.* (1971).

altitudes above 70 km. These integrated intensity measurements may be more fully understood from the height profile of the OH emission measured in the twilight (Vallance Jones *et al.*, 1971). The height profiles shown in Figure 4 provide a comparison of the observed 1.75 μ OH profile with the $O_2(^1\Delta_g)$ profile obtained on the same flight; the similarity above 80 km is direct evidence for both excitation mechanisms occurring through the atmospheric O_3 content. The agreement between the theoretical and observed profiles (Figure 5) suggests that the basic model atmosphere is correct although the ratio of the amounts of OH emission predicted above and below 80 km is different. It has been suggested that the height variation of the OH emission during the day and twilight could be determined from a measurement of the rotational intensity distribution. However, the problem of simply observing the emission may preclude the determination of useful rotational temperatures. Recently

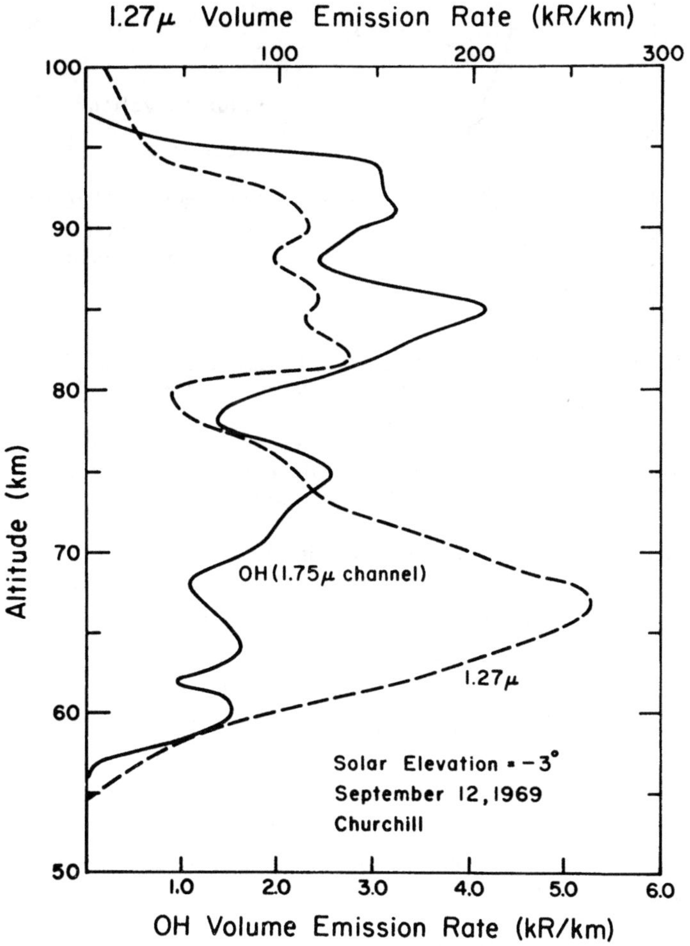

Fig. 4. The twilight altitude distributions for the 1.27 μ O emission and the OH emission near 1.75 μ (Vallance Jones *et al.*, 1971).

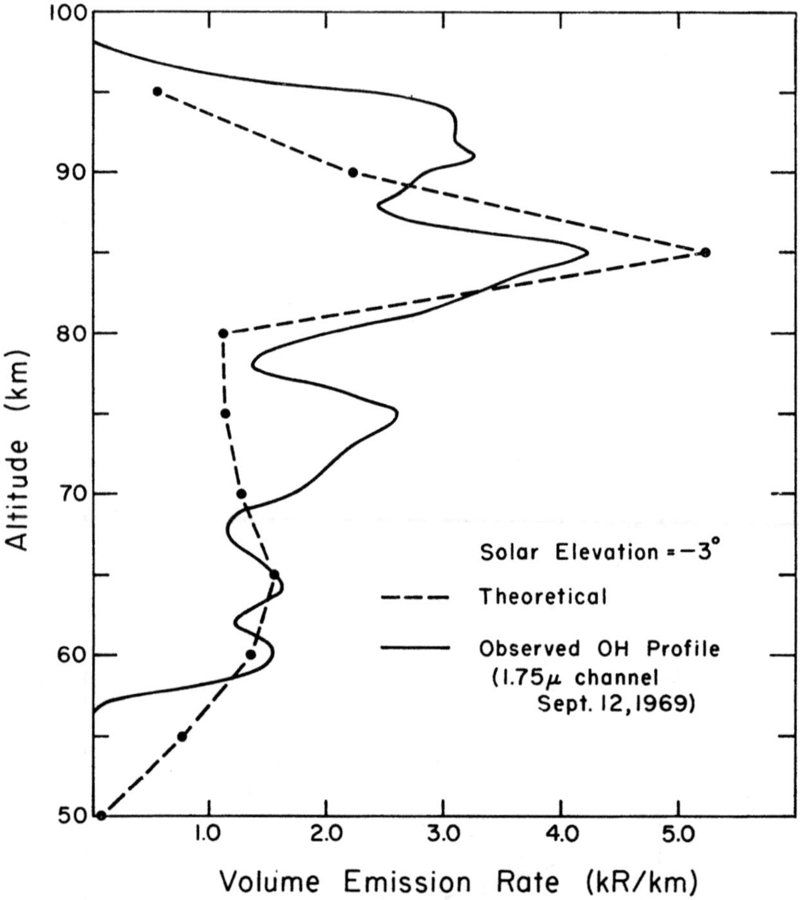

Fig. 5. The comparison of the observed 1.75 μ OH altitude profile with that predicted from a model atmosphere.

there have been other attempts to calculate the diurnal variation of the OH emission (Hesstvedt, 1970; Shimazaki and Laird, 1970). The simplified calculation of Hesstvedt appears to show an OH height profile which does not change greatly between night and day while the calculation of Shimazaki and Laird indicates a profile which has a significant diurnal variation. It should be noted that the excited OH molecules produced at very low altitude during the daytime probably undergo collisional quenching (Moreels *et al.*, 1970) so that the order of magnitude of the day-night intensity ratio is approximately unity. However, above 55 km there is little quenching. The morning twilight decrease is due to the decrease in O_3 through photolysis while the evening variation is due to the loss of daytime H at the lower altitudes before the high altitude O_3 has been reestablished.

B. INFRARED ATMOSPHERIC SYSTEM OF OXYGEN

The IR O_2 emissions at 1.27 μ and 1.58 μ are excited through the photolysis of O_3 in the Hartley band (Vallance Jones and Gattinger, 1963)

$$O_3 + h\nu \rightarrow O_2(^1\Delta_g) + O(^1D).$$

This emission, unlike many others, was first detected in the dayglow and has since been extensively studied from balloons, rockets and aircraft and with ground based methods. The majority of the earlier investigations (Noxon, 1968) were made at Saskatoon with a ground based scanning spectrometer which detected the 0–1 band at 1.58 μ in the twilight (Vallance Jones and Harrison, 1958; Vallance Jones and Gattinger, 1963; Gattinger and Vallance Jones, 1966). The latter authors were only able to obtain agreement between theory and observation through the introduction of collisional quenching of the excited state by O. Subsequent balloon borne observations (Evans *et al.*, 1967) showed that the evening twilight decay of the emission was much slower than indicated by ground based measurement. In fact even a simple O_3 model revealed a decay which was faster than that observed, in particular for the winter observations. However, in each case the theoretical calculations were restricted by the unknown quantum efficiency of the photolysis mechanism although recent measurements by Gauthier and Snelling (1970), suggest that $O_2(^1\Delta_g)$ is produced with unit efficiency at 2537 Å. Even allowing for this there is still an excess emission in the late twilight over that predicted on the basis of measured O_3 profiles and their extrapolation to 100 km (Evans, 1967). A measurement of the altitude profile of the (0–0) 1.27 μ emission during the day (Evans *et al.*, 1968) showed an upper layer near 85 km; subsequent observations of the dayglow profile (Evans and Llewellyn, 1970; Wood *et al.*, 1970) have also revealed this second layer. Evans *et al.* (1968) considered a large number of possible mechanisms for this upper layer but in the absence of further information were unable to determine the exact mechanism. Recent ground based observations (Evans *et al.*, 1970a) have shown that this upper layer will satisfactorily explain the observed 'excess' twilight emission if the excitation mechanism is cut off at UV sunset. These ground based observations have used a two channel photometer which detects the 1.27 μ emission in the twilight; an extensive calculation of the absorption of the 1.27 μ band by ground state O_2 using a Voigt line profile has shown that the atmospheric transmission is 6% (Evans *et al.*, 1970b) in agreement with that proposed by Lowe (1969). It has been suggested that the seasonal variation of the excess twilight emission (Evans and Llewellyn, 1970), is due to a seasonal variation of the O_3 layer above 80 km. It is assumed in deriving this variation that the nightglow emission of approximately 100 kR is not enhanced during the day; for some mechanisms it is possible that the nightglow is suppressed during the day. Unfortunately other observations of the altitude profile (Haslett *et al.*, 1969) have not revealed the second layer, as the rocket apogee was insufficient to provide the complete profile, but have indicated an intensity that is in good agreement with that obtained from balloon measurements (Evans *et al.*, 1969). Recently, Gattinger (1969a) has detected the 0–1 band at 1.58 μ around local noon using a solar subtraction technique. The intensity determined in this way was some 29 MR for the 0–0 band, using the photon intensity band ratio of 46 determined by Findlay (1969).

A balloon observation by Pick *et al.* (1971) has indicated that this ratio is approximately 70 so that Gattinger's measurement implies an intensity of some 45 MR; this intensity is twice that observed by some investigators although the method of solar subtraction should give a decreased intensity rather than a very large one (Hunten, 1970). For the auroral zone altitude profiles of the $O_2(^1\Delta_g)$ the contribution of a significant emission through particle precipitation may be neglected as the flights were made under quiet conditions. The possibility of a contribution from hot electrons may also be omitted as other measurements (Evans *et al.*, 1970c) suggest that this will be above the altitude of present observations.

C. SODIUM EMISSION

The first measurements of the Na emission in the daytime were made by Blamont and Donahue (1961) using the Zeeman photometer although the polarization effects were neglected (Gadsen *et al.*, 1966). More recently there have been further measurements by Gadsden and Purdy (1970) and Albano *et al.* (1970) which have considered the polarization effect although the contributions from the Ring effect (Grainger and Ring, 1962) are neglected. However, Albano *et al.* do suggest that the contribution may not be insignificant. It is possible to estimate the contribution from the Ring effect if it is assumed that it occurs through a flat continuum which is a few percent of the sky brightness (Noxon, 1968). Figure 6 shows the resulting apparent Fraunhofer line profile in the skylight spectrum for both a 5% and 10% continuum. The false emission line which has a width equal to the Fraunhofer line width is included in the figure. For a Zeeman field of 4 kG, approximately 280 mK wave number shift (as marked by the arrow), the change in the intensity of the false emission line is 1.5% (for a 10% Ring effect). If the Na line width is 0.05 Å the false intensity is 5 kR for a 10% Ring effect and 2.5 kR for a 5% effect; these intensities are a significant fraction of the measured emission intensity. However, Hunten (1970) has shown that the Ring effect emission may exhibit a diurnal variation as sometimes observed by Noxon (1968) so that the existing Na observations cannot be reduced by an arbitrary amount; the matter is further complicated by local ground and haze conditions. Despite this it would appear reasonable to reduce the observed intensities by approximately a factor of 1.5 at local noon and lesser factors at other times. Even with this reduction the dayglow observations (Albano *et al.*, 1970) show a Na abundance enhancement of greater than two between day and twilight. There have been recent rocket observations of Na dayglow (Donahue and Meier, 1967; Hunten and Wallace, 1967) which indicate daytime enhancements although one other flight showed no enhancement. The enhancement observed by Hunten and Wallace was 1.45, in approximate agreement with our suggested ratio, while that observed by Meier and Donahue was a factor of 5. The volume emission rate derived by Hunten and Wallace (1967) for the enhanced daytime flight showed that the peak emission occurs at the same altitude as that for the twilight emission observed during the period around the flight. Hunten and Wallace have explained their results in terms of aerosols rather than a simple photochemical theory which predicts that the daytime enhancement will be lower

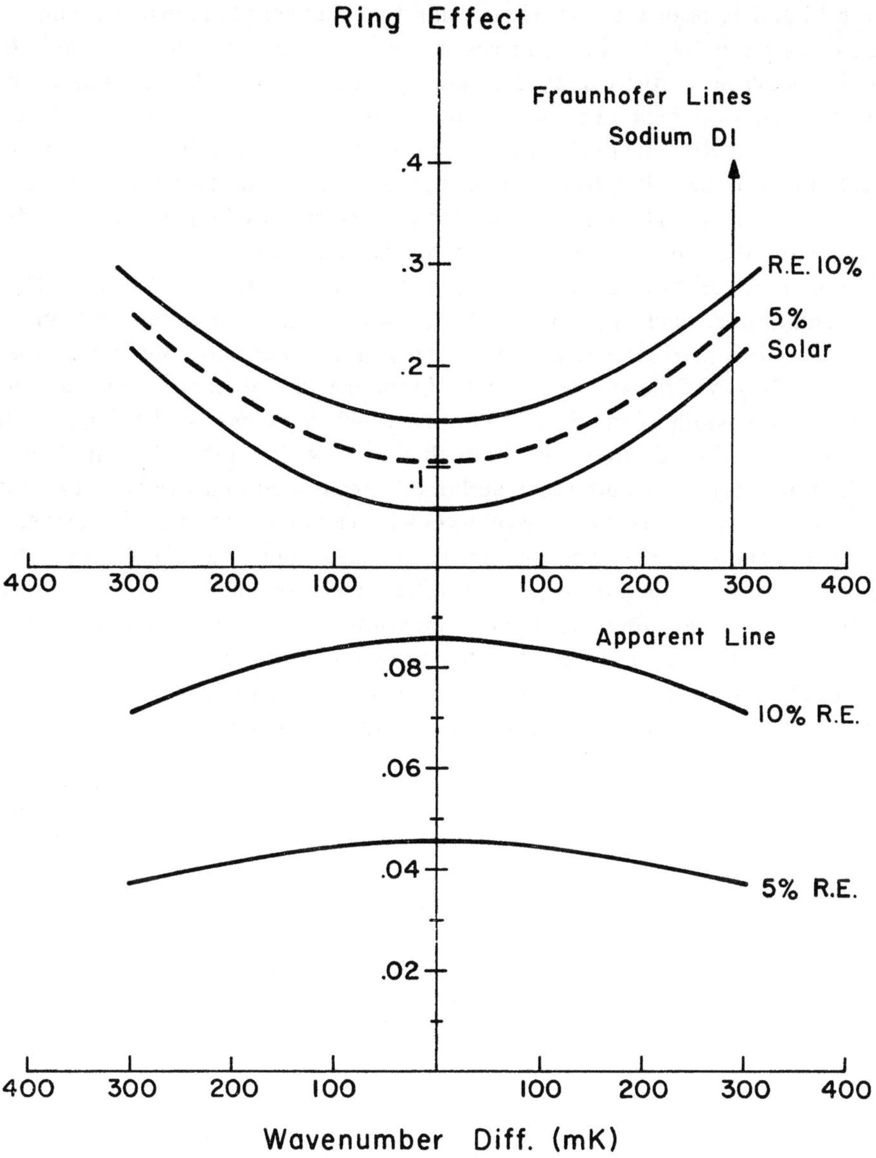

Fig. 6. The apparent sky-light Fraunhofer spectrum for the Na D_1 line with Ring effect contributions of 5% and 10%. The false emission line is shown in the lower part of the figure.

than the twilight layer (Blamont and Donahue, 1964). Hunten (1971) has reviewed the recent twilight Na observations. However, before the rocket and ground based observations can be directly compared the full significance of the Ring effect should be determined.

D. LITHIUM AND POTASSIUM

Hunten and Wallace (1967) also attempted to measure the daytime height profiles of the Li and K resonance emissions using two channel photometers similar to that used to measure Na. However, in neither case were the measurements able to give definite evidence for the existence of an emission although the Li observation may suggest an emitting region near 80 km in agreement with that measured in the twilight by Sullivan and Hunten (1964). More recent measurements of the twilight Li emission (Gault and Rundle, 1969) have shown that the peak of the emission profile is at the same altitude as for Na although other measurements by Sullivan (1970) indicate a seasonal and diurnal variation in the peak emission altitude.

For K, the recent twilight measurements (Gault and Rundle, 1969) suggest a peak emission altitude which is below that for Na and Li. In an attempt to measure the K dayglow emission Albano *et al.* (1970) have used a Zeeman photometer operating at the wavelength of the D_1 line (7699 Å). In this case the observations are complicated by the narrow width of the Fraunhofer line which requires that the Doppler shift of the line be considered. However, for such a narrow Fraunhofer line the Ring effect will be extremely important as the authors demonstrated in a series of careful experiments. In this case a Zeeman field of 4 kG will scan across most of the false emission profile and the intensity of an assumed resonance line would be 10 kR. This is a factor of 2000 larger than the measured twilight emission intensity (Gault and Rundle, 1969) and it appears that the daytime measurement is a direct observation of the Ring effect (Albano *et al.*, 1970). As noted previously the intensity of the false emission line should exhibit a diurnal variation which is symmetric about noon and the observations (Figure 7) are in good agreement with this conclusion. It is apparent that

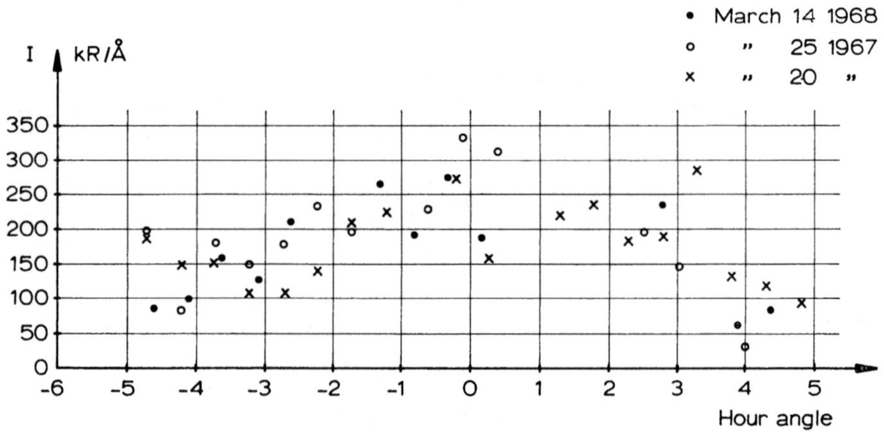

Fig. 7. The diurnal variation of the emission intensity at 7699 Å (Albano *et al.*, 1970).

the observations of the alkali metals in the daytime are not definitive and there is the need for further experiments which are capable of unique interpretation. It should be noted that a recent study by Kvifte and Wallace (1970) has shown that measurements of the Na emission in the twilight can also provide a measure of the Na abundance in the daytime.

E. VISIBLE ATOMIC OXYGEN EMISSIONS

O I(5577 Å). There have been a large number of observations of the green line in the daytime using rocket borne photometers and spectrometers in addition to the ground based observations of Noxon (1963) using the polarimeter. Unfortunately these measurements are unable to provide clear information on the diurnal variation of the dayglow although the observations suggest that the integrated intensity is probably an order of magnitude larger than the nightglow. Although a direct comparison of the integrated intensities is of little value, useful information can be obtained from a comparison of the height profiles (Figure 8). A measurement of the green line night-

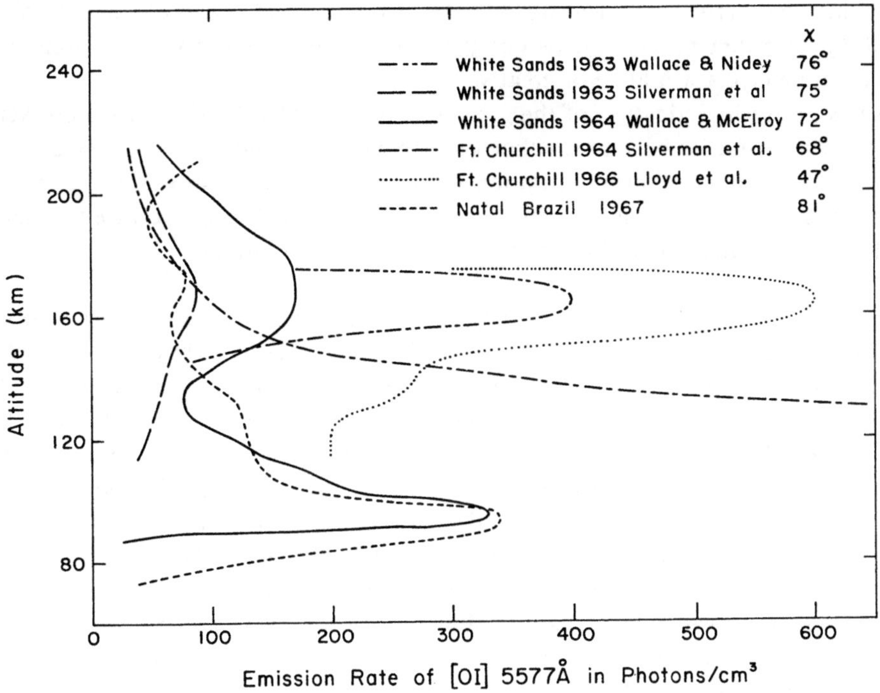

Fig. 8. Rocket observations of the dayglow altitude distributions of the O green line ($\lambda = 5577$ Å) (Dandekar, 1969).

glow profile by Gulledge *et al.* (1968) permits a direct comparison with the observations of Wallace and McElroy (1966); the integrated nighttime intensity is in close agreement with that observed by Gulledge *et al.* If it is assumed that the low altitude emission is due to a photochemical mechanism then the emission intensity is proportional to $n^3(0)$ and the O concentration near 95 km is approximately 10^{12} cm^{-3}. This concentration represents a 60% increase over the nighttime value and is in agreement with that suggested by Evans and Llewellyn (1970). However, there is a large emission (~1.5 kR) during the day near 170 km which cannot arise through a

Chapman mechanism and must be due to some other process. There is a corresponding nighttime emission ($\sim$30 R) although the altitude appears to be somewhat greater ($\sim$240 km). Wallace and McElroy (1966) suggested that this high altitude daytime emission was due to photoelectron excitation although Dandekar (1969) has considered that photodissociation and dissociative recombination could contribute significant emission on the basis of measured structure in the altitude profile. However, a consideration of the instrumentation used suggests that it may have been polarization sensitive and that a significant fraction of the structure could have been from this effect or from the effects of stray light. There is little doubt that dissociative recombination will contribute to the observed dayglow emission but the determination of the magnitude will probably require a combined experiment similar to that of Donahue *et al.* (1970) to measure both positive ions and the emission intensity. It would appear that measurements designed to measure the diurnal variations of the two emitting layers would be of great value.

An extensive investigation of the green line in the twilight has been undertaken by Deehr (1969) and he has shown that the irregular observations of an enhancement can be fitted with a photodissociation mechanism alone. However, these enhancements appear to be associated with magnetic activity which may produce the changes in the O_2 concentration at 200 km required by the observations.

OI(6300 Å). Early estimates of the dayglow emission indicated an intensity of some 50 kR (Brandt, 1958), which should be readily detectable from both the ground and high altitude. The most complete measurements are due to Noxon (1964, 1968) although there have been other measurements by Jarrett and Hoey (1963, 1967) and by Bens *et al.* (1965). The latter measurements have used a high resolution Fabry-Pérot etalon system and are probably influenced by the Ring effect while those of Jarrett and Hoey are undoubtedly affected by the Fraunhofer structure and the resulting interference pattern (Cogger and Shepherd, 1965). The measurements of Noxon (1964, 1968) have revealed that the emission intensity may exhibit large fluctuations over short time periods although during solar minimum the intensity has remained relatively stable, typically 5 to 10 kR. However, with the onset of another solar maximum the intensity fluctuations have again been observed (Noxon, 1968); it is necessary, therefore, to consider the period of the solar cycle when comparing results.

The available height profile measurements are shown in Figure 9. The integrated intensities for the measurements of Nagata *et al.* (1968) and Wallace and McElroy (1966) differ by a factor of 2; it is probable that part of this discrepancy may have been due to scattered light as the Japanese measurements were made with a vehicle which had a large cone angle (25°). The apparent agreement of these latter measurements with calculation is also somewhat doubtful as the $O(^1D)$ quenching coefficient adopted is smaller than that recently measured (Noxon, 1970). It is doubtful that the excess emission observed at high altitude is due to 'hot' electrons, as suggested by Noxon (1964, 1968) for the rapid fluctuations in the observed integrated intensity.

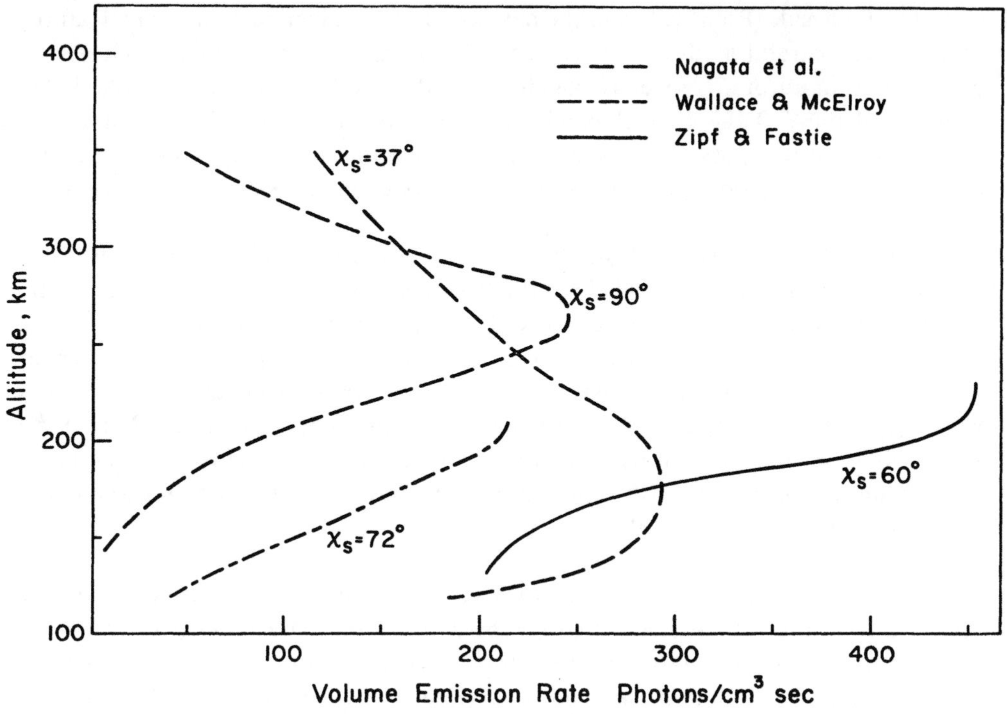

Fig. 9. Rocket observations of the dayglow altitude distribution of the O red line ($\lambda = 6300$ Å).

Wallace and McElroy (1966) compared their observations with the emission intensities expected from dissociative recombination, photodissociation and photoelectron excitation and were able to show that the production due to dissociative recombination provides only a small contribution to the observed intensity. Considerable support for this conclusion is provided by the twilight observation of Nagata *et al.* (1968) which indicates the decay of the low altitude emission during the flight; such a rapid decay also implies an extremely rapid excitation mechanism such as photodissociation.

It is the production through photodissociation of $O_2 + h\nu \rightarrow O(^1D) + O(^3P)$, which leads to the true twilight enhancement of the 6300 Å emission. The posttwilight enhancement is due to dissociative recombination where the O_2^+ ions are produced by charge transfer from O^+; it is obvious that the absence of ions in the morning will lead to an asymmetry in the emission intensities from such a mechanism in the morning and evening twilights. It was shown by Cole (1965) that the morning twilight, or predawn, enhancement of the 6300 Å emission was due to conjugate point photoelectrons; however, for Alaska the conjugate point photoelectrons are unable to reach the emission region and so provide no morning enhancement (Deehr, 1969).

F. OTHER ATOMIC OXYGEN EMISSIONS

O I(7774 Å). Recently Wallace and Broadfoot (1969) have made a spectrometer survey of the weak IR emissions in the dayglow and the identification of any features attests to the careful analysis undertaken. The O emission at 7774 Å, which is well documented in aurora, is severely blended with the 0–0 band of the atmospheric O_2 system and the 2–0 N_2^+ Meinel band. However, as the 1–0 N_2^+ Meinel band was also observed it was possible to estimate the O emission intensity as 1.5 kR ± 0.5 kR; this intensity is in good agreement with that estimated by Green and Barth (1967).

O I(8446 Å). This emission is also well established in the auroral spectrum but was not reported in the airglow until Shefov (1962) made a successful measurement in the twilight; the observed intensity was 13 R. The emission was obviously present in the spectra recorded at 85 km (Wallace and Broadfoot, 1969) although the observed intensity (1.1 kR) is significantly larger than that calculated by Dalgarno *et al.* (1969) who suggest an intensity of 300 R from photoelectron excitation and 50 R from the Bowen fluorescence mechanism. The preliminary emission associated with the latter mechanism will be at 1.13 μ where it will be extremely difficult to detect as the maximum intensity may only be 600 R.

O I(4368 Å). The allowed transition which produces this emission terminates in the upper state of the 1304 Å line which has been measured on a number of occasions (Fastie and Crosswhite, 1964; Fastie, 1968). There is only one reported dayglow observation of this visible oxygen emission (Krassovsky and Shefov, 1965) and the measured intensity is extremely large, approximately 10 kR. This large intensity should be compared with the spectra obtained by Wallace and Nidey (1964) which do not reveal a feature at the required wavelength; for this reason the observation must be questioned. Recently Tinsley (1970) has been able to observe the evening twilight emission using a spectrometer. The intensity is approximately 1 R and decays during the evening twilight period to 0.2 R.

G. O_2 ATMOSPHERIC SYSTEM

The Atmospheric system of O_2 ($b\,^1\Sigma_g^+ - X\,^3\Sigma_g^-$) has been measured from rockets by Wallace and Hunten (1968) and from the ground by Noxon (1968). The rocket observations have provided measurements of both the 0–0 and 0–1 bands while the ground based measurements can only detect the 0–1 band at 8645 Å as the 0–0 band is absorbed by the ground state molecules in the atmosphere below the emission region. Thus the integrated zenith intensity will be smaller for the (0–0) band at altitudes below which the absorption exceeds the ratio of the band brightnesses. However, Wallace and Hunten (1968) were able to compare both band intensities with theory by calculating the absorption of a Doppler array of lines by a Doppler array in which the emission region has an optical depth greater than zero; a similar method for Voigt line shapes has been given by Evans *et al.* (1970b). The observations have been explained in terms of resonant scattering, transfer from $O(^1D)$ and O_3 photolysis; this latter mechanism was necessary to explain the excess emission near 50 km. A recent laboratory observation by Gauthier and Snelling (1970) shows that the $O_2(^1\Sigma_g^+)$ is not produced in O_3 photolysis so that the low altitude emission must arise through energy transfer. This problem has been considered by Rundle (1971) who shows that a satisfactory agreement between theory and observation is possible if the O_2 concentration above 100 km is reduced from that used by Wallace and Hunten. The contribution of the chemiluminescent nightglow mechanism to the dayglow intensity is not directly apparent in the rocket observations and it would be of value to measure the $O_2(^1\Sigma_g^+)$ concentration profile during the late twilight.

H. NI (5200 Å)

The N doublet at 5200 Å has been studied in the twilight (Dufay, 1953; Blackwell *et al.*, 1960; Weill, 1969) but the dayglow observations are restricted to the rocket measurements of Wallace and Nidey (1964), Wallace and McElroy (1966). The observations of Weill have shown that the contribution to the morning twilight intensity from conjugate point photoelectrons is less than 2 R. The integrated daytime intensity is approximately 100 R but even with the high throughput of the photometer it is difficult to measure the emission height profile. These latter investigators considered three possible excitation mechanisms,

$$NO^+ + e \rightarrow N(^2D) + O$$
$$N_2^+ + e \rightarrow N(^2D) + N$$
$$N_2^+ + O \rightarrow NO^+ + N(^2D),$$

although their observations did not permit a determination of the contribution from each source. It has been noted by Noxon (1968) that none of these mechanisms directly involve the N content so that it is not possible to provide a measurement of the N concentration through observations of the 5200 Å emission. It is to be anticipated that the atomic nitrogen emission at 10400 Å, the analog of the O green line, which may be present in the spectra of Wallace and Broadfoot (1969) cannot provide this information either. However, there are observations of the N emission at 1200 Å (Fastie *et al.*, 1964) which was detected through the recorded asymmetry in the Lyα emission line shape and there have been subsequent observations by Pearce (1970); the 1200 Å line occurs through an allowed transition terminating in the ground state. This may cause the N emission to have an optical thickness greater than unity, although it will be significantly less than the 1304 Å O line due to the decreased N column concentration. It is probable, therefore, that a measurable fraction of the observed 1200 Å emission occurs through resonant scattering and could be used to determine the N concentration. We have assumed that the oscillator strength for this emission is approximately 0.1, by comparison with the derived oscillator strength for the 1304 Å O line (Strickland and Donahue, 1970), and that the decrease in the overhead intensity between 150 and 200 km is due to the decrease in the column concentration of N. The required N concentration in this region is 5×10^7 cm^{-3}, almost two orders of magnitude larger than that estimated by Barth (1961) but in closer agreement with recent calculations (Strobel *et al.*, 1970). It would appear, therefore, that the observed emission may occur either through photoelectron excitation, as for the 1304 Å line, or through resonance.

There have also been attempts to observe airglow X-rays (Grader *et al.*, 1968) which could provide further information on the atomic N content.

Acknowledgments

Much of the new work reported here has been through our collaboration with Dr. A. Vallance Jones which we are pleased to acknowledge. The work has been supported with grants from N.R.C. (Canada) and DRB (Canada).

References

Albano, J., Blamont, J. E., Chanin, M. L., and Petitdidier, M.: 1970, *Ann. Geophys.* **26**, 151.
Barth, C.: 1961, in *Chemical Reactions in the Lower and Upper Atmosphere* (Proceedings of Stanford Research Institute international symposium) Interscience Publishers, New York, p. 303.
Bens, A. R., Cogger, L. L., and Shepherd, G. G.: 1965, *Planetary Space Sci.* **13**, 551.
Blackwell, D. E., Ingham, M. F., and Rundle, N.: 1960, *Astrophys. J.* **131**, 15.
Blamont, J. E. and Donahue, T. M.: 1961, *J. Geophys. Res.* **66**, 1407.
Blamont, J. E. and Donahue, T. M.: 1964, *J. Geophys. Res.* **69**, 4093.
Brandt, J. C.: 1958, *Astrophys. J.* **128**, 718.
Cogger, L. L. and Shepherd, G. G.: 1965, *Planetary Space Sci* **13**, 1163.
Cole, K. D.: 1965, *Ann. Geophys.* **21**, 156.
Coulson, K., Dave, S., and Sekera, Z.: 1960, *Tables Related to Radiation Emerging From a Planetary Atmosphere With Rayleigh Scattering*, University of California Press.
Dalgarno, A., McElroy, M. B., and Stewart, A. I.: 1969, *J. Atmospheric Sci.* **4**, 753.
Dandekar, B. S.: 1969, *Planetary Space Sci.* **17**, 1609.
Deehr, C. S.: 1969, *Ann. Geophys.* **25**, 881.
Donahue, T. M. and Meier, R. R.: 1967, *J. Geophys. Res.* **72**, 2803.
Donahue, T. M., Zipf, E. C., and Parkinson, T. D.: 1970, *Planetary Space Sci.* **18**, 171.
Dufay, M.: 1953, *Ann. Phys.* **8**, 813.
Evans, W. F. J.: 1967, Ph.D. Thesis, University of Saskatchewan.
Evans, W. F. J., Hunten, D. M., Llewellyn, E. J., and Vallance Jones, A.: 1968, *J. Geophys. Res.* **73**, 2885.
Evans, W. F. J. and Llewellyn, E. J.: 1970, *Ann. Geophys.* **26**, 167.
Evans, W. F. J., Llewellyn, E. J., Haslett, J., and Megill, L. R.: 1970c, *J. Geophys. Res.* **75**, 6425.
Evans, W. F. J., Llewellyn, E. J., and Vallance Jones, A.: 1967, *Nature* **213**, 352.
Evans, W. F. J., Llewellyn, E. J., and Vallance Jones, A.: 1969, *Planetary Space Sci.* **17**, 933.
Evans, W. F. J., Wood, H. C., and Llewellyn, E. J.: 1970a, *Planetary Space Sci.* **18**, 1065.
Evans, W. F. J., Wood, H. C., and Llewellyn, E. J.: 1970b, *Can. J. Phys.* **48**, 747.
Fastie, W. G.: 1968, *Planetary Space Sci.* **16**, 929.
Fastie, W. G. and Crosswhite, H. M.: 1964, *Planetary Space Sci.* **12**, 1021.
Fastie, W. G., Crosswhite, H. M., and Heath, D. F.: 1964, *J. Geophys. Res.* **69**, 4129.
Findlay, F. D.: 1969, *Can. J. Phys.* **47**, 687.
Gadsden, M., Blamont, J. E., and Donahue, T. M.: 1966, *J. Geophys. Res.* **71**, 5047.
Gadsden, M. and Purdy, C. M.: 1970, *Ann. Geophys.* **26**, 43.
Gattinger, R. L.: 1969a, *Can. J. Phys.* **47**, 367.
Gattinger, R. L.: 1969b, *Ann. Geophys.* **25**, 825.
Gattinger, R. L. and Vallance Jones, A.: 1966, *Planetary Space Sci.* **14**, 1.
Gault, W. A. and Rundle, H. N.: 1969, *Can. J. Phys.* **47**, 85.
Gauthier, M. and Snelling, D. R.: 1970 *Chem. Phys. Letters* **5**, 93.
Grader, R. J., Hill, R. W., and Seward, F. D.: 1968, *J. Geophys. Res.* **73**, 7149.
Grainger, J. F. and Ring, J.: 1962, *Nature* **193**, 762.
Green, A. E. S. and Barth, C. A.: 1967, *J. Geophys. Res.* **72**, 3975.

Gulledge, I., Packer, D. M., Tilford, S. G., and Vanderslice, J. T.: 1968, *J. Geophys. Res.* **73**, 5535.
Haslett, J. C., Megill, L. R., and Schiff, H. I.: 1969, *Can. J. Phys.* **47**, 2351.
Hesstvedt, E.: 1970, *J. Geophys. Res.* **75**, 2337.
Hughes, J. V.: 1964, *Appl. Opt.* **3**, 1135.
Hunten, D. M.: 1967, *Space Sci. Rev.* **6**, 493.
Hunten, D. M.: 1970, *Astrophys. J.* **159**, 1107.
Hunten, D. M.: 1971, this volume, p. 3.
Hunten, D. M. and Wallace, L.: 1967, *J. Geophys. Res.* **72**, 69.
Isaksen, I., Llewellyn, E. J., and Vallance Jones, A.: 1971, to be published.
Jarrett, A. M. and Hoey, M. E.: 1963, *Planetary Space Sci.* **11**, 1251.
Jarrett, A. M. and Hoey, M. E.: 1967, *Planetary Space Sci.* **15**, 591.
Krassovsky, V. I.: 1956, in *The Airglow and the Aurorae* (ed. by E. B. Armstrong and A. Dalgarno), Pergamon Press, London, p. 86.
Krassovsky, V. I. and Shefov, N. N.: 1965, *Space Sci. Rev.* **4**, 176.
Kvifte, G. I. and Wallace, L.: 1970, *Planetary Space Sci.* **18**, 623.
Lloyd, J. W. F., Silverman, S., Nardone, L. J., and Cochrun, B. L.: 1965, *Appl. Opt.* **4**, 1602.
Lowe, R. P.: 1969, *Phil. Trans. Roy. Soc. London* **A264**, 163.
Lytle, E. A. L. and Hampson, J.: 1964, *Nature* **202**, 76.
Moreels, G., Evans, W. F. J., Blamont, J. E., and Vallance Jones, A.: 1970, *Planetary Space Sci.* **18**, 637.
Nagata, T., Tohmatsu, T., and Ogawa, T.: 1968, *J. Geomagnet. Geoelec. Kyoto* **20**, 315.
Noxon, J. F.: 1963, *J. Atmospheric Terrest. Phys.* **25**, 637.
Noxon, J. F.: 1964, *J. Geophys. Res.* **69**, 3245.
Noxon, J. F.: 1968, *Space Sci. Rev.* **8**, 92.
Noxon, J. F.: 1970, *J. Chem. Phys.* **52**, 1852.
Pearce, J. B.: 1970, private communication.
Pick, D., Llewellyn, E. J., and Vallance Jones, A.: 1971, *Can. J. Phys.* **49**, in press.
Rundle, H. N.: 1971, this volume, p. 90.
Shefov, N. N.: 1962, *Poljarnye sijania i syecenie nocnogo neba*, serija Rezul'taty MGG **9**, 55.
Shefov, N. N.: 1969, *Planetary Space Sci.* **17**, 1629.
Shimazaki, T. and Laird, A. R.: 1970, *J. Geophys. Res.* **75**, 3221.
Strickland, D. J. and Donahue, T. M.: 1970, *Planetary Space Sci.* **18**, 661.
Strobel, D. F., Hunten, D. M., and McElroy, M. B.: 1970, *J. Geophys. Res.* **75**, 22.
Sullivan, H. M.: 1970, *Ann. Geophys.* **26**, 161.
Sullivan, H. M. and Hunten, D. M.: 1964, *Can. J. Phys.* **42**, 937.
Tinsley, B. A.: 1970, *J. Geophys. Res.* **75**, 3932.
Vallance Jones, A. and Gattinger, R. L.: 1963, *Planetary Space Sci.* **11**, 961.
Vallance Jones, A. and Harrison, A. W.: 1958, *J. Atmospheric Terrest. Phys.* **13**, 45.
Vallance Jones, A., Evans, W. F. J., Wood, H. C., and Llewellyn, E. J.: 1971, to be published.
Wallace, L. and Broadfoot, A. L.: 1969, *Planetary Space Sci.* **17**, 975.
Wallace, L. and Hunten, D. M.: 1968, *J. Geophys. Res.* **73**, 4813.
Wallace, L. and McElroy, M. B.: 1966, *Planetary Space Sci.* **14**, 677.
Wallace, L. and Nidey, R.: 1964, *J. Geophys. Res.* **69**, 471.
Weill, G. H.: 1969, in *Atmospheric Emissions* (ed. by B. M. McCormac and A. Omholt) Van Nostrand Reinhold Corporation, New York, p. 449.
Wood, H. C., Evans, W. F. J., Llewellyn, E. J., and Vallance Jones, A.: 1970, *Can. J. Phys.* **48**, 862.
Zipf, E. C. and Fastie, W. G.: 1963, *J. Geophys. Res.* **68**, 6208.

INSTITUTE DISCUSSION

It was suggested that the dust layer may be responsible for the minimum in the 8–6 OH band as shown in Figure 1. E. Llewellyn stated that the photochemical calculations predict such a minimum and that the dust layer is not responsible.

F. Bunn asked if some of the structure in the OH Meinel intensity observations could be due to CH_4. E. Llewellyn thought that this was possible and that the data should be carefully compared with the spectra of H. Gush and F. Bunn.

DAYGLOW NITROGEN BAND SYSTEMS

A. L. BROADFOOT
Kitt Peak National Observatory, Tucson, Ariz., U.S.A.

1. Introduction

The band systems of N_2, 1P and 2P, and of N_2^+, Meinel and 1N, are dominant features of the dayglow spectrum (cf., Hunten, 1967; Noxon, 1968). Models of the sunlit atmosphere can be adjusted to agree satisfactorily with the twilight-dayglow observations, and the emissions seem to be well understood. With our present knowledge of the emission from these band systems it is possible to assess the excitation mechanisms which lead to the emission and thereby remove some of the assumptions inherent in the atmospheric modeling process.

This review attempts to illustrate how the signature of the excitation mechanism is carried into the measured intensity, and also into the vibrational and rotational intensity distributions. Observations are suggested which would assist in determining the importance of each excitation mechanism to the overall emission.

2. Excitation

Efficient excitation of the N_2 bands can result only from electron bombardment since a spin change is required in the transition. (An energy level diagram is given in Figure 1.)

$$N_2(X\,^1\Sigma_g^+) + e \rightarrow N_2(C\,^3\Pi_u, B\,^3\Pi_g, A\,^3\Sigma_u^+) + e\,. \tag{1}$$

The emission is somewhat complicated since molecules in the $C\,^3\Pi_u$ state cascade into the $B\,^3\Pi_g$ state which in turn cascade into the $A\,^3\Sigma_u^+$ state through emission of the 2P and 1P, respectively. The natural lifetime of the $A\,^3\Sigma_u^+$ state is of the order of 2 sec (Shemansky, 1969), a sufficient time for radiationless deactivation or further excitation by fluorescence, electron collision, or resonance scattering. High vibrational levels of the $A\,^3\Sigma_u^+$ state can collisionally cross or even radiate to the $B\,^3\Pi_g$ state followed by radiation of the 1P to lower $A\,^3\Sigma_u^+$ state vibrational levels (Campbell and Thrush, 1967; Shemansky and Broadfoot, 1971). In any case the most significant source of excitation would be due to the initial Reaction (1); the theory is reasonably tractable (cf., Broadfoot and Hunten, 1964; Shemansky and Broadfoot, 1971).

Nitrogen ions are produced in all of the ion states by either XUV solar radiation,

$$N_2(X\,^1\Sigma_g^+) + h\nu\,(\lambda < 796\,\text{Å}) \rightarrow N_2^+(A\,^2\Pi_u, B\,^2\Sigma_u^+, X\,^2\Sigma_g^+) + e \tag{2}$$

or by electron bombardment,

$$N_2(X\,^1\Sigma_g^+) + e \rightarrow N_2^+(A\,^2\Pi_u, B\,^2\Sigma_u^+, X\,^2\Sigma_g^+) + 2e\,. \tag{3}$$

B. M. McCormac (ed.), The Radiating Atmosphere, 34–44. *All Rights Reserved.*

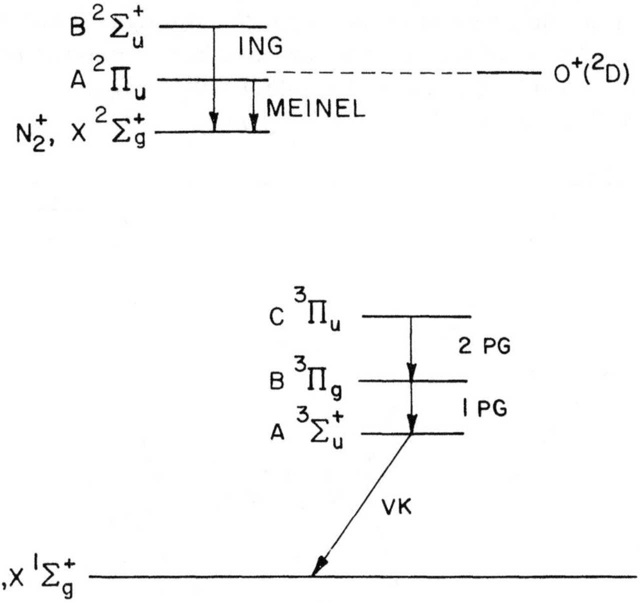

Fig. 1. Energy level diagram, N_2, $N_2{}^+$.

A more selective ionizing reaction was suggested by Dalgarno and McElroy (1966),

$$N_2 (X\ {}^1\Sigma_g^+) + O^2 ({}^2D) \rightarrow N_2^+ (A\ {}^2\Pi_u, v' = 1) + O\,. \tag{4}$$

The resonance between the $O^+ ({}^2D)$ energy and the $v' = 1$ vibrational level of the $A\ {}^2\Pi_u$ state was discussed by Omholt (1959) as a possible source of auroral excitation. Zipf (1969) has reported that N_2 quenches O^+ $({}^2D)$ with a rate coefficient of 3×10^{-10} cm^2 sec^{-1}, about ten times smaller than the rate coefficient suggested by Dalgarno and McElroy (1966) for Reaction (4).

When N_2 molecules are simultaneously ionized and excited to the $A\ {}^2\Pi_u$ and $B\ {}^2\Sigma_u^+$ states they immediately radiate the Meinel and 1N band systems and enter the N_2^+ $(X\ {}^2\Sigma_g^+)$ ground state. Ions in the $X\ {}^2\Sigma_g^+$ state illuminated by sunlight will resonantly scatter about 0.11 photons ion^{-1} sec^{-1} through each of the Meinel and 1N band systems:

$$N_2^+ (X\ {}^2\Sigma_g^+) + h\nu \rightleftarrows N_2^+ (A\ {}^2\Pi_u, B\ {}^2\Sigma_u^+)\,. \tag{5}$$

3. Emission Intensity

In the twilight an observer takes advantage of the movement of the earth's shadow through an emitting layer to measure the integrated emission intensity as a function of solar depression. A rocket, on the other hand, measures the integrated emission intensity as a function of altitude. In either case the chain of reactions leading to emission is complicated. A measure of our understanding of the emission is obtained by comparison of the products of models of the atmosphere with the measured intensity curves. Models of the twilight sky have been generated by Broadfoot and Hunten (1966) and subsequently by Broadfoot (1968) (Figure 3). The dayglow was modeled by Wallace and McElroy (1966) (Figure 2).

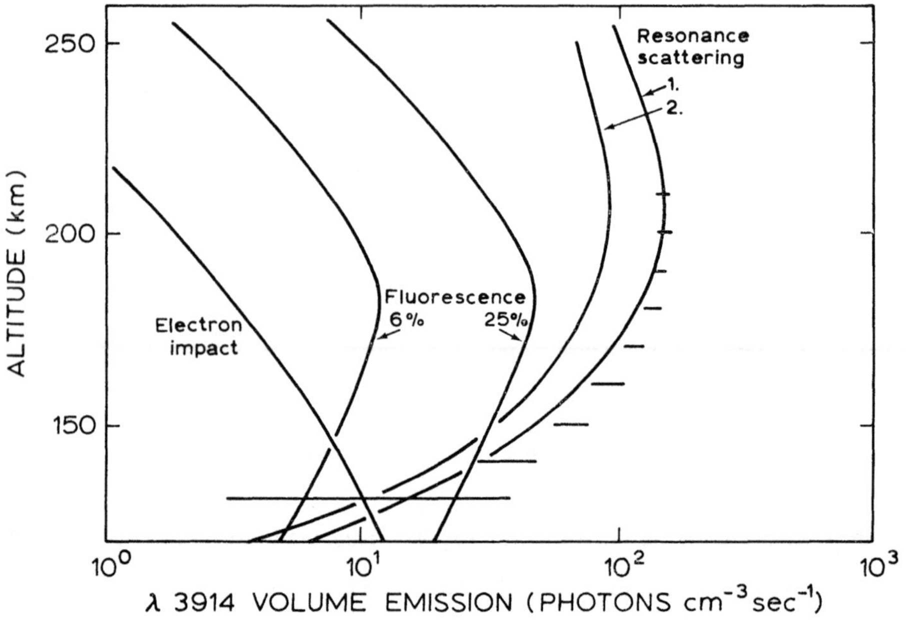

Fig. 2. Theoretical model showing the contribution of three excitation mechanisms to the emission of the 3914 Å band of N_2^+ (Wallace and McElroy, 1966).

Rocket measurements by several groups Zipf and Fastie, 1964; Wallace and Nidey, 1964; Nagata *et al.*, 1965; Holmes *et al.*, 1965; Wallace and McElroy, 1966; Hoffman *et al.*, 1969) on the N_2^+ emission and ion density fit the daytime models of the atmosphere satisfactorily. Model predictions are fairly insensitive to the origin of the emission. Figure 2 shows that separation of the processes would be impractical by intensity measurements alone.

The twilight model predictions are also very insensitive to the origin of the emission. When the assumed conditions are combined and the effects of tropospheric scattering are included (Broadfoot, 1968) then the observations and calculations agree nicely (Figure 4). There is no doubt that the emission is predominantly a high altitude feature, but confirmation that the model is correct in detail will have to come from other methods of detection.

It is possible to determine the importance of the photo-electron flux in the dayglow by measurement of the emission intensity of the N_2 band systems by either the twilight or rocket methods. Distribution of energy among photoelectrons has been the subject of several papers in the past few years (Dalgarno *et al.*, 1969; Nagy *et al.*,

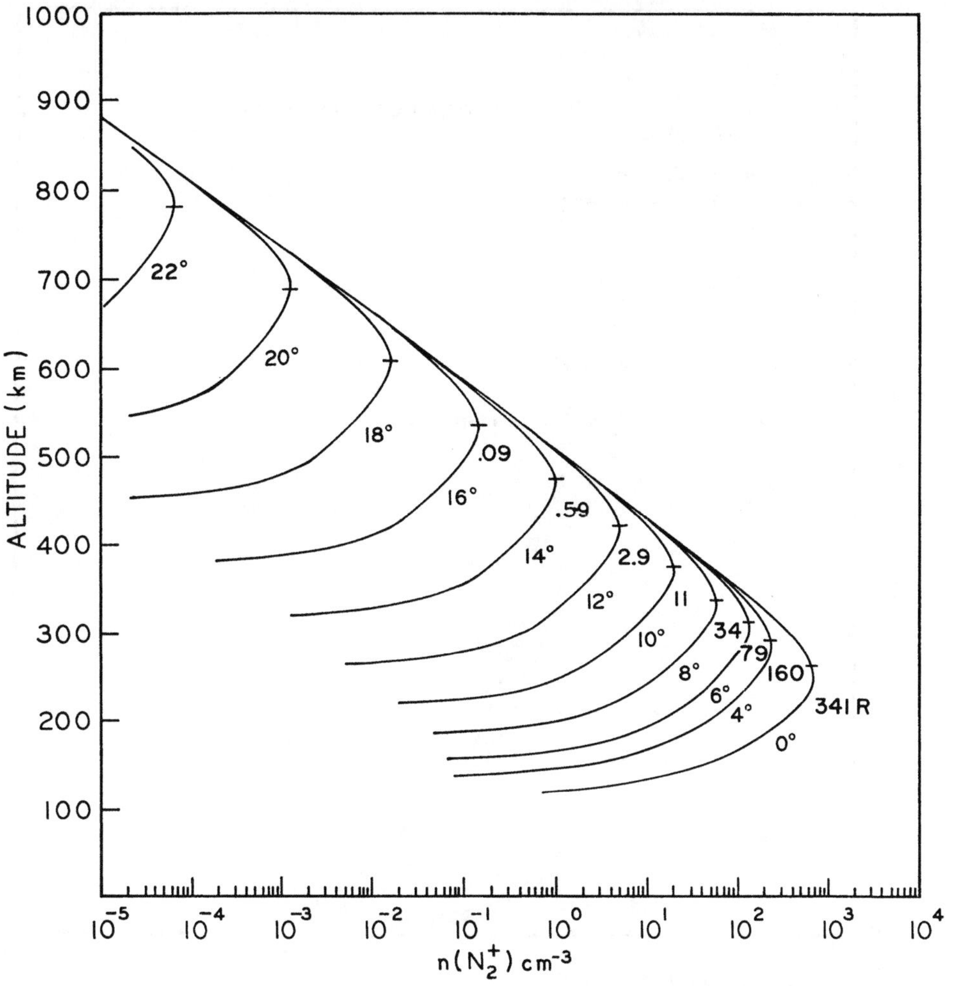

Fig. 3. Calculated N_2^+ ion concentration at various solar depressions. The centroid of the layer is indicated as well as the resulting emission rate (Broadfoot, 1968).

1969; Doering *et al.*, 1970; Stewart, 1970; Nagy and Banks, 1970). Collisional cross sections for the bands of interest here have recently been revised by Shemansky and Broadfoot (1971).

4. Vibrational Distribution of Intensity

The intensity of a band is given by

$$I_{v'v''} = N_{v'} A_{v'v''},$$

where $N_{v'}$ is the population of the upper vibrational level and $A_{v'v''}$ is the probability for spontaneous transition to the lower state v'' level. Good sets of $A_{v'v''}$ values are

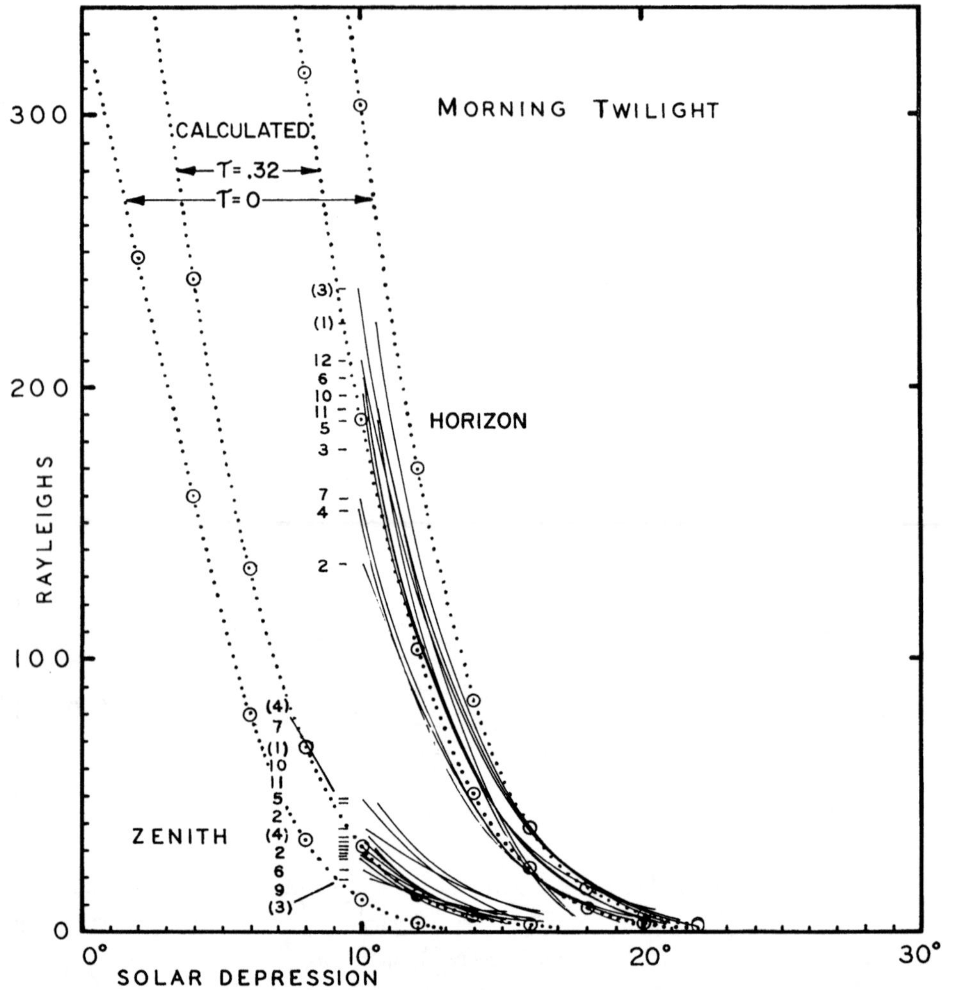

Fig. 4. Morning twilight intensity. The intensity of the 3914 Å band is given as a function of solar depression. Two calculated curves are shown by dots, with ($\tau = 0.32$) and without ($\tau = 0$) tropospheric scattering (Broadfoot, 1968).

now available for all of the band systems discussed here (Shemansky and Broadfoot, 1971). With moderate resolution it is possible to measure the relative intensity of several bands in a sequence, for instance, the 0–1, 1–2, 2–3, 3–4 bands of the 1N. Since these bands are from different vibrational levels, $v' = 0$, 1, 2, 3, the relative populations among these levels can be determined.

The N_2 systems can only be excited efficiently by electron bombardment. The upper state vibrational distribution reflects the Boltzmann distribution among the ground state molecules weighted by the transition probability which can be assumed proportional to the Franck-Condon factor. Although complicated to some extent by cascading, the distribution is predictable (Broadfoot and Hunten, 1964; Shemansky and Broadfoot, 1971).

Three sources of N_2^+ emissions are considered: simultaneous ionization and excitation from the N_2 ground state, resonance scattering by the N_2^+ ion ground state, and ionization by the $O^+(^2D)$ reaction. Each of these sources results in a distinctive vibrational distribution among the excited states. The previous work on this subject (Broadfoot, 1967) is updated to include the new absolute transition probabilities (Shemansky and Broadfoot, 1971).

Besides the differences in vibrational distribution, the sources of emission are different in another way. If the ion lifetime, τ, is very short ($\tau < 10^{-3}$ sec) as in the aurora or very low altitude dayglow, resonance scattering cannot contribute significantly; the vibrational distribution in the excited states arises from a Boltzmann distribution in the $N_2(X\,^1\Sigma_g^+)$ state weighted by the Franck-Condon factors of the ionizing transition. On the other hand, if the ion lifetime is very long ($\tau > 10^3$ sec) as at high altitude, resonance scattering will dominate; in this case an equilibrium condition is set up among all of the vibrational levels in the ion states. The lifetime of the ion is used as a variable parameter to illustrate the effect of mixing these emission sources (Figures 5, 6, and 7).

The lifetime of the ion can easily be extracted from the model atmosphere calculations:

$$\tau = \frac{\text{ion concentration}}{\text{ion destruction rate}}.$$

Broadfoot (1968) has included tropospheric scattering in the calculations of the observed emission at twilight and gives the integrated ion lifetimes listed in Table I. Ground based observations are only possible when the solar depression exceeds 10°.

Although the calculated apparent lifetimes increase to greater solar depression the change in vibrational distribution expected is small in Figure 5. A measured twilight distribution (Broadfoot, 1967) is plotted in Figure 5 and is in good agreement with

TABLE I

Model twilight emission 3914 Å band

Emission predicted at a zenith angle of 70° toward and in the azimuth of the sun (Broadfoot, 1968)

Solar depression (observer)	Centroid (km)	I (R)	Apparent lifetime (sec)
10	298	189	13
12	318	103	16
14	342	51	20
16	369	23	26
18	400	9.5	35
20	436	3.6	45
22	474	1.2	60

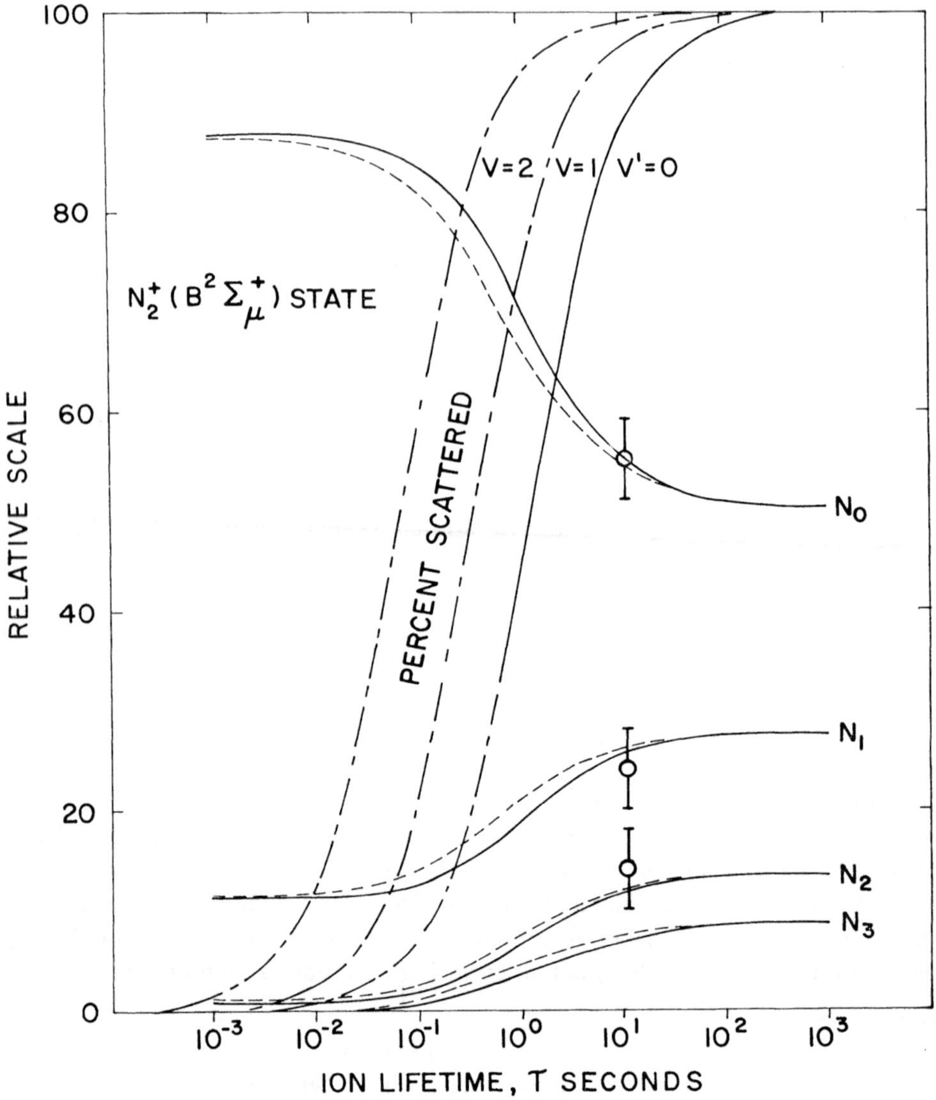

Fig. 5. The calculated dependence of the relative populations among the vibrational levels in the N_2^+ ($B\,^2\Sigma_u^+$) state on the ion lifetime in a sunlit atmosphere. Dashed curve includes the O^+ (^2D) reaction (see text). Measured relative populations from Broadfoot (1967) are indicated.

the model calculations. If shorter lifetimes were detected this would indicate a contribution from lower altitudes due to fast photoelectrons.

The Meinel system has not been detected in the twilight; in any case the expected measurement error would be large and tend to obscure apparent lifetime variations.

This technique of separating the excitation processes may be more fruitful with

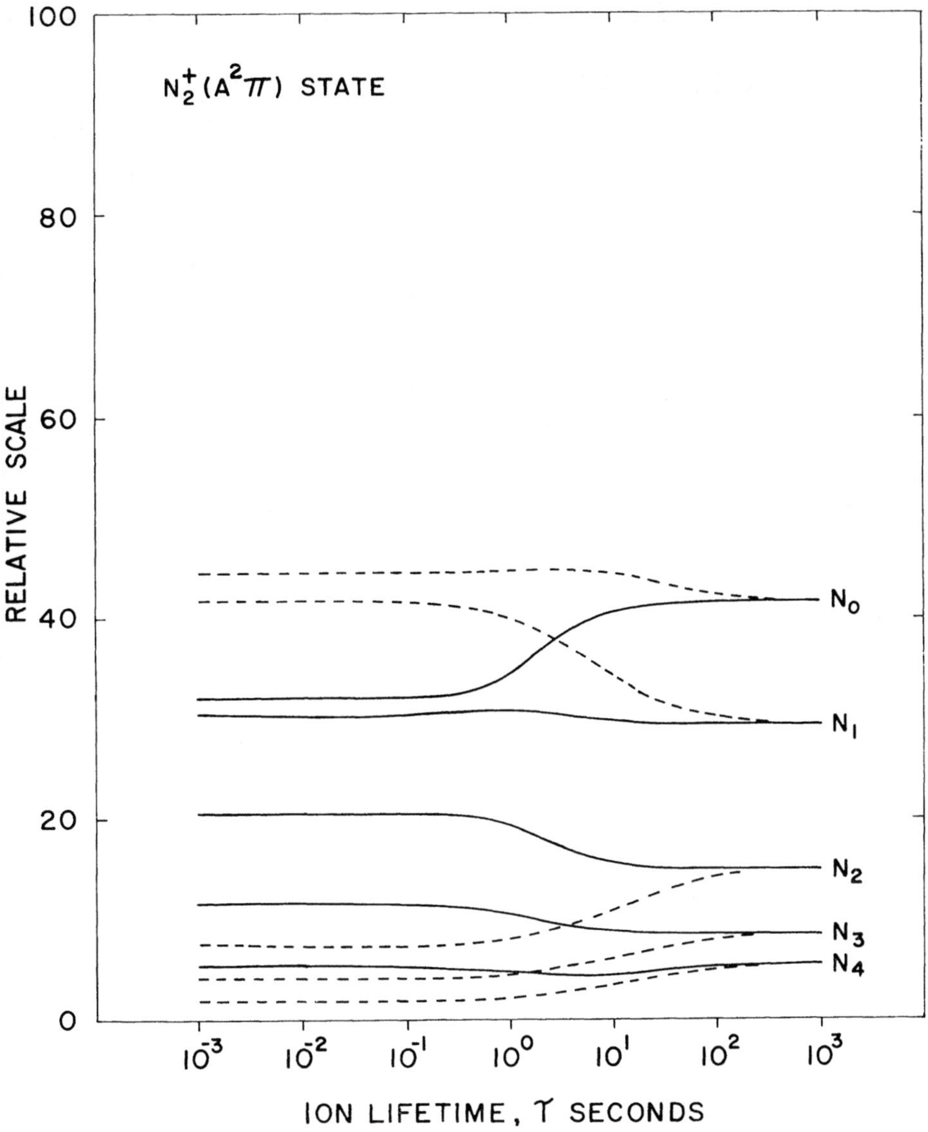

Fig. 6. The calculated dependence of the relative populations among the vibrational levels in the N_2^+ ($A\ ^2\Pi$) state on the ion lifetime in a sunlit atmosphere. Dashed curve includes the O^+ (^2D) reaction (see text).

rocket investigations of the dayglow. A typical apparent lifetime (1 sec) of the ionized layer was estimated from Wallace and McElroy (1966) by Wallace and Broadfoot (1969). As a rocket penetrates the ion layer (cf., Figure 2) the dominant source of emission rapidly becomes resonance scattering. This should be detected as a change in the vibrational distribution and should be consistent with an increase in the

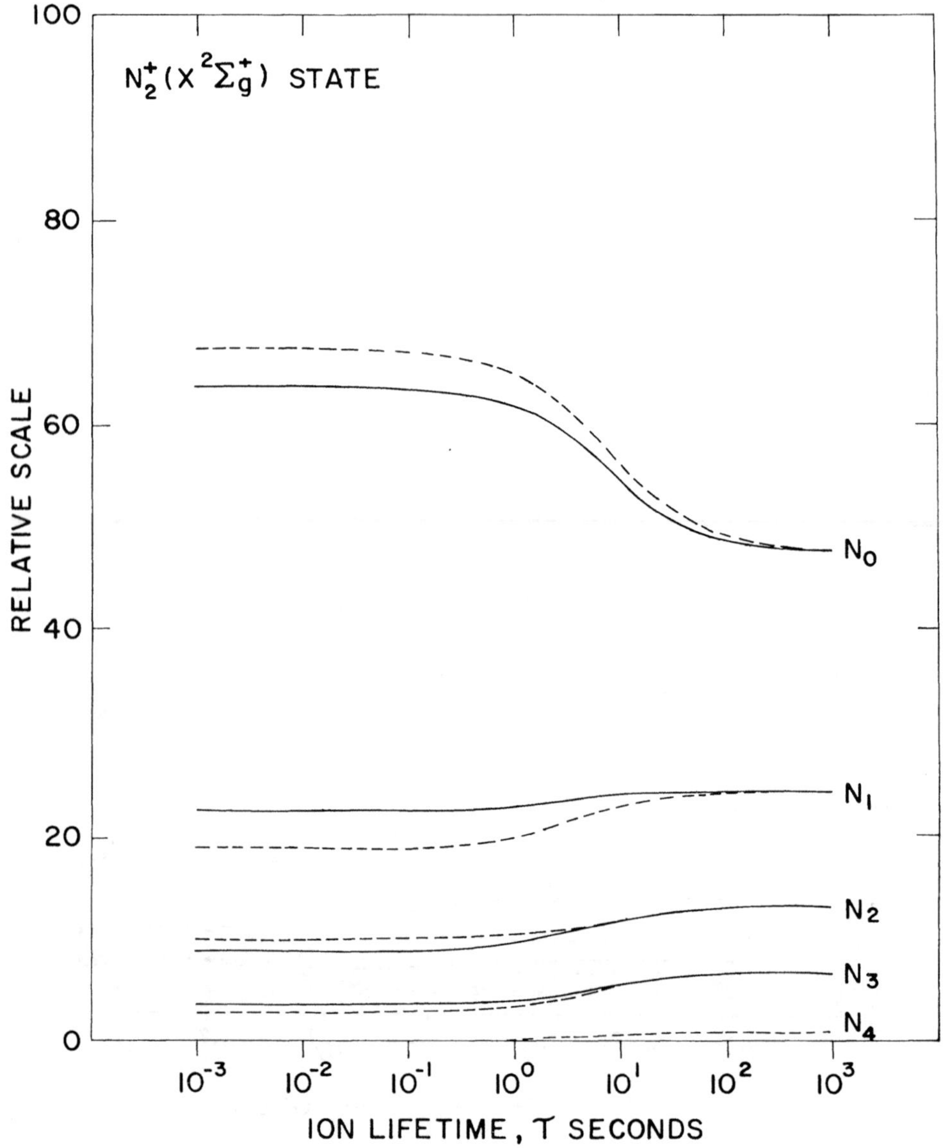

Fig. 7. The calculated dependence of the relative populations among the vibrational levels in the N_2^+ ($X\,^2\Sigma_g^+$) state on the ion lifetime in a sunlit atmosphere. Dashed curve includes the O^+ (^2D) reaction (see text).

apparent lifetime of the ion. Ahmed reported at this meeting that he had dayglow measurements which do show this characteristic.

The third source of ionization, the O^+ (^2D) reaction, may be distinguishable in the vibrational distribution of the N_2^+ band systems. It is not known how the energy is distributed among the ion vibrational levels of the ion states. The energy available from the $O^+(^2D)$ ion is in very near resonance with the $A\,^2\Pi_u$, $v'=1$ vibrational level (Omholt, 1959). Higher levels than this cannot be excited. However, Moore and Doering (1969) report that low energy ion-neutral reactions, which excite the ionized states of N_2, tend to distribute the ion products more uniformly among the vibrational levels than predicted by the Franck-Condon principle. For the purpose of illustration, the assumption was made that half the N_2^+ ions were produced by the $O^+(^2D)$ reaction and that the ions entered the $A\,^2\Pi_u$ state equally through the $v'=0$ and 1 levels; dashed curves in Figure 5, 6, and 7 represent the resulting distributions. The effect on the 1N is small but the change in the Meinel vibrational distribution may be detectable. A rocket measurement of Meinel bands from the $v'=1$ and 2 levels could indicate the importance of the $O^+(^2D)$ reaction.

5. Rotational Distribution of Intensity

The 3914 Å band of N_2^+ is undoubtedly the most useful band for studies of rotational energy distribution. It is the most intense band in the N_2 and N_2^+ systems and has a relatively simple structure to observe or calculate. The other systems have innumerable branches and intermeshing of the rotational lines. The three sources of emission mentioned above also have distinguishing characteristics detectable in the rotational line intensity.

Simultaneous ionization and excitation of the N_2^+ $(B\,^2\Sigma_g^+)$ state is followed by spontaneous emission of the 1N. Emission takes place in about 10^{-8} sec which precludes any possibility of rotational redistribution in the excited state. Emission arising in this manner reflects the rotational distribution of the $N_2(X\,^1\Sigma_g^+)$ ground state which is Boltzmann in thermal equilibrium with the atmosphere. Once the ion is formed a few collisions will allow rotational redistribution but there would be little difference from the neutral ground state.

The feature which distinguishes resonance scattering is called the 'Swings Effect' (cf., Hunten, 1963). The scattering rate of each rotational line depends on the solar flux at that wavelength and therefore the Fraunhofer absorption modifies the rotational line intensities significantly. This effect was first detected in the 3914 Å band emission by Vallance Jones and Hunten (1960); it is also prominent in comets.

The $O^+(^2D)$ reaction may extend the rotational distribution significantly. Broadfoot and Hunten (1966) obtained a 3914 Å band spectrum which fitted a synthetic spectrum fairly well, except that the observed band appeared to have an extended rotational development. This observation led Dalgarno and McElroy (1966) to suggest that the $O^+(^2D)$ reaction could be important since it could perturb the rotational distribution in the observed manner. Moore and Doering (1968) have since shown that the rotational development of the N_2^+ bands is substantially extended by using low energy ions as the excitation particle. We can only speculate what the distribution might be for a zero energy reaction such as the $O^+(^2D)$. Moore and Doering (1968) found the effect increased with decreasing collision energy; their lowest energy ions produced a tail which was comparable to a rotational temperature of 3500 K.

Although a quantitative study has not been undertaken the evidence suggests that the rotational distribution may be useful also in separating the three sources of emission outlined above.

References

Broadfoot, A. L.: 1967, *Planetary Space Sci.* **15**, 1801.
Broadfoot, A. L.: 1968, *Planetary Space Sci.* **16**, 693.
Broadfoot, A. L. and Hunten, D. M.: 1964, *Can. J. Phys.* **42**, 1212.
Broadfoot, A. L. and Hunten, D. M.: 1966, *Planetary Space Sci.* **14**, 1303.
Campbell, I. M. and Thrush, B. A.: 1967, *Proc. Roy. Soc.* **A296**, 201.
Dalgarno, A. and McElroy, M. B.: 1966, *Planetary Space Sci.* **14**, 1321.
Dalgarno, A., McElroy, M. B., and Stewart, A. I.: 1969, *J. Atmospheric Sci.* **26**, 753.
Doering, J. P., Fastie, W. G., and Feldman, P. D.: 1970, *J. Geophys. Res.* **75**, 4787.
Hoffman, J. H., Johnson, C. Y., Holmes, J. C., and Young, J. M.: 1969, *J. Geophys. Res.* **74**, 6281.
Holmes, J. C., Johnson, C. Y., and Young, J. M.: 1965, *Space Res.* **5**, 756.
Hunten, D. M.: 1963, *Planetary Space Sci.* **10**, 37.
Hunten, D. M.: 1967, *Space Sci. Rev.* **6**, 493.
Moore, J. H., Jr. and Doering, J. P.: 1968, *Phys. Rev.* **174**, 178.
Moore, J. H., Jr. and Doering, J. P.: 1969, *Phys. Rev.* **177**, 218.
Nagata, T., Tahmatsu, T., and Ogawa, T.: 1965, *Planetary Space Sci.* **13**, 1273.
Nagy, A. F. and Banks, P. M.: 1970, Paper given at the A.G.U. Meeting, Washington.
Nagy, A. F., Fontheim, E. G., Stolarski, R. S., and Beutler, A. E.: 1969, *J. Geophys. Res.* **74**, 4667.
Noxon, J. F.: 1968, *Space Sci. Rev.* **8**, 92.
Omholt, A.: 1959, *Geophys. Publ.* **20**, No. 11.
Shemansky, D. E.: 1969, *J. Chem. Phys.* **51**, 689.
Shemansky, D. E. and Broadfoot, A. L.: 1971, *J. Chem. Phys.*, submitted.
Stewart, A. I.: 1970, Paper given at the A.G.U. Meeting, Washington.
Vallance Jones, A. and Hunten, D. M.: 1960, *Can. J. Phys.* **38**, 458.
Wallace, L. and Broadfoot, A. L.: 1969, *Planetary Space Sci.* **17**, 975.
Wallace, L. and McElroy, M. B.: 1966, *Planetary Space Sci.* **14**, 677.
Zipf, E. C.: 1969, *Can. J. Phys.* **47**, 1863.
Zipf, E. C. and Fastie, W. G.: 1964, *J. Geophys. Res.* **69**, 2357.

INSTITUTE DISCUSSION

L. Thomas emphasized that L. Broadfoot's statement that "an evening enhancement of 3914 Å seen in the winter hemisphere as a result of conjugate region photoelectrons" is of considerable interest in ionispheric studies. He asked if the observations of resonant scattering of metastable He at 10 828 Å or in another line at the same level also provide evidence of the existence of high energy photoelectrons from the conjugate point. L. Broadfoot was unaware of any reported observations but felt that the He line at 3889 Å followed the 3914 Å band activity very well.

It is especially important to note that resonance scattering of solar radiation will seriously modify the rotational temperature.

SOME OBSERVATIONS OF THE EARTH'S ULTRAVIOLET DAYGLOW

J. B. PEARCE

Laboratory for Atmospheric and Space Physics University of Colorado, Boulder, Colo., U.S.A.

1. Introduction

Scanning UV spectrometers of the kind described by Fastie (1952a, b, 1963, 1967) aboard sounding rockets and satellites have been used to measure the earth's UV dayglow. This paper will describe some recent observations and comment upon their interpretation.

2. Experimental Detail

Observations of the upper atmosphere made at zenith angles greater than 90° and from an altitude above the observed layer are characterized by an increased optical signal compared to those made from below the layer at a smaller zenith angle for optically thin situations. This 'limb brightening' can result in a signal gain of as much as 40 for problems of atmospheric interest on the earth.

A set of observations utilizing this ability was made from Aerobee rocket above White Sands Missile Range on June 13, 1969. The instrument used to make the observations was a $\frac{1}{4}$ m focal length Ebert-Fastie scanning the spectral range 1100 to 2000 Å with a 10 Å resolution and 2000 to 4300 Å with 20 Å resolution in 3 sec. The instrument was equipped with a $\frac{1}{4}$ m focal length reflecting telescope which was used to limit the field of view to 5×50 km at the maximum viewing distance of ~ 1100 km. The instrument was oriented such that the 5 km dimension limited the field in the vertical. This instrument was identical in its characteristics to those observing Mars from Mariners 6 and 7.

During the time that the rocket was above 100 km a sun locked attitude control system caused the instruments viewing zenith distance to twice sweep the region from 90° to 100° and back to 90°. This allowed the altitude of the minimum ray height, that altitude from which the significant part of the optical signal comes, to sweep from the vehicle altitude down through the atmosphere to approximately 100 km below the vehicle and back, this sequence occurring in the 60 sec just before and the 60 sec just after apogee at 172 km. Thus, 40 spectra were obtained of the atmosphere in the altitude range between 60 and 170 km. Figure 1 is a spectrum typical of these taken when the instrument was at 171 km and viewing the atmosphere at 105 km at a 98° zenith angle.

3. Dayglow Observations

With the exception of the Lyα line of H at 1216 Å all of the features in this spectrum are due to N and O in various chemical states.

B. M. McCormac (ed.), The Radiating Atmosphere, 45–48. *All Rights Reserved.*

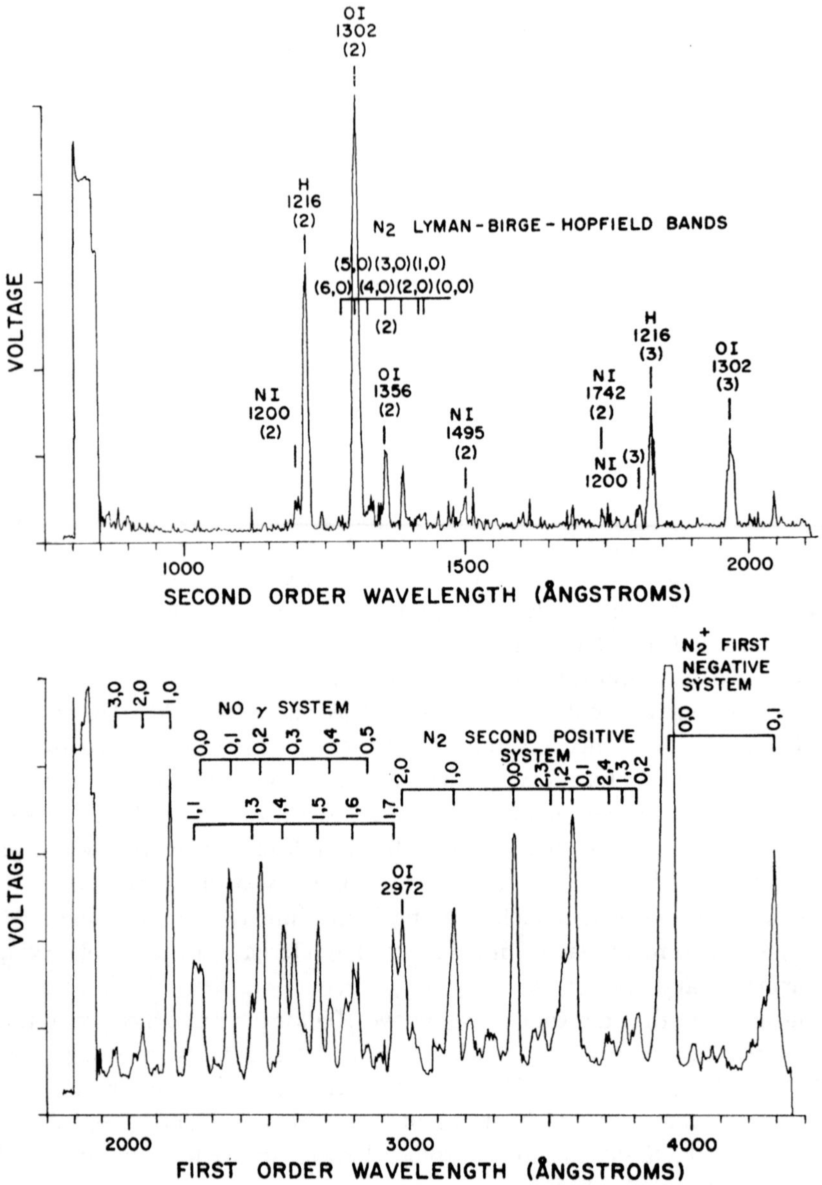

Fig. 1. The spectrum of the earth's UV dayglow from 1100 to 2000 Å (above) and 1900 to 4300 Å (below). The features at the left of each segment are wavelength fiducials. In the upper segment each feature is labelled with a numeral in parenthesis indicating its diffraction order.

A. OXYGEN

Three features due to O are found in this spectrum: the 1302 Å, 1305 Å, and 1306 Å resonance triplet ($^3P-{}^3S$); the 1356 Å line ($^3P-{}^5S$); and the 2972 Å line ($^3P-{}^1S$). The resonance triplet and the 1356 Å line in the dayglow have been measured and studied thoroughly in the past years (cf., Fastie *et al.*, 1964; Green and Barth, 1967; Dalgarno *et al.*, 1969). The triplet is caused, in part, at low altitudes by the resonance scattering of sunlight. Since the oscillator strength of this transition is relatively large, the observations are difficult to interpret in terms of an O density because the intensity of the radiation depends sensitively upon a number of parameters. The 1356 Å line is optically forbidden, however, and resonance scattering fails to account for the measured intensity by a factor of about 100 (Donahue, 1965). It is most likely that both of these lines are excited at low altitudes by photoelectron impact on ground state O. The 2972 Å line is also optically forbidden. As such, it may also be excited by electron impact; however, the relatively low (4.17 eV) energy of its upper state make it accessible to a number of photochemical processes as well.

B. NITROGEN

Several lines of N: the 1200 Å resonance triplet ($^4S-^4P$); the 1493 Å triplet ($^2D-^2P$); the 1744 Å quartet ($^2P-^2P$); and the 3466 Å forbidden line ($^4S-^2P$) are observed. The situation with regard to the 1200 Å feature (Fastie *et al.*, 1964) is somewhat analogous to its counterpart in O, the 1302 Å, 1305 Å, and 1306 Å triplet, except that it is not been studied as completely. The 1493 Å, 1744 Å, and 3466 Å lines are believed to be due to photoelectron excitation of N and N_2 at these altitudes, and, in the case the 3466 Å line, with a possible photochemical contribution. The production of the 1200 Å, 1493 Å, and 1744 Å lines by photoelectron impact on N_2 has recently been measured in the laboratory (Ajello, 1970).

In addition, the Lyman-Birge-Hopfield (LBH), Vegard-Kaplan (V-K) and 2P bands of N_2, as well as the 1N bands of N_2 ions are observed. The LBH ($a\,^1\Pi_g-X\,^1\Sigma_g^+$) bands appear to be excited by photoelectron impact, i.e., the relative intensities of the bands observed are predicted by the Franck-Condon principle. The V-K ($A\,^3\Sigma_u^+-X\,^1\Sigma_g^+$) and 2P ($C\,^3\Pi_u-B\,^3\Pi_g$) both require an optically forbidden transition from the ground state for excitation, and hence are most likely excited by photoelectrons (cf., Dalgarno *et al.*, 1969). For a discussion of the 1N($B\,^2\Sigma_u^+-X\,^2\Sigma_g^+$) system see Dalgarno *et al.* (1969) and Broadfoot (1971).

C. NITRIC OXIDE

Also present in the spectrum are the γ bands ($A\,^2\Sigma^+-X\,^2\Pi$) of NO (Barth, 1964). This system is observed in resonance fluorescence (Barth, 1964) and thus, given a knowledge of the intensity of the solar flux at the appropriate wavelengths and the band strengths, the number of molecules in the field of view may be calculated (Pearce, 1969a) and the density of NO determined (Barth, 1966a, b; Pearce, 1969b; Meira, 1971).

Acknowledgments

The experiment described in this paper was planned and executed under the direction of Professor C. A. Barth. I wish to thank him and Mr. K. K. Kelly for their vital contributions. This work was supported under NASA Grant No. NGL-06-003-052.

References

Ajello, J. M.: 1970, *J. Chem. Phys.* **53**, 1156.
Barth, C. A.: 1964, *J. Geophys. Res.* **69**, 3301.
Barth, C. A.: 1966a, *Ann. Geophys.* **22**, 198.
Barth, C. A.: 1966b, *Planetary Space Sci.* **14**, 623.
Broadfoot, A. L.: 1971, this volume, p. 34.
Dalgarno, A., McElroy, M. B., and Stewart, A. I.: 1969, *J. Atmospheric Sci.* **26**, 753.
Donahue, T. M.: 1965, *Planetary Space Sci.* **13**, 871.
Fastie, W. G.: 1952a, *J. Opt. Sci. Amer.* **42**, 641.
Fastie, W. G.: 1952b, *J. Opt. Sci. Amer.* **42**, 647.
Fastie, W. G.: 1963, *J. Quant. Spectr. Radiative Transfer.* **3**, 507.
Fastie, W. G.: 1967, *Appl. Opt.* **6**, 397.
Fastie, W. G., Crosswhite, H. M., and Heath, D. F.: 1964, *J. Geophys. Res.* **69**, 4129.
Green, A. E. S. and C. A., Barth: 1967, *J. Geophys. Res.* **72**, 3975.
Meira, L. G.: 1971, *J. Geophys. Res.* **76**, 202.
Pearce, J. B.: 1969a, *J. Quant. Spectr. Radiative Transfer.* **9**, 1593.
Pearce, J. B.: 1969b, *J. Geophys. Res.* **74**, 853.

PART II

ATMOSPHERIC PROCESSES

INTERPRETATION OF AIRGLOW IN TERMS OF EXCITATION MECHANISMS

D Region Nightglow

R. L. GATTINGER

Astrophysics Branch National Research Council of Canada, Ottawa, Canada

Abstract. The behavior of the OH emissions and the IR Atmospheric O_2 bands in the night airglow is discussed with the aid of a model which simulates an oxygen-hydrogen atmosphere. The evaluation of a number of coefficients which enter the model is discussed. The effects of atmospheric mixing on the brightness variations of the emissions are considered.

1. Introduction

Numerous attempts have been made to explain the observed behavior of the night airglow emissions (Chamberlain, 1961; Shefov, 1969). The present discussion is limited to consideration of the excitation mechanisms involved in producing the OH emissions and the IR Atmospheric band system of O_2. The processes involved are characterized in terms of an oxygen-hydrogen atmosphere. Bates and Nicolet (1950) laid the framework for the understanding of the important processes occurring in an oxygen-hydrogen atmosphere, and many aspects of the behavior of the OH emissions in this type of atmosphere have been dealt with since then (e.g., Wallace, 1962). A significant advancement towards the solution of the problem was presented by Hunt (1966) who performed a numerical integration of the appropriate simultaneous differential equations. Subsequent models have been developed which include vertical diffusion (Hesstvedt, 1968; Shimazaki and Laird, 1970). The present model is an extension of one which has been described previously (Gattinger, 1968, 1969); the same reaction numbering system has been used. It incorporates recent measurements of reaction rates and solar photodissociation coefficients.

It is necessary to remember that all models of this type have a number of serious limitations. The most obvious one is the complete lack of experimental information on some reaction rates. Another limitation is the difficulty of mathematically simulating realistic boundary conditions in the model since this requires a knowledge of general atmospheric circulation patterns to establish the correct flux conditions. Also, a limitation on the accuracy of the solution may be introduced by the numerical integration techniques employed since computational time must usually be kept as short as possible.

2. Model

A. REACTION RATES

The reactions considered when deriving the model were the same as those used previously (Table IA, Gattinger, 1968) with the exception of those listed in Table IB.

B. M. McCormac (ed.), The Radiating Atmosphere, 51–63.

TABLE IA

Reactions (Gattinger, 1968)

Number	Reaction	Rate coefficient
1*a*	$O_2 + h\nu \rightarrow O(^3P) + O(^3P)$	J_{2H} (Herzberg)
1*b*	$O_2 + h\nu \rightarrow O(^1D) + O(^3P)$	J_{2S} (Schumann-Runge)
2*a*	$O_3 + h\nu \rightarrow O_2(^3\Sigma_g^-) + O(^3P)$	J_{3C} (Chappuis)
2*b*	$O_3 + h\nu \rightarrow O_2(^1\Delta_g) + O(^1D)$	J_{3H} (Hartley)
3	$O(^3P) + O_2 + M \rightarrow O_3 + M$	$k_3 = 1.5 \times 10^{-34} \exp(445/T)$
4	$O(^3P) + O_3 \rightarrow 2O_2$	$k_4 = 5.6 \times 10^{-11} \exp(-2850/T)$
5*a*	$O(^3P) + O(^3P) + M \rightarrow O_2 + M$	$k_{5a} + k_{5b} = 2.7 \times 10^{-33}$
5*b*	$O(^3P) + O(^3P) + M \rightarrow O_2(^1\Delta_g) + M$	$k_{5b} = k_{5a}/4$
6	$O(^1D) + O_3 \rightarrow 2O_2$	$k_6 = 10^{-11}$
7	$O(^1D) + M \rightarrow O(^3P) + M$	$k_7 = 10^{-12}$
8	$H + O_3 \rightarrow OH + O_2$	$k_8 = 1.5 \times 10^{-12} \sqrt{T}$
9	$OH + O(^3P) \rightarrow H + O_2$	$k_9 = 3 \times 10^{-12} \sqrt{T}$
10	$OH + O_3 \rightarrow HO_2 + O_2$	$k_{10} = 5 \times 10^{-13}$
11	$OH + OH \rightarrow H_2O + O(^3P)$	$k_{11} = 5 \times 10^{-12} \sqrt{T} \exp(-1000/T)$
12	$OH + HO_2 \rightarrow H_2O + O_2$	$k_{12} = 1.5 \times 10^{-12} \sqrt{T}$
13	$H + HO_2 \rightarrow H_2 + O_2$	$k_{13} = 2 \times 10^{-13}$
14	$H + HO_2 \rightarrow 2OH$	$k_{14} = 1 \times 10^{-11}$
15	$O(^3P) + HO_2 \rightarrow OH + O_2$	$k_{15} = 1.5 \times 10^{-12} \sqrt{T}$
16	$H + O_2 + M \rightarrow HO_2 + M$	$k_{16} = 3.3 \times 10^{-33} \exp(800/T)$
17	$H + OH + M \rightarrow H_2O + M$	$k_{17} = 2.5 \times 10^{-31}$
18	$H + H + M \rightarrow H_2 + M$	$k_{18} = 2.6 \times 10^{-32}$
19	$HO_2 + HO_2 \rightarrow H_2O_2 + O_2$	$k_{19} = 5 \times 10^{-12} \sqrt{T} \exp(-1000/T)$
20	$OH + H_2O_2 \rightarrow H_2O + HO_2$	$k_{20} = 4 \times 10^{-13}$
21*a*	$O(^3P) + H_2O_2 \rightarrow OH + HO_2$	$k_{21a} = 1 \times 10^{-15}$
21*b*	$O(^3P) + H_2O_2 \rightarrow O_2 + H_2O$	$k_{21b} = 1.5 \times 10^{-13} \sqrt{T} \exp(-2000/T)$
		$k_{21a} + k_{21b} = 2 \times 10^{-15}$
22*a*	$H + H_2O_2 \rightarrow H_2 + HO_2$	$k_{22a} = 1 \times 10^{-13}$
22*b*	$H + H_2O_2 \rightarrow OH + H_2O$	$k_{22b} = 5 \times 10^{-12} \sqrt{T} \exp(-3000/T)$
23	$O(^1D) + H_2 \rightarrow OH + H$	$k_{23} = 1 \times 10^{-11}$
24	$O(^1D) + H_2O \rightarrow 2OH$	$k_{24} = 1 \times 10^{-11}$
25	$O(^3P) + OH + M \rightarrow HO_2 + M$	$k_{25} = 1.4 \times 10^{-31}$
26	$HO_2 + O_3 \rightarrow OH + 2O_2$	$k_{26} = 1 \times 10^{-14}$
27	$H_2O + h\nu \rightarrow OH + H$	J_w
28	$H_2O_2 + h\nu \rightarrow 2OH$	J_{HP}
29	$H + O_2 \rightarrow OH + O$	$k_{29} = 3.6 \times 10^{-9} \sqrt{T} \exp(-12000/T)$
30	$H_2 + OH \rightarrow H_2O + H$	$k_{30} = 5 \times 10^{-12} \sqrt{T} \exp(-3000/T)$
31	$H + OH \rightarrow H_2 + O$	$k_{31} = 2 \times 10^{-13} \sqrt{T} \exp(-3400/T)$
32	$HO_2 + h\nu \rightarrow OH + H$	J_P
33	$H + O_3 \rightarrow HO_2 + O$	$k_{33} = 1.5 \times 10^{-11} \sqrt{T} \exp(-2000/T)$
34	$H_2 + O \rightarrow OH + H$	$k_{34} = 5 \times 10^{-13} \sqrt{T} \exp(-4400/T)$
35	$O_2(^1\Delta_g) \rightarrow O_2(^3\Sigma_g^-) + h\nu$	$k_{35} = 2.6 \times 10^{-4}$
36	$O_2(^1\Delta_g) + M \rightarrow O_2(^3\Sigma_g) + M$	$k_{36} = 1.7 \times 10^{-11} \sqrt{T} \cdot F$
37	$O_2(^1\Delta_g) + O(^3P) \rightarrow O_2(^3\Sigma_g^-) + O(^3P)$	k_{37}
38	$O_2(^1\Delta_g, {}^1\Sigma_g^-) + O_3 \rightarrow O_2(^3\Sigma_g^-) + O_3$	$k_{38} = 1.7 \times 10^{-14}$

Rate k_7 has been changed to 6×10^{-11} cm^3 sec^{-1} in accordance with a number of recent laboratory measurements (Zipf, 1969). Rate k_{24} has been varied, but, unless indicated otherwise, a value of 3×10^{-11} cm^3 sec^{-1} (Biedenkapp *et al.*, 1970) has been assumed. Snelling and Hampson (1969) have discussed the interpretation of several

TABLE IB

Modifications and additions to the reactions considered in the previous model (Gattinger, 1968).

Number	Reaction	Rate coefficient	Reference
1b	$O_2 + h\nu \rightarrow O(^1D) + O(^3P)$	J_{2S}	Text
2b	$O_3 + h\nu \rightarrow O_2(^1\Delta_g) + O(^1D)$	J_{3H}	Text
2c	$O_3 + h\nu \rightarrow O_2(^1\Delta_g) + O(^1S)$	J_{3HS}	Text
5b	$O + O + M \rightarrow O_2(^1\Delta_g) + M$	$k_{5b} = k_{5a}$	Text
7	$O(^1D) + M \rightarrow O(^3P) + M$	$k_7 = 6 \times 10^{-11}$	Zipf (1969)
24	$O(^1D) + H_2O \rightarrow 2OH$	$k_{24} = 3 \times 10^{-11}$	Biedenkapp *et al.* (1970)
27	$H_2O + h\nu \rightarrow OH + H$	J_W	Text
36	$O_2(^1\Delta_g) + M \rightarrow O_2(^3\Sigma_g^-) + M$	$k_{36} = 1.7 \times 10^{-11} \sqrt{T} \cdot F$	Text
39	$H + O_3 \rightarrow OH^* + O_2(^1\Delta_g)$	k_{39}	Text
40	$O(^1S) + O_2(^3\Sigma_g^-) \rightarrow O(^3P) + O_2(^3\Sigma_g^-)$	$k_{40} = 2 \times 10^{-13}$	Zipf (1969); Text
41	$O(^1S) + H_2O \rightarrow 2OH$	$K_{41} = 3 \times 10^{-10}$	Stuhl and Welge (1969)

measurements of this reaction rate. It is important that rates k_7 and k_{24} be known since the O_3 concentration is partially controlled as an indirect result of these two reactions.

Reaction 39 was considered as a possible source of $O_2(^1\Delta_g)$ molecules. In order to produce enough $O_2(^1\Delta_g)$ molecules to account for the observed brightness of the IR Atmospheric band system in the night airglow (Evans *et al.*, 1967) k_{39} would have to be about 10^{-12} cm^3 sec^{-1}. However, Schiff (1970) has found that the upper limit to k_{39} is several orders of magnitude lower, and consequently, the reaction has been neglected.

Reaction 2c in conjunction with Reaction 41 has been suggested by Hampson (1969) as a possible source of hydrogen species which would be effective in removing O_3 near the stratopause. The evaluation of the solar photodissociation coefficient, J_{3HS}, is discussed in the next section. A significant portion of the $O(^1S)$ atoms that are produced is removed by Reaction 40. Very little is known about the products of this reaction. If $O_2(^1\Sigma_g^+)$ and $O(^1D)$ are formed, then these products have very little effect on the atmosphere since the same products are formed indirectly and at a much faster rate by photodissociation of O_3 in the Hartley band (Wallace and Hunten, 1968; Noxon, 1970). For the present model the reaction products were assumed to be $O(^3P)$ and $O_2(^3\Sigma_g^-)$. The effect of including Reactions 2c, 40, and 41 will be discussed later.

Reaction rate k_{5b} was selected to provide enough $O_2(^1\Delta_g)$ molecules to attempt to explain the observed brightness of the IR Atmospheric band system in the night airglow (Evans *et al.*, 1967).

A value of 1.5×10^{-9} has been used for F in Reaction 36 (Findlay *et al.*, 1969).

The assumed atmospheric temperature profiles were extensions of those discussed previously (Gattinger, 1968).

The simultaneous differential equations which describe the rate of change of the constituents were numerically integrated using the finite difference method employed by Hunt (1966). The solutions were obtained for thirty-six hour periods with local noon taken as the starting point.

B. SOLAR PHOTODISSOCIATION COEFFICIENTS

The photodissociation coefficients are very similar to those used previously (Gattinger, 1968). It has been assumed that the efficiency of production of $O_2(^1\Delta_g)$ molecules in Reaction 2b is 75%. Approximately 50% of the photodissociations could lead to $O_2(^1\Sigma_g^+)$ as discussed by Wallace and Hunten (1968). However, Gauthier and Snelling (1970) have found that $O_2(^1\Sigma_g^+)$ may not be produced. If it is assumed that the efficiency of Reaction 2b is much less than 50% then it becomes impossible to explain the observed brightness of the IR Atmospheric bands in the day airglow (Evans *et al.*, 1967) since the rate of removal of $O_2(^1\Delta_g)$ molecules via Reaction 36 has been measured (Findlay *et al.*, 1969).

The method of evaluation of the dissociation coefficient for $O_2(^3\Sigma_g^-)$ in the region of the Schumann-Runge continuum has remained unchanged, but it has been modified considerably for the Schumann-Runge band region. Previous models have usually been developed with the simplifying assumption that the absorption cross section for the Schumann-Runge bands can be expressed as a series of averaged values over a number of specified wavelength intervals all of which are assumed to be independent of the path length of the absorber. This is a poor assumption if the individual wavelength intervals chosen contain rotational lines with widely differing absorption cross sections. Hudson *et al.* (1969) have measured the effective absorption cross sections as a function of path length for the Schumann-Runge bands. The photodissociation coefficients which they calculated for the Schumann-Runge band system were used in the present analysis.

Incorporating a variable absorption cross section with path length in the region of the Schumann-Runge bands has a very significant effect on the calculated photodissociation coefficient for water, J_W (Reaction 27). Anderson (1970) has given the effective absorption cross sections which he obtained from Hudson for the Schumann-Runge bands. These results have been adapted for numerical processing by fitting curves for σ (cross section in cm^2) of the form

$$\sigma = \sigma_0 e^{-A \ln N}$$

to the measured cross sections for various wavelength intervals as specified in Table II. The values of the constants σ_0 and A which are also listed in Table II specify the calculated values of the intercept and slope respectively of a straight line on a log-log plot. The variable N refers to the column density of $O_2(^3\Sigma_g^-)$ in molecules per cm^2. Anderson (1970) considered this effect of the variation of cross section with path length when he evaluated J_W, but it appears as though he overestimated J_W in the 50 to 60 km region when his values are compared with those given in Table III. Shimazaki and Laird (1970) have neglected absorption by the Schumann-Runge bands, and have assumed instead that absorption in the Herzberg continuum is more important in this region. This leads to a serious overestimate of J_W for the region

TABLE II

Constants used to evaluate absorption cross sections in the Schumann-Range band system

Wavelength	σ_0	A
1750–1782.6	1.014×10^{-9}	-0.5427
1782.6–1830.9	2.664×10^{-13}	-0.3767
1830.9–1864.1	7.097×10^{-14}	-0.3685
1864.1–1925.2	3.162×10^{-15}	-0.3256
1925.2–1973.4	4.618×10^{-18}	-0.2023

from about 30 to 70 km which results in an excessively high production of H and OH via Reaction 27. This in turn causes the O_3 concentration to be significantly reduced as will be shown later. The portion of J_W which arises in the region of the Schumann-Runge continuum and the effects of solar Lyα have been calculated in the same manner as was done previously (Gattinger, 1968).

The photodissociation coefficient, J_{3HS}, in Reaction 2c has been calculated using the O_3 cross sections given by Watanabe (1958) and the cross sections for absorption in the Schumann-Runge bands as discussed above. The value of J_{3HS} as a function of altitude is given in Table III.

TABLE III

Values of J_W and J_{3HS} (in sec^{-1}) for an overhead sun

Height – km	J_W	J_{3HS}
40		4.0×10^{-7}
50	1.0×10^{-8}	1.7×10^{-6}
60	9.3×10^{-8}	3.1×10^{-6}
70	7.0×10^{-7}	4.6×10^{-6}
80	4.0×10^{-6}	5.6×10^{-6}

C. EDDY DIFFUSION

Atmospheric mixing processes are not well understood, and they are therefore difficult to simulate mathematically. The technique used here is similar to that described by Hesstvedt (1968); it is based on the Fickian diffusion equation (Lettau, 1951). The effect is to modify the concentrations of the minor constituents as derived by the solution of the photochemical equations in such a way as to cause each of them to tend towards a constant number density mixing ratio with height relative to the total atmospheric density at that height. This type of solution is applicable only to small scale motions when diffusion coefficients, such as those given by Johnson and Wilkins (1965), are used as was the case in this model.

D. BOUNDARY CONDITIONS

As mentioned previously it is very difficult to establish an accurate set of boundary conditions since this implies that all the pertinent parameters are known. Shimazaki and Laird (1970) assumed that diffusive equilibrium was achieved at 150 km and they then varied the eddy diffusion coefficient to achieve various solutions. The technique employed here was to assume that photochemical equilibrium exists at the lower and upper boundaries of 30 and 95 km respectively. The concentrations of the minor constituents at the boundaries as derived in this way are generally adequately realistic for purposes of isolating the important processes.

A special problem which must be considered is that of the effects of the escape of H from the exosphere on the distribution of hydrogen compounds in the mesosphere. The problem becomes even more complicated when the lateral flow of H at the base of the exosphere is considered. The H flow may actually be downwards below 120 km on some occasions (Donahue, 1969). A general solution of the global H circulation patterns would have to be included along with a solution on a global basis of the model discussed here. This would form a logical extension of the current models.

3. Evaluating Importance of Various Reactions

A. REACTION CHAINS INVOLVING HYDROGEN COMPOUNDS

It has been recognized that there are several reaction chains involving H compounds which are effective in removing O and O_3 (e.g., Hampson, 1965). Reactions 10 and 26 are capable of removing significant amounts of O_3 below the stratopause. Reactions 8 and 9 remove a considerable amount of O and O_3 in the upper part of the mesosphere. Most of the H, OH, and HO_2 is formed directly or indirectly by Reactions 24, 27, and 41. The relative importance of these three reactions as determined using the model is discussed in the next section. The technique used was to obtain a standard solution in which representative values of all the reaction rates and photodissociation coefficients were used, and then to obtain solutions for which one of the parameters under investigation had been changed.

B. SOURCES OF ATOMIC HYDROGEN, HYDROXYL AND PERHYDROXYL MOLECULES

The importance of Reaction 24 in the removal of O and O_3 was evaluated by varying k_{24} from 2×10^{-10} cm^3 sec^{-1} to the measured value of about 3×10^{-11} cm^3 sec^{-1} which was used in the standard solution. The results given in Table IV indicate that this reaction is significant. If the value of 2×10^{-10} cm^3 sec^{-1} is used the predicted O_3 concentration is reduced so that agreement with observations (Johnson *et al.*, 1952; Miller, 1969) is better at some heights than if a rate of 3×10^{-11} cm^3 sec^{-1} is used since the O_3 concentration in the latter case increases by about 25% in the lower part of the mesosphere. If the rate is further reduced the disagreement is increased. Since this reaction has a definite influence on the O_3 concentration it should be the subject of further laboratory investigations.

The rate of Reaction 41 was varied from the standard value of 3×10^{-10} cm^3 sec^{-1} (Stuhl and Welge, 1969) to 1×10^{-11} cm^3 sec^{-1} to check the importance of the reaction. The results (Table IV) indicate that this mechanism may be important in the lower part

TABLE IV

Comparison between observed (mid-latitude) and predicted (58.8°N, Aug. 1) O_3 concentrations (cm^{-3}) at sunset

Height (km)	Johnson *et al.* (1952)	Miller (1969)	Standard solution 58.8°N, Aug. 1	$k_{24} = 2 \times 10^{-10}$	$k_{41} = 1 \times 10^{-11}$	J_W – Shimazaki and Laird (1970)	$J_W = 0$	Standard solution 58.8°N, Dec. 21
30	2.6×10^{12}		3.0×10^{12}	2.9×10^{12}	3.0×10^{12}	7.6×10^{11}	3.0×10^{12}	3.1×10^{12}
35	1.2×10^{12}		1.8×10^{12}	1.7×10^{12}	1.8×10^{12}	5.1×10^{11}	1.8×10^{12}	1.5×10^{12}
40	4.8×10^{11}		1.0×10^{12}	9.4×10^{11}	1.1×10^{12}	2.2×10^{11}	1.0×10^{12}	6.3×10^{11}
45	1.5×10^{11}		3.7×10^{11}	3.0×10^{11}	4.6×10^{11}	5.3×10^{10}	3.7×10^{11}	2.9×10^{11}
50	5.8×10^{10}	8.0×10^{10}	8.8×10^{10}	6.9×10^{10}	1.1×10^{11}	9.3×10^{9}	8.9×10^{10}	1.3×10^{11}
55	2.1×10^{10}	2.5×10^{10}	2.5×10^{10}	1.9×10^{10}	3.1×10^{10}	1.9×10^{9}	2.7×10^{10}	5.4×10^{10}
60	7.6×10^{9}	1.2×10^{10}	7.1×10^{9}	5.9×10^{9}	8.1×10^{9}	5.4×10^{8}	9.3×10^{9}	2.1×10^{10}
65	2.5×10^{9}	6.6×10^{9}	1.2×10^{9}	1.2×10^{9}	1.3×10^{9}	1.4×10^{8}	2.4×10^{9}	4.4×10^{9}
70	6.3×10^{8}	3.0×10^{9}	2.3×10^{8}	2.3×10^{8}	2.3×10^{8}	4.4×10^{7}	5.7×10^{8}	6.3×10^{8}
75		5.0×10^{8}	3.2×10^{7}	3.2×10^{7}	3.2×10^{7}	1.1×10^{7}	7.0×10^{7}	5.3×10^{7}
80		8.6×10^{7}	2.0×10^{7}	2.0×10^{7}	2.0×10^{7}	1.2×10^{7}	2.5×10^{8}	1.3×10^{8}
85		1.0×10^{8}	1.2×10^{8}	1.2×10^{8}	1.2×10^{8}	9.0×10^{7}	3.4×10^{8}	1.0×10^{8}
90		1.1×10^{8}	8.2×10^{7}	8.2×10^{7}	8.2×10^{7}	7.5×10^{7}	1.1×10^{8}	3.3×10^{7}
95			1.6×10^{7}	1.6×10^{7}	1.6×10^{7}	1.6×10^{7}	1.7×10^{7}	6.7×10^{6}

of the mesosphere. Laboratory investigations are required to determine experimentally the coefficient for Reaction 2c. Also, if $O(^1S)$ atoms are formed via Reaction 2c it is possible that they may influence the atmospheric photochemistry in another way.

The photodissociation coefficient, J_W, in Reaction 27 was varied in two different ways as compared with its standard value to determine its effects on the O and O_3 concentrations. One solution was obtained with the assumption made by Shimazaki and Laird (1970) that solar energy in the 1750 to 2000 Å region was not absorbed by the Schumann-Runge bands but by the much weaker Herzberg continuum. The results (Table IV) clearly indicate that this mechanism is extremely effective as would be expected. However, since the absorption cross sections for the Schumann-Runge bands have been measured in the laboratory this assumption can no longer be considered as a realistic one. A second solution was obtained in which J_W was set to zero to determine the heights at which the photodissociation of water vapor has the greatest effect. The results (Table IV) show that in the 75 km region there is a significant difference between this solution and the standard solution even though the initial concentrations for both were assumed to be those existing in an oxygen-hydrogen atmosphere. This result stresses again the importance of assumptions involved in evaluating the photodissociation coefficient for H_2O. A comparison between the standard solutions for winter and summer at high latitudes indicates that the O_3 concentration between 50 and 80 km should be considerably higher during the winter. This is caused mainly by the decrease in the rate of photodissociation of water vapor at these levels during the winter. The solution for which J_W was set equal to zero for summer time conditions indicated a similar trend in the 50 to 80 km region. The O_3 concentration in the 40 km region should be smaller in winter according to the solution. This is a result of a decrease in solar energy which is capable of forming odd forms of oxygen.

4. Results and Discussion

A. OH EMISSIONS

Many observations of the diurnal variation of the intensity of the OH emission bands have been made (e.g. Berthier, 1956; Huppi and Stair, 1968; Snelling and Hampson, 1969; Moreels *et al.*, 1970). The results indicate that the brightness of the emission at midday, which decreases towards sunrise and sunset, is approximately equal to the nighttime value, that a twilight enhancement often occurs, that the brightness changes during the night are highly variable, and that a sudden decrease in brightness occurs at dawn. It is possible to predict some of these features, namely, the midday brightness, the twilight enhancement, and the sudden decrease at dawn using photochemical models (e.g., Hunt, 1966; Gattinger, 1969). One feature which has not been dealt with very much is the extreme variability of the emission during the night. The models generally predict that after the twilight enhancement the brightness decreases through the night (Figure 1). Berthier (1956) frequently found that the twilight enhancement was followed by a decay which lasted until about midnight, and then the brightness increased steadily until just before sunrise.

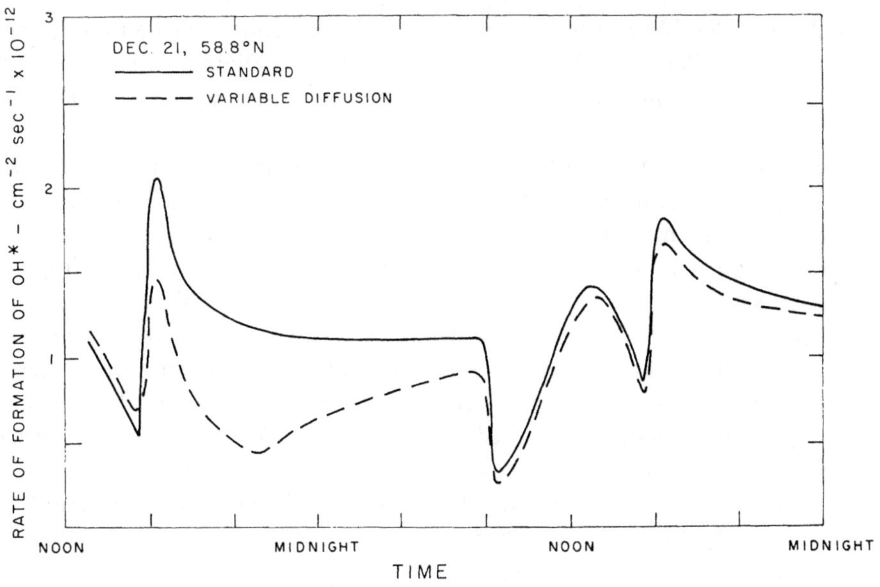

Fig. 1. Diurnal brightness variations of the OH emission.

Ballif and Venkateswaran (1963) have found that it is possible for the OH emission brightness to increase from sunset to sunrise if a very low H concentration is assumed to exist near the mesopause. This effectively causes the peak of the twilight enhancement to occur much later during the night, if it occurs at all, which does not agree with Berthier's (1956) observations. The likelihood of occurrence of this mechanism was checked by obtaining a solution for December 21, 58.8° N, with one of the initial conditions being that a very low H concentration of 10^6 atoms per cm^3 existed from 60 to 95 km. The results showed that a brightness increase may occur over a considerable portion of the first night, but during the second night the normally predicted twilight enhancement followed by a slow decrease in brightness are the dominant features. This occurs because the H concentration is increased very considerably during the sunlit hours above the very low initial values as a result of the photodissociation of water vapor. It is unlikely that the H_2O vapor mixing ratio will become so small as to make the photodissociation process unimportant unless upward mixing below 60 km is very slow as suggested by Hunten (1970). It is relevant to note that an attempt to explain the observed vertical profile of the OH emission near sunset (Llewellyn and Evans, 1971) using the current model indicated that the water vapor mixing ratio assumed to exist above 60 km may be too high. The average hydrogen production rate over a twenty-four hour period is about 10^9 cm^{-2} sec^{-1} which considerably exceeds the hydrogen escape rate (Donahue, 1969) so that the assumed boundary condition of photochemical equilibrium at 95 km should give realistic values in the upper mesosphere since photochemical loss processes for hydrogen should considerably exceed the escape rate. These results indicate that the mechanism suggested by Ballif and Venkateswaran (1963) for explaining an increase of OH emission brightness with time during the night may not be physically realizable in the atmosphere.

One process which must not be overlooked when dealing with OH brightness variations is that of atmospheric mixing. When O is transported downwards in the 80 km region during the night it releases H via Reaction 9 and forms O_3 via Reaction 3 which in turn results in the formation of OH molecules via Reaction 8. The importance of atmospheric mixing was assessed by obtaining a solution which assumed normal initial conditions with the exception that mixing was assumed to be ineffective until about 6 hr after the first sunset at which time the full effect of mixing was introduced. The results (Figure 1) show that this mechanism may well be responsible for the type of OH brightness variations observed by Berthier (1956). It remains to be shown experimentally that such changes occur in the rate of atmospheric mixing.

The predicted height profile of the OH emission in the night airglow (Hunt, 1966; Hesstvedt, 1968; Gattinger, 1969; Shimazaki and Laird, 1970) agrees reasonably well with the observations (Packer, 1961; Tarasova, 1963). A solution obtained using the current model gave similar results (Figure 2). It would be very instructive to measure the height profile of the emission near sunset, midnight, and sunrise on a night which

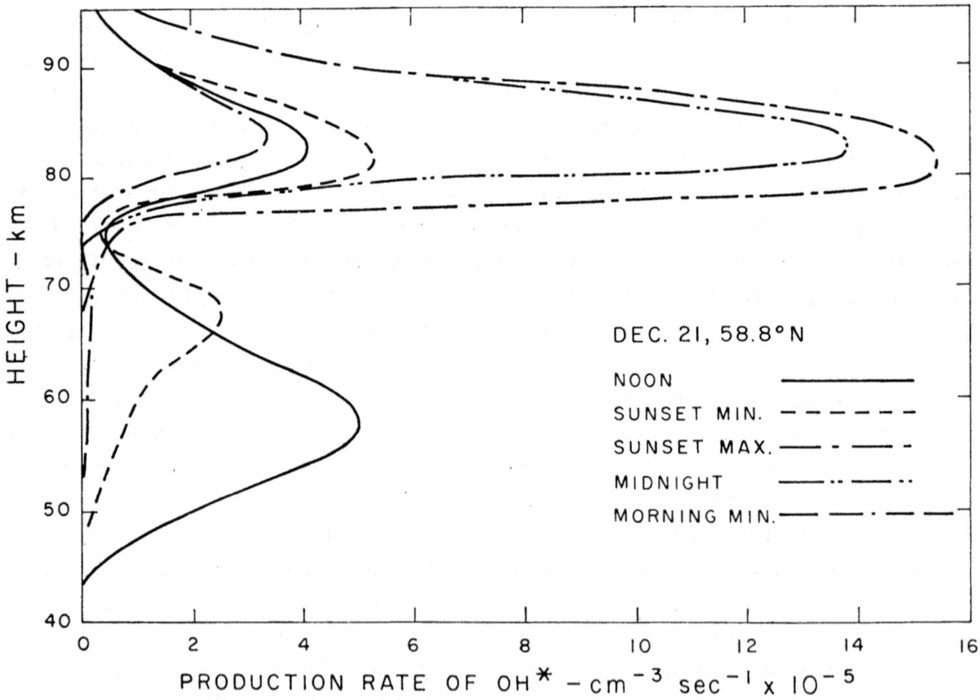

Fig. 2. Variation of the height profile of the OH emission.

exhibited the characteristic brightness variation observed by Berthier (1956) since it may then be possible to observe both time and spatial effects.

B. IR ATMOSPHERIC O_2 BAND EMISSIONS

The results of a number of measurements indicate that the brightness of the 1.27 μ IR Atmospheric O_2 band in the night airglow is relatively constant and has a value of about 75 kR (Gush and Buijs, 1964; Evans *et al.*, 1967). As mentioned previously the source of the 1.27 μ band in the night airglow was assumed to be Reaction 5b. It would seem reasonable to expect that the 1.27 μ band should exhibit a seasonal brightness variation if the emission is caused by Reaction 5b since the brightness at a given height must then depend upon the square of the O concentration which in turn should possess some type of seasonal dependence. Both atmospheric circulation and the amount of solar energy available to photodissociate O_2 and H_2O should have a significant effect on the O concentration.

The change in the rate of atmospheric mixing suggested when dealing with the OH night airglow emissions also affected slightly the brightness of the 1.27 μ band as expected. The solution indicated that the brightness increased by about 20% from the time that mixing was assumed to be effective until the following morning which

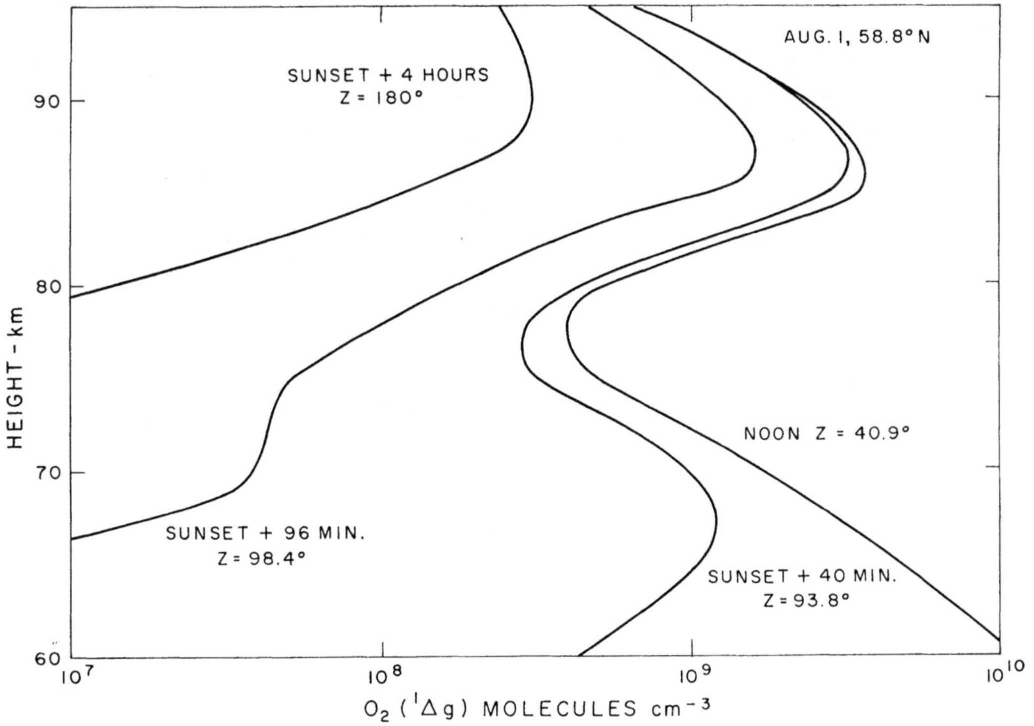

Fig. 3. Variation of the height profile of the 1.27μ O_2 band.

was about 10 hr. Such small changes are very difficult to observe with current IR detectors, but improved detection systems may make it possible to check the predicted correlation.

The predicted height profile of the 1.27 μ band emission is given in Figure 3. A higher concentration of O in the 90 km region is required to explain the brightness observed at night; downward motion from above 100 km may be important. There have been no definitive measurements of the 1.27 μ emission profile in the night airglow as yet. A measurement of the height profile of the O concentration at the same time would make the 1.27 μ band height profile much more useful.

5. Conclusions

Incorporation of a number of recently measured reaction rates and photodissociation coefficients into the model of an oxygen-hydrogen atmosphere had led to some difficulties in predicting the observed height profile of the O_3 concentration in the lower mesosphere. The rate of Reaction 24 should be investigated further since it is very effective at that altitude. The rate of photodissociation of H_2O must be carefully evaluated in any model which deals with OH emissions since it has a very significant effect on the O_3 and H concentrations in the upper mesosphere. Reaction 41 as suggested by Hampson (1969) may also be of importance in the determination of the O_3 concentration. Vertical motions of hydrogen species should be considered more fully.

The mechanism suggested by Ballif and Venkateswaran (1963) for producing an increase in the brightness of the OH emissions throughout the night appears to be unacceptable when a more complete time-dependent solution is obtained. A more satisfactory solution is obtained if it is assumed that there is a moderate change in the rate of atmospheric mixing during the night.

If the 1.27 μ O_2 emission arises from Reaction 5b then the peak of the emission should be at about 90 km. A simultaneous measurement of the height profiles of the 1.27 μ band brightness and the O concentration would form a good test of the possible existence of this mechanism. The model indicates that the 1.27 μ band brightness in the night airglow is dependent upon the rate of atmospheric mixing which suggests that there may be a brightness correlation between the 1.27 μ band and the OH emission.

References

Anderson, J. G.: 1970, University of Colorado, Ph.D. Thesis.
Ballif, J. and Venkateswaran, S. V.: 1963, *J. Atmospheric Sci.* **20**, 1.
Bates, D. R. and Nicolet, M.: 1950, *J. Geophys. Res.* **55**, 301.
Berthier, P.: 1956, *Ann. Geophys.* **12**, 113.
Biedenkapp, D., Hartshorn, L. G., and Bair, E. J.: 1970, *Chem. Phys. Letters* **5**, 379.
Chamberlain, J. W.: 1961, in *Physics of the Aurora and Airglow*, Academic Press, N.Y.
Donahue, T. M.: 1969, *J. Geophys. Res.* **74**, 3717.
Evans, W. F. J., Llewellyn, E. J., and Vallance Jones, A.: 1967, *Nature* **213**, 352.
Findlay, F. D., Fortin, C. J., and Snelling, D. R.: 1969, *Chem. Phys. Letters* **3**, 204.

Gattinger, R. L.: 1968, *Can. J. Phys.* **46**, 1613.
Gattinger, R. L.: 1969, *Ann. Geophys.* **25**, 825.
Gauthier, M. and Snelling, D. R.: 1970, *Chem. Phys. Letters* **5**, 93.
Gush, H. P. and Buijs, H.: 1964, *Can. J. Phys.* **42**, 1037.
Hampson, J.: 1965, *Chemiluminescent Emissions Observed in the Stratosphere and Mesosphere, Canadian Armament Research and Development Establishment*, T.N. 1690/65.
Hampson, J.: 1969, *Colloquium – Processes of Importance to the Stratosphere and Mesosphere*, Quebec City.
Hesstvedt, E.: 1968, *Geophys. Norv.* **27**, 1.
Hudson, R. D., Carter, V. L., and Breig, E. L.: 1969, *J. Geophys. Res.* **74**, 4079.
Hunt, B. G.: 1966, *J. Geophys. Res.* **71**, 1385.
Hunten, D. M.: 1970, private communication.
Huppi, E. R. and Stair, A. T. Jr.: 1969, in *Atmospheric Emissions* (ed. by B. M. McCormac and A. Omholt), Van Nostrand Reinhold Company, New York, p. 471.
Johnson, F. S. and Wilkins, E. M.: 1965, *J. Geophys. Res.* **70**, 1281, 4063.
Johnson, F. S., Purcell, J. D., Tousey, R., and Watanabe, K.: 1952, *J. Geophys. Res.* **57**, 157.
Lettau, H.: 1951, in *Compendium of Meteorology* (ed. by T. F. Malone), Waverly Press, Inc., Baltimore, p. 320.
Llewellyn, E. J. and Evans, W. F. J.: 1971, this volume, p. 17.
Miller, D. E.: 1969, private communication.
Moreels, G., Evans, W. F. J., Blamont, J. E., and Vallance Jones, A.: 1970, *Planetary Space Sci.* **18**, 637.
Noxon, J. F.: 1970, *J. Chem. Phys.* **52**, 1852.
Packer, D. M.: 1961, *Ann. Geophys.* **17**, 67.
Schiff, H.: 1970, private communication.
Shefov, N. N.: 1969, *Planetary Space Sci.* **17**, 797, 1629.
Shimazaki, T. and Laird, A. R.: 1970, *J. Geophys. Res.* **75**, 3221.
Snelling, D. and Hampson, J.: 1969, *University of Illinois Aeronomy Report No. 32*, p. 223.
Stuhl, F. and Welge, K. H.: 1969, *Can. J. Chem.* **47**, 1870.
Tarasova, T. M.: 1963, *Planetary Space Sci.* **11**, 437.
Wallace L.: 1962, *J. Atmospheric Sci.* **19**, 1.
Wallace, L. and Hunten, D. M.: 1968, *J. Geophys. Res.* **73**, 4813.
Watanabe, K.: 1958, *Adv. Geophys.* **5**, 153.
Zipf, E.: 1969, *Can. J. Chem.* **47**, 1863.

INSTITUTE DISCUSSION

There were no answers to A. T. Stair's request for an explanation of why there is no decrease in O_2 $(^1\Delta)$ in the morning twilight.

L. Thomas commented that the reaction $O(^1D) + H_2O \rightarrow 2\,OH$ does not significantly decrease the H_2O concentration at 80 km so one cannot explain Narcissi's observations of a sharp decrease in water attached cluster ions at that altitude with this reaction.

INTERPRETATION OF *F* REGION NIGHTGLOW

J. F. NOXON

Blue Hill Observatory, Harvard University, Cambridge, Mass., U.S.A.

1. Introduction

Nightglow emission from the *F* region (150 to 500 km) appears to be the result primarily of reactions involving charged species. We shall give major emphasis to the optical emission resulting from (a) the various reactions which remove O^+, the dominant positive ion, and (b) very low energy electron excitation. We do not discuss those nighttime features, such as emission from H and He, which are essentially long lived twilight phenomena. We are also unable to do more than mention the important thermal emission from high altitude; examples are the fine structure emission from O (see Kockarts and Peetermans, 1970) and the possibility of IR emission from CO_2.

2. O^+ Reactions

The familiar reactions of O^+ with N_2 and O_2, and the subsequent dissociative recombination of NO^+ and O_2^+, are still thought to be the major source of $O(^1D)$, $O(^1S)$ and $N(^2P)$ in the nighttime *F* region; these are the states which, respectively, radiate at 6300 Å, 5577 Å and 5200 Å. These features have been exhaustively discussed during the previous two meetings; we shall therefore comment primarily upon recent developments. On the other hand it has only become clear within the last year that there may be other O^+ reactions which contribute to *F* region nightglow; these are the radiative recombination of O^+ with an electron and the mutual neutralization of O^+ and O^-. In this case the work is so new that only a preliminary account can be presented at this time.

A. ION MOLECULE REACTIONS

O_2^+ and NO^+ are formed in the *F* region by reaction of O^+ with O_2 and with N_2. The dissociative recombination of O_2^+ with electrons then yields $O(^1D)$ and $O(^1S)$ in about a 10/1 ratio; the yield for $O(^1D)$ is very close to unity (Zipf, 1970). Collisional quenching of $O(^1D)$ is important up to 300 km; $O(^1S)$, on the contrary, does not suffer this fate except below 100 km. $N(^2D)$ is presumably formed by dissociative recombination of NO^+; the state is quenched at all altitudes up to at least 500 km by electrons and below about 250 km by O_2. Quenching of $N(^2D)$ by O has been invoked (Weill, 1969) but the evidence is not conclusive. The latest computed values for electron quenching of $N(^2D)$ suggest that this process dominates the loss rate in the *F* region above about 250 km.

Above 200 km the dominant ion is O^+ and the rate of production of the metastables is governed by the slower first step, the reaction of O^+ with O_2 or N_2. The rate of

B. M. McCormac (ed.), The Radiating Atmosphere, 64–72.

dissociative recombination, in the second step, becomes unimportant; what matters for the nightglow is only the efficiency with which a given state is produced. The volume emission rates then become

$$R\,(6300\text{ Å}) = \frac{\varepsilon_1 k_1 n\,(e)\,n\,(O_2)}{[1 + k_3 n\,(N_2)/A_1]} \sim n\,(e) < 300\text{ km};\; \sim n\,(e)\,n\,(O_2) > 300\text{ km} \tag{1}$$

$$R\,(5577\text{ Å}) = \varepsilon_2 k_1 n\,(e)\,n\,(O_2) > 200\text{ km} \tag{2}$$

$$R\,(5200\text{ Å}) = \frac{\varepsilon_3 k_2 n\,(e)\,n\,(N_2)}{[1 + (k_4 n\,(O_2) + k_5 n\,(e))/A_2]} \sim n\,(e) < 250\text{ km};\; \sim n\,(N_2) > 250\text{ km}. \tag{3}$$

The asymptotic behavior is noted in the limit ot high and low altitude on the right. Current values for the various coefficients are given in Table I.

TABLE I

Reaction rates

$O^+ + O_2 \rightarrow O + O_2^+$	$k_1 = 1.5 \times 10^{-11}$ cm^3 sec^{-1}	Ferguson (1969)
$O^+ + N_2 \rightarrow N + NO^+$	$k_2 = 7 \times 10^{-13}$ cm^3 sec^{-1}	Ferguson (1969)
$O(^1D) + N_2 \rightarrow O(^3P) + N_2$	$k_3 = 9 \times 10^{-11}$ cm^3 sec^{-1}	Noxon (1970)
$N(^2D) + O_2 \rightarrow NO + O$	$k_4 = 6 \times 10^{-12}$ cm^3 sec^{-1}	Black *et al.* (1969)
		Lin and Kaufman (1969)
$N(^2D) + e \rightarrow N(^4S) + e$	$k_5 = 3 \times 10^{-9}$ cm^3 sec^{-1}	Henry and Williams (1968)
$\varepsilon_1 = 1$; $\varepsilon_2 = 0.1$	Zipf (1970)	
$A_1 \sim 10^{-2}$ sec^{-1}		
$A_2 \sim 10^{-5}$ sec^{-1}		

Before turning to a discussion of the individual emissions we note that the ratio 5577 Å/6300 Å should be relatively independent of altitude above 300 km but vary in proportion to $n(N_2)$ below this height. 5200 Å/6300 Å should vary as $[n(N_2)/n(O_2]^2$ at low altitude and hence be relatively insensitive to height below about 250 km where quenching of $N(^2D)$ tends to be dominated by O_2 rather than by electrons. At a higher altitude the ratio should vary as $n(N_2)/n(e)$ and should therefore decrease markedly with increasing altitude. These estimates of altitude regimes depend, of course, on the choice of neutral atmosphere and, for 5200 Å, on the electron density profile. Nevertheless one has the prediction that 5200 Å should follow 6300 Å when the *F* region is low and both features are strong; this is the case at low latitude (Weill, 1969). While both 6300 Å and 5577 Å should be enhanced when the *F* region moves to lower altitude the effect should be greater for 5577 Å.

The only published rocket measurement does not appear to show any significant variation in the 6300 Å/5577 Å ratio over the height range where one would have expected it. However the authors' conclusion that quenching of $O(^1D)$ must therefore be small has been questioned (see Peterson, 1970; Gulledge-Packer *et al.*, 1970). rom low latitude observations Weill (1969) finds a good correlation between 5200 Å and 6300 Å but a much poorer one between 5577 Å and 6300 Å. He suggests that the failure for the 5577 Å/6300 Å ratio to increase when the *F* region is low may be accounted for by additional production of $O(^1D)$ at low altitude through quenching of $N(^2D)$ by O. The position on intensity ratios, particularly when observed from the ground, is not entirely clear and the need for direct high altitude measurements is evident. More surprises may well be in store as satellite measurements become available. The recent report by Donahue *et al.* (1970), that the scale height of 5577 Å on the topside is several times larger than predicted by Equation (2), may be a portent of things to come.

1. 6300 Å

It is well established that this feature is ionospherically controlled. The Barbier formula, discussed at length during previous Institutes, is of occasional aid in interpreting the behavior of 6300 Å, but its use often seems simply to lead to confusion. It may be applied only when the shape of the *F* layer is preserved during vertical

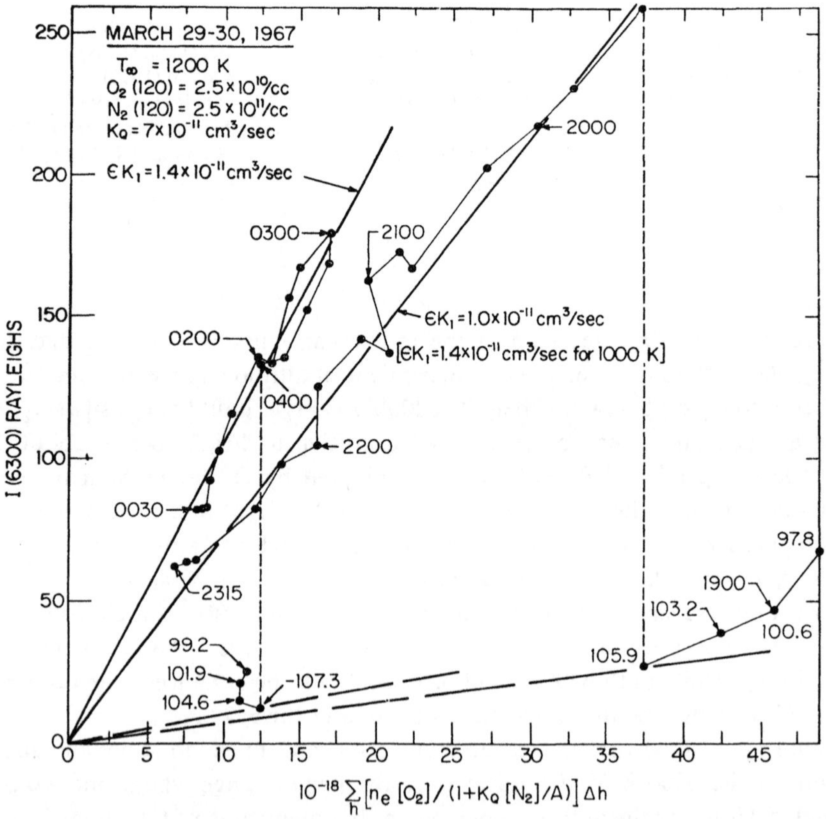

Fig. 1. Plot of observed 6300 Å nightglow intensity against theoretical value obtained by integrating Equation (1) over height (Noxon and Johanson, 1970). Local time and solar zenith angle are indicated. Intensities for $K < 105°$ are reduced by 10. The pre-dawn enhancement is due to ionic recombination.

movement and when quenching of $O(^1D)$ is ignored. The latter assumption is almost never true and attempts to infer the neutral particle scale height using the formula will inevitably lead to an overestimate.

The more recent approach is to use measured electron density profiles and numerically integrate Equation (1). Peterson and Van Zandt (1969) have applied this approach to low latitude 6300 Å and have shown that the variation of 6300 Å with time is very well explained only if quenching is rapid. Noxon and Johanson (1970) have applied the method to mid-latitude 6300 Å and reach the same conclusion; they also show that the quantitative agreement between theory and observation is good where all quantities entering into Equation (1) are obtained from experimentally measured values. Figure 1 shows one of their plots; the rate coefficients are those shown in Table I and the model atmosphere is derived from recent rocket measurements and an assumed exospheric temperature. One can note the departure of observed 6300 Å intensity from the prediction based upon ionospheric recombination for solar zenith angle less than 105°. This is the result of additional production of $O(^1D)$ from photodissociation of O_2 in the Schumann-Runge continuum; the magnitude is correctly predicted as a function of solar zenith angle by an exact theoretical calculation which employs recent measurements of the absolute solar brightness in the 1500 Å to 1700 Å region.

One may infer from such studies that the 6300 Å nightglow is almost entirely the result of ionic recombination since the intercept on the ordinate axis is negligible; this point is discussed by Noxon and Johanson. If additional processes of any significance exist they must mimic the ionic recombination rate rather well; a process such as the production of $O(^1D)$ by excitation transfer from $N(^2D)$ might do this. Nevertheless such analyses have their limits, largely imposed by uncertainties in the electron density profile. They need confirmation by rocket measurements which simultaneously determine both the volume emission intensity and plasma density.

Some recent Doppler temperature measurements on 6300 Å nightglow have been reported (Armstrong, 1969; Nagy *et al.*, 1971). These provide further evidence that (a) the temperatures correspond very roughly to the exospheric temperature inferred from satellite drag; (b) interesting and rather rapid changes in the temperature exist; and (c) that horizontal winds in the *F* region may be detected through distortions in the line profile.

2. 5577 Å

The existence of an *F* layer component in nightglow 5577 Å is now well established; yet, as mentioned earlier, it is by no means clear that the observations support the notion of ionic recombination of O_2^+ as the sole source. Ground based detection of the upper layer is generally possible only at low latitude, where excursions of the *F* region to low altitude occur; one may then detect a correlation between 6300 Å and the total 5577 Å intensity. (A recent study has been presented by Chiplonkar and Tillu (1970).) At mid-latitude the 100 km layer of 5577 Å dominates the weak high altitude component. While rocket and satellite observations will doubtless clarify the position a ground based method of separating the high and low altitude components appears to be yielding interesting results. Shepherd (1969) has reviewed the method; it involves a careful analysis of the 5577 Å line profile in order to resolve it into three components. A narrow (~200 K) core comes from the 100 km layer while the high altitude component is composed of (a) a thermalized component at 1000 K, and (b) a broader skirt resulting from emission from the partly thermalized $O(^1S)$ which still retains some of the excess kinetic energy resulting from the dissociative recombination of O_2^+. Recent work using this approach (Hernandez, 1970) shows considerable promise. The outstanding problem at present involves the curious dependence of 5577 Å upon altitude reported by Donahue *et al.* (1970).

3. 5200 Å

Weill (1969) has shown how well 5200 Å correlates with 6300 Å in the nightglow. Hernandez and Turtle (1969) have demonstrated conclusively that the feature is present in mid-latitude nightglow and have noted the problem of contamination by OH even at high resolution. There is general agreement that the 5200 Å/6300 Å ratio is of the order of 3%; although Weill concluded that the existence of 5200 Å implied an abnormally large rate of O^+ reaction with N_2 this does not follow if proper account is taken of $O(^1D)$ quenching. If one forms the 5200 Å/6300 Å ratio, using Equations (1) and (3) and the rate coefficients in Table I, there results (for altitudes <250 km)

$$\frac{I(5200\,\text{Å})}{I(6300\,\text{Å})} = \frac{\varepsilon_3}{\varepsilon_1} \cdot \left[\frac{n(N_2)}{n(O_2)}\right]^2 \cdot \frac{k_2 k_3 A_2}{k_1 k_4 A_1}. \tag{4}$$

It follows that with $n(N_2)/n(O_2) \sim 10$ (CIRA, 1965) $\varepsilon_3/\varepsilon_1$ needs only to be about 0.3. More recent model atmospheres have relatively less O_2 although this may be somewhat offset by the effect of electron quenching which was ignored in deriving Equation (4). Although the rate of reaction of O^+ with N_2 can be greatly accelerated if the latter is vibrationally excited there would appear at present to be no need to appeal to such a device to account for the 5200 Å intensity. The combined effect of O_2 and electron quenching of $N(^2D)$ is such as to yield an effective lifetime of ~20 min in the nightglow between 200 and 300 km. Weill (1969) reports a somewhat shorter lifetime from his measurements of the time delay between 5200 Å and 6300 Å; a longer value is favored by Hernandez and Turtle (1969). In any event there appears to be no compelling reason, at present, to doubt that the 5200 emission is due to NO^+ recombination and that the intensity does not conflict with laboratory measurements of the rate at which O^+ reacts with N_2. The importance of $N(^2D)$ quenching by O remains unclear; a laboratory measurement would be desirable.

Although the intensity of 5200 Å is low the concentration of $N(^2D)$ is appreciable, due to the very small emission probability. It appears that the concentration is such that electron quenching may be a significant means for heating the ambient electrons in the nighttime *F* region. The nightglow intensity suggests that ε_3 approaches unity and a similar conclusion has been reached by Norton and Barth (1970) in their discussion of NO production. Assuming $\varepsilon_3 = 1$ the energy supplied per ambient electron via $N(^2D)$ is

$$Q = \frac{2.5\, k_2 n(e)\, n(N_2)\, k_4}{(A_2 + k_3 n(O_2) + k_4 n(e))} \text{ eV/electron} \cdot \text{sec}. \tag{5}$$

Local electron cooling is dominated by O (Dalgarno and Degges, 1968) between 250 and 350 km; the cooling per electron is $4 \times 10^{-15}\, n(O)$ eV/sec. If the heating and cooling rates are equated one finds

$$(T_e - T_n) = \frac{2.5\, k_2 k_4 n(N_2)\, n(e)}{4 \times 10^{-15} n(O)\, [A_2 + k_3 n(O_2) + k_4 n(e)]}$$

With $n(e) \sim 3 \times 10^5$ cm^{-3} and a 1000 K (CIRA, 1965) atmosphere the value for $T_e - T_n$ at 300 km is ~100 K. It is obvious that appropriate changes in the assumed particle distribution will reduce the excess temperature to trivial values but the process is clearly one that cannot be neglected in any discussion of the nighttime electron temperature in the *F* region. If it is confirmed that ε_3 is indeed ~1, and unless the N_2/O ratio is seriously overestimated, then measurable effects on $T_e - T_n$ exist.

B. RADIATIVE RECOMBINATION OF O^+

Until very recently it was thought that the only significant reactions of O^+ in the *F* region leading to airglow were those with O_2 and N_2. The discovery of 1304 Å and 1356 Å emission from O in the nightglow, however, led Hanson (1969) to suggest radiative recombination of O^+ and electrons as the source. Knudsen (1970) has suggested the additional process of mutual neutralization of O^+ with O^-. He notes that the two-body formation of O^- should be accompanied by a continuum emission in the 7000 to 8000 Å region and that recent satellite measurements may have detected such a feature. Tinsley (1970) describes a twilight measurement of emission from O at 4368 Å; the line, like the UV lines, originates from a high level of excitation in atomic oxygen, and the decay appears consistent with Hanson's mechanism. Although these two reactions are expected to yield primarily highly excited states of O there may be some direct production of low singlet states from the mutual neutralization reaction. It is still undetermined as to whether there are significant effects on the 5577 Å line and whether other O emissions from high level states may be found to result from such processes. This work has been stimulated by satellite photometry of *F* region nightglow; the majority of the observations are as yet unpublished but will be of the greatest interest. Whether the mechanisms so far proposed are quantitatively adequate remains to be seen.

3. Low Energy Electron Excitation

Under certain conditions, not strictly described as auroral, one may expect significant excitation of $O(^1D)$ by low energy electrons in the *F* region. This arises (a) where a flux of photoelectrons is incident from the magnetic conjugate point, and (b) when conductive heat flow from the protonosphere leads to an appreciable rise in the electron temperature in the F region. The second sometimes may be a consequence of the first, but it is also possible when other processes heat the plasma in the protonosphere.

The conjugate flux is thought to be responsible, at least in part, for the 'pre-dawn enhancement' of 6300 Å which occurs at certain seasons well before the onset of local photoelectron production. A discussion of recent work is given by Noxon and Johanson (1970). It appears that the direct excitation of O (^1D) is sensitive to the local ionospheric conditions, in particular, upon the ratio $n(e)/n(O)$; when the latter is large a major fraction of conjugate photoelectron energy is deposited in the local electron gas and less goes to O. Some of the pre-dawn enhancement, at some locations, is a direct result of a lowering of the F layer and the consequent enhancement of ionic recombination. Depending upon local conditions the enhancement due to direct excitation of O by conjugate electrons may vary from 10 to 100 R. At some locations the conjugate point remains sunlit throughout the night with a corresponding contribution to nightglow 6300 Å.

The mid-latitude 'red arc' in 6300 Å (discussed at previous Institute meetings), is now believed to be simply the result of thermal electron excitation of O(^1D). Direct evidence is provided by the satellite measurements reported by Norton and Findlay (1969) in which T_e was found to exceed 4000 K at 1000 km over a red arc. It is not difficult to show that this boundary condition implies a conductive heat flow into the F region which is adequate to produce the arc. Such conditions are unusual except at high latitudes and conductive heat flow, although always present, is seldom large enough to affect nightglow 6300 Å at mid and low latitude. It may well be a significant and permanent feature at high latitude. It may also be that the polar nightglow ultimately depends upon a soft electron flux, if only through its effect in maintaining the F region.

Acknowledgment

This work is supported by a grant to Harvard University from the National Science Foundation, GP 4293.

Note added in proof: In the six months following the Institute a number of papers related to the UV tropical airglow have appeared. The oxygen lines at 1304 Å and 1356 Å were measured by Barth and Schaffner (1970) and by Hicks and Chubb (1970). Hanson (1970) has discussed the relative importance of his and Knudsen's mechanisms. It appears that the measured intensity of the lines is considerably larger than would be predicted by either recombination reaction; either the rates have been underestimated or an additional excitation process must exist. The possibility of excitation by a flux of low energy electrons has been considered but Barth and Schaffner note that flux measurements made on their satellite indicate this process to be unimportant. Heikkila's (1971) detection of an intense flux of trapped electrons and protons in the equatorial region and the broadband photometric observations of Sparrow *et al.* (1970) may have some relevance to the UV airglow problem which still awaits a definitive interpretation.

References

Armstrong, E. B.: 1969, *Planetary Space Sci.* **17**, 957.
Barth, C. A. and Schaffner, S.: 1970, *J. Geophys. Res.* **75**, 4299.
Black, G., Slanger, T. G., St. John, G. A., and Young, R. A.: 1969, *J. Chem. Phys.* **51**, 116.
Chiplonkar, M. W. and Tillu, A. D.: 1970, *Ann. Geophys.* **26**, 213.
CIRA: 1965, Cospar International Reference Atmosphere compiled by Working Group IV, COSPAR, North-Holland Publishing Company, Amsterdam, Holland.
Dalgarno, A. and Degges, T. C.: 1968, *Planetary Space Sci.* **16**, 128.
Donahue, T. M., Guenther, B., and Blamont, J. E.: 1970, Post Deadline paper at 1970 Washington meeting of Am. Geophys. Union.
Ferguson, E. E.: 1969, *Ann. Geophys.* **25**, 819.
Gulledge-Packer, I., Packer, D. M., Tilford, S. G., and Vanderslice, J. T.: 1970, *J. Geophys. Res.* **75**, 2616.
Hanson, W. B.: 1969, *J. Geophys. Res.* **74**, 3720.
Hanson, W. B.: 1970, *J. Geophys. Res.* **75**, 4343.
Heikkila, W.: 1971, *J. Geophys. Res.* **76**, 1076.
Henry, R. J. W. and Williams, R. E.: 1968, *Publ. Astron. Soc. Pacific* **80**, 669.
Hernandez, G.: 1970, personal communication.
Hernandez, G. and Turtle, J. P.: 1969, *Planetary Space Sci.* **17**, 675.
Hicks, G. T. and Chubb, T. A.: 1970, *J. Geophys. Res.* **75**, 6233.
Knudsen, W. C.: 1970, *J. Geophys. Res.* **75**, 3862.
Kockarts, G. and Peetermans, W.: 1970, *Planetary Space Sci.* **18**, 271.
Lin, C. L. and Kaufman, F.: 1969, DASA Symposium, Stanford Research Institute, Menlo Park, Calif.
Nagy, A. F., Hays, P. B., and McWatters, K. D.: 1971, *Space Res.* **XI**, to be published.
Norton, R. B. and Barth, C. A.: 1970, *J. Geophys. Res.* **75**, 3903.
Norton, R. B. and Findlay, J. A.: 1969, *Planetary Space Sci.* **17**, 1867.
Noxon, J. F.: 1970, *J. Chem. Phys.* **52**, 1852.
Noxon, J. F. and Johanson, A. E.: 1970, *Planetary Space Sci.*, **18**, 1367.
Peterson, V. L.: 1970, *J. Geophys. Res.* **75** 2613.
Peterson, V. L. and Van Zandt, T. E.: 1969, *Planetary Space Sci.* **17**, 1725.
Shepherd, G. G.: 1969, in *Atmospheric Emissions* (ed. by B. M. McCormac and A. Omholt), Van Nostrand Reinhold Company, New York, p. 411.
Sparrow, J. G., Ney, E. P., Stoddart, J. W., and Burnett, G. B.: 1970, *J. Geophys. Res.* **75**, 5475.
Tinsley, B. A.: 1970, *J. Geophys. Res.* **75**, 3932.
Weill, G.: 1969, in *Atmospheric Emissions* (ed. by B. M. McCormac and A. Omholt), Van Nostrand Reinhold Company, New York, p. 449.
Zipf, E. C.: 1970, *Bull. Am. Phys. Soc.* **15**, 418.

INSTITUTE DISCUSSION

Below about 250 km altitude, N_2 is a major sink for vibrational excitation.

Although pre-dawn enhancement may only be a consequence of lowering the *F* region, R. Smith stated that in England this is not the case.

M. Isherwood discussed the dissociative recombination in the *F* region and the fit to the Barbier formula from 6300 Å data taken at Exeter. D. Hunten stated that M. Isherwood's analysis proves that Barbier's formula is not based on the correct physics, but it is just a four parameter fit to the data.

Some of the 6300 Å observations obtained from OGO 4 and OGO 6 were discussed by J. Blamont. A large amount of data has been obtained from OGO 4, and OGO 6 is still functioning. Most of the data is yet to be analyzed and presented.

A vertical photometer with a 10° field of view was on OGO 4. Both OGO 4 and OGO 6 were carrying a photometer which scans from the horizon down 30°. On OGO 6 are two vertical Fabry-Pérot interferometers spectrally scanning at 6300 Å and 6200 Å. From the line shape the temperature to ± 20K is obtained.

The temperature observations do not agree with the Jacchia 1970 model. For an altitude of 275 km the high latitude regions show a temperature of 1300 to 1500 K while the mid-latitudes and equatorial regions, have a temperature of about 1000 K.

In general, there is no conjugacy between the southern and northern tropical red arcs. There is a large variation in the arcs as a function of UT and LT. Sometimes there is no arc in the south when there is a very strong arc in the north. At other times the southern arc is much weaker than the northern one. Sometimes there are strong arcs in both the north and the south, but no conjugacy.

The OGO satellites also measured the $O(^3S)$ radiation at 1304 Å. There is no correlation of $O(^3S)$ and $O(^1D)$ between $\pm 40°$ latitude. An example was shown where there was a northern tropical arc of 6300 Å and 1304 Å. There was a southern arc of 1304 Å but no 6300 Å.

THE RELATIONSHIP OF THE PLASMASPHERE AND THE STABLE AURORAL RED ARCS IN THE MAGNETIC STORM OF OCTOBER 29 TO NOVEMBER 7, 1968

C. R. CHAPPELL, K. K. HARRIS, and G. W. SHARP
Lockheed Palo Alto Research Laboratory, Palo Alto, Calif., U.S.A

1. Introduction

The densities of ambient H^+ and He^+ ions in the plasmasphere were measured by the Lockheed light ion mass spectrometer on OGO 5 during the magnetic storm of October 29 to November 7, 1968. Three occurrences of red arcs were reported during this storm period. In one satellite pass during the storm the ion densities were measured on an L shell at the base of which an intense red arc was being simultaneously observed by a ground station.

It is our purpose here to report these recent measurements of the ambient H^+ ions which show the general change in size of the plasmasphere associated with this period of SAR arc observation. We will discuss these changes and correlate them with ground based measurements of 6300 Å SAR arc emissions and with the magnetic activity over this period.

2. Instrumentation and Data Presentation

The details of the design and calibration of the Lockheed ion spectrometer can be found in previous works (Harris and Sharp, 1969, 1971). During this storm period the OGO 5 orbit was such as to allow the measurement of ambient ion densities from a perigee height of 4600 km to an apogee of approximately 23 R_E. Particular orbits through the plasmasphere considered here cover magnetic latitudes of about $-40°$ inbound to about $+40°$ outbound. The L parameter will be used to organize the ion density and to compare these densities with ground based observations made at similar L shells. Figure 1 shows the trajectory of the satellite on an inbound pass presented as a function of L and magnetic latitude. The field lines which define the location of the SAR arc are also drawn in the Figure. It will be noted that the H^+ density is measured in the plasmasphere in flux tubes which intersect the ionosphere at and around the position of SAR arcs observed from the ground.

Because of a change in the instrument ranging logic during April 1968, measurements of H^+ ion densities after that date are limited to concentrations less than about 200 ions cm^{-3}. However, H^+ densities of values $\gtrsim 200$ ions cm^{-3} can be estimated by using the measurements of the He^+ ions and the fairly consistent ratio of H^+/He^+ within the plasmasphere between about $L=2$ and the plasmapause.

3. Observations During the Storm of October 29 to November 7, 1968

The sequence of events and the times of the different observations during the magnetic

B. M. McCormac (ed.), The Radiating Atmosphere, 73–80.

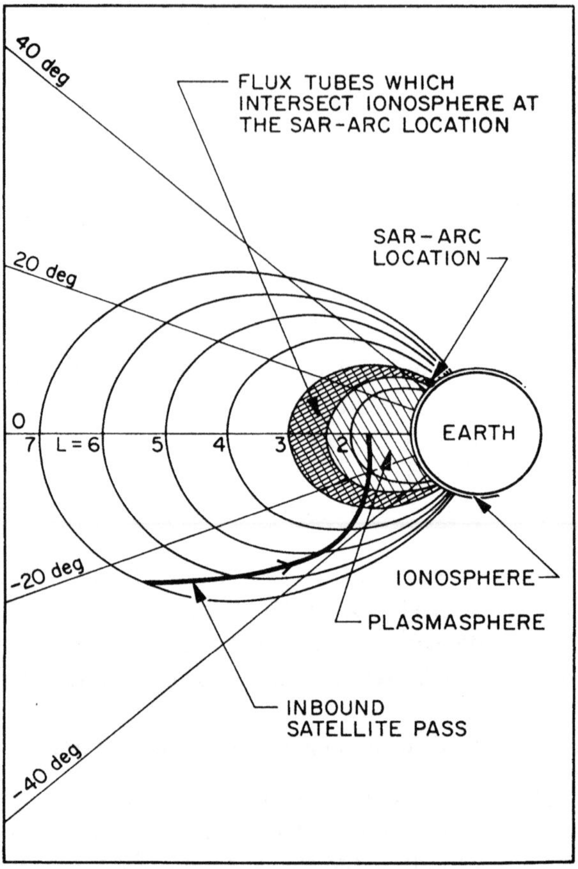

Fig. 1. An inbound pass presented as a function of *L* and geomagnetic latitude. This pass shows that the H^+ densities are measured in the plasmasphere in flux tubes which intersect the ionosphere at the SAR arc locations.

storm of October 29 to November 7, 1968, are best shown on the plot of the Dst (*H*) and K_p indices vs. time presented in Figure 2. The SC began at 0200 UT on October 29 with the initial injection of ring current particles and the consequent decrease in Dst occurring at 1700 UT on the same day.

The first satellite pass through the plasmasphere occurred following this injection (as shown by the arrows marking the UT of the satellite entrance into and exit from the plasmasphere, about 2100 and 2200 UT, respectively). There were continuing injection events at 2100 UT on October 29, 1100 UT on October 30, and about 0000 UT on October 31. During this latter injection event, a SAR arc was observed by Roble (1969) at Ann Arbor and by Hoch (1970) at Richland. Another very large injection of ring current particles occurred at 1400 UT on October 31. This injection caused a main phase decrease of $-212\ \gamma$. During the following 17 hr rapid recovery phase of this event another SAR arc was observed by Roble (1969) at Ann Arbor for 11 hr, by

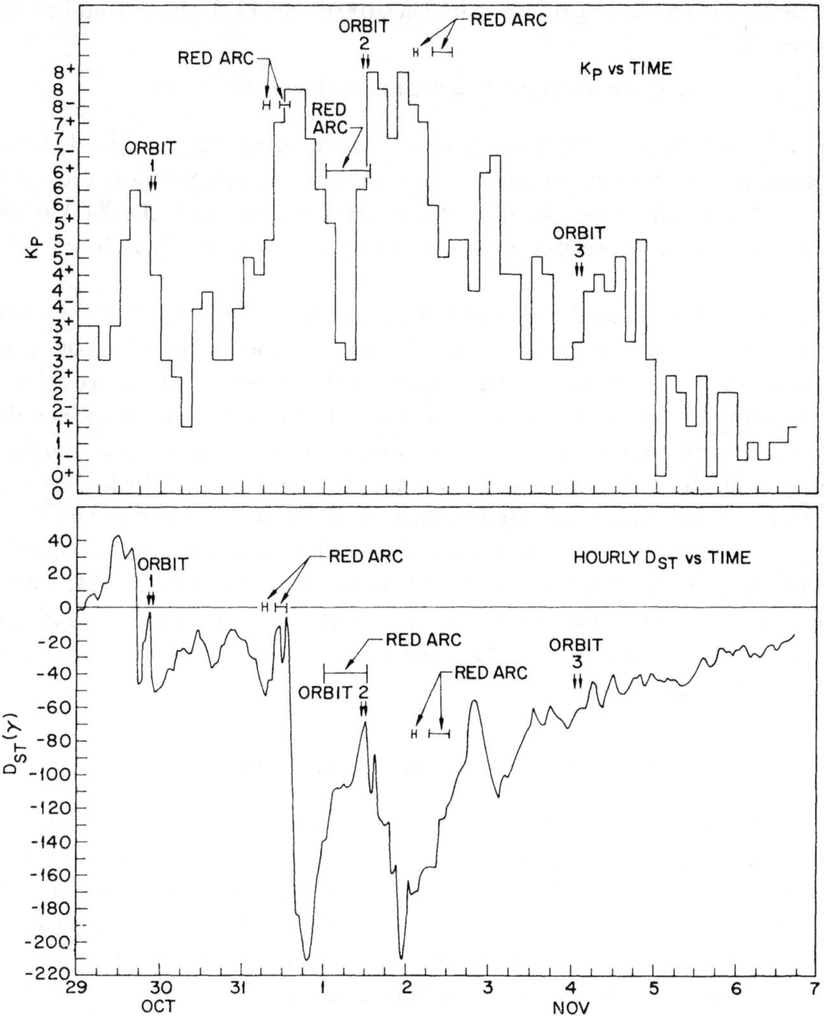

Fig. 2. Shows the K_p and Dst (*H*) indices during the October 29 to November 7, 1968, magnetic storm. The times of the satellite passes through the plasmasphere and the SAR arc observations are also shown.

Hoch at Richland for $1\frac{1}{2}$ hr, and briefly by Marovich (1970) at the Fritz Peak. During the ground observations of the arc at Richland, a second satellite pass at 1100 to 1200 UT measured the properties of the plasmasphere above the location of the arc. At about 1300 UT on November 1 there was another large injection of ring current particles which drove the magnetic field to $-210\ \gamma$ by 2300 UT of that day. Again, the 21 hr rapid recovery phase following this injection was accompanied by an observation of a SAR arc by Marovich (1970) at Fritz Peak. A smaller injection event then occurred at 2100 UT on November 2 and was followed by the slow 5 day recovery of the magnetic field to its undisturbed values. During the slow recovery phase of the storm the plasmasphere characteristics were again measured at 0100 to 0230 UT on November 4 by the OGO 5 vehicle.

4. Changes in the Plasmasphere During the Storm

Measurements of the plasmasphere were made at three different stages of this storm. The inbound passes of the spacecraft measured the H^+ density profiles in the midnight to dawn LT sector and the outbound passes sampled the profiles in the 0900 to 1000 LT region. In each of the three passes the plasmasphere was greatly reduced below its average size.

Orbits 1 and 3 measured the plasmapause position at about $L=3$. Orbit 2 showed steep gradient locations at $L=2$ and $L=2.3$ on the inbound and outbound passes, respectively. These very low L value locations for the plasmapause probably result from convection events which occur during the storm. These events reduce the size of the plasmasphere both by driving the flux tubes to lower L values near midnight and by peeling away the outer portions of the plasmasphere around dusk. The presence of these convection events is indicated by the high K_p values shown in Figure 2.

This general decrease in plasmapause radius with increasing magnetic activity has been shown previously by several authors (Corcuff and Delaroche, 1964; Carpenter, 1967; Taylor *et al.*, 1968, and others). In this large storm, however, the reduction in size is greatly enhanced and the plasmapause is located at the very low L values characteristic of SAR arcs.

5. Plasmapause Location vs. SAR Arc Location

The inbound pass on November 1, 1968 (orbit 2) measured the H^+ ion densities in flux tubes which intersected the ionosphere at the position of a SAR arc which was observed simultaneously by the Richland station. This orbit is shown in detail in Figure 3. In this figure the heavy solid line shows the orbit as a function of L and LT with a time sequence of satellite positions shown by the points A, B, C and D with their corresponding UT also listed in the figure. In the insets at the base of the figure the H^+ ion density profiles along the orbit are shown. (The solid line portion of these profiles is the measured H^+ concentration, and the dotted portion represents values of H^+ concentration extrapolated from the measured He^+ concentration and measured ratio of H^+/He^+ at $L=3$.) Each of these profiles can be mated to the appropriate section of the orbit by matching the points A, B, C, and D with the corresponding points of the orbit. The figure also shows a partial qualitative sketch of the plasmasphere shape (shaded area) as determined by drawing connecting contour lines through the 1000 ions cm^{-3}, 100 ions cm^{-3}, and 10 ions cm^{-3} positions of the inbound and outbound profiles. The dashed line in the Figure represents the approximate position of the plasmapause during periods of quiet to moderate activity as determined by plotting the steep density gradient locations in some 75 orbits of OGO 5.

The inbound density profile shown in the left hand inset has a double knee characteristic which was probably caused by a partial corotation of the plasmasphere bulge

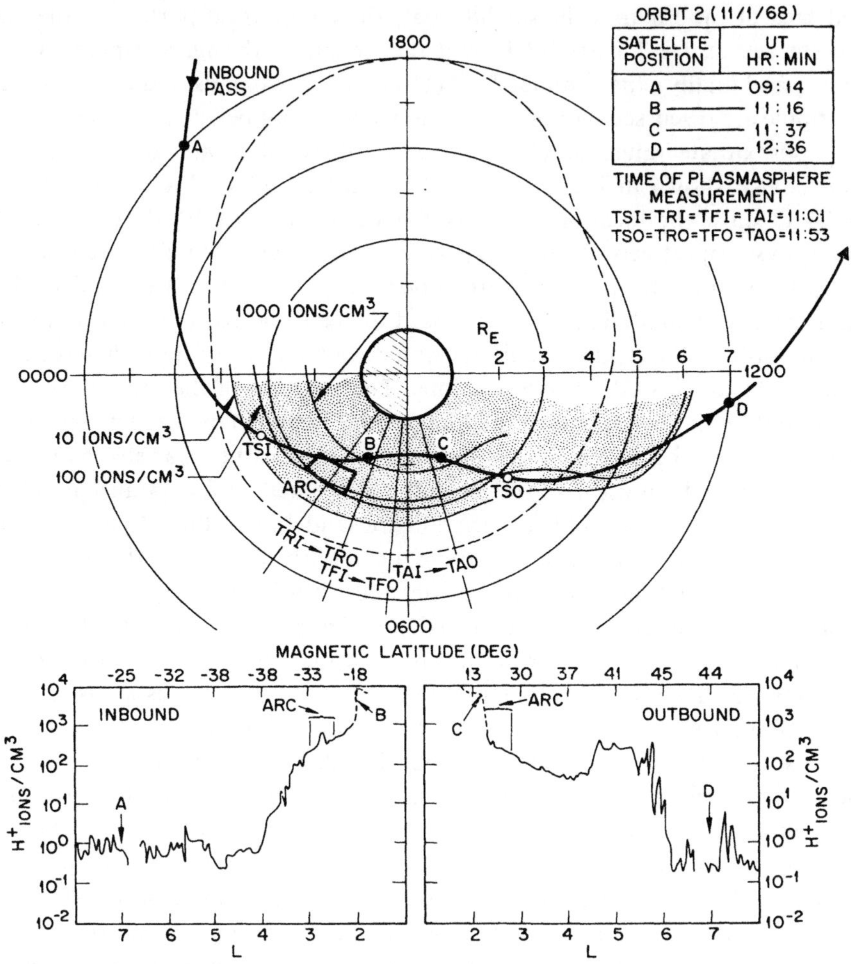

Fig. 3. Shows the satellite pass of November 1, 1968, as a function of L and LT. The heavy solid line represents the satellite path and the dashed line shows the average plasmapause location. The shaded area shows a partial sketch of the plasmasphere shape on this particular day with the 10, 100, and 1000 ions cm^{-3} contour lines also shown. The sectors labeled TRI→TRO, TFI→TFO, and TAI→TAO show the LT location of SAR arc ground stations during the satellite pass through the plasmasphere. The rectangular shaped area shows the position of a simultaneously observed SAR arc. The insets at the bottom show the H^+ ion density profiles along the inbound and outbound passes.

into this LT region during the 6 hr period of reduced $K_p(\sim 3^-)$ which preceded the profile measurement (Figure 2). The steep gradient located at $\sim L=2$ on this inbound pass probably represents the plasmapause position during the period of extremely high $K_p(\sim 8^-)$ which occurred on the preceding day (0600 to 1800 UT on October 31).

The peculiar outbound profile may well be an example of a longitudinal ripple effect in the plasmasphere surface which has been discussed by Carpenter *et al.* (1969), and Park and Carpenter (1970). The ripple effect is shown at about 1000 LT in the shaded area of the Figure. The satellite exits the compressed portion of the plasmasphere surface at $L=2.5$ (0800 LT) and then reenters the plasmasphere at $L=4.5$ (1000 LT) and finally exits at about $L=6$(1100 LT). The details of the probable causes of this ripple are discussed in more detail elsewhere (Chappell *et al.*, 1971).

The approximate range of LT coverage of three SAR arc ground observation stations is also shown in Figure 3. The three sectors, TRI→TRO, TFI→TFO, and TAI→TAO show the location in LT of the Richland, Fritz Peak, and Ann Arbor observatories, respectively, during the period of time that the satellite travels from the point TSI entering the plasmasphere inbound, to the point TSO and exiting the plasmasphere outbound. The satellite passed through the Richland LT sector during a SAR arc observation and through the Ann Arbor LT sector immediately following a period of 11 hr of SAR arc observations. The position of the arc observed by the Richland station at the time of the satellite passage is shown by the rectangular shaped area in the figure. The arc position is shown as an area rather than just a line because the several hundred km width of the arc in the ionosphere at $L=3$ translates along the magnetic field lines to a width of several thousand km in the magnetic equatorial plane. Therefore, we include a spread of about ± 0.3 L shells around the arc location. This spread of ± 0.3 L also encompasses any uncertainties in the exact L shell location of the arc as determined by the ground station and any distortions of the field lines due to the ring current. With this spread, the arc is located approximately between $L=2.4$ and $L=3.0$. As shown in the inbound inset in Figure 3, this corresponds to a density range of about 100 to 500 ions cm^{-3} near the equatorial plane in the plasmasphere in flux tubes which enter the ionosphere near the SAR arc position.

The outbound portion of this same orbit also passes very close to the position of the SAR arc that was observed by the Ann Arbor station for an 11 hr period which ended just 1 hr before this measurement was made. The arc position at its last point of observation was $L=2.3$ (Roble, 1969). Using the outbound density profile shown in the right hand inset in Figure 3, and taking into account the ± 0.3 L spread around $L=2.3$ due to arc width, we find that the arc position occurred at plasmasphere densities of 100 to 1000 ions cm^{-3}. This agrees quite well with the densities characteristic of the arc position measured during the inbound pass.

Both of the above passes indicate that the SAR arc appears to be located at the base of flux tubes which are positioned near the outer portion of the plasmasphere. These flux tubes have an equatorial H^+ density of 100 to 1000 ions cm^{-3}. This connection between SAR arc location and plasmapause location is further supported by the characteristic movement of the SAR arc during the night. Roble (1969) has shown that the arc position decreased in L value from about $L=3.0$ at 20:30 hr LT on October 31 to $L=2.3$ at 05:45 LT on November 1. This decrease in L value of the SAR arc from dusk to dawn generally agrees with the decrease in L value of the plasmapause location through this same LT region. The general decrease in plasmapause position through the nightside is shown by the dashed line in Figure 3. This decrease has also been shown previously by Carpenter (1966), Binsack (1967), Taylor *et al.* (1970) and others.

6. Discussion

From these observations there appear to be at least two ingredients necessary for the production of a SAR arc. First, there must be an injection of ring current particles which is probably the source of energy for the arcs. This idea was first suggested by Cole (1964, 1965). This injection causes a decrease in the Dst (H) index – a decrease which *is* observed during the SAR arc observations in this storm.

Second, the plasmasphere must be greatly reduced in size to the L values characteristic of the location of the SAR arcs. These size reductions are probably due to the series of convection events which both compress and peel away the outer portion of the plasmasphere during the storms.

With both of the above ingredients present the SAR arcs seem to occur during the *rapid recovery* phase of the storm, i.e., during the initial loss of the newly injected ring current particles in the 10 to 20 hr following injection. The SAR arc is generally located in L value near the position of the plasmapause. The arc is found to occur in the ionosphere at the base of flux tubes which have ambient ion densities of 100 to 1000 ions cm^{-3} near the magnetic equatorial plane.

The agreement between plasmapause and SAR arc location is further supported by the general trend of movement of the arcs. As the LT progresses from dusk to dawn the arcs move to lower L values – a movement which is consistent with the decrease in L value of the plasmapause across the nightside due to its asymmetrical shape. Also, data from polar satellites have shown the presence of a density trough both for H^+ ions and electrons above SAR arcs (Norton and Findlay, 1969; Chandra *et al.*, 1970). This ionospheric trough is most likely the effect of the equatorial plasmapause on the ionosphere, thereby again connecting the SAR arc position with the plasmapause position. Finally, Cole (1971) has also shown this agreement between plasmapause and SAR arc location indirectly through the common relationship of both to the distribution of geomagnetic disturbances.

The SAR arc formation mechanisms of Cole (1965), Cornwall *et al.* (1970), and Eather and Carovillano (1970) predict that the arc will be found at the point of decreasing ambient ion density in the magnetosphere – the plasmapause. Each of these mechanisms employs the ring current particles as the energy source and transfers this energy to the ambient electrons of the plasmasphere. This energy transfer is accomplished through coulomb collisions in the Cole (1965) mechanism and through pitch angle diffusion resulting from an ion cyclotron resonance in the Cornwall *et al.* (1970), and Eather and Carovillano (1971) mechanisms. Although additional information is needed to establish the exact process through which the suggested energy transfer is accomplished, the data shown here do support the predicted connection between the plasmapause and the SAR arc position, and reveal the importance of the plasmapause location in determining the positions of the SAR arc.

Acknowledgments

The authors would like to thank Mr. A. J. Fernandez for his work in processing the data which are presented herein. We are indebted to Dr. Richard J. Hoch and Mr. Edward Marovich for furnishing unpublished data on the SAR arc positions during the Halloween storm. We are also indebted to Dr. R. M. Thorne and Dr. A. F. Nagy for helpful discussions.

This work was performed under NASA Contract NAS 5-9092.

References

Binsack, J. H.: 1967, *J. Geophys. Res.* **72**, 5231.
Carpenter, D. L.: 1966, *J. Geophys. Res.* **71**, 693.
Carpenter, D. L.: 1967, *J. Geophys. Res.* **72**, 2969.
Carpenter, D. L., Park, C. G., Taylor, H. A. Jr., and Brinton, H. C.: 1969, *J. Geophys. Res.* **74**, 1837.
Chandra, S., Maier, E. J. R., Troy, B. E. Jr., and Narasinga Rao, B. C.: 1970, presented at Spring meeting Amer. Geophys. Union.
Chappell, C. R., Harris, K. K., and Sharp, G. W.: 1971, *J. Geophys. Res.* **76**, 2357.
Cole, K. D.: 1964, *J. Geophys. Res.* **69**, 3595.
Cole, K. D.: 1965, *J. Geophys. Res.* **70**, 1689.
Cole, K. D.: 1970, *J. Geophys. Res.* **75**, 4216.
Corcuff, Y. and Delaroche, M.: 1964, *Compt. Rend.* **258**, 650.
Cornwall, J. M., Coroniti, F. V., and Thorne, R. M.: 1970, *J. Geophys. Res.* **75**, 4699.
Eather, R. H. and Carovillano, R. L.: 1971, *Cosmical Electrodynamics*, to be published.
Harris, K. K. and Sharp, G. W.: 1969, *Trans. IEEE Geosci.* **2**, 93.
Harris, K. K. and Sharp, G. W.: 1971, *J. Geophys. Res.*, to be published.
Hoch, R. J.: 1970, private communication.
Marovich, E.: 1970, private communication.
Norton, R. B. and Findlay, J. A.: 1969, *Planetary Space Sci.* **17**, 1867.
Park, C. G. and Carpenter, D. L.: 1970, *J. Geophys. Res.* **75**, 3825.
Roble, R. G.: 1969, Ph.D. Thesis, University of Michigan.
Taylor, H. A., Brinton, H. C., and Deshmukh, A. R.: 1970, *J. Geophys. Res.* **75**, 2481.
Taylor, H. A., Brinton, H. C., and Pharo, M. W.: 1968, *J. Geophys. Res.* **73**, 961.

RECENT OBSERVATIONS OF RED ARCS AND MID-LATITUDE AURORA

H. LAUCHE

Max-Planck-Institut für Aeronomie, Institut für Ionosphärenphysik, Lindau, West Germany

Two meridian photometers have been used to measure the red and green emission of O since October 1968. One instrument is at Lindau, Germany, and the other in the Sierra Nevada mountains near Granada, Spain. A third photometer of the same type was erected at Tsumeb, Southwest Africa in April 1970.

Due to the different latitudes of these three stations
(Lindau: 51.50 N; 10.00 E; $L = 2.36$ at 100 km,
Granada: 37.00 N; 3.50 W; $L = 1.45$ at 100 km, and
Tsumeb: 19.72 S; 17.72 E; $L = 1.37$ at 100 km),
the photometric results differ substantially from each other. At Lindau we find typical mid-latitude phenomena as reported by Roble *et al.* (1970) and Hoch and Clark (1970). Red arcs and auroras appear at times of large magnetic disturbances ($K_p > 5$). Since 1968 red arcs have been measured on three occasions (Table I).

TABLE I

Date	Time (UT)	Magnetic indices K_p
October 31, 1968	20.00–22.00	7
November 1, 1968	20.00–23.00	8+
July 9 to 10, 1970	10.45–02.00	6–7

The red arc of November 1, 1968, and July 9, 1970, displayed a double structure.

Several auroras have been observed. The brightest was on March 8, 1970 from 18.30 hr to 24.00 UT. The zenith intensities for the red O line were 5 kR and 6 kR for the green line.

At Sierra Nevada, Spain, mid-latitude auroras and red arcs were observed on March 23, 1969, from 21.00 to 24.00 hr UT when $K_p = 8$ and on May 14, 1969 from 23.00 to 23.30 hr UT when $K_p = 7$. Several tropical red arcs were observed at times of very low magnetic activity.

At Tsumeb we find in most cases phenomena conjugate to those which we observe in Sierra Nevada. A comparison of the data obtained at these low latitude stations would seem to indicate that, occasionally, only one tropical arc extending far to the north exists, rather than two on both sides of the equator.

References

Hoch, R. S. and Clark, K. C.: 1970, *J. Geophys. Res.* **75**, 2511.
Roble, R. G., Hays, P. B., and Nagy, A. F.: 1970, *Planetary Space Sci.* **18**, 431.

B. M. McCormac (ed.), The Radiating Atmosphere, 81. *All Rights Reserved.*

FLUCTUATIONS IN THE INTENSITY AND EXCITATION TEMPERATURE IN THE OH AIRGLOW (8–3) BAND

G. VISCONTI
Istituto di Fisica, Università, L'Aquila, Italy
F. CONGEDUTI and G. FIOCCO
European Space Research Institute, Frascati, Italy

1. Introduction

Since the identification of the OH rotational vibrational bands in the nightglow spectrum by Meinel (1950) these emissions have been studied by several workers. Measurements of the excitation temperature of the OH obtained by a comparison of the intensity of the rotational lines have been discussed by Kvifte (1961), Wallace (1961, 1962) and Krassovsky (1963). Some uncertainty exists in regard to the relative importance of the possible excitation process and whether the rotational states of OH are in thermodynamic equilibrium with the upper mesospheric ambient.

The OH airglow is an important indicator of the mesospheric conditions. Theoretical models of the O-H atmosphere (Wallace, 1962; Ballif and Venkateswaran, 1963; Hunt, 1966; Hesstvedt, 1970; Shimazaki and Laird, 1970) have been used to predict seasonal and diurnal variation of the OH intensity. Experiments, however, may not only yield the total integral intensity in a number of bands, but also may resolve lines in bands and give ratios between them. In this way excitation temperature of rotational and vibrational levels can be measured. Because of the possibility of contamination by other emissions as well as the difficulties in measuring atmospheric transmission and checking the instrumental calibration with continuity, integral intensity data, obtained with wide band photometers, may be difficult to interpret and often affected by errors. Temperature variations obtained from the ratios between lines are less likely to be affected by such uncertainties although instrumentation of adequate sensitivity and resolution is required.

2. Description of the Experiment

The measurements reported here have utilized a tilting interference filter photometer. The filter, which is temperature stabilized, is slowly tilted, to vary the wavelength of peak transmission from 7403 to 7300 Å. Its pass band is 3 Å at 7403 Å. The entire scan from 7403 to 7300 Å is covered by 49 channels.

Under the assumption of thermal equilibrium the intensity of rotational lines within the same band is given by

$$I = C\nu^4 S_J \exp\left[- B_v J\,(J + 1)\, hc/kT\right]$$

where S_J is the line strength, B_v the effective rotational constant, J the rotational quantum number, and T the rotational temperature.

B. M. McCormac (ed.), The Radiating Atmosphere, 82–89. *All Rights Reserved.*

Besides the rotational temperature a doublet temperature based on the relative intensities of the branch originating from the $^2\pi_{1/2}$, $^2\pi_{3/2}$ levels is defined. In most cases this temperature has been found to be lower than the corresponding rotational temperature (Meinel, 1950).

For each night two successive scans were combined and from each combined new spectrum the dark current and sky background were subtracted. A synthetic spectrum was then constructed, assuming the transfer function of the instrument to be represented by a Gaussian curve: the amplitude and width of the Gaussian were a function

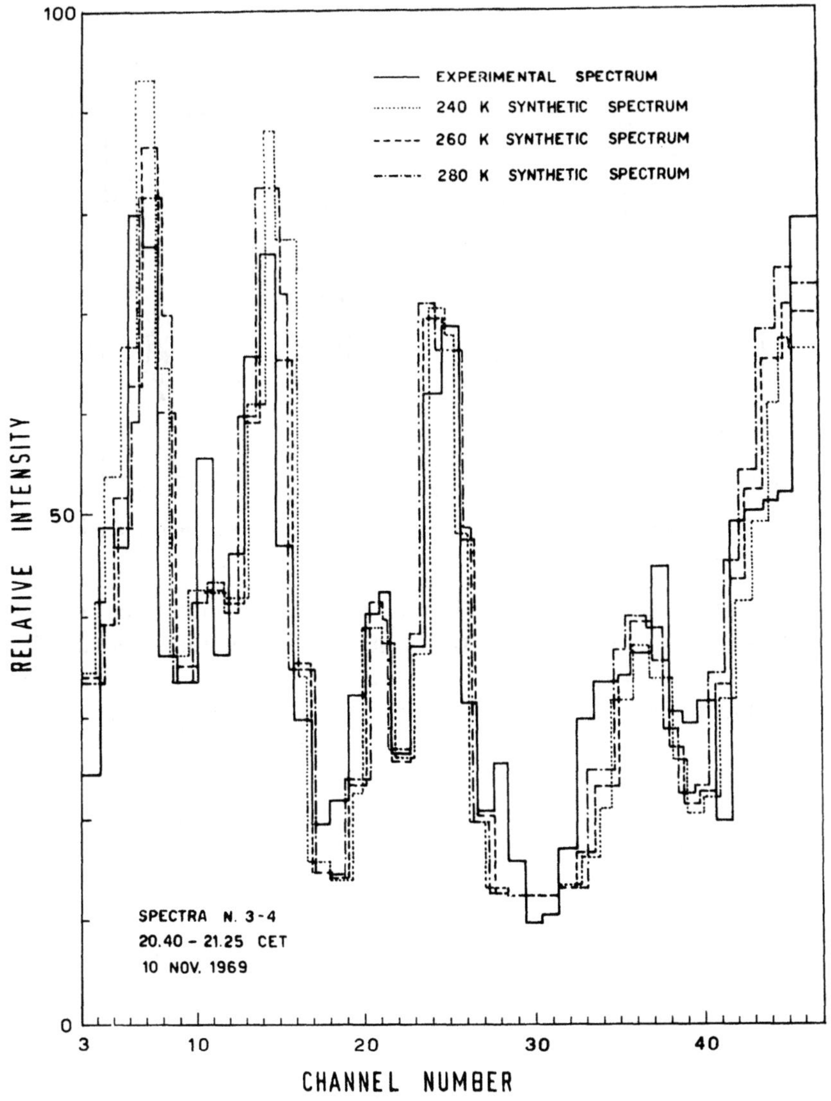

Fig. 1. Comparison of typical experimental spectrum with synthetic spectra for three different values of rotational temperature.

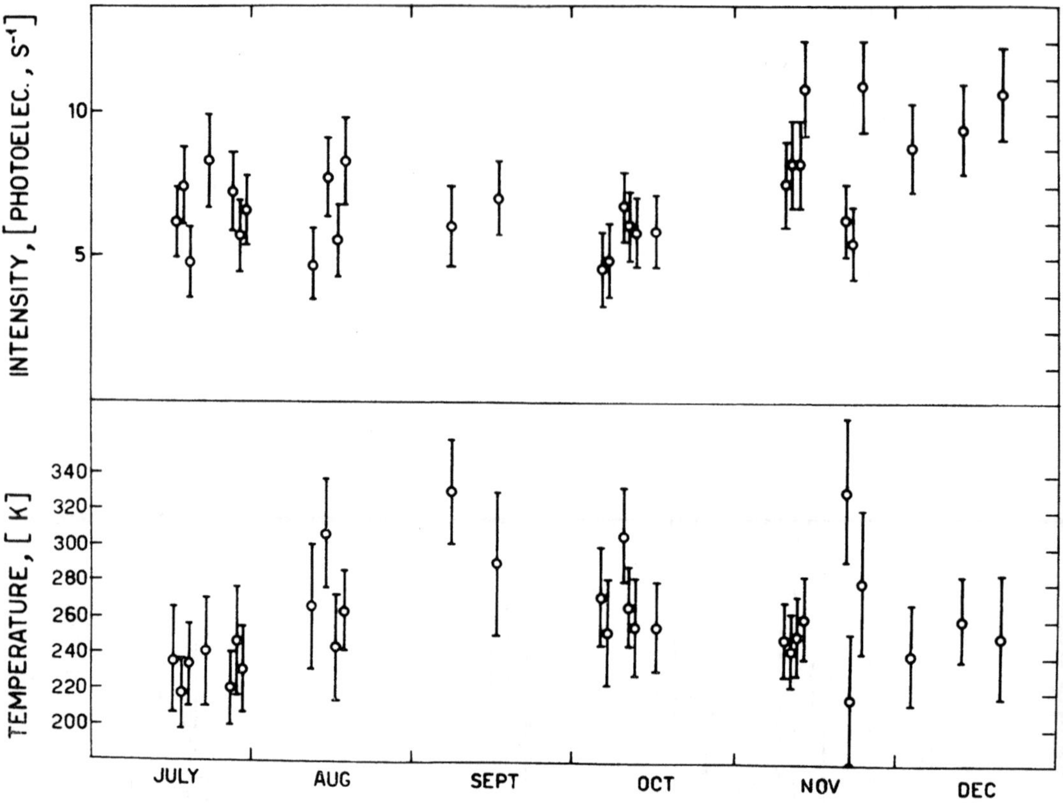

Fig. 2. Seasonal variations of the intensity and rotational temperature.

of the wavelength, according to the calibrations. The contribution of all lines of the (8 – 3) OH band was considered and the excitation temperature was varied in steps of 20 K from 120 to 400 K. The ratio between the P_1 and P_2 branches was fixed assuming that the doublet temperature coincided with the rotational temperature. A fit was then carried out between the synthetic spectra and the real spectra using the χ^2 criterion. Finally the result was interpolated to obtain the most probable value of temperature. The rotational constants were taken from Wallace (1960). Figure 1 is an example of the raw data obtained summing up two scans with the background subtracted, and of a set of synthetic spectra. The probable error on each channel reading was taken as the square root of the counts. The probable error on the temperature was obtained by the χ^2 fitting considering the change in the value of temperature corresponding to an increase in χ^2 of an amount $\sqrt{N}$ with N the number of channels (49). The value of χ^2 varied between 30 and 150 indicating a good fit.

3. Results

The results of 22 nights and 276 spectra taken during 1969 and of 7 nights and 72 spectra taken during July to December 1968 are reported. Measurements reported were taken only within a few days of a new moon since it was found that the presence of the moon substantially increased the background and affected the measurements. The instrument was operated at two locations near L'Aquila (42° lat. 23° long.): at Campo Imperatore (30 km from L'Aquila, altitude 2200 m) until August 1969 and, because of logistic difficulties and poor meteorological conditions, at Preturo (10 km from L'Aquila, altitude 700 m) for the remaining period.

The nights during which data were taken are indicated in Figure 2. The excitation temperature obtained by averaging the results of each night between the third and sixth hr after sunset are shown as ordinates. The diagram also indicates the night average, in the same period, of the intensity for each night. However, since calibrations were carried out only occasionally, there is no assurance that the sensitivity of the instrument was constant during the entire period; only a crude account could be taken of changes in atmospheric transmission, which meant discarding entire sets of data. While little can be said with regard to the seasonal fluctuations, we notice, as expected, a tendency in the temperature to increase from summer to fall which is apparent also in the results of Wallace (1961).

At present we are mainly concerned with the night variations of intensity and temperature. These were studied at 1 hr intervals. For short periods of time (one night) the calibration can be assumed to be constant and the intensity fluctuation can be regarded as meaningful. In order to obtain a pattern we studied the temperature variation after sunset for each night. This was done by considering the difference in temperature at successive hourly intervals for each night. The diagram in Figure 3 shows a decrease of about 20° over a period of about 8 hr. Behavior with respect to LT is also given in Figure 3.

Figure 3 also shows the hourly variation of the intensity of all the P_1 lines, relative to the minimum recorded in each night. Cross correlations between hourly values of the relative intensities and of the inverse temperature fluctuations were evaluated and the results are shown in the diagrams of Figure 4. Plots of relative intensity are shown and lines of best fit for the correlation coefficient are indicated for monthly averages, together with the value of the coefficient and the number of points utilized.

The correlation coefficient was always negative and varied between -0.14 and -0.72. Considering that the error expected in these estimates is about ± 0.2 we believe that the correlation coefficient indicates a real effect, namely, that changes of temperature and intensity are in phase.

While reports exist of diurnal intensity fluctuations for the OH airglow we have found no studies of these types of fluctuations for the rotational temperature. In fact, knowledge of the diurnal fluctuation of mesospheric kinetic temperature is also extremely limited and a relationship between OH excitation temperature and kinetic temperature has not been established. Below a certain altitude, approximately 80 km, it is likely that the two temperatures coincide in view of the expected collision rate and the lifetime of the excited states.

Hunt (1966) has provided estimates for the rates of formation of OH with height

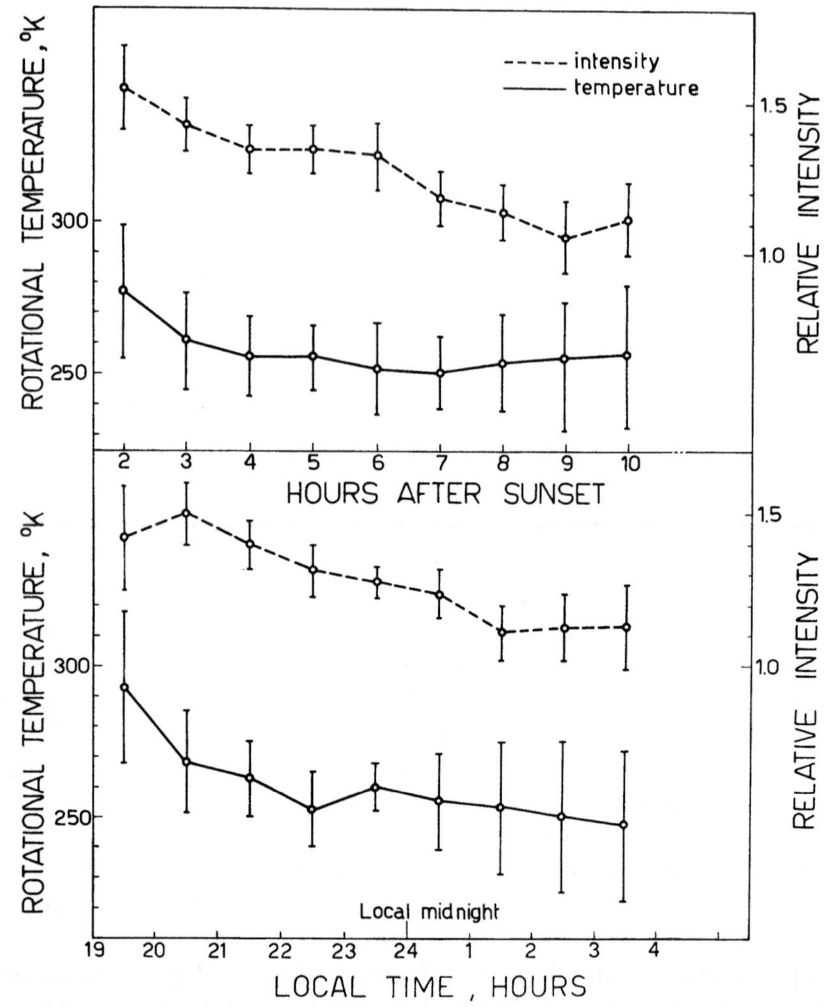

Fig. 3. Nocturnal variation of relative intensity and rotational temperature; time variations are referred to local time and time after sunset. (Fiocco *et al.*, 1970.)

for daytime and nighttime conditions. According to his results the rate of OH formation and, thus, the radiation at night would be high only in the higher mesosphere and lower thermosphere.

In the daytime one would essentially add contributions of the stratosphere and the lower mesosphere where, however, quenching is effective. Hesstvedt (1970), considering diffusion and quenching in his recent model, gives a vertical distribution for the OH* emission which shows a trend similar to that found by Hunt (1966). Because of quenching, Hesstvedt's model involves a much smaller range of altitudes for the diurnal change of OH* emission, and obtains a much smaller variation.

Our intensity data are in reasonable agreement with the diurnal intensity variation

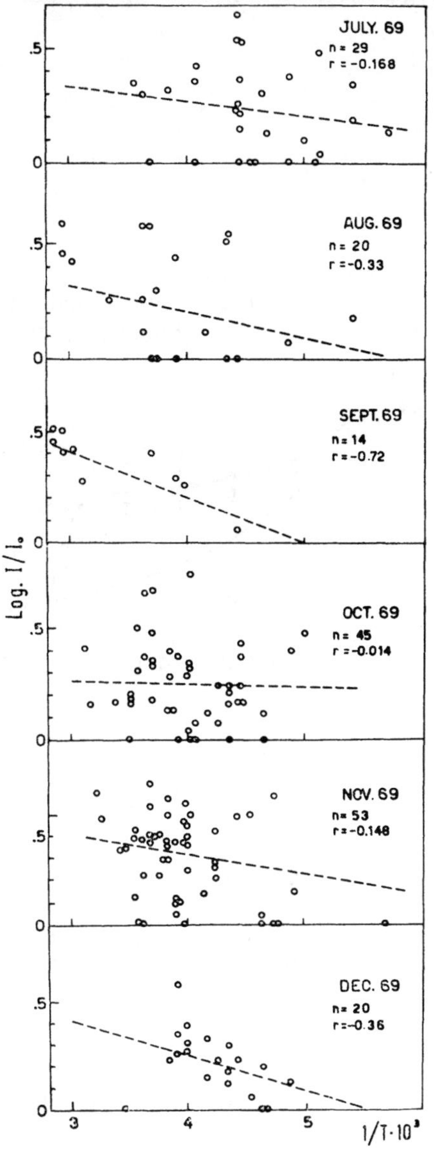

Fig. 4. Correlation between log I/I_0 and inverse rotational temperature. I_0 represents the minimum value of intensity as recorded during each night. Also indicated are the number of points (*n*), the correlation coefficients (*r*), and the lines of best fit.

given by Hesstvedt (1970), and can be explained with the change in the concentration of H and O_3, species entering in the formation of OH^* according to the reaction $H+O_3 \rightarrow OH^*+O_2$. Similar results are obtained by Shimazaki and Laird (1970); the variation that they obtain at night is somewhat longer than the one we observe (Figure 5).

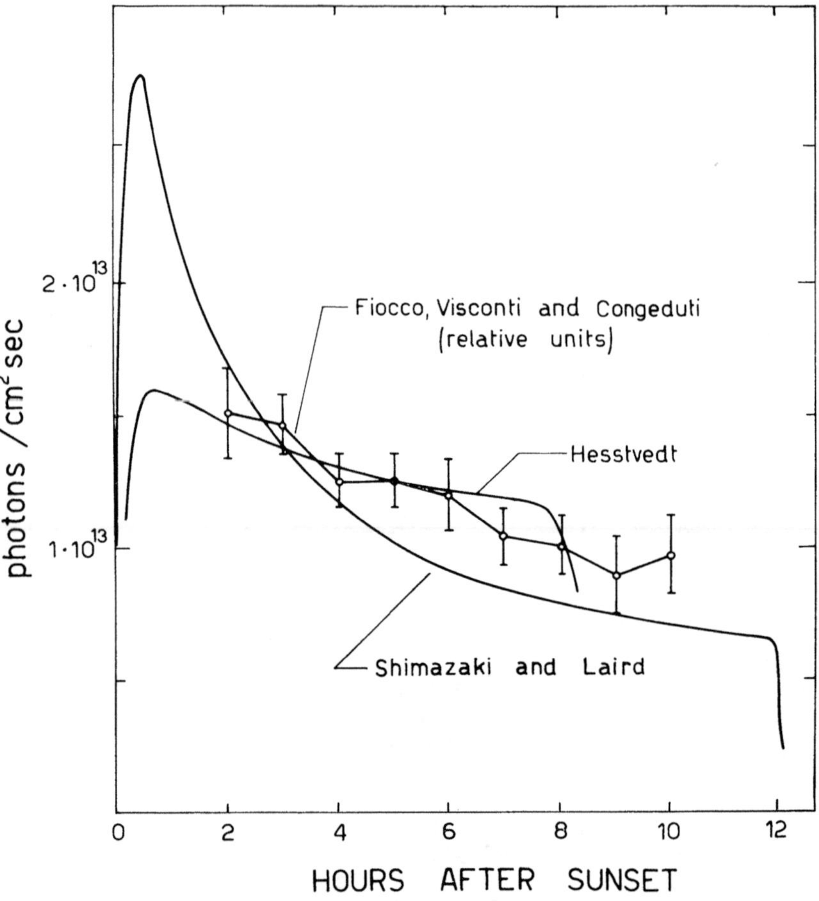

Fig. 5. Comparison between the nocturnal variation experimentally determined (relative units), and the variation obtained by the models of Shimazaki and Laird (1970) and Hesstvedt (1970).

A discussion of the temperature variation should take into account also the formation process and the mechanisms for populating the levels. A change in rotational temperature may reflect changes in the excitation process and in the height where the process takes place, as well as changes in ambient kinetic temperature. According to Hesstvedt's (1970) model, the change in intensity from sunset to midnight is accompanied by a change in the bottom height of the layer of approximately only 1 km. If the temperature was stationary this change could account for a variation of only a few degrees. The excursion that we notice appears to be larger: if Hesstvedt's model is valid, it could point either to a large variation in ambient temperature or to substantial differences between rotational temperature and ambient temperature.

Acknowledgment

We thank Dr. A. D'Altorio for his contribution to the initial part of this work and Messrs. O. Consorte and V. Galli for assistance in carrying out the measurements. This research was supported in part by the Consiglio Nazionale delle Ricerche.

References

Ballif, J. R. and Venkateswaran, S. V.: 1963, *J. Atmospheric Sci.* **20**, 1.
Fiocco *et al.*: 1970, *Nature* **228**, 1079.
Hesstvedt, A.: 1970, *J. Geophys. Res.* **75**, 2337.
Hunt, B. G.: 1966, *J. Geophys. Res.* **71**, 1385.
Krassovsky, V. I.: 1963, *Planetary Space Sci.* **10**, 7.
Kvifte, G.: 1961, *Planetary Space Sci.* **5**, 158.
Meinel, A. B.: 1950, *Astrophys. J.* **111**, 555; **112**, 120.
Shimazaki, T. and Laird, A. R.: 1970, *J. Geophys. Res.* **75**, 3221.
Wallace, L.: 1960, *Astrophys. J.* **132**, 894.
Wallace, L.: 1961, *J. Atmospheric Terrest. Phys.* **20**, 85.
Wallace, L.: 1962, *J. Atmospheric Sci.* **19**, 1.

DAYGLOW AND TWILIGHT EXCITATION MECHANISMS FOR AIRGLOW

H. N. RUNDLE

Institute of Space and Atmospheric Studies, University of Saskatchewan

An attempt is made to survey the present ideas on the phenomena responsible for dayglow and twilight emissions starting with constituents prevalent at higher heights. Comprehensive reviews are Chamberlain (1961), Hunten (1967) and Noxon (1968).

1. HI and HeI

Emissions of nighttime and daytime Lyα and nighttime Hα are established as resonant scattering and fluorescence from Lyβ absorption, respectively, by H extending out to several R_E. The Lyα radiative transfer problem has been solved (Donahue, 1966); however, it is difficult to obtain H density profiles from observations since the optical depth is 3 to 10 at 120 km. The H density profile is the result of diffusion of H from the turbopause up to the geocorona. Diurnal variations of this profile may be the result of thermospheric diurnal temperature variations of about 100 K as well as injection of H_2 at the turbopause from the homosphere (Patterson, 1966; Tinsley, 1969a) with subsequent dissociation (Bates and Nicolet, 1965). The dissociation of H_2 is accomplished by collision with 0 at temperatures above 800 K (Bates and Nicolet, 1965). The behavior of the H emissions is consistent with the ratio of H densities at exospheric heights to those below 250 km being increased at night over the daytime ratio.

The 10830 Å emission observed at twilight is established as resonance scattering of solar radiation by the metastable He(2^3S) state of orthohelium. Production of this state (McElroy, 1965) is mainly by photoelectron impact on He(1^1S) and requires 19.7 eV. The lifetime of He(2^3S) is about 10 min due to photoionization and less for heights below 500 to 600 km by the Penning reactions, He(2^3S)+X→He(1^1S)+X^++e, where X=0, O_2 and N_2. Diffusion of metastable atoms will modify the effective lifetimes (McElroy, 1965; Patterson, 1967). Daytime intensities should be less than twilight intensities due to faster cooling of photoelectrons.

During years of high solar activity, the intensities of the He and H emissions increase and decrease, respectively, (Krassovsky *et al.*, 1966; Shefov, 1968; Tinsley, 1969b) which is what is expected by atmospheric temperature control on He and H densities. The He emission exhibits a maximum intensity during winter (Shefov, 1968; Harrison, 1969; Tinsley, 1969b) and probably is a result (Tinsley, 1968) of increased He densities during the winter (Jacchia, 1969). Hα exhibits a maximum intensity during summer due to the geometry of the earth and sun; however, an increase of H densities is indicated during the fall (Tinsley, 1967). The excess winter He

B. M. McCormac (ed.), The Radiating Atmosphere, 90–104.

has been interpreted as a result of a height variation of the turbopause (Jacchia, 1969). Krassovsky (Shefov, 1968) invoked transport of He^+ to the winter hemisphere and subsequent recombination. The polar wind (Banks, 1969) and thermal escape should also play a role. Geomagnetic variations (Shefov, 1969) and daily fluctuations of the He and H emissions will reflect variations of atmospheric structure. The semi-annual variation in the thermospheric and exospheric densities (Cook, 1969) and H chemistry in the homosphere may also be factors.

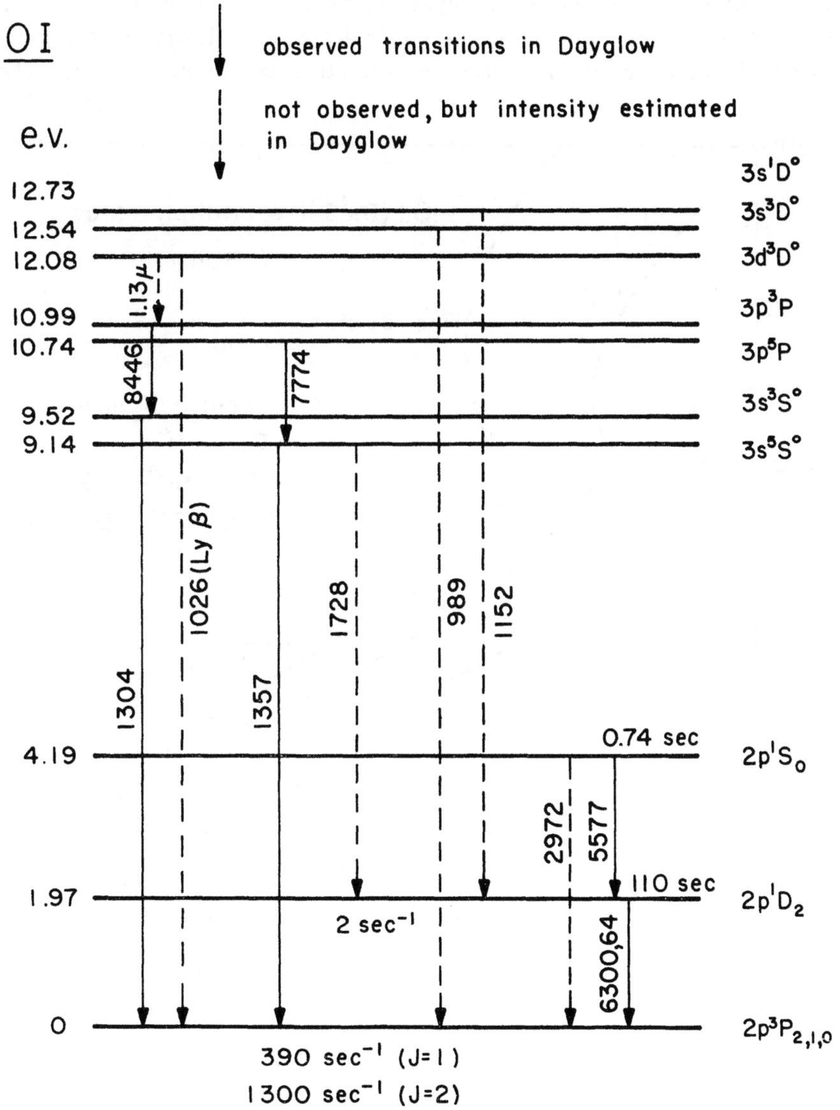

Fig. 1. Atmospheric O I transitions and energy levels. The transition wavelengths are in Ångstroms except where units are indicated. Some transition probabilities are indicated.

2. OI

A. UV EMISSIONS

The OI energy levels above 9 eV (Figure 1) are populated mainly by photoelectron impact (Green and Barth, 1967; Dalgarno *et al.*, 1969). The allowed $2p^3P - 3s^3S^\circ$ transition at 1305 Å exhibits resonance scattering where the optical depth becomes at least 10^4. The difficult radiative transfer problem has been solved (Donahue, 1965; Strickland and Donahue, 1970) and they show that the measurements of Fastie (1968) are consistent with photoelectron impact being the dominant mechanism around 150 to 200 km and resonant scattering above 400 km as illustrated in Figure 2. Absorption of Lyβ (Bowen fluorescence) and radiative recombination of O^+ (Donahue

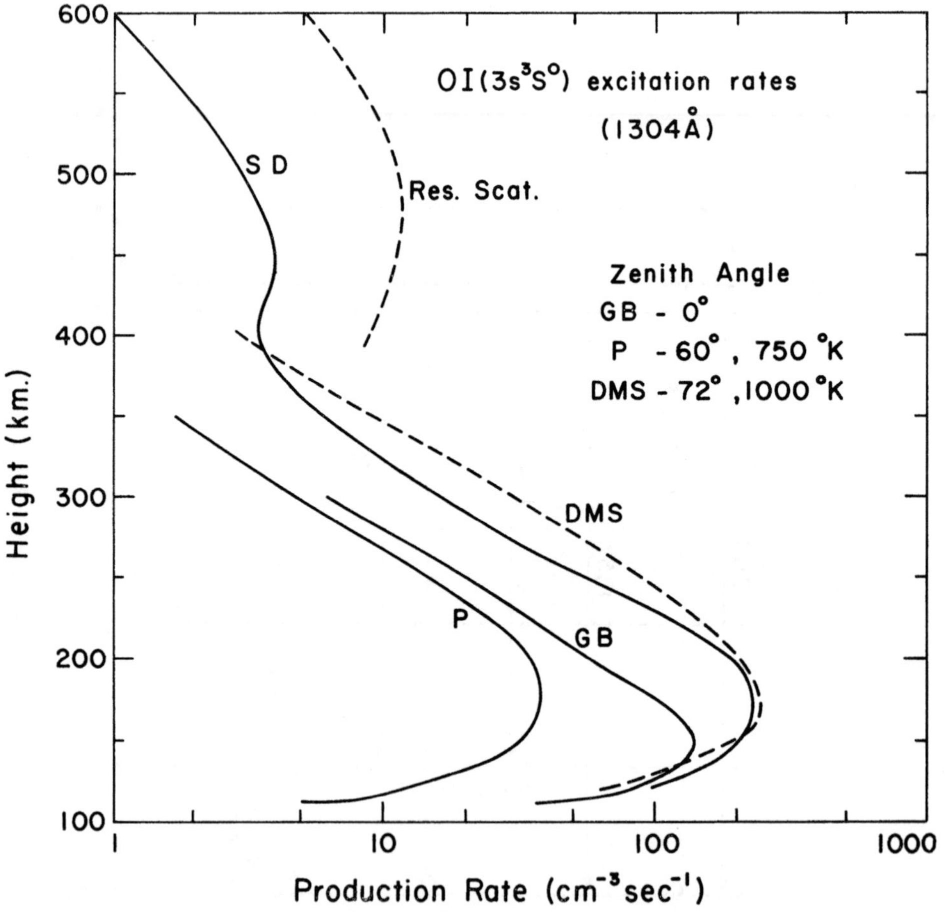

Fig. 2. Initial source functions or excitation rates due to photoelectron impact before radiation imprisonment are considered for OI($3s^3S^\circ$) which gives rise to the 1304 Å triplet. GB – Green and Barth (1967); P – Prasad (1969); DMS – Dalgarno *et al.* (1969); SD – Strickland and Donahue (1970). The latter curve is derived from observations of Fastie (1968). The resonance scattering curve is as calculated by Strickland and Donahue (1970).

and Fastie, 1964; Hanson, 1969) populate the $3d^3D^0$ level and are minor daytime sources. Bowen fluorescence will contribute about 50 R to the 8446 Å and 1.13 μ lines (Dalgarno *et al.*, 1969). Photoelectron impact accounts for several hundred rayleighs of 8446 Å and 7774 Å emission and about 1/10th of this value for the 1.13 μ emission (Green and Barth, 1967; Dalgarno *et al.*, 1969). The 8446 Å and 7774 Å lines have been measured in the dayglow at 1 kR and 1.6 kR, respectively (Wallace and Broadfoot, 1969). Russian observers have measured the twilight 8446 Å line for several years (Shefov, 1968) and have explained the results by Bowen fluorescence.

The forbidden transition $^3P - {}^5S^0$ gives about 350 R at 1356 Å and 1358 Å which is explained by photoelectron excitation of the upper state. The intensity profile indicates an escape of locally produced photoelectrons (Prasad, 1969). Other transitions at 989 Å and 1152 Å have been predicted (Green and Barth, 1967) with 900 R and 1100 R, respectively, due to excitation by photoelectron impact. The 989 Å line will also be emitted by resonance scattering since it is an allowed line involving the ground state.

B. VISIBLE O I EMISSION

The red and green lines (5577 Å and 6300 Å) are a result of formation of the $2p^1S$ and $2p^1D$ levels. The production mechanisms are Schumann-Runge dissociation of O_2 (Hinteregger *et al.*, 1965; Wallace and McElroy, 1966; Deehr, 1969), electron impact (Wallace and McElroy, 1966) and dissociative recombination of O_2^+ (Zipf, 1969). The Chapman reaction (Schiff, 1969) is thought to enhance 5577 Å something like a factor of 4 during the daytime (Wallace and McElroy, 1966; Dandekar, 1969) due to an increase in O. Another possible source is $O(^3P)+N(^2D)\rightarrow N(^4S)+O(^1D)$ (Peterson *et al.*, 1966). If the rate is 2×10^{-13} cm^3 sec^{-1} (Weill, 1969), then this mechanism is as important as dissociative recombination of O_2^+ above 200 km. The quenching of $O(^1S)$ is accomplished by O, O_2 and H_2O. $O(^1D)$ is quenched by an N_2 or O_2 molecule about equally well (Noxon, 1970). The rates are summarized (Zipf, 1969). The lifetimes of $O(^1S)$ and $O(^1D)$ are equal to their radiative lifetimes at heights of about 85 km and 250 km, respectively.

Photoelectron impact above 200 km and photodissociation below are the important mechanisms for 3 kR total emission of redline dayglow (Wallace and McElroy, 1966). As the solar elevation angle decreases, the contribution from photodissociation below 200 or 150 km height decreases (Noxon, 1968) due to the increased slant path of solar emission. Deehr (1969) explained the twilight decay of the red line with photodissociation, as well as dissociative recombination of O_2^+ and neglected the less important electron impact mechanisms along the lines of earlier treatments (Chamberlain, 1961; Hunten, 1967) Deehr used a deactivation rate of 10^{-11} cm^3 sec^{-1} for $O(^1D)$ which seems low; hence, the additional mechanisms (discussed above) could be significant during the twilight also. A summary of green line dayglow measurements (Dandekar, 1969) indicates the layer due to the Chapman reaction just below 100 km and emission higher up which was explained by photoelectron excitation (Wallace and McElroy, 1966). However, dissociative recombination of O_2^+ can equally explain the higher altitude results as pointed out by Wallace and McElroy themselves. The twilight decay of 5577 Å which must be observed against the relatively bright night sky green line emission has been explained by photodissociation (Deehr, 1969). Again, the electron impact mechanism during twilight has been neglected. Large values up to 50 kR of 6300 Å dayglow (Noxon, 1968) supposedly result from excitation of $O(^1D)$ atoms by superthermal electrons or the energetic electrons associated with electron temperatures over 3000 K. Such temperatures have been measured, e.g. during March, 1966 (Brace *et al.*, 1969). Lower temperatures occurred during 1964 (Evans, 1967; Evans and Mantas, 1968; Nagy *et al.*, 1969). The high temperatures and superthermal electrons can be caused by photoelectrons or interactions of the protonosphere and of the trapped radiation belts with the ionosphere (Knudsen, 1968; Nagy *et al.*, 1969). The green line has not been monitored during daytime, but should exhibit similar less pronounced variations. Abnormally large twilight intensities of the green line (Megill, 1960; Deehr, 1969) may be due to variations in atmospheric O_2 densities (Deehr, 1969.) Heating of ionospheric electrons by conjugate photoelectrons has been invoked to explain the pre-dawn enhancement of 6300 Å emission (Cole, 1965; Evans and Gastman, 1970).

3. N_2

Excited N_2 states are a result of photoelectron impact. The intensities expected are 15 kR for the 2P system and an intensity ratio for 1P/2P of about two (Green and Barth, 1967) which indicates that the $C^3\pi_\mu$ and the $B^3\pi_g$ states are equally excited by photoelectron impact. Dalgarno *et al.* (1969) calculated somewhat lower intensities. Further discussion is given by Broadfoot (1971). Separate measurements of the 2P and 1P bands (Barth and Pearce, 1966; Wallace and Broadfoot, 1969) show 6.4 kR for the 2P system and 1.5 for the intensity ratio. The Vegard-Kaplan bands should be a bit more intense than the 1P system (Green and Barth, 1967; Dalgarno *et al.*, 1969) and quenching proceeds by collision with O (Zipf, 1969). The higher vibrational levels may be further quenched by collisions with N_2 (Dalgarno *et al.*, 1969). The intensities of the singlet band systems have been calculated assuming photoelectron excitation, however, most of the cross sections are not known.

4. N_2^+

The N_2^+ 1N and Meinel bands are a result of resonant scattering (Broadfoot, 1967a) and electron impact (Wallace and McElroy, 1966; Dalgarno *et al.*, 1969). Photo-ionization results in 30%, 60% and 10% of N_2^+ $X^3\Sigma$, $A^2\pi$ and $B^2\Sigma$, respectively (Schoen, 1969). From 150 to 300 km heights, charge transfer between O^+ (^2D) and N_2 yields high rotational development of N_2^+ $(A^2\pi, v'=1)$ which persists for the resulting N_2^+ $(X^3\Sigma)$. The role of this mechanism is still in some doubt. Quenching of the excited state is considered not important.

Destruction of N_2^+ (Broadfoot and Hunten, 1966) is by dissociative recombination (Wallace and McElroy, 1966; Biondi, 1969) and collision with O and O_2 (Ferguson

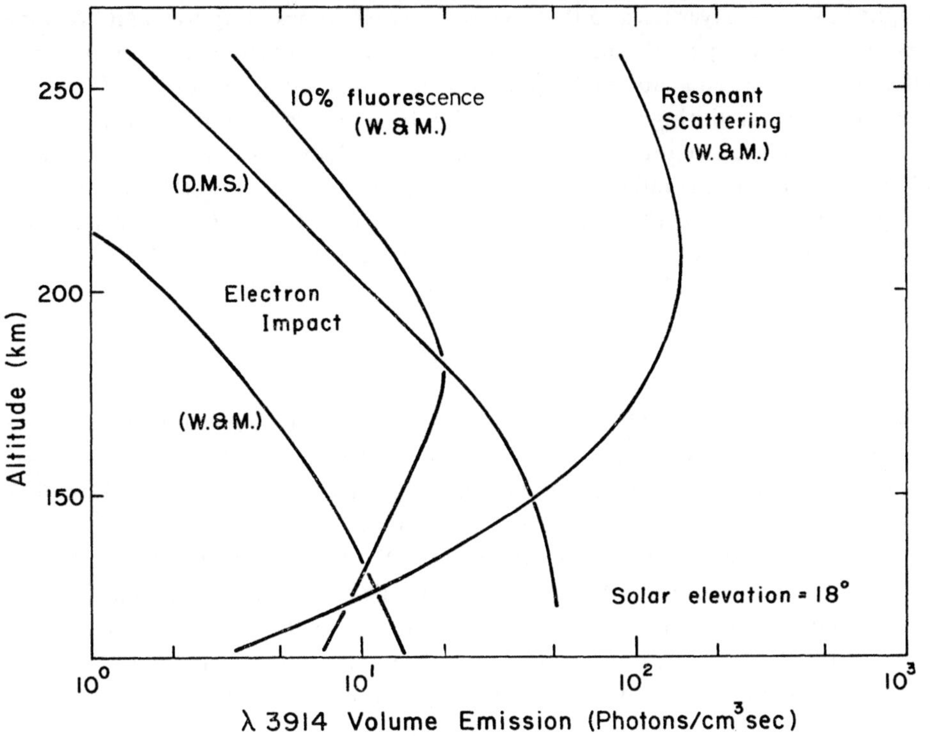

Fig. 3. The contribution of various excitation mechanisms to excitation of $N_2^+(B^2\Sigma_u^+)$ which gives rise to the N_2^+ 1N 0-0 band at 3914 Å. W and M is after Wallace and McElroy (1966) and DMS is after Dalgarno *et al.* (1969). 10% fluorescence means that 10% of N_2 photoionization contributes to the $N_2^+(B^2\Sigma_u^+)$ state.

1969). Thus, N_2^+ is rapidly destroyed below 150 km by O_2 or O. The lifetime of N_2^+ becomes about 1 sec at 150 km and 85 sec at about 500 km height. The population of different vibrational levels and the relative contribution of each production mechanism is dependent on the lifetime of the ion. Long lifetimes (over 1 sec) correspond to increased resonant scattering and are discussed by Broadfoot (1967a, 1971).

Wallace and McElroy (1966) assumed the first three of the above production mechanisms for the occurrence of the daytime $0-0$ 1N band. The results of their calculation are shown in Figure 3. The resulting N_2^+ ion densities compare favorably with mass spectrometer ion measurements taken by Brinton *et al.* (1969) and Goldberg and Blumle (1970).

Integrated intensity measurements of the Meinel dayglow bands (Wallace and Broadfoot, 1969) are approximately in agreement with the 1N measurements. The measured ratio of 1.5 to 2.5 of the intensities of the $(1-0)$ Meinel and $(0-0)$ 1N bands suggests that change transfer between $O^+(^2D)$ and N_2 is unimportant. However, the two bands were measured at different times.

Twilight observations (Broadfoot and Hunten, 1966) are difficult to interpret due to tropospheric scattering (Broadfoot, 1968). As the solar depression increases, the height of the N_2^+ layer rises so that the ions have lifetimes of up to 60 sec. Resonance scattering will then be the main excitation process as demonstrated by the Swings effect (Broadfoot and Hunten, 1966) with photoionization by solar UV controlling the density of N_2^+ during twilight. Ionization by photoelectron impact is unimportant except for a large evening twilight enhancement at Kitt Peak by impact of conjugate photoelectrons. The rotational temperature appears to be larger than the exospheric temperature so that perhaps charge exchange between $O^+ (^2D)$ and N_2 cannot be neglected as suggested by the present dayglow measurements. Further discussion is given by Broadfoot (1971) on N_2^+ mechanisms.

5. NI

The 5200 Å doublet arises from the $N(^2D^0)$ level which has a natural lifetime of 26 hr. The production of $N(^2D^0)$ occurs from the destruction of N_2^+ by dissociative recombination and ion exchange with O and also by dissociative recombination of NO^+ (Biondi, 1969). The loss of $N(^2D^0)$ is thought to be by $N(^2D)+O_2(X^3\Sigma_g)\rightarrow N(^4S)+$ $+O_2(^1\Delta_g, {}^1\Sigma_g^+)$ or $N+NO$ (Zipf, 1969), $N(^2D)+O\rightarrow N+O(^1D)$ (Weill, 1969) and $N(^2D)+e\rightarrow N+e$ (Seaton, 1955; Henry and Williams, 1968). The lifetime of $N(^2D)$ is about 15 min at 175 km, 200 km, and 300 km due to the above three mechanisms, respectively. The first loss reaction is important for NO production (Norton and Barth, 1970) and the second may be important for $O(^1D)$ production (Weill, 1969).

The 90 R dayglow intensity (Wallace and McElroy, 1966) can be explained by any of the above production mechanisms with destruction by collision with O_2 as the majority of the emission was below 200 km. Wallace and Broadfoot (1969) identified the 10404 Å line from $N(^2P)$ which can only be produced by dissociative recombination of N_2^+. The NO chemistry has been explained by production of $N(^2D)$ mainly by dissociative recombination of NO^+ below 150 km with $N_2^+ +O\rightarrow N(^2D)+NO^+$ of secondary importance (Norton and Barth, 1970). The slowly decaying evening twilight 5200 Å is due to dissociative recombination of NO^+ since N_2^+ is rapidly lost after sunset and $N(^2D)$ has only a 15 min lifetime.

6. O^+ and O_2^+

Emissions from these ions have not been detected in dayglow or twilight; however, the $O^+(^2D^0_{3/2,5/2}-{}^2P^0_{3/2,5/2})$ transition is expected to give about 100 R for each of the four lines involved during the day and about 10 R for a solar depression of 10^0 (Dalgarno and McElroy, 1965, 1966). The production should be by photoionization to excited O^+ states with subsequent cascading. Since the natural lifetime of the $O^+(^2P^0)$ state is a few sec, quenching by N_2 and electrons may be important but not severe. Quenching of the $O^+(^2D^0)$ states by N_2 (Zipf, 1969) should be severe enough to effectively give a trivial intensity to the $(^4S^0-{}^2D^0)$ transition and was noted above as a possible source of N_2^+. Charge exchange with O_2 and quenching by electron collisions are also important. Several weak O_2^+ bands are expected as a result of photoionization and ionization by photoelectron impact; however, most of the cross sections are not known with any certainty (Dalgarno and McElroy, 1965; Green and Barth, 1967; Dalgarno *et al.*, 1969).

7. NO

The NO γ bands (Barth, 1966; Pearce, 1969; Meira, 1971) are due to resonance scattering. The latter measurement indicates about 4×10^7 NO molecules cm^{-3} between 100 and 70 km. The problem is then one of explaining the presence of NO although more measurements are needed along with accurate solar fluxes, f values and branching ratios.

NO densities are a result of the chemistry of the N–O atmosphere (Norton and Barth, 1970; Strobel *et al.*, 1970). Above 150 km, the major production and loss of NO are by $N+O_2 \rightarrow NO+O$ (k_1) and $NO+N \rightarrow N_2^*+O$ (k_2) (Nicolet, 1965; Schiff, 1969). At high enough heights diffusion is important. Chemical equilibrium probably exists at 200 km where $[N] \sim 5 \times 10^6$ cm^{-3} since the lifetime of NO is about 2 hr. If the temperature is about 1000 K, $[NO] \sim 2 \times 10^6$ cm^{-3} also. Due to the large activation energy of k_1, the production of NO becomes too slow to account for NO densities measured below 150 km. $O^+ + N_2 \rightarrow NO + NO^+$ is not much faster (Ferguson *et al.*, 1965; Hunten and McElroy, 1968). However, vibrationally excited N_2 may increase the reaction rate (Barth, 1966). The use of $O_2\,(^1\Delta)$ in reaction k_1 (Hunten and McElroy, 1968) does not help since the reaction rate is still about a factor of 100 too low (Clark and Wayne, 1969). If $N\,(^2D)$ is used in reaction k_1 (Norton and Barth, 1970; Strobel *et al.*, 1970) the reaction rate of 5×10^{-12} cm^3 sec^{-1} (Black *et al.*, 1969; Lin and Kaufman, 1969) is adequate. The chemistry of $N\,(^2D)$ and $N\,(^4S)$ presented by Norton and Barth (1970), and Strobel *et al.* (1970) results in $[N\,(^4S)] \sim 10^6$ cm^{-3} at 100 km. Thus, the chemical lifetime of NO is about 12 hr at this height and longer at lower heights so that diffusion controls NO densities below about 95 km. The downward flux of NO molecules in the absence of a loss process is $\phi = n_1 K(H^{-1} - H_1^{-1})$ where n_1 and H_1 are the density and scale height, respectively, of NO. K is the eddy diffusion coefficient and H is the atmospheric scale height. Since H_1 appears to be large (Meira, 1971), $\phi \sim n_1 K/H \sim 10^8$ molecules cm^{-2} sec^{-1}. A chemical sink will exist below 70 km, say, and may be distributed over some height range so that both ϕ and n_1 will eventually decrease to zero. An interesting large variation in NO densities can occur by relatively small variations in the production rate of $N\,(^2D)$ and $N\,(^4S)$ from dissociative recombination of NO^+ and dissociation of N_2 by photoelectron impact, respectively, (Strobel *et al.*, 1970).

8. OH

The nightglow vibration-rotation Meinel bands of OH persist throughout the daytime and are usually considered to be a result of $H+O_3 \rightarrow O_2+OH^*$ $(^2\pi)$ although $H+O_2\,(v'>4) \rightarrow OH^*\,(^2\pi)+O$ is still popular with the Russian school (Krassovsky, 1969). The O_2 reaction is considered most unlikely from laboratory work (Kaufman,

1969) and the O_3 reaction yields 80 to 85% of OH products in the $v''>4$ levels with increasing population up to $v''=9$ (Anlauf *et al.*, 1968).

Twilight and dayglow observations support the O_3 reaction (Llewellyn and Evans, 1971). The chemistry of H and O involves some 40 reactions along with eddy diffusion (Hunt, 1966; Gattinger, 1968, 1969, 1971; Shimazaki and Laird, 1970).

The relative population of the OH vibrational levels has a seasonal variation at Zvenigorod and has been used to interpolate a diurnal variation which indicates enhanced lower level population when sunlight is present (Shefov, 1969). Such a phenomenon can be explained by two mechanisms producing excited OH (supposedly the O_3 and O_2 ones) or else by a redistribution of the population of vibrational levels before emission. Little is known about quenching of excited OH; however, the reaction rate of $OH^* + O_3 \rightarrow HO_2 + O_2$ or $H + 2O_2$ depends strongly on the vibrational level of OH^* (Kaufman, 1969) and H_2O appears to be an efficient quenching agent (Kaufman, 1964). Redistribution of vibrational populations by $O_2(X^3\Sigma_g) + OH(v') \rightarrow \rightarrow O_2(^1\Delta_g) + OH(v'', v' - v'' \sim 1 \text{ eV})$ has been suggested (Evans and Llewellyn, 1970).

9. O_2

The atmospheric $b^1\Sigma_g^+ - X^3\Sigma_g^-$ and IR atmospheric $a^1\Delta_g - X^3\Sigma^-$ bands are prominent dayglow emissions. Production of the forbidden atmospheric band is accomplished by resonant scattering (Wallace and Hunten, 1968) and $O(^1D) + O_2(^3\Sigma_g) \rightarrow O(^3P) + + O_2(^1\Sigma)$ which competes with N_2 quenching of $O(^1D)$. The occurrence of $O_2(^1\Delta)$ as an end product is thought to be negligible (Noxon, 1970). The $O(^1D)$ is produced by Schumann-Runge dissociation of O_2 and Hartley dissociation of O_3. Hartley dissociation of O_3 may yield $O_2(^1\Sigma)$ for $2000 \text{ Å} < \lambda < 2660 \text{ Å}$ (Wallace and Hunten, 1968). However, at 2537 Å, the reaction appears to give only $O_2(^1\Delta_g)$ (Gauthier and Snelling, 1970). The nightglow chemiluminescent reaction persists during the daytime at 90 km. Apparently, $O + O + M \rightarrow O_2(^1\Sigma) + M$ is too slow according to present laboratory measurements which are not understood (Wallace and Hunten, 1968; Young, 1969). Redistribution of OH vibrational populations by O_2 collision as mentioned earlier may also produce $O_2(^1\Sigma)$. $O + O_3 \rightarrow O_2(^1\Sigma) + O_2$ is another possibility (Wallace and Hunten, 1968).

The contribution of these mechanisms at various heights has been calculated (Wallace and Hunten, 1968) with the assumption that O_3 photodissociation is important and that N_2 is 5 times more effective per collision than O_2 in $O(^1D)$ quenching. The picture is quite consistent. However, if O_3 photolysis does not give $O_2(^1\Sigma)$, and N_2 and O_2 are about equally effective $O(^1D)$ quenching agents per collision, the picture is as shown in Figure 4. The additional production of $O_2(^1\Sigma)$ at 50 km, that was explained by O_3 photodissociation, can now be replaced by a greater production of $O_2(^1\Sigma)$ by $O(^1D)$ deactivation. However, the production rate above 100 km height is now overestimated but could be reduced to the amount measured by Wallace and Hunten (1968) if the O_2 density is reduced by almost a factor of 2. This may not be too unreasonable since a seasonal variation of O_2 densities at 100 to 120 km seems to

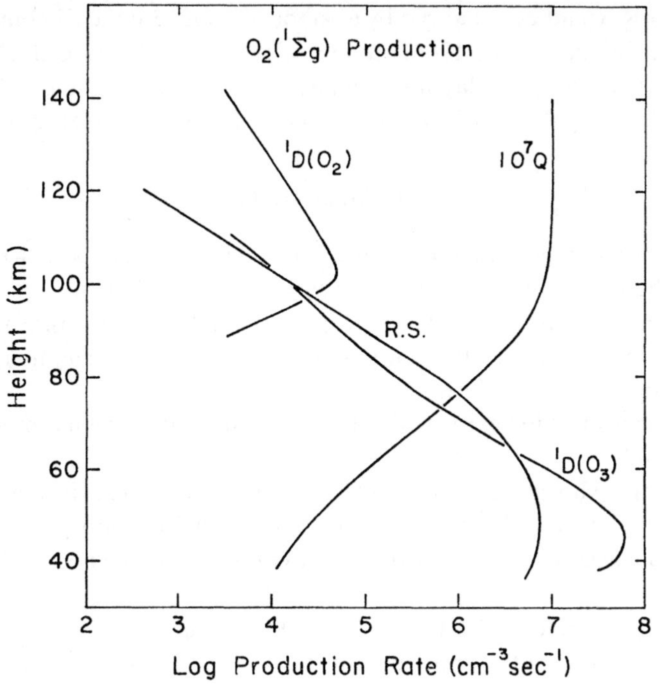

Fig. 4. Production rates of $O_2(^1\Sigma_g)$. Resonance scattering (RS) is as calculated by Wallace and Hunten (1968). The production by deactivation of $O(^1D)$ by O_2 collision is $O_2(^1D)$ and $O_3(^1D)$ where ^1D is a result of photolysis of O_2 and O_3, respectively. These are four times the rates given by Wallace and Hunten. Q is the quenching factor as calculated by Wallace and Hunten.

occur with a minimum during the winter (Norton and Warnock, 1968). The chemiluminescent mechanism probably supplies as much $O_2(^1\Sigma)$ as the revised $O_3(^1D)$ mechanism does at 90 to 95 km heights (Wallace and Hunten, 1968). Quenching of the $O_2(^1\Sigma)$ level is thought to be accomplished mainly by N_2 (Zipf, 1969). The loss, $Q=A/(A+k_{N_2})$ where A is the Einstein coefficient, is shown in Figure 4 as calculated by Wallace and Hunten (1968). The quenching lifetime is equal to the radiative lifetime at a height of about 85 km.

Photolysis of O_3 by the Hartley band is the principal daytime production mechanism of $O_2(^1\Delta)$ and as seen above, the efficiency is probably unity. Quenching of the $O_2(^1\Delta)$ state occurs by N_2 or O_2 (Findlay *et al.*, 1969; Steer *et al.*, 1969). The radiative lifetime is equal to the quenching lifetime at about 70 to 75 km. 1.4 MR out of 27 MR is contributed from above 80 km heights (Evans *et al.*, 1968, 1969). Twilight observations (Gattinger, 1968; Evans *et al.*, 1969; Evans and Llewellyn, 1970) suggest that the emission above 80 km is somewhat complex. The evening twilight decay of $O_2(^1\Delta)$ is longer ($\sim$60 min) in the winter than in the summer ($\sim$35 min) (Evans and Llewellyn, 1970). Hence, the layer above 80 km appears to be enhanced during the winter. Several possibilities have been suggested for $O_2(^1\Delta)$ production above 80 km. Conversion of $O_2(^1\Sigma)$ to $O_2(^1\Delta)$ by collision may supply $\sim$250 kR and the nightglow of about 100 kR would be enhanced by a factor of about 3 if the Chapman mechanism is responsible (Evans *et al.*, 1968). If $OH(v')+O_2 \rightarrow O_2(^1\Delta)+OH(v'')$ is the main nighttime mechanism, this daytime enhancement would not exist. The remaining 50 to 70% of the $O_2(^1\Delta)$ production would then be due to O_3 photolysis.

10. Alkali Metals

The presence of resonance radiation from Ca II, Na, Li and K has been firmly established during twilight and during daytime for the case of Na. Ca II ions are a result of meteoric deposition between 100 and 250 km heights (Broadfoot, 1967b). Sodium densities have been extensively studied for almost two decades and a summary of the more important properties is as follows:

(a) A seasonal variation exists with a maximum in winter which is more pronounced at higher latitudes.

(b) Variations with periods of about a month are common during winter at high latitudes (e.g. 52° N at Saskatoon) and are associated with stratospheric warmings.

(c) Diurnal variations occur with the daytime densities enhanced over twilight values.

(d) Any height variations for the Na layer at 90 km are small.

(e) The scale height of the upper part of the layer is 3 to 4 km which is considerably less than that of the atmosphere.

Less is known about Li and K. However, the seasonal variation of Li consists of a low background with a winter maximum and spikey enhancements during the winter (Gault and Rundle, 1969). Potassium seasonal variations appear less well defined. Lithium seems to be at about the same height as Na while K seems to be at a lower height. The ratio of the densities of Na:K:Li ~ 1000:5:1. However, such figures vary during the year and involve a guess as to the amount oxidized and ionized. The possibility of contamination by artificial atmospheric Li injections has been noted (Gadsden, 1964; Kellogg, 1964). However, the first Li measurements were made during August and September 1957 before high altitude artificial injections occurred. There is speculation that a previous large explosion at ground level succeeded in injecting Li into the upper atmosphere (Martell, 1970). The behavior of the alkali layers has not yet been decisively explained. Photochemistry, meteoric ablation, and aerosols have been invoked and will now be discussed.

Photochemistry of Na was first discussed by Chapman in 1939 and is summarized by Hunten (1967). The important reactions are $Na+O_3 \rightarrow NaO+O_2$ and $NaO+O \rightarrow \rightarrow Na+O_2$ with rates k_1 and k_2 respectively. The lifetime of a Na atom at 90 km ~ 15 min by k_1, and k_2 gives NaO a lifetime of about 1 sec. Equilibrium is then prevalent so that

$$\frac{[Na]}{[NaO]} = \frac{k_2[O]}{k_1[O_3]} \sim 7 \frac{[O]}{[O_3]}.$$

Oxygen and O_3 depend significantly on atmospheric dynamics; thus, photochemical control of [Na] is attractive except for the facts that enhanced Na layers should be lower if one assumes that the total Na is conserved, and the scale height of the layer should be that of the atmosphere. Sodium is photoionized at a rate of 2×10^{-5} sec^{-1} but daytime recombination must be rapid since no evening-morning difference is observed in Na densities, and, compared to Na densities, Na^+ densities measured by ion mass spectrometers are low. Recombination can be by dissociative recombination of oxide ions which is formed by oxidation of Na^+. Charge exchange with negative ions is another possibility. To explain the small topside scale height during the daytime, Hanson and Donaldson (1967) invoked upward eddy diffusion of Na atoms with subsequent photoionization. The Na ion is then forced downward by westward winds across the magnetic field where the ion can recombine by dissociative recombination of the ion oxide. However, the small topside scale height has been measured during the night (Sandford and Gibson, 1970) which indicates the presence of another mechanism.

Ablation of meteors between 90 to 110 km as a source of free alkali metal layers has been discussed (Gadsden, 1968, 1970) with the additional processes of ionization and eddy diffusion to a lower chemical sink. The topside scale height and the essentially fixed height of the Na layer can be explained by varying the parameters. The observed meteor fluxes (Keay, 1963; Millman and McKinley, 1963) show no relationship to Na densities and ionization does not decrease the evening twilight densities; therefore, Gadsden assumes that eddy diffusion is the overwhelming controlling mechanism. The problem of Na density variations in time is then replaced by the problem of explaining a vertical eddy diffusion coefficient with interesting seasonal and diurnal properties. The correlation between Na density enhancements and stratospheric warmings would need to be explained by a reduction in the vertical eddy diffusion (or mixing in general) at 90 km heights when one really would expect the opposite. The eventual fate of alkali atoms locked up in the mesosphere sink is neglected.

The aerosol hypothesis has several variations. Those that identify the aerosols with micrometeorites settling in the atmosphere are unsatisfactory (Gadsden, 1968). Moreover, if such aerosols provide a source of Na atoms, the atmospheric Na density should correlate with meteor activity and one has the same problem from which the meteor ablation hypothesis suffers. One is led to the conclusion that there is atmospheric control of the Na densities; thus, the supposed aerosols are resident in the mesosphere for a time which is thought to be 5 yr. All sources of aerosols are easily accommodated. The terminal velocity at 95 km altitudes is ~ 100 cm sec^{-1} for a particle of radius 0.01 μ. Since eddy diffusion velocities are about the same or less, one is led to consider aerosols with radii $\leqslant 0.01$ μ at 95 km. One could call such particles macromolecules containing $\leqslant 10^4$ atoms. Large scale atmospheric motions may very well carry larger particles upward but 0.1 μ radii particles will fall 5 km in about 15 min at 95 km and will tend to exist below 70 to 80 km after the large scale atmospheric motion ceases. Atmospheric aerosols will act as a source and perhaps as a sink for Na atoms (as well as other constituents) and the smaller ones may even change their radii in the process so that their scale height is varied. One must consider $Ma' \rightleftarrows Ma + Na$ where Ma stands for a macromolecule. If equilibrium can be assumed for Na production and loss by the above process plus oxidation and reduction,

$$[\text{Na}] = \frac{k_2[\text{NaO}][\text{O}] + p[\text{Ma}]}{k_1[\text{O}_3] + l[\text{Ma}]}$$

where p and l stand for the production and loss rates of Na due to macromolecules which will depend on temperature and the past history of the particle. It appears that the macromolecules terms influence the Na densities as much as the O or O_3 terms. The introduction of macromolecules into the mechanism has several attractive features and the initial work (Donahue and Meier, 1967; Hunten and Wallace, 1967; Gadsden, 1968) should be extended.

Acknowledgments

I wish to thank Dr. D. M. Hunten for a critical reading of the manuscript. This research was supported by the Canadian National Research Council.

References

Anlauf, R. G., MacDonald, R. G., and Polanyi, J. C.: 1968, *Chem. Phys. Letters* **1**, 619.
Banks, P. M.: 1969, in *Atmospheric Emissions* (ed. by B. M. McCormac and A. Omholt), Van Nostrand Reinhold Company, New York, p. 533.
Barth, C. A.: 1966, *Ann. Geophys.* **22**, 198.
Barth, C. A. and Pearce, J. B.: 1966, *Space Res.* **6**, 381.
Bates, D. R. and Nicolet, M.: 1965, *Planetary Space Sci.* **13**, 905.
Biondi, M. A.: 1969, *Can. J. Chem.* **47**, 1711.
Black, G., Stanger, T. C., St. John, G. A., and Young, R. A.: 1969, *J. Chem. Phys.* **51**, 116.
Brace, L. H., Mayo, H. G., and Findlay, J. A.: 1969, *J. Geophys. Res.* **74**, 2952.
Brinton, H. C., Pharo, III, M. W., Mayr, H. G., and Taylor, Jr., H. A.: 1969, *J. Geophys. Res.* **74**, 2941.
Broadfoot, A. L.: 1967a, *Planetary Space Sci.* **15**, 1801.
Broadfoot, A. L.: 1967b, *Planetary Space Sci.* **15**, 503.
Broadfoot, A. L.: 1968, *Planetary Space Sci.* **16**, 693.
Broadfoot, A. L.: 1971, this volume, p. 34.
Broadfoot, A. L. and Hunten, D. M.: 1966, *Planetary Space Sci.* **14**, 1303.
Chamberlain, J. W.: 1961, in *Physics of the Aurora and Airglow*, Academic Press.
Clark, I. D. and Wayne, R. P.: 1969, *Chem. Phys. Letters* **3**, 405.
Cole, K. D.: 1965, *Ann. Geophys.* **21**, 156.
Cook, G. E.: 1969, *Ann. Geophys.* **25**, 451.
Dalgarno, A. and McElroy, M. B.: 1965, *Planetary Space Sci.* **13**, 947.
Dalgarno, A. and McElroy, M. B.: 1966, *Planetary Space Sci.* **14**, 1321.
Dalgarno, A., McElroy, M. B., and Stewart, A. I.: 1969, *J. Atmospheric Sci.* **26**, 753.
Dandekar, B. S.: 1969, *Planetary Space Sci.* **17**, 1609.
Deehr, C. S.: 1969, *Ann. Geophys.* **25**, 881.
Donahue, T. M.: 1965, *Planetary Space Sci.* **13**, 871.
Donahue, T. M.: 1966, *Ann. Geophys.* **22**, 175.
Donahue, T. M. and Fastie, W. G.: 1964, *Space Res.* **4**, 304.
Donahue, T. M. and Meier, R. R.: 1967, *J. Geophys. Res.* **72**, 2803.
Evans, J. V.: 1967, *Planetary Space Sci.* **15**, 1557.
Evans, J. V. and Gastman, I. J.: 1970, *J. Geophys. Res.* **75**, 807.
Evans, J. V. and Mantas, G. P.: 1968, *J. Atmospheric Terrest. Phys.* **30**, 563.
Evans, W. F. J., Hunten, D. M., Llewellyn, E. J., and Vallance Jones, A.: 1968, *J. Geophys. Res.* **73**, 2885.

Evans, W. F. J. and Llewellyn, E. J.: 1970, *Ann. Geophys.* **26**, 167.
Evans, W. F. J., Llewellyn, E. J., and Vallance Jones, A.: 1969, *Planetary Space Sci.* **17**, 933.
Fastie, W. G.: 1968, *Planetary Space Sci.* **16**, 929.
Ferguson, E. E.: 1969, *Ann. Geophys.* **25**, 814.
Ferguson, E. E., Fehsenfeld, F. C., Goldan, P. D., and Schmeltekopf, A. L.: 1965, *J. Geophys. Res.* **70**, 4323.
Findlay, F. D., Fortin, C. J., and Snelling, D. R.: 1969, *Chem. Phys. Letters* **3**, 204.
Gadsden, M.: 1964, *Ann. Geophys.* **20**, 383.
Gadsden, M.: 1968, *J. Atmospheric Terrest. Phys.* **30**, 151.
Gadsden, M.: 1970, *Ann. Geophys.* **26**, 141.
Gattinger, R. L.: 1968, *Can. J. Phys.* **46**, 1613.
Gattinger, R. L.: 1969, *Ann. Geophys.* **25**, 825.
Gattinger, R. L.: 1971, this volume, p. 51.
Gault, W. A. and Rundle, H. N.: 1969, *Can. J. Phys.* **47**, 85.
Gauthier, M. and Snelling, D. R.: 1970, *Chem. Phys. Letters* **5**, 93.
Goldberg, R. A. and Blumle, L. J.: 1970, *J. Geophys. Res.* **75**, 133.
Green, A. E. S. and Barth, C. A.: 1967, *J. Geophys. Res.* **72**, 3975.
Hanson, W. B.: 1969, *J. Geophys. Res.* **74**, 3720.
Hanson, W. B. and Donaldson, J. S.: 1967, *J. Geophys. Res.* **72**, 5513.
Harrison, A. W.: 1969, *Can. J. Phys.* **47**, 1377.
Henry, R. J. W. and Williams, R. E.: 1968, *Publ. Astron. Soc. Pacific* **80**, 669.
Hinteregger, H. E., Hull, L. A., and Schmidtke, O.: 1965, *Space Res.* **5**, 756.
Hunt, B. G.: 1966, *J. Geophys. Res.* **71**, 1385.
Hunten, D. M.: 1967, *Space Sci. Rev.* **6**, 493.
Hunten, D. M. and McElroy, M. B.: 1968, *J. Geophys. Res.* **73**, 2421.
Hunten, D. M. and Wallace, L.: 1967, *J. Geophys. Res.* **72**, 69.
Jacchia, L. G.: 1969, *Space Res.* **9**, 478.
Kaufman, F.: 1964, *Ann. Geophys.* **20**, 106.
Kaufman, F.: 1969, *Can. J. Chem.* **47**, 1917.
Keay, C. S. L.: 1963, *Monthly Notices Roy. Astron. Soc.* **126**, 165.
Kellog, W. W.: 1964, *Space Sci. Rev.* **3**, 275.
Knudsen, W. C.: 1968, *J. Geophys. Res.* **73**, 841.
Krassovsky, V. I.: 1969, *J. Geophys. Res.* **74**, 3064.
Krassovsky, V. I., Shefov, N. N., and Vaisberg, O. L.: 1966, *Ann. Geophys.* **22**, 203.
Lin, C. L. and Kaufman, F.: 1969, DASA Symposium on the Physics and Chemistry of the Upper Atmosphere, Stanford Research Institute, Menlo Park, California.
Llewellyn, E. J. and Evans, W. F. J.: 1971, this volume, p. 17.
Martell, R. A.: 1970, private communication with D. M. Hunten.
McElroy, M. B.: 1965, *Planetary Space Sci.* **13**, 403.
Megill, L. R.: 1960, *J. Atmospheric Terrest. Phys.* **17**, 276.
Meira, L. G., Jr.: 1971, Ph.D. Thesis, University of Colorado; *J. Geophys. Res.,* **76**, 202.
Millman, P. M. and McKinley, D. W. R.: 1963, in *The Moon, Meteorites and Comets* (ed. by B. M. Middlehurst and G. P. Kuiper), The Solar System, vol. IV, University of Chicago Press, p. 674.
Nagy, A. F., Fontheim, E. G., Stolarski, R. S., and Beutler, A. E.: 1969, *J. Geophys. Res.* **74**, 4667.
Nicolet, M.: 1965, *J. Geophys. Res.* **70**, 679.
Norton, R. B. and Barth, C. A.: 1970, *J. Geophys. Res.* **75**, 3903.
Norton, R. B. and Warnock, J. M.: 1968, *J. Geophys. Res.* **73**, 5798.
Noxon, J. F.: 1968, *Space Sci. Rev.* **8**, 92.
Noxon, J. F.: 1970, *J. Chem. Phys.* **52**, 1852.
Patterson, T. N. L.: 1966, *Planetary Space Sci.* **14**, 417.
Patterson, T. N. L.: 1967, *Planetary Space Sci.* **15**, 1219.
Pearce, J. B.: 1969, *J. Geophys. Res.* **74**, 853.
Peterson, V. L., Van Zandt, T. E., and Norton, R. B.: 1966, *J. Geophys. Res.* **71**, 2255.
Prasad, S. S.: 1969, *J. Geophys. Res.* **74**, 4772.
Sandford, M. C. W. and Gibson, A. J.: 1970, *J. Atmospheric Terrest. Phys.* **32**, 1423.
Schiff, H. I.: 1969, *Can. J. Chem.* **47**, 1903.
Schoen, R. I.: 1969, *Can. J. Phys.* **47**, 1879.

Seaton, M. J.: 1955, in *The Airglow and Aurora* (ed. by A. B. Armstrong and A. Dalgarno), Pergamon Press, London, p. 289.
Shefov, N. N.: 1968, *Planetary Space Sci.* **16**, 1103.
Shefov, N. N.: 1969, *Planetary Space Sci.* **17**, 1629.
Shimazaki, T. and Laird, A. R.: 1970, *J. Geophys. Res.* **75**, 3221.
Steer, R. P., Ackerman, R. A., and Pitts, J. N.: 1969, *J. Chem. Phys.* **51**, 843.
Strickland, D. J. and Donahue, T. M.: 1970, *Planetary Space Sci.* **18**, 661.
Strobel, D. F., Hunten, D. M., and McElroy, M. B.: 1970, *J. Geophys. Res.* **75**, 4307.
Tinsley, B. A.: 1967, *Planetary Space Sci.* **15**, 1757.
Tinsley, B. A.: 1968, *Planetary Space Sci.* **16**, 91.
Tinsley, B. A.: 1969a, *Planetary Space Sci.* **17**, 769.
Tinsley, B. A.: 1969b, in *Atmospheric Emissions* (ed. by B. M. McCormac and A. Omholt), Van Nostrand Reinhold Company, New York, p. 539.
Wallace, L. and Broadfoot, A. L.: 1969, *Planetary Space Sci.* **17**, 975.
Wallace, L. and Hunten, D. M.: 1968, *J. Geophys. Res.* **73**, 4813.
Wallace, L. and McElroy, M. B.: 1966, *Planetary Space Sci.* **14**, 677.
Weill, G. M.: 1969, in *Atmospheric Emissions* (ed. by B. M. McCormac and A. Omholt), Van Nostrand Reinhold Company, New York, p. 449.
Young, R. A.: 1969, *Can. J. Chem.* **47**, 1927.
Zipf, E. C.: 1969, *Can. J. Chem.* **47**, 1863.

INSTITUTE DISCUSSION

E. Zipf stated that recent experimental measurements from a rocket in an 80 kR aurora showed that the concentration of NO is approximately equal to that of O_2 and the ratio of NO^+ to O_2^+ is large.

According to B. McCormac it is important to know the total inventory for each alkali metal, atomic as well as molecular. It is necessary to know what fraction of the total is being observed in order to develop a model.

H. Sullivan mentioned that he had observed sharp enhancements of Li during the summer, one case being July 30, 1970.

AIRGLOW OBSERVATIONS AND THE INTERPRETATION OF IONOSPHERIC DISTURBANCES ASSOCIATED WITH MAGNETIC STORMS

L. THOMAS

S.R.C., Radio and Space Research Station, Ditton Park, Slough, Bucks.

1. Introduction

The disturbances of the ionosphere associated with magnetic storms have been recognized for many years. However, it has not been possible to describe the effects in any detail and as a result no satisfactory explanation is available. It seems likely that changes in neutral composition play an important role, and some evidence for such changes has been provided by airglow observations. These observations have also indicated changes in other parameters, such as neutral gas temperature, which can have important consequences.

The principal ionospheric effects associated with magnetic storms are outlined in Section 2. It will be seen that the most pronounced effects occur in the $F2$ and D regions. In Sections 3 and 4 the processes controlling the normal behavior of these regions are briefly described and the current ideas concerning the changes occurring during magnetic storms are indicated; the contributions of airglow studies to such ideas are emphasized.

There is increasing evidence that the disturbances revealed by ionospheric and airglow observations at mid- and low latitudes at the time of or following magnetic storms are associated with disturbances originating at higher latitudes. The implications of these disturbances for the more general question of magnetospheric storms are indicated in Section 5.

2. Ionospheric Disturbances

Radio wave observations carried out over more than four decades have shown marked changes in the ionosphere associated with the disturbances in the geomagnetic field which characterize magnetic storms. At mid- and low latitudes the major changes occur in the $F2$ and D regions (Ratcliffe and Weekes, 1960), and disturbances at E and $F1$ region heights are rather small (Sato, 1957). This is illustrated in Figure 1 which shows the effects occurring at $F2$, E and D region heights during and following the magnetic storm which occurred at 0417 UT on May 16, 1956. The full curves represent the variations of (a) the peak electron density of the $F2$ region ($NmF2$) at Slough, England, (b) the peak electron density of the E region (NmE) at De Bilt, Holland, and (c) the changes in reflection height of 16 kHz wave transmissions from Rugby to Cambridge, England, as reported by Belrose and Thomas (1968). The E region electron densities were not measurable at nighttime, and the shorter interrup-

B. M. McCormac (ed.), The Radiating Atmosphere, 105–115.

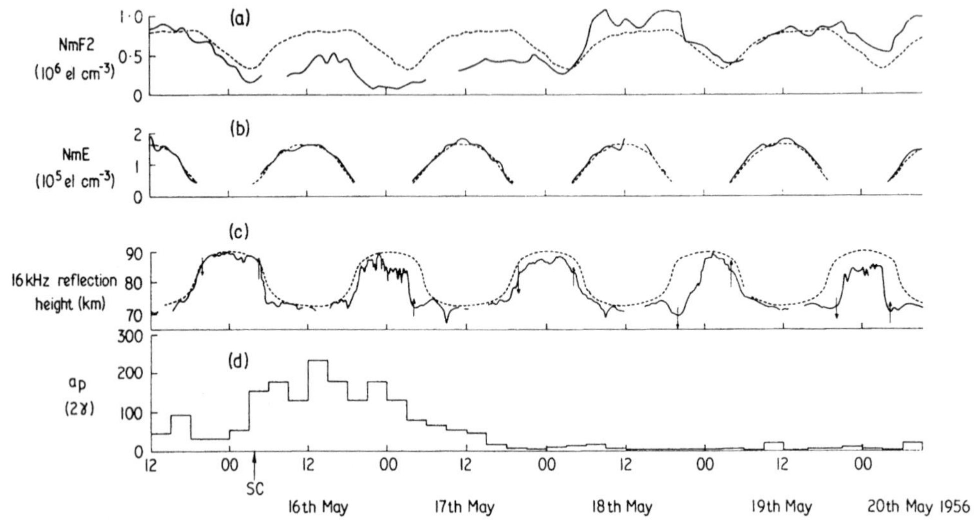

Fig. 1. Variations of ionospheric parameters in Europe and of the planetary magnetic field disturbance index, a_p, during and following the magnetic storm occurring at 0417 UT on May 16, 1965 (a) the peak electron density of the *F*2 region at Slough, England; (b) the peak electron density of the *E* region at De Bilt, Holland; (c) the *D* region reflection height of 16 kHz waves from G.B.R Rugby as deduced from measurements at Cambridge, England (Belrose and Thomas, 1968); and (d) the a_p index.

tions of the curves for *NmF*2 and 16 kHz reflection heights are due to a variety of causes. In each case the broken curves show the mean variation for normal days. Also shown by (d) in Figure 1 is the variation of the 3 hourly planetary magnetic disturbance index, a_p.

It is seen that for the *F*2 region the major disturbance was a reduction in *NmF*2 during May 16 and 17, and the effects on May 19, when the geomagnetic field had recovered, were small. Very little disturbance occurred in the daytime *E* region as shown by the *NmE* results. The observations of 16 kHz wave transmissions showed a very marked *D* region disturbance on the morning of May 17, a partial recovery with the termination of the storm on the night of May 17 to 18, and very pronounced effects from late afternoon on May 18 to the morning of May 20. The deviation from the normal variation is particularly marked at sunset on May 18 and 19, and sunrise on May 20.

The *F* region changes associated with magnetic storms have been subject to considerable study (Obayashi, 1964; Warren, 1969). It is known that the development of the disturbance depends on the local time of the storm, season, and latitude. In general, marked decreases in electron concentration occur during summer months in mid-latitudes and increased concentrations can occur during winter. Corresponding changes in total electron content are often observed (Lawrence *et al.*, 1963). The changes in the height of maximum ionization are generally small (Thomas and Robbins, 1958).

The information available on *D* region disturbances at mid-latitudes associated with magnetic storms has been derived from changes in characteristics of radio wave propagation (Bracewell *et al.*, 1951; Lauter and Sprenger, 1952; Lauter and Knuth, 1967; Belrose and Thomas, 1968). As already noted from Figure 1, these studies have shown that the greatest changes in ionization occur not at the time of maximum magnetic activity but 2 to 4 days later, and these changes sometimes persist for 10 days or longer. It has been found from observations of radio wave transmissions for frequencies up to 300 kHz that the delayed effects are most pronounced at twilight and nighttime and are usually absent at noon, as shown in Figure 1. However, observations at higher frequencies reveal such effects during daytime also (Belrose and Thomas, 1968; Bourne and Hewitt, 1968).

3. The Use of Airglow Observations in the Interpretation of F2 Region Disturbances During Storms

It is customary to consider changes in electron concentration of the $F2$ region in terms of the continuity equation:

$$\frac{\partial N}{\partial t} = Q - L - \operatorname{div}(NV) \tag{1}$$

where Q and L represent the electron production and loss per unit volume, respectively, and V the effective velocity of the plasma.

On the basis of current ideas, electron production during daytime occurs by the photoionization of O atoms, and electron loss is by dissociative recombination of O_2^+ and NO^+ ions formed by reactions of the primary O^+ ions with O_2 and N_2 gases:

$$O^+ + O_2 \rightarrow O_2^+ + O \tag{2}$$

$$O^+ + N_2 \rightarrow NO^+ + N. \tag{3}$$

The loss of electrons is determined by the rate of formation of these molecular ions, i.e., $k_2\,[O_2]+k_3\,[N_2]$, where k_2, k_3 represent the rate coefficients of Reactions (2) and (3).

Three contributions to the plasma velocity are recognized:

$$V = V_D + V_W + V_E \tag{4}$$

where V_D represents the part due to ambipolar diffusion, V_W the effect of atmospheric winds, and V_E the effect of electromagnetic drift.

Observations of reductions in $F2$ electron concentrations during magnetic storms have been interpreted in terms of decreases in electron production (Taylor, 1965) and increases in electron loss (Seaton, 1956; King, 1962). Changes in daytime arising from thermal expansion alone are too small to produce the decrease in $NmF2$ of 50% sometimes observed (Thomas, 1966). However, if changes in the ratio of O to O_2 or N_2 concentrations are invoked many of the features of this type of F region disturbance are reproduced (King, 1962; Chandra and Herman, 1969). Airglow observations of the forbidden multiplet of O^+ ($^2D-{}^2P$) at 7319 to 7330 Å could provide a useful monitor of the electron production rate. Furthermore, the twilight effect in 6300 Å ($^3P-{}^1D$) emission of O (Barbier, 1959) could help to indicate changes in molecular concentrations giving rise to increases in electron loss rate. Hunten (1967) has pointed out that an analysis of this effect in terms of the photodissociation of O_2 can be used to give the O_2 concentration at heights where N_2 quenching is unimportant, and the N_2 concentrations at a lower height.

Attempts have also been made to explain the reduction of $F2$ region electron densities in terms of increases in rate coefficients of Reactions (2) and (3) by atmospheric heating (Yonezawa, 1963) and by vibrational excitation of the N_2 gas (Thomas and Norton, 1966). It is found that increases in the temperature of the neutral atmosphere indicated by observations of the atmospheric drag on earth satellites (Jacchia *et al.*, 1967) or by airglow observations (Hays *et al.*, 1969; Armstrong and Bell, 1970) are probably insufficient to account for a 50% reduction in $NmF2$ during daytime, but might be important at night.

Thomas and Norton (1966) proposed that vibrational excitation of N_2 might occur during conditions of high electron temperature existing during storms (Evans, 1965; Rao, 1968). Walker *et al.* (1969) have also suggested that excitation might occur during reactions of N and NO or by quenching of $O(^1D)$ atoms by N_2 molecules; there are airglow observations indicating that both sources might be enhanced during storms. For instance, Krassovsky (1968) has postulated an increased concentration of atmospheric NO during storms based on the observed increased intensity of the airglow continuum, and an increased concentration of $O(^1D)$ atoms is believed to give rise to the mid-latitude red arcs (Roach and Roach, 1963). The enhanced electron temperatures associated with these arcs (Norton and Findlay, 1969) will also contribute to the vibrational excitation of N_2.

King and Roach (1961) and Norton and Findlay (1969) have noted reduced electron densities associated with mid-latitude arcs. It is to be expected from the lifetimes against quenching of vibrationally excited N_2 by electrons or by other N_2 molecules (Walker *et al.*, 1969) that the excited molecules will diffuse to considerable distances before quenching. Thus the enhanced electron loss rates might extend either side of the arcs, as is seen for some events.

Considerable effort has been made to explain F region disturbances in terms of electromagnetic drifts caused by electric fields set up in the ionosphere during magnetic storms (e.g., Maeda and Sato, 1959). However, at mid-latitudes the vertical component of drifts is reduced by frictional drag between ions and air molecules when the electric field varies during a time comparable with a day (Kohl, 1963). Such drifts might still be important near the equator where because of particular geometrical facts this frictional effect is thought to be less important. It is well known that E-W electric fields cause upward movements of ionization which play an important part in the daytime equatorial ionosphere (Bramley and Peart, 1965). Balsley and Woodman (1969) have found a reversal in the sense of the vertical drift of ionization near the equator between day and night; this implies a corresponding reversal in the zonal electric field. Studies of F region parameters and enhanced emissions of 6300 Å radiation have indicated the importance of movements of the layer at nighttime. Furthermore, observations at Tamanrassett (22.8°N, 5.3°E) have shown that the intensity of 6300 Å emission is enhanced during magnetic storms (Barbier, 1962; Weill and Christophe-Glaume, 1967) and it would appear that further observations of this emission might help to identify storm effects attributable to electric fields.

Theoretical studies by Kohl and King (1967), Rishbeth (1967), Kohl *et al.* (1968) and others have shown that thermospheric winds can have a serious influence on the distribution of ionization in the F layer. The frictional drag between air molecules and ions can cause a movement of ionization but this is confined to the direction of the magnetic field. The change in height resulting from the vertical component of the movement is then accompanied by a change in the effective electron loss rate $(k_2[O_2]+k_3[N_2])$. To date, information on atmospheric winds has been derived from models of the thermospheric temperature using solutions of the equation of motion. These refer to mean normal conditions and no attempt has been made to consider theoretically the wind system during disturbed periods.

Biondi and Feibelmann (1968) have described a technique for estimating winds from the Doppler shift of 6300 Å emission in the nightglow. Since the $O(^1D)$ atoms have a radiative lifetime of 110 sec they will acquire the gross motion of the wind, and the frequency shift will be proportional to the velocity component in the direction of the observer. In a development of this technique described by Armstrong (1969), scanning Fabry-Pêrot interferometer observations from opposite directions in the sky have been used to deduce winds. The only measurements for storm conditions have been reported by Hays (1970) which indicated meridional winds of speed 100 to 400 m sec^{-1}.

Observations of the atmospheric drag on earth satellites have indicated that the atmospheric heating during storms increases with increasing latitude (Jacchia *et al.*, 1967). It would be expected therefore that the equatorwards winds predicted theoretically (Geisler, 1967; Kohl and King, 1967) and found experimentally for normal nighttime conditions (Vasseur, 1969) would be enhanced during storms. The reduced electron loss resulting from the upward movement of ionization could help explain the increased electron content sometimes observed at mid-latitudes during disturbed conditions. Observations of the Doppler shift of 6300 Å could provide wind measurements for quantitative studies of particular events.

4. D Region Disturbances Associated with Magnetic Storms and Information Provided by Airglow Observations

Although there have been major developments in ideas concerning the formation and behavior of the D region in recent years, no satisfactory model is yet available. Consequently, only qualitative explanations of the disturbances associated with magnetic storms can be attempted.

It has long been realized that the relatively large ambient pressures existing at mesospheric heights permit rapid three-body collision processes and, in particular, the formation of negative ions:

$$e + X + M \rightarrow X^- + M. \quad (5)$$

The process in which X and M are both O_2 molecules is known to be fast (Chanin *et al.*, 1959).

The continuity equation for electrons can then be written as:

$$\partial N/\partial t = Q - L - (A - D) \quad (6)$$

where Q represents the electron production, L the loss by recombination, and A and D represent the effects of electron attachment and various detachment processes, respectively.

Only a limited part of the solar spectrum can penetrate to D region heights. Electron production in the middle D region results from the photoionization of NO by Lyα radiation (Nicolet, 1945), of $O_2\,(^1\Delta_g)$ by radiation in the range 1027 to 1118 Å (Hunten and McElroy, 1968; Huffman *et al.*, 1971), and of all gases by 2 to 8 Å X-rays. At the lower heights ionization is produced by galactic cosmic rays. It is evident that changes in concentration of NO during magnetic storms, as indicated by the airglow continuum emission (Krassovsky, 1968), or of $O_2\,(a^1\Delta_g)$, could contribute to the D region disturbance. Observations of the infra-red atmospheric band system $O_2\,(a^1\Delta_g - X^3\Sigma_g^-)$ could reveal marked changes in concentration of the metastable molecule.

The chemical loss L arises from dissociative recombination of water cluster ions $H^+ \cdot (H_2O)_n$ or of O_2^+ and NO^+ ions, depending on altitude (Narcisi and Bailey, 1965; Reid, 1970). Reid (1970) has pointed out that the effective recombination coefficient in the D region might be sensitive to temperature changes if the equilibrium distribution among the various cluster ions depends on temperature (Ferguson and Fehsenfeld, 1969) and the more complex clusters have larger recombination coefficients. Thus changes in temperature might contribute to D region effects during magnetic storms. Furthermore, the formation of these cluster ions $H^+ \cdot (H_2O)_n$, depending on reactions involving H_2O (Ferguson and Fehsenfeld, 1969) and O (Ferguson, 1971), will be modified by changes in the height distributions of these minor constituents during storms. Changes in temperature or in concentration of these constituents could be revealed by suitable airglow observations.

In view of the pronounced effects observed during twilight periods following magnetic storms (Figure 1), it seems likely that changes in the negative ion chemistry are involved. This chemistry is again dominated by the presence of minor neutral constituents, the day-to-night changes in concentrations of O, O_3, NO and NO_2 being particularly important. Lauter and Knuth (1967) have proposed, however, that the twilight effects are associated with anomalous ionization caused by photodetachment from negative ions produced by an influx of high energy particles.

Changes in neutral composition at mesospheric heights during and following magnetic storms have been indicated by airglow observations. Weill and Christophe-

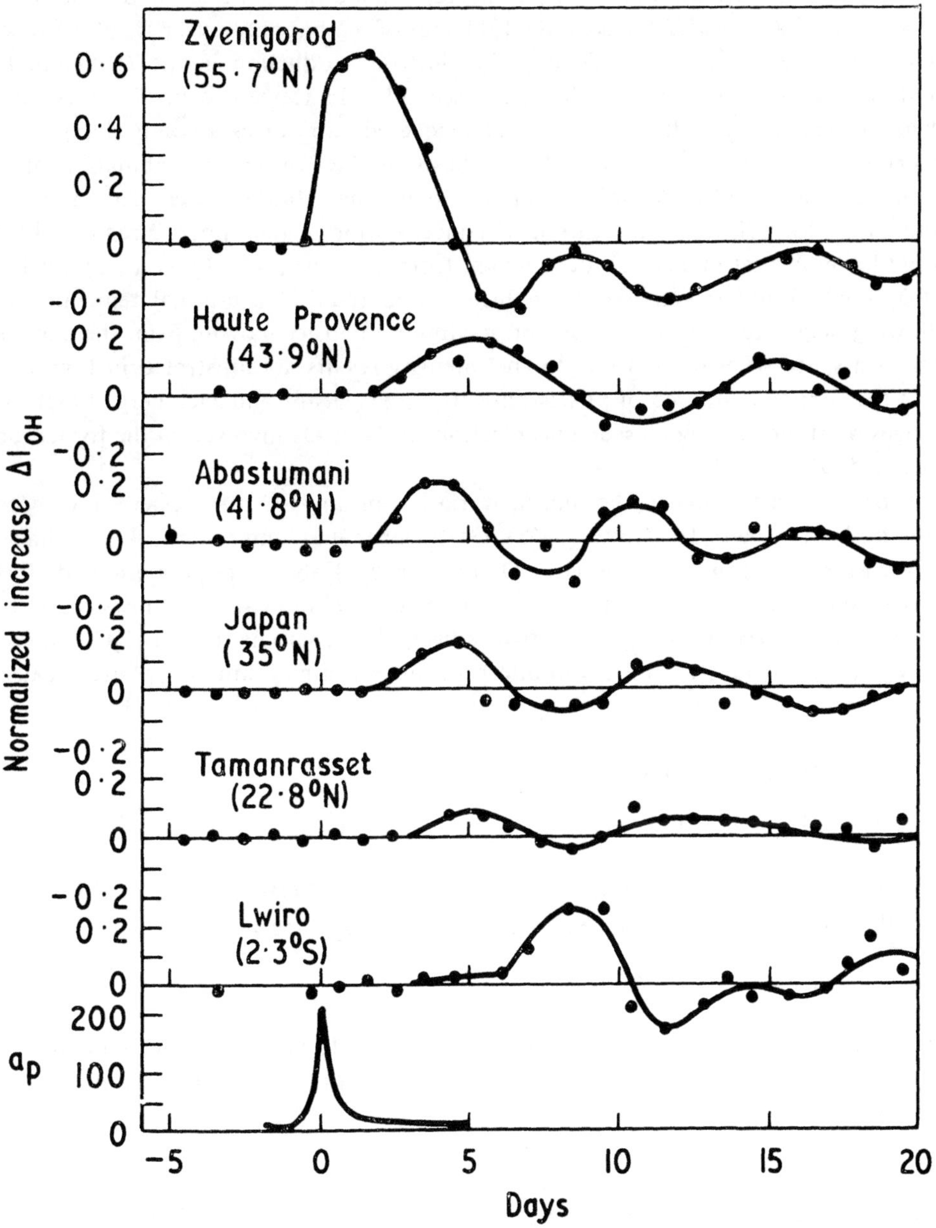

Fig. 2. Mean variations of the increments in intensity of OH emission (normalized to mean seasonal values) at different latitudes during and following geomagnetic storms (Shefov, 1969).

Glaume (1967) found that tne 5577 Å, O_2 bands, and continuum at Haute Provence (43.9°N, 5.7°E) were all enhanced during storms, and the OH and Na doublet were reduced. No change was detected in the 1N bands of N_2^+ but only a very small enhancement is required to alter significantly the electron production in the *D* region. In addition, the 5577 Å emission showed a second enhancement several hours later when the geomagnetic field had almost recovered; no corresponding change was observed in the other emissions. On the basis of the 3 atom recombination of O proposed to account for the 5577 Å emission (Chapman, 1931), the late enhancement would seem to indicate an increase in O concentration in the upper *D* region which would be important in electron detachment from O_2^- ions ($O_2^- + O \rightarrow O_3 + e$). Shefov (1968, 1969) found increased OH intensities and reduced rotational temperatures following magnetic storms, the zone of maximum intensity moving from the auroral zone at a speed of 5 to 10 m sec^{-1}. The intensity results are illustrated in Figure 2. Since the radiative lifetime of OH is only 10^{-2} sec (Heaps and Herzberg, 1952) the changes must indicate increased concentrations of H or O_3 involved in the formation of excited OH radicals.

Theoretical treatments of the diurnal variation of an H-O atmosphere have been made by Hunt (1966), Hesstvedt (1968), Gattinger (1969), and Shimazaki and Laird (1970). Marked diurnal variations of O_3 are expected above about 50 km and of H below about 80 km. The disappearance of O below 80 km at sunset allows the H to be converted to OH and HO_2; these two radicals then combine to give H_2O and O_2. However, the introduction of O will interrupt this recombination and, instead, cause the reactions:

$$HO_2 + O \rightarrow OH + O_2 \tag{7}$$

$$OH + O \rightarrow H + O_2 . \tag{8}$$

The H produced can then react with O_3 to give excited OH radicals. Furthermore the presence of O also enhances the O_3 concentration:

$$O + O_2 + M \rightarrow O_3 + M . \tag{9}$$

Thus, an enhancement of O at heights below about 100 km could account for the delayed effect in airglow intensities observed by Weill and Christophe-Glaume (1967) and Shefov (1969). Attention has already been drawn to the importance of O in the negative ion chemistry of the *D* region.

The enhanced OH intensities could also imply increased atmospheric heating since the reaction between H and O_3 and the reactions shown above are all exothermic. As already indicated, changes of temperature could have a direct effect on the electron loss by dissociative recombination of water cluster ions. This effect arises from the temperature dependence of three-body reactions involved in the formation of these cluster ions; other effects on the positive and negative ion chemistry could also arise from the temperature dependence of such reactions.

5. Relationship to High Latitude and Magnetospheric Phenomena

Norton and Findlay (1969) have suggested that mid-latitude red arcs might be related to the 'trough' observed in the latitude variations of *F*2 region electron densities (Dayharsh and Farley, 1965; Muldrew, 1965) and ion densities (Sharp, 1966). The latitude of this trough is known to move to mid-latitudes with moderate increases of magnetic activity (Muldrew, 1965) and this feature could clearly contribute to the reduced *F*2 region electron densities sometimes observed at nighttime during disturbed conditions.

Rycroft and Thomas (1970) have suggested that the geomagnetic field line terminating at the trough also coincides with the plasmapause, i.e., the region at a few R_E in the magnetosphere where there is a sudden decrease in electron density (Carpenter, 1963). An association between mid-latitude red arcs and the plasmapause has also been claimed. For example, Glass *et al.* (1970) have suggested that the changes with time of the geocentric distance of lines of force linking arcs in the Northern and Southern Hemispheres, as inferred from movements and changes in orientation of the arcs, are similar to those of the plasmapause.

It would seem, therefore, that at least part of the storm time decrease of *F* region electron densities at mid-latitudes during nighttime is associated with the plasmapause. Of the mechanisms proposed to explain the enhanced 6300 Å emission in the associated arc, collisional excitation by hot electrons is believed to be the most likely (Walker and Rees, 1968) and the high electron temperatures observed over the arcs (Norton and Findlay, 1969) are consistent with this mechanism. Vibrational excitation of N_2, either by electron collisions or quenching of $O(^1D)$ atoms, could enhance the electron loss within and perhaps either side of the arc. Walker and Rees (1968) conclude that conduction from above is the most likely source of heat and some experimental support for this has been provided by Roble *et al.* (1970).

Attempts have been made to attribute *F* region disturbances at mid-latitudes to compositional changes originating in effects occurring near the auroral zone. The effects of atmospheric waves set up by auroral currents (Chimonas and Hines, 1970) have been emphasized by King (1966) and experimental evidence for such waves seems to have been found (Hunsucker and Tveten, 1967; French, 1968). It seems likely that the enhanced mixing near the turbopause caused by such waves could lead to many of the features of *F* region disturbances (Chandra and Herman, 1969) and might also lead to changes in minor constituents at mesospheric heights and, thereby, account for the airglow and *D* region effects.

Alternative explanations of ionospheric disturbances associated with magnetic storms have involved changes in neutral composition arising out of storm time circulations of particular neutral constituents. The increased intensities of OH and $O_2(^1\Delta_g)$ emissions observed by Brown and Belon (1969) and Noxon (1970), respectively, confirm neutral atmosphere changes at mesospheric heights in the auroral zone. Since the radiative lifetimes of the excited OH radicals and the $O_2(^1\Delta_g)$ molecules are only 10^{-2} sec and 60 min, respectively, their transport to mid-latitudes cannot be important.

Maeda and Aikin (1968) have suggested that under particular conditions of particle precipitation enhanced O concentrations could result at auroral latitudes from the dissociation of O_2 and O_3. Oxygen atoms produced at heights above 90 km will have lifetimes in excess of a week. Furthermore, enhanced NO concentrations (Krassovsky, 1968) could persist for more than a week at middle mesospheric heights. Hence, transport of either O atoms or NO molecules from auroral to mid-latitudes is possible and the arrival of these minor constituents could be important in explaining the *D* region effects following magnetic storms. A meridional transport of O atoms has also been invoked to explain *F* region disturbances (Duncan, 1969).

Acknowledgment

This paper is published by permission of the Director of the Radio and Space Research Station of the Science Research Council.

References

Armstrong, E. B.: 1969, *Planetary Space Sci.* **17**, 957.
Armstrong, E. B. and Bell, J. A.: 1970, *Planetary Space Sci.* **18**, 784.
Balsley, B. B. and Woodman, R. F.: 1969, *J. Atmospheric Terrest. Phys.* **31**, 865.
Barbier, D.: 1959, *Ann. Geophys.* **15**, 179.
Barbier, D.: 1962, *J. Phys. Soc. Japan* **17**, 255.
Belrose, J. S. and Thomas, L.: 1968, *J. Atmospheric Terrest. Phys.* **30**, 1397.
Biondi, M. A. and Feibelmann, W. A.: 1968, *Planetary Space Sci.* **16**, 431.
Bourne, I. A. and Hewitt, L. W.: 1968, *J. Atmospheric Terrest. Phys.* **30**, 1381.
Bracewell, R. N., Budden, K. G., Ratcliffe, J. A., Straker, T. W., and Weekes, K.: 1951, *Proc. Inst. Elect. Engrs.* **98**, 221.
Bramley, E. N. and Peart, M.: 1965, *J. Atmospheric Terrest. Phys.* **27**, 1201.
Brown, N. B., and Belon A. E.: 1969, *Trans. Amer. Geophys. Union* **50**, 258.
Carpenter, D. L.: 1963, *J. Geophys. Res.* **68**, 1675.
Chandra, S. and Herman, J. R.: 1969, *Planetary Space Sci.* **17**, 841.
Chanin, L. M., Phelps, A. V., and Biondi, M. A.: 1959, *Phys. Rev. Letters* **2**, 344.
Chapman, S.: 1931, *Proc. Roy. Soc.* **A132**, 353.
Chimonas, G. and Hines, C. O.: 1970, *Planetary Space Sci.* **18**, 565.
Dayharsh, T. I. and Farley, W. W.: 1965, *J. Geophys. Res.* **70**, 5361.
Duncan, R. A.: 1969, *J. Atmospheric Terrest. Phys.* **31**, 59.
Evans, J. V.: 1965, *J. Geophys. Res.* **70**, 2726.
Ferguson, E. E.: 1971, *Proc. ESRIN/ESRO Symposium on Mesospheric Models and Related Experiments*, to be published by Van Nostrand Reinhold Company, New York.
Ferguson, E. E. and Fehsenfeld, F. C.: 1969, *J. Geophys. Res.* **74**, 5743.
French, A. G.: 1968, *Planetary Space Sci.* **16**, 993.
Gattinger, R. L.: 1969, *Ann. Geophys.* **25**, 825.
Geisler, J. E.: 1967, *J. Atmospheric Terrest. Phys.* **29**, 1469.
Glass, N. W., Wolcott, J. H., Miller, L. W., and Robertson, M. M.: 1970, *J. Geophys. Res.* **75**, 2579.
Hays, P. B.: 1970, private communication.
Hays, P. B., Nagy, A. F., and Roble, R. G.: 1969, *J. Geophys. Res.* **74**, 4162.
Heaps, H. S. and Herzberg, G.: 1952, *Z. Physik* **133**, 48.
Hesstvedt, E.: 1968, *Geophysiske Publikasjoner* **27**, 1.
Huffman, R. E., Paulsen, D. E., Larrabee, J. C., and Cairns, R. B.: 1971, *J. Geophys. Res.*, in press.
Hunsucker, R. D. and Tveten, L. H.: 1967, *J. Atmospheric Terrest. Phys.* **29**, 909.
Hunt, B. G.: 1966, *J. Geophys. Res.* **71**, 1385.
Hunten, D. M.: 1967, *Space Sci. Rev.* **6**, 493.

Hunten, D. M. and McElroy, M. B.: 1968, *J. Geophys. Res.* **73**, 2421.
Jacchia, L. G., Slowey, J., and Verniani, F.: 1967, *J. Geophys. Res.* **72**, 1423.
King, G. A. M.: 1962, *Planetary Space Sci.* **9**, 95.
King, G. A. M.: 1966, *J. Atmospheric Terrest. Phys.* **28**, 957.
King, G. A. M. and Roach, F. E.: 1961, *J. Nat. Bur. Std.* **65D**, 129.
Kohl, H.: 1963, *The Ionosphere*, The Institute of Physics and the Physical Society, London, p. 198.
Kohl, H. and King, J. W.: 1967, *J. Atmospheric Terrest. Phys.* **29**, 1045.
Kohl, H., King, J. W., and Eccles, D.: 1968, *J. Atmospheric Terrest. Phys.* **30**, 1733.
Krassovsky, V. I.: 1968, *Nature* **217**, 1136.
Lauter, E. A. and Knuth, R.: 1967, *J. Atmospheric Terrest. Phys.* **29**, 411.
Lauter, E. A. and Sprenger, K.: 1952, *Zeit. Met.* **6**, 161.
Lawrence, R. S., Posakony, D. J., Garriott, O. K., and Hall, S. C.: 1963, *J. Geophys. Res.* **68**, 1889.
Maeda, K. and Aikin, A. C.: 1968, *Planetary Space Sci.* **16**, 371.
Maeda, K. I. and Sato, T.: 1959, *Proc. IRE* **47**, 232.
Muldrew, D. B.: 1965, *J. Geophys. Res.* **70**, 2635.
Narcisi, R. S. and Bailey, A. D.: 1965, *J. Geophys. Res.* **70**, 3687.
Nicolet, M.: 1945, *Mem. Inst. Roy. Met. Belg.* **19**, 162.
Norton, R. B. and Findlay, J. A.: 1969, *Planetary Space Sci.* **17**, 1867.
Noxon, J. F.: 1970, *J. Geophys. Res.* **75**, 1879.
Obayashi, T.: 1964, *Res. Geophys.* **1**, 335.
Rao, P. B.: 1968, *J. Geophys. Res.* **73**, 1661.
Ratcliffe, J. A. and Weekes, K.: 1960, *Physics of the Upper Atmosphere* (ed. by J. A. Ratcliffe), Academic Press, London, p. 378.
Reid, G. C.: 1970, *J. Geophys. Res.* **75**, 2551.
Rishbeth, H.: 1967, *J. Atmospheric Terrest. Phys.* **29**, 225.
Roach, F. E. and Roach, J. R.: 1963, *Planetary Space Sci.* **11**, 523.
Roble, R. G., Hays, P. B., and Nagy, A. F.: 1970, *Planetary Space Sci.* **18**, 431.
Rycroft, M. J. and Thomas, J. O.: 1970, *Planetary Space Sci.* **18**, 65.
Sato, T.: 1957, *J. Geomag. Geoelect. (Kyoto)* **9**, 57.
Seaton, M. J.: 1956, *J. Atmospheric Terrest. Phys.* **8**, 122.
Sharp, G. W.: 1966, *J. Geophys. Res.* **71**, 1345.
Shefov, N. N.: 1968, *Nature* **218**, 1238.
Shefov, N. N.: 1969, *Planetary Space Sci.* **17**, 797.
Shimazaki, T. and Laird, A. R.: 1970, *J. Geophys. Res.* **75**, 3221.
Taylor, G. N.: 1965, *J. Atmospheric Terrest. Phys.* **27**, 735.
Thomas, L.: 1966, *J. Geophys. Res.* **71**, 1357.
Thomas, L. and Norton, R. B.: 1966, *J. Geophys. Res.* **71**, 227.
Thomas, J. O. and Robbins, A. R.: 1958, *J. Atmospheric Terrest. Phys.* **13**, 131.
Vasseur, G.: 1969, *J. Atmospheric Terrest. Phys.* **31**, 397.
Walker, J. C. G. and Rees, M. H.: 1968, *Planetary Space Sci.* **16**, 915.
Walker, J. C. G., Stolarski, R. S., and Nagy, A. F.: 1969, *Ann. Geophys.* **25**, 831.
Warren, E. S.: 1969, *Proc. IEEE* **57**, 1029.
Weill, G. M. and Christophe-Glaume, J.: 1967, *Ann. Geophys.* **23**, 317.
Yonezawa, T.: 1963, *The Ionosphere*, The Institute of Physics and the Physical Society, London, p. 128.

INSTITUTE DISCUSSION

The relationship of the plasmasphere to the *F* region was discussed. R. Chappell has shown that there is an association between *M* region arcs and the plasmapause. L. Thomas pointed out that the work of Norton and Findlay and also of Rycroft and Thomas has shown associations between the *F* region 'through' the *M* region arc, and the plasmapause. It is evident that this system contributes to the *F* region disturbance at latitudes as far south as $L = 2.5$ during magnetic storms.

The accuracy of the models for neutral winds at times of storms was discussed. L. Thomas stated that the theoretical models worked out by Geisler, Kohl, King, Lindzen, and others are based on models of temperature distributions, such as those of Jacchia. These relate to mean conditions and will not be relevant for storm times. Armstrong and Hays have shown how winds for disturbed conditions can be deduced from the Doppler shift of 6300 Å emissions.

ATMOSPHERIC DYNAMICS AND AIRGLOW FLUCTUATIONS

ROBERT D. SEARS
Lockheed Palo Alto Research Laboratory, Palo Alto, Calif. U.S.A.

1. Introduction

The airglow signal emitted from the upper atmosphere contains information about the fluid dynamical properties of the emitting region as well as about the photochemical processes leading to emission. The temporal and spatial variations of airglow volume emission rates are measures of corresponding variations of the density of the radiating excited states, hence, of the neutral atmospheric density fluctuations as a whole in the airglow emitting region. The statistical properties of such fluctuations are related to the state of turbulence in the emitting region, its energy spectrum, and characteristic scale lengths and times.

2. Experimental Method

The technique for measuring height resolved atmospheric density fluctuations is based upon four fundamental premises:

(a) A triangulation method defines a remote experimental volume.

(b) The observed airglow signal consists of two components: a temporally constant average level and a randomly fluctuating component, i.e.,

$$I(t) = \langle I \rangle + \Delta I(t). \tag{1}$$

(c) The atmospheric density fluctuations, $\Delta\varrho$, have a finite spatial and temporal correlation length which may be described by

$$\begin{aligned} R_{ij}(\Delta x_i, \tau) &= \langle \Delta\varrho(x_i, t)\, \Delta\varrho(x_i + \Delta x_i, t + \tau)\rangle \\ &\rightarrow \mathrm{O} \text{ for } |\Delta x_i| \gg L, |\tau| \gg T \\ &\rightarrow \langle \Delta\varrho^2(x_i, t)\rangle \text{ for } |\Delta x_i| \ll L, |\tau| \ll T \end{aligned} \tag{2}$$

where: $\langle\ \rangle$ indicates temporal average; x_i are spatial coordinates; t is time; L is a correlation scale length, and T is a correlation scale time.

(d) The fractional fluctuations in density of all species are linearly proportional

$$\frac{\Delta[M]}{\langle[M]\rangle} = \frac{\Delta[N]}{\langle[N]\rangle} = \frac{\Delta\varrho}{\langle\varrho\rangle}. \tag{3}$$

Two geometrical methods of isolating an experimental volume at a well defined altitude in an airglow emitting region are illustrated in Figure 1 (Krause, 1967). The ground based technique (Figure 1a) was utilized by Wolff (1966) in his attempt to determine O altitude profiles by correlation methods similar to those described herein.

B. M. McCormac (ed.), The Radiating Atmosphere, 116–122.

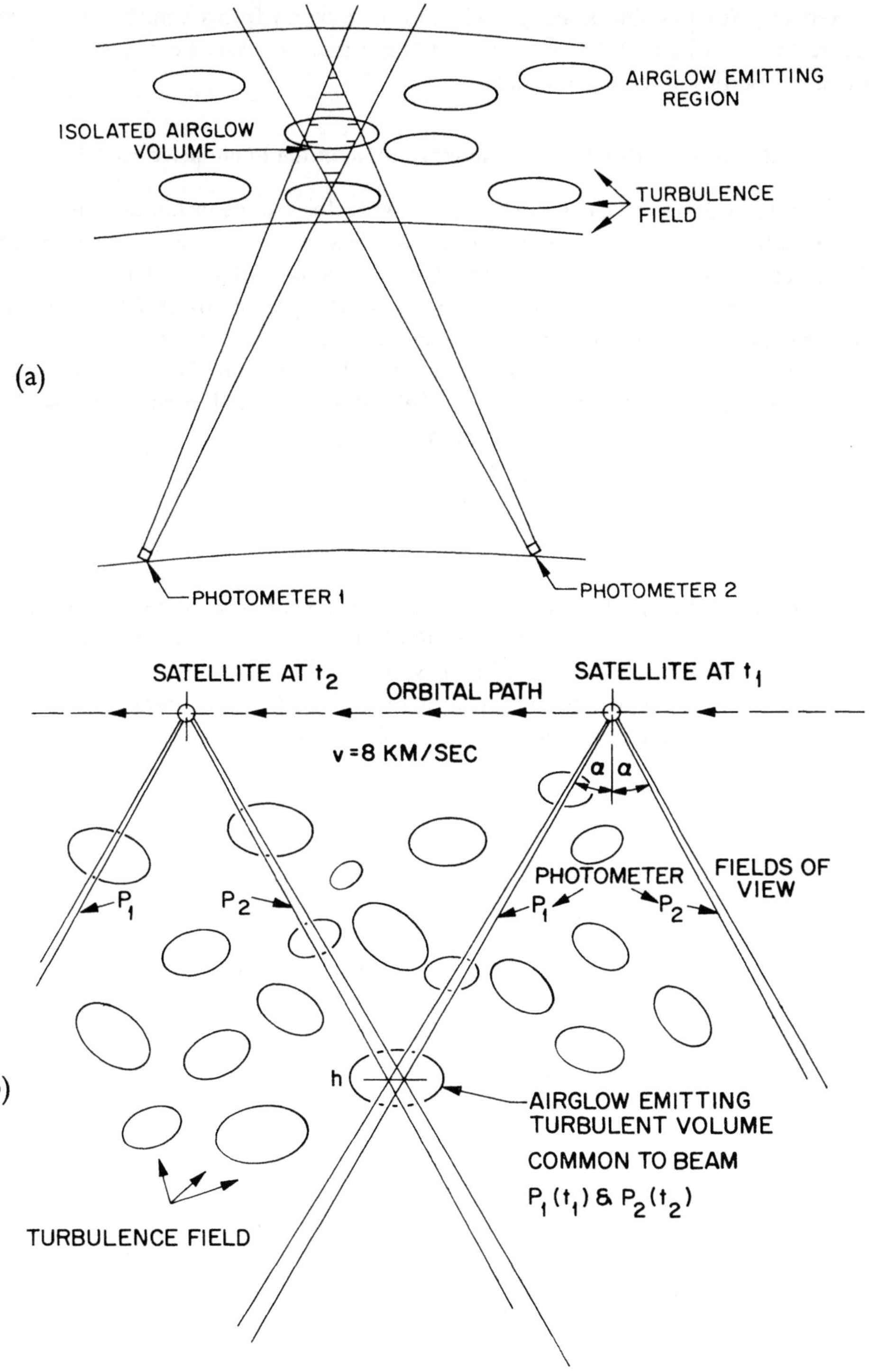

Fig. 1a–b. Triangulation methods for isolating a turbulent airglow emission region (a) ground based technique; (b) satellite borne technique.

A geometry for satellite based investigation of auroral forms equivalent to that in Figure 1b was used by Johnson *et al.* (1967) but their analysis did not utilize the statistical properties of the auroral signals.

3. Statistical Analysis and Interpretation of the Photometer Signals

The measured signals at each photometer (Figure 1a) are the instantaneous values of airglow intensity vs. detector orientation, field of view, wavelength, and time. A region of intersection or of closest separation for the two beams is defined by their orientations. The fluctuating component of the airglow signal is separated from the instantaneous value by subtraction of the temporal average value as suggested by Equation (1). The measured airglow intensity fluctuation at each photometer is the column integral of all volume emission fluctuations in the field of view which is assumed to be of unit dimensions. This column integral for photometer 1 is:

$$\Delta I_1(t) = \int_0^\infty \Delta\zeta(x_i, t)\, dx_1 \tag{4}$$

where $\Delta\zeta(x_i, t)$ is the volume emission fluctuation, x_i are spatial coordinates chosen so that x_1 and x_2 are along the lines of sight of photometers 1 and 2, respectively, and x_3 is perpendicular to their plane of intersection.

Following the analytical development of Fisher and Krause (1967), we form a two beam covariance product which is a measurable quantity,

$$\begin{aligned} G_{12}(x_i, \tau) &= \langle \Delta I_1(t)\, \Delta I_2(t+\tau)\rangle \\ &= \frac{1}{t_0} \int_0^\infty \Delta\zeta_1(x_i, t)\, dx_1 \int_0^\infty \Delta\zeta_2(x_i, t)\, dx_2\, dt\,. \end{aligned} \tag{5}$$

By rewriting Equation (3) and substituting into Equation (5) after changing the order of integration we obtain

$$\begin{aligned} G_{12}(x_i, t) &= \left(\frac{\langle I\rangle}{\langle \varrho\rangle}\right)_0^2 \int_0^\infty dx_1 \int_0^\infty dx_2 \frac{1}{t_0} \int_0^{t_0} \Delta\varrho(x_i, t)\, \Delta\varrho(x_i, t+\tau)\, dt \\ &= \left(\frac{\langle I\rangle}{\langle \varrho\rangle}\right)_0^2 L_1 L_2 R_{12}(x_3, \tau)\,. \end{aligned} \tag{6}$$

The quantity $(\langle I\rangle/\langle\varrho\rangle)^2$ is the ratio squared of the temporally averaged local airglow volume emission rate to the temporally averaged local atmospheric density. This value is assumed herein to be constant or only slowly varying over distances $\pm L$ from the beam intersection altitude. Large changes in $\langle I\rangle/\langle\varrho\rangle$ over the scale length of the inhomogeneity L would result in the computation of erroneous values of L_1 and L_2, the scale lengths of the inhomogeneities along the photometer axes. The quantity $R_{12}(x_3, \tau)$ is the covariance of the density fluctuation which equals $\langle \Delta\varrho^2 \rangle$ if $\tau \cong 0$.

The scale time, T, of the density fluctuations may be found from the value of delay time, τ, at which $G_{12}(x_3, \tau)$ reaches a predetermined value. For exponentially varying density fluctuations it is convenient to set

$$T = \tau \text{ at } \frac{G_{12}(x_3, \tau)}{G_{12}(x_3, 0)} = e^{-1}.$$

The scale length, L_3, of the density inhomogeneity in the x_3 dimension can be found by introducing a beam separation, Δx_3. From the functional behavior of $R_{12}(x_3 + \Delta x_3, 0)$ expressed in Equation (2), we see that a scale length $\Delta x_3 = L_3$ may be defined as that which is necessary to cause a predetermined drop in the value of $G_{12}(\Delta x_3, 0)$, e.g., to e^{-1} of $G_{12}(\Delta x_3 = 0, \tau = 0)$. Consequently, if spatially isotropic density fluctuations are assumed, $L_1 = L_2 = L_3 = L$ and $\langle \Delta\varrho^2 \rangle$ may be determined from the experimental data. The consequences of spatially inhomogeneous density fluctuations are greatly increased complexities in the computations and are a requirement to determine 2 or 3 dimensional values of scale lengths.

The preceding analytical development was based upon the premise of spatially stationary density fluctuations and stationary photometer beams. The effects of horizontal transport of density fluctuations upon this experiment can be deduced if we assume a uniform wind velocity, v_0, along coordinate axis, x_3. Then, for a beam separation, Δx_3, the peak value of $G_{12}(\Delta x_3, \tau)$ would appear at $\tau = \pm v_0 \Delta x_3$.

A much more flexible experimental technique involving scanning of both beam allows simultaneous measurements of both transport velocities and density fluctuations over the entire range of altitude of the emitting region. The analytical methods involved are analogous to those summarized herein, except for a change of variables to relate delay time, τ, to the altitude of the beam intersection.

In addition to the statistical correlation scale lengths and mean square value of density fluctuations, the power spectral density of the atmospheric fluctuations is an important tool for understanding upper atmospheric turbulence. The power spectral density of the mass density fluctuations is found from the Fourier transform of the 2 beam covariance product

$$P_w(k) = \frac{1}{2\pi} \int_{-\infty}^{\infty} e^{-ik\tau} R_{12}(\tau) \, dt. \tag{7}$$

If, instead of the delay time, τ, a covariance product is formed for the beam separation distance, Δx_3, the spatial power spectral density, $P_{x3}(k)$, can be computed. This functional quantity can be interpreted in terms of the type of turbulence present in the emitting region as was pointed out by Barat and Blamont (1969). If $P_{x3}(k) \propto k^{-5/3}$, an isotropic inertial range or Kolmogorov type of turbulence is present, whereas, if $P_{x3}(k) \propto k^{-7/5}$, a buoyant turbulence exists and it may be anisotropic.

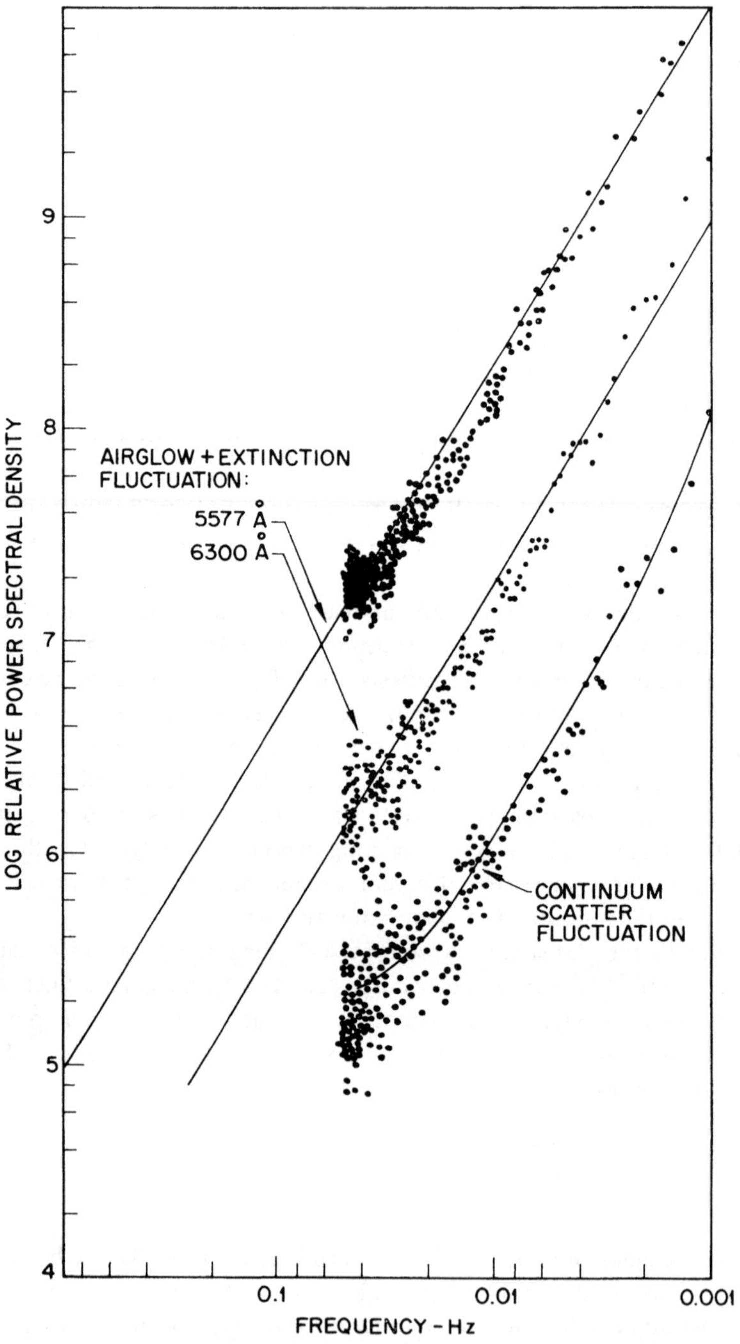

Fig. 2. Power spectral density curves of airglow emission plus atmospheric extinction fluctuations at 5577 Å and 6300 Å and atmospheric scatter fluctuations alone.

4. Measured Airglow Fluctuation Power Spectra

In order to test the general feasibility of measuring airglow intensity fluctuations and of computing their power density spectra, an experiment was conducted using a single beam photometer operating at the O airglow wavelengths 5577 Å and 6300 Å. Because in a single beam experiment the sources of extraneous fluctuations in the data cannot be eliminated by the cross covariance process, care must be taken to identify those remaining. The photometer signal may be described as a sum of uncorrelated fluctuation sources

$$\begin{aligned}\Delta I\,(\text{obs}) = {} & \Delta I\,(\text{airglow}) + \Delta I\,(\text{shot noise}) \\ & + \Delta I\,(\text{atmospheric extinction noise}) \\ & + \Delta I\,(\text{statistical photon noise}) \\ & + \Delta I\,(\text{other sources}) .\end{aligned} \tag{8}$$

All extraneous sources were eliminated or made small except the noise caused by fluctuations in atmospheric extinction in the troposphere. From expression (8) the remaining observed noise sources were:

$$\Delta I\,(\text{obs}) = \Delta I\,(\text{airglow}) + \Delta I\,(\text{extinction}) . \tag{9}$$

It is easy to show that

$$R_{11}\,(\tau)\,(\text{obs}) = R_{11}\,(\tau)\,(\text{airglow}) + R_{11}\,(\tau)\,(\text{extinction}) \tag{10}$$

and

$$P_{11}\,(k) = P_{11}\,(k)\,(\text{airglow}) + P_{11}\,(k)\,(\text{extinction}) . \tag{11}$$

That is, the power spectral densities of uncorrelated random processes are additive. A measure of $P_{11}(k)$ (extinction) was obtained by monitoring the intensity fluctuations of moonlight scattered from the atmospheric aerosols (which contribute most of the extinction in the visible wavelengths) in a wavelength band adjacent to the airglow emission lines. The intensity of the scattered moonlight signal was low enough not to contribute appreciably to the airglow signal measurement itself. Figure 2 illustrates the results of power spectral computations of 5577 Å and 6300 Å airglow signals and those for scattered moonlight. Most airglow power spectral density computations show steeper functional dependence upon k than that of the background scattered moonlight at least over part of the range of periods 10^{-1} to 10^{-3} sec.

The relatively wide field of view of the photometers (3°) discriminates against detection of any inertial spectral fluctuations in the 100 km region because the 5 km spatial resolution of the instrument is much larger than the appropriate inertial scale length indicated by Barat and Blamont (1969). Extinction fluctuations in the troposphere would be expected to be in the inertial range, hence, the airglow spectrum should be slightly steeper than $k^{-5/3}$ whereas the background extinction power spectral density should approximate the $k^{-5/3}$ behavior. This is consistent with the preliminary observations. Unfortunately not enough data exist to date to allow adequate determination of the exact functional form of the airglow intensity fluctuation power spectral density alone, or to validly compare power spectra of 5577 Å and 6300 Å fluctuations.

5. Conclusions

In this paper we have outlined a method for obtaining height resolved data related to the turbulent fluid dynamical properties of the atmosphere. Preliminary data from a single photometer indicate that the fluctuations in airglow intensity can be detected, separated from extraneous sources of noise, and its statistical properties computed.

Acknowledgments

The author acknowledges the contributions of R. W. Deuel who assisted with the experimental measurements and of R. F. Kamm, Jr., who programmed the power spectral analysis routine. This work was partially supported by the Lockheed Independent Research Program.

References

Barat, J. and Blamont, J.: 1969, in *Atmospheric Emissions* (ed. by B. M. McCormac and A. Omholt), Van Nostrand Reinhold Company, New York, p. 507.
Fisher, M. J. and Krause, F. R.: 1967, *J. Fluid Mech.* **28**, 705.
Johnson, R. G., Meyerott, R. E., and Evans, J. E.: 1967, in *Aurora and Airglow* (ed. by B. M. McCormac), Reinhold Publishing Corporation, New York, p. 169.
Krause, F.: 1967, *Trans. Amer. Geophys. Union* **48**, 111.
Wolff, M.: 1966, *Trans. Amer. Geophys. Union* **47**, 75.

PART III

AURORA

AURORAL MORPHOLOGY

C. S. DEEHR* and A. EGELAND

The Norwegian Institute of Cosmic Physics
The University of Oslo, Oslo, Norway

and

F. SØRAAS

The Institute of Physics
University of Bergen, Bergen, Norway

1. The General Location of Auroral Activity

A. INTRODUCTION

'Auroral morphology' has grown to include far more than the mapping of visible auroral forms. Satellites, rockets, and airplanes with sophisticated particle and radiation detectors now locate auroral particles and their effects. However, the simplest of observations has not yet been made extensively enough in time and space to adequately describe even one aspect of the auroral phenomena. It is still quite common that even the more sophisticated experiments involving a great deal of data are not correlated with any other observations of a single event. Thus, many reviews of auroral morphology have been based on comparison and mapping of statistical data taken over a period of years at different levels of activities. An attempt will be made here to show that most auroral effects are closely associated with the auroral oval, and then present some new satellite data as examples of the processes occurring there.

B. THE STATIC OVAL

The statistical auroral distribution has been mainly documented by all-sky camera photographs (Feldstein, 1969; Lassen, 1970). Spectrographic (Sandford, 1968; Weins and Vallance Jones, 1969) and geomagnetic (Fukushima and Nagata, 1967; Mishin *et al.*, 1970) studies have also been used. Effects of particle precipitation have been used to expand this picture of auroral activity (Hartz and Brice, 1967; Hartz, 1971). It is generally agreed that most of the aurora lies in a relatively narrow band in the shape of an oval whose center lies a few degrees from the magnetic pole along the nightside of the solar meridian. The earth rotates underneath this pattern which is fixed with respect to the sun, resulting in a characteristic diurnal variation for a ground based observer. Some differences in this pattern occur between the hemispheres (Sandford, 1968) but the austral oval seems to behave in much the same manner as the boreal oval (Chubb and Hicks, 1970).

While the existence of the oval is generally accepted, its exact form is not agreed upon. An analysis of high latitude all-sky camera data from Greenland (Lassen, 1970)

* on leave from The Geophysical Institute, College, Alaska.

B. M. McCormac (ed.), The Radiating Atmosphere, 125–149.

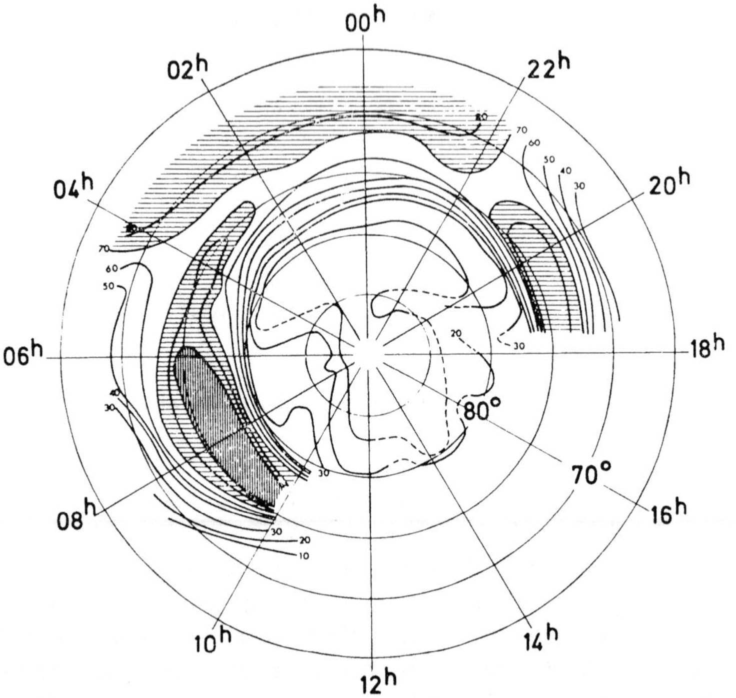

Fig. 1. Polar graph showing distribution of occurrence frequency of auroras for $K_p = 2$ for 1964–67. Coordinates are corrected geomagnetic latitude and time. Isoauroral lines are based on zenithal frequency and drawn for multiples of 10% (Figure from Lassen, 1970.)

has indicated that the oval of percentage occurrence is actually two horseshoe-shaped regions, one on the dayside and one on the nightside (Figure 1). The prongs of the horseshoes lie in the dusk and dawn and meet as K_p increases. The oval is then almost complete at medium activity, but there are still zones of decreased statistical occurrence near 0600 LT and 0018 LT.

In order to understand what these results mean, one must consider the limitations imposed on these data by the observing instrument, the all-sky camera. It has been shown that the threshold for detection of an aurora is between 500 R and 1 kR of 5577 Å emission, depending on background sky brightness, position in the sky, and the condition of the camera (Davis *et al.*, 1959). Thus, on the basis of a 75% occurrence value, for example, one cannot conclude that 25% of the time there is no aurora in that sector, because the luminosity can be below the detection level of the camera.

An excellent method of observing oval activity is by the use of a high speed jet aircraft (Akasofu, 1968b). By flying westward along the oval, the entire magnetic day passes above the airplane in a relatively short period of UT. The results have shown the oval to be complete under medium to low magnetic activity (Akasofu, 1970a; Buchau, 1970). There was no 'fork in the road' or missing portion although there were regions where the aurora was not as bright as others. Thus, we suggest that the oval is complete, although there may be regions of reduced flux near dusk and dawn.

C. POLAR CAP AURORA

The relation of polar cap auroras to the aurora occurring in the oval region is not yet entirely clear. 'Auroral oval type' auroras have been seen over the magnetic pole on certain occasions (Feldstein, 1970). Figure 1 shows two tongues of increased frequency occurrence which extend from the auroral oval into the polar cap in the morning and afternoon sectors (Lassen, 1970). These auroras are the polar cap auroras observed from Thule and Vostok; their frequency and position are independent of K_p except in the early afternoon when the frequency decreases with increasing K_p. The latter effect (anticorrelation with K_p) has produced some confusion in the past and is probably a result of the aurora moving equatorward with increasing K_p. As K_p increases the afternoon auroras in the polar cap move equatorward, but the morning distribution remains until $K_p > 5$. However, the geometry is such that the sun-earth alignment which is usually given as the main characteristic of polar cap aurora could result from the polar cap auroras being split from the auroral oval where it is approximately sun-earth aligned (Akasofu, 1970a; Gustafsson, 1970a, b).

D. THE AURORAL ZONES

The spatial distribution of precipitating electrons of energy less than 50 keV observed by the 1965 Lockheed satellite (Sharp and Johnson, 1968) showed a relatively complicated pattern on the dayside. The hard and soft electron zones observed have been used as support for a number of proposals about the existence of several auroral 'zones'. The documentation of many effects of particle precipitation by Hartz and Brice (1967) and Hartz (1971) led to the mapping of two regions of auroral activity.

A fairly complicated picture is proposed by Mishin *et al.* (1970). On the basis of statistical mapping of magnetic activity it is shown that there could exist two almost circular zones, one at 67° and one at 77° IN Lat. Many of the arguments proposed by Mishin *et al.* (1970) are the same as used by Hartz and Brice in their development of two main regions of precipitation.

Recent ESRO satellite data on the distribution of low energy auroral particle precipitation are interpreted by Riedler and Borg (1970) to fit the Hartz and Brice model of an auroral oval and an auroral zone. However, the authors document no less than 4 soft electron zones and 3 hard electron zones. As Feldstein (1969) points out, there is nothing in the arguments for multiple zones which does not exclude the possibility of a single, closed oval, changing size and shape with magnetic activity.

E. THE DYNAMIC OVAL

Generally speaking, the auroral oval widens and shifts equatorward with increasing magnetic activity (Feldstein and Starkov, 1967; Sandford, 1968; Lassen, 1970). Recent satellite results have confirmed this tendency. Chubb and Hicks (1970) plotted the latitude of the *equatorward maximum* of the UV aurora as a function of K_p and found that the rate of equatorward motion of the aurora with increasing K_p was much greater on the dayside than on the nightside. This is in agreement with the mapping of magnetic disturbances by Fukushima and Nagata (1967). Thus, the oval approaches circular shape as K_p increases.

Perhaps most important to the discussion here is that the UV aurora observed from the satellite matched the oval distribution very well found from ground based all-sky camera data (Feldstein, 1969; Chubb and Hicks, 1970). In view of this it is difficult to reconcile the existence of several auroral zones presented in the previous section. The following will include a discussion of recent observations indicating the possible origin of these zones.

F. DAYSIDE AURORA

The proposed auroral 'zones' are separated from the average auroral oval in the polar cap and on the dayside where the extent of observations is least. Statistical averages have shown that there exists a region of softer electrons poleward of the region where the harder electrons predominate and that the proton precipitation takes place in one broad region (Sharp *et al.*, 1969). There is also a statistically significant softer poleward edge to the nighttime aurora (see Section 3E). The latter observation is supported by the average enhancement of 6300 Å emission relative to other emissions just poleward of the main aurora on the night found by Eather (1969). The association of this 6300 Å enhancement with short lived sun aligned arcs in the polar cap led Eather and Akasofu (1969) to suggest that the low energy electron 'bursts' observed by Hoffman and Evans (1968) in that region may have been spatial rather than temporal effects.

A graphic example of the soft energy spectrum on the poleward side of the morning and dayside auroral oval has been shown by Akasofu (1970b). Colored all-sky camera photographs show a broad region of red aurora (high altitude) poleward of a bright green band (lower altitude) at the equatorward side of the dayside oval. Photometric observations by Romick (1970), Whalen (1970), and Eather and Mende (1971), confirm this interpretation.

Considering the above discussion, it is not difficult to see how the different concepts of the auroral 'zones' come about. Most of the early zones of maximum occurrence were simply the locus of the maximum activity. The auroral zone was circular in geomagnetic space at approximately latitude 68°. The concept of an oval (Feldstein 1969) showed that the instantaneous position of the aurora can be quite different from the diurnal path of its maximum effect. Thus, the studies which led to the concepts of the zones (Section 1D) were carefully plotted with regard to the statistical diurnal (longitudinal) variation of the various phenomena.

It has now become apparent, however, that both UT and LT variations in the geomagnetic field are linked to the instantaneous shape of the auroral oval or zone of auroral precipitation. Thus, the mapping of a precipitation zone averaged without regard for the prevailing geomagnetic activity leads to a chart of the locus of maximum latitudinal effect of the particular data involved. This may well be the case regarding the auroral 'zone' of Hartz and Brice (1967). The only reference to visible auroral emissions associated with the 'zone' is 'mantle aurora' (Sandford, 1968). This phenomenon is apparently the only direct auroral emission which is a 'drizzle phenomenon.' However, because the mantle aurora has been documented only by spectrographic exposures of 1 hr or more, it is hardly fair to say that it cannot be 'burst-like' and must be slowly varying. All of the particle studies upon which the existence of this zone were based took no account of the effect of different magnetic conditions prevailing at the time of the observations (cf., Hartz *et al.*, 1963; McDiarmid and Burrows, 1964; Sharp and Johnson, 1968; Riedler and Borg, 1970). Therefore, because the particle spectrum tends to become harder with decreasing latitude (Figures 2 and 5) (Sharp *et al.*, 1969) the hard particles will tend to maximize at lower latitude than the soft and 2 zones emerge with their position dependent on the handling of these data. Perhaps an analogus situation occurs in the case of the double zones of Mishin *et al.* (1970); one can only point out that other treatments of the magnetic data (Fukushima and Nagata, 1967; Feldstein, 1969) are in agreement with the oval concept.

Thus, the picture which emerges is that most auroral activity takes place along the instantaneous auroral oval. Apparently the dayside aurora does not seem to be as confined as that on the nightside, although under quiet magnetic conditions the dayside oval can consist of a single arc (Buchau *et al.*, 1969). However, the gradient of characteristic particle energy with latitude is less and the rate of expansion with magnetic activity is greater on the dayside. As a result, the association of the auroral activity there with the instantaneous oval is more difficult than on the nightside. Obviously, observations of the day and twilight regions are needed and they must be interpreted in terms of magnetic time, latitude, and activity.

2. Auroral Heights

A. CHANGE OF HEIGHT WITH LATITUDE

Boyd (1969) measured the height of maximum luminosity in 5577 Å for quiet auroral arcs at 4 stations between 64.5 and 70.2° dp lat. From 24 nights during one winter he scaled off 1100 points. The results show a steady increase in average height, from about 110 km at 65° to 165 km at 72°. Figure 2 shows his results plotted together with the average of all measured auroral points of Størmer (Egeland and Omholt, 1966). Statistically, at least, the altitude seems to be roughtly constant equatorward of 65°, but poleward of that point, the altitude changes markedly with latitude.

This relatively abrupt average altitude change may result in the observed 'coastal effect'. Nadobovich (1963) noted the tendency of auroral arcs to follow the coastline. Andriyenko (1965) analyzed the height of aurora as a function of distance from the coastline, and found that the average height of the aurora increased over the sea. He also found the same effect in Vegard's data from northern Norway. Cursory examination of the location of the 67° IN Lat circle shows that the only two coastlines in the world which are approximately aligned with this latitude region are the coasts of Siberia and northern Norway (Evans *et al.*, 1969). Because the change in altitude

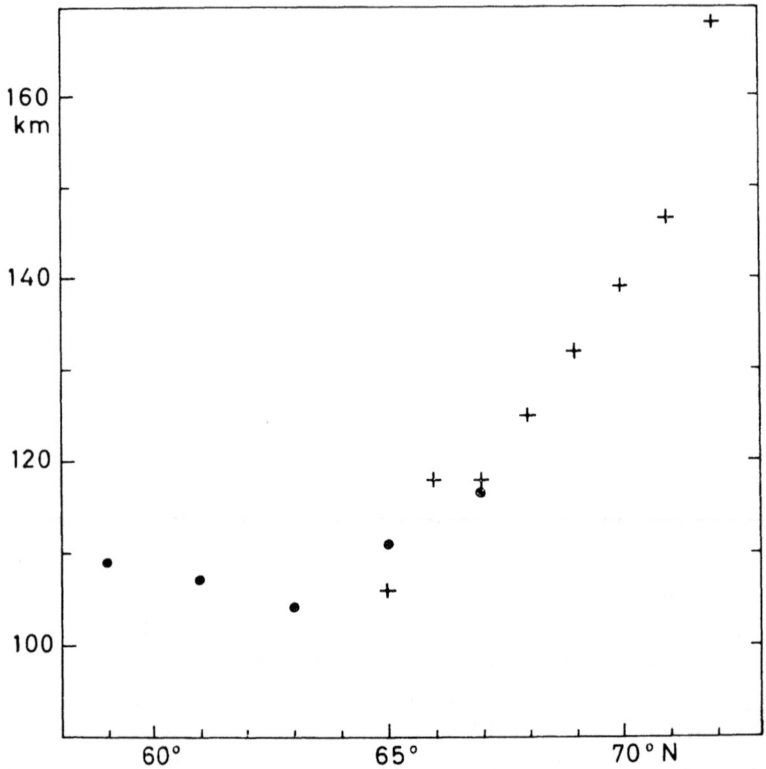

Fig. 2. Altitude of the aurora as a function of geomagnetic latitude. The dots are lower borders from Størmer's data (Egeland and Omholt, 1966). The crosses are maximum intensity from Boyd (1969) (Figure from Omholt, 1971).

shown in Figure 2 was observed over continental North America, we may conclude that perhaps the change in altitude observed by Andriyenko is related to magnetospheric rather than lithospheric processes.

B. HEIGHT OF POLAR CAP AURORAS

The altitude of the polar cap auroas has been the subject of two investigations giving quite different results. Photographic measurements from Spitzbergen (73 to 76° IN Lat) by Starkov (1968) showed a lower border altitude between 145 and 180 km. The observations of Lassen (1969) from the same latitude interval in Greenland showed an altitude distribution like those observed in the auroral oval.

Lassen (1969) suggested that perhaps the difference between these height measurements may have resulted from a difference in solar activity. Close examination of Lassen's (1970) latest data indicates that this is probably the case. Figure 1 shows that the greater part of the aurora measured during quiet sun years at 75° IN Lat was undoubtedly oval aurora. Lassen (1970) reports that the oval moves equatorward with K_p, leaving the polar cap type aurora over the station during periods of higher magnetic activity. During the maximum activity years, when the Starkov (1968) data were collected, the oval probably never moved poleward far enough to cover the 75° region and more polar cap type aurora were observed then than oval aurora.

It is interesting to note that the altitudes of polar cap auroras measured by Starkov (1968) fall near the Størmer-Boyd curve shown in Figure 2. Lassen's (1969) points do not fall on this curve by a considerable margin. The 'knee' of the curve in Figure 2 is at the latitude of the average nighttime oval. Satellite particle observations indicate that the position of the 'knee' would follow the oval to a higher latitude on the dayside. Thus because Lassen's IQSY altitudes in the latitude region 70 to 75° IN Lat pertained mostly to dayside oval aurora, the average altitude should be less than that observed by Starkov. This interpretation suggests that much of the scatter in auroral morphology data would be removed if the data were referred to the instantaneous latitude of the auroral oval. Hopefully, some 'invariant line' can be found in the auroral display which could be used as a reference point. Such singularities as the 'knee' mentioned above, or the most poleward conjugate arc (Davis *et al.*, 1971) are possibilities.

3. The Description of Individual Auroral Events

A. THE AURORAL SUBSTORM

It is evident from the preceding discussions that although the auroral oval may be well defined on a magnetospheric scale its actual location at any given time is certainly complicated by the level of magnetospheric activity and processes occurring within the boundaries of the oval. The latter processes have been documented to a certain extent in an exceptional work on the dynamics of the aurora by Akasofu (1968a). Individual events in the oval were shown to follow a pattern which Akasofu calls the polar substorm. From spectrographic and all-sky camera observations, the electron and proton aurora have been described in terms of the auroral substorm by Montbriand (1971). Because the polar substorm reflects large scale activity in the geomagnetic field called the magnetospheric substorm, it is of fundamental interest to document auroral observations within this framework. Attempts will be made in the following sections to interpret these satellite data in this manner.

B. THE ESRO 'AURORAE' SATELLITES

Although the preceding review of some of the major problems in auroral morphology is by no means complete, many other aspects will be presented in other parts of this volume. The sections to be presented here will deal with specific examples of auroral observations from the ESRO 'Aurorae' Satellites. The data are mainly from the photometer observations by the Norwegian Institute of Cosmic Physics and observations of protons over 100 keV by the Physics Institute at the University of Bergen in cooperation with the Danish Space Research Institute at Lyngby.

ESRO IA was launched October 3, 1968, while ESRO IB went up on the 1st of October 1 yr later. Because both satellites were magnetically stabilized, the photometers were aligned to view the aurora along the magnetic lines of force. Because half of the Northern Hemisphere is illuminated by sunlight, obscuring the aurora, the photometric data were restricted to the near midnight meridian between 55° and 85° north latitude. The perigee and apogee for the IA satellite were 275 and 1,500 km, respectively, while ESRO IB had an almost circular orbit of about 340 km. ESRO IA had a midnight-noon orbit during the period of interest, while IB was located near 0300 LT.

The two photometers measured the intensity of Hβ and N_2^+ emissions in the aurora. Unlike the other detectors aboard ESRO I, which make *in situ* measurements, the photometers integrate all the light which is emitted or scattered into the 10° field of view from the satellite down to the earth. The optical filtering in the photometers has been chosen such that only auroral emissions from Hβ (4861 Å) and N_2^+ (4278 Å) are observed by the satellite in the absence of scattered light from the sun or moon. The H emission indicates the extent of proton precipitation because the excited H atoms are produced by the interaction of fast protons with atmospheric constituents. The N_2^+ emission indicates the total energy deposition and the total ionization rate. A relatively detailed technical description of the photometers was given by Deehr *et al.* (1970a).

The particle detectors were designed to measure protons in the energy range 100 keV to 5 MeV. This experiment employed three totally depleted semiconductor detector, looking along, normal and 45° to the local magnetic field line. A detailed description of the satellite instrument is given by Sørass *et al.* (1969). Results on the precipitation pattern of these protons and its relation to geomagnetic activity have been reported by Aarsnes *et al.* (1970).

C. RELATION OF LUMINOSITY TO PARTICLE PRECIPITATION

Figure 3 shows the intensities of the 4278 Å N_2^+ band and the Hβ line, together with simultaneous data for 6 keV precipitating electrons and 6 keV protons with pitch angles 10° and 80° (Riedler and Borg, 1970; Deehr *et al.*, 1970b). It is clearly seen that there is correspondence between the electron precipitation and the 4278 Å emission. In this figure the optical emission should be shifted $\frac{1}{4}$ to $\frac{3}{4}$° poleward in order to compensate for the curvature of the magnetic field lines; the photometers are aligned with the magnetic field at the satellite. The major part of the N_2^+ emission therefore is associated with the 6 keV electron precipitation zone, although the 1 keV zone at the edges of the main zone seems to broaden the main region of luminosity.

The Hβ filter passband was shifted to the red allow for the Doppler shifted emission from H atoms receding from the satellite. This filter choice tends to minimize contamination due to the Vegard-Kaplan (2.15)N_2 band which is the most probable contaminator of terrestrial observations with filters shifted to the blue. Thus, the N_2^+ signal associated with the electron spike in 6 keV precipitation represented an increase of 400% while the associated Hβ increase was less than 10%. It is clear, therefore, that the photometric data provide a very good indication of the extent of the total electron and proton precipitation regions. Typical Hβ intensities in the main proton precipitation zone were 50 to 200 R.

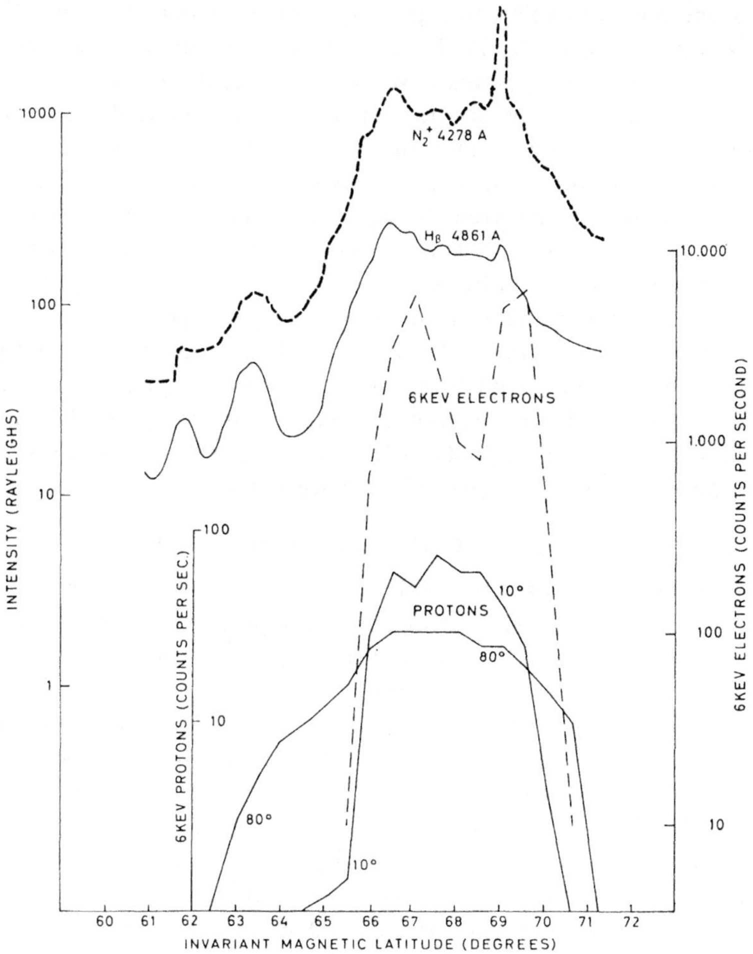

Fig. 3. Photometric and particle data. Pass number 308. The IN Lat is that of the field line passing through the satellite.

There is a remarkable zone of high pitch angle protons which extends some degrees of latitude to the equatorwards side of the apparent precipitation boundary (Figure 3). This depleted loss cone type of anisotropy is described by Hultqvist *et al.* (1970) and is associated with an Hβ intensity of 10 to 50 R. The ratio, 3, of the N_2^+ to Hβ emission in this area is what would be expected for pure proton bombardment. This low ratio is found in many other passes to be associated with the equatorward depleted loss cone region. The emission cannot be connected with scattered light from various geographical features or with some airglow contamination. The observed intensity (10 to 50 R) is also reasonable when one considers that the intensity is reduced from that in the main precipitation region because of a greatly reduced proton flux, mainly due to reflection of the protons below the satellite (back up the field line) before the charge exchange process occurs.

There is one possible explanation to this emission region which should be tested more rigorously. The satellite, in this case, was observing at an altitude of about 400 km. At this height the protons have begun charge exchange collisions with the atmospheric constituents, whereby fast H atoms are produced. The atoms travel considerable distances along straight lines before they are reionized and attached to another field line. This diffusion effect is important (Omholt, 1956; Davidson, 1965) because the mean distance between charge exchange collisions at 400 km altitude for a horizontally moving H atom is 2000 km. The travel in horizontal direction depends on the pitch angle, θ, and the altitude, h, the former factor being proportional to sin θ and the latter to e^{-h} according to the density of the atmosphere.

In the equatorward direction, H atoms travel horizontally or slightly into the atmosphere if they have high pitch angles, because the magnetic field lines have an inclination of 70 to 80°. Hence the atoms are readily reionized, and a diffusion distance of some hundred km is expected. In the poleward direction 80° atoms travel horizontally or upwards and most likely escape out into space again.

D. EFFECT OF PITCH ANGLE ANISOTROPY ON H SPECTRAL PROFILE

Hultqvist *et al.* (1970) reported the depleted loss cone anisotropy (Section 3C) along with another type of pitch angle anisotropy in the proton data called a field aligned anisotropy. That is on certain occasions the 0° proton flux premodinates usually near the center of the main precipitation region.

Because the Doppler profile of the H emission is dependent on the velocity of the emitting atom with respect to the observer, it is expected that these anisotropic pitch angle distributions would affect the profile observed from the ground. It is usually assumed that the direction of observation (parallel or perpendicular to the field lines) is the dominant factor in forming the spectral profile. However, because a significant amount of the luminosity may be associated with particles with an anisotropic pitch angle distribution which varies considerably with latitude, the latitude of the observer must also be considered along with the angle of the observation to the field lines. That is, a zenith observation just equatorward of the main precipitation zone would produce the relatively narrow symmetric profile usually associated with a horizon observation. This may explain some of the 'narrow line' observations which have been made in the zenith (Montalbetti, 1959; Galperin, 1963).

Another effect which has been observed is the 'red shift' in the H spectral profile observed in the zenith from the ground (Eather, 1966). This shift is probably caused by a certain percentage of the H atoms traveling back up the field lines as they emit. It would be accentuated by a pitch angle anisotropy of either the depleted loss cone or field aligned type. The tilt of the magnetic field should result in a greater red shift poleward of the main precipitation region than equatorward. To conclude, the variations in the Doppler profile of H emission observed from the ground (Johansen and Omholt, 1963) are indicative of the reality of the pitch angle anisotropies and are probably more dependent on the latitude of the observer and the type of aurora than was previously thought.

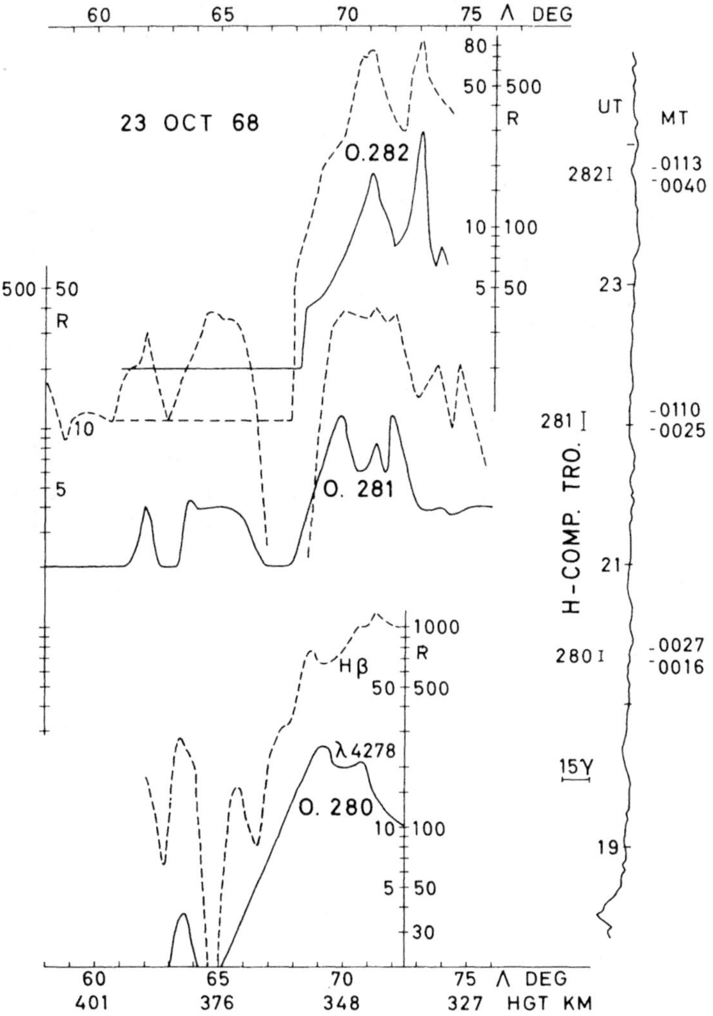

Fig. 4. Photometric data from orbits 280, 281, and 282 as a function of IN Lat. The dashed trace represents the Hβ intensities and the solid line N_2^+ (4278 Å) emission. The larger values on the ordinate apply to the latter emission, while the smaller ones represent the former. The trace of the *H* component of the geomagnetic field at Tromsø is recorded on the right hand side in UT. The geomagnetic time of the satellite passage over northern auroral latitudes on the nightside is indicated along this trace.

E. THE LOCATION OF THE OVAL DURING QUIET MAGNETIC CONDITIONS

During all passes of the satellite over the auroral region some aurora was observed even during the quietest of magnetic conditions. According to Feldstein (1970) the midnight portion of the auroral oval lies near 70° IN Lat. The intensities shown in Figure 4 are typical for this condition and the lack of all-sky camera data (Stringer and Belon, 1967) is understandable because the probable intensity of the 5577 Å O green

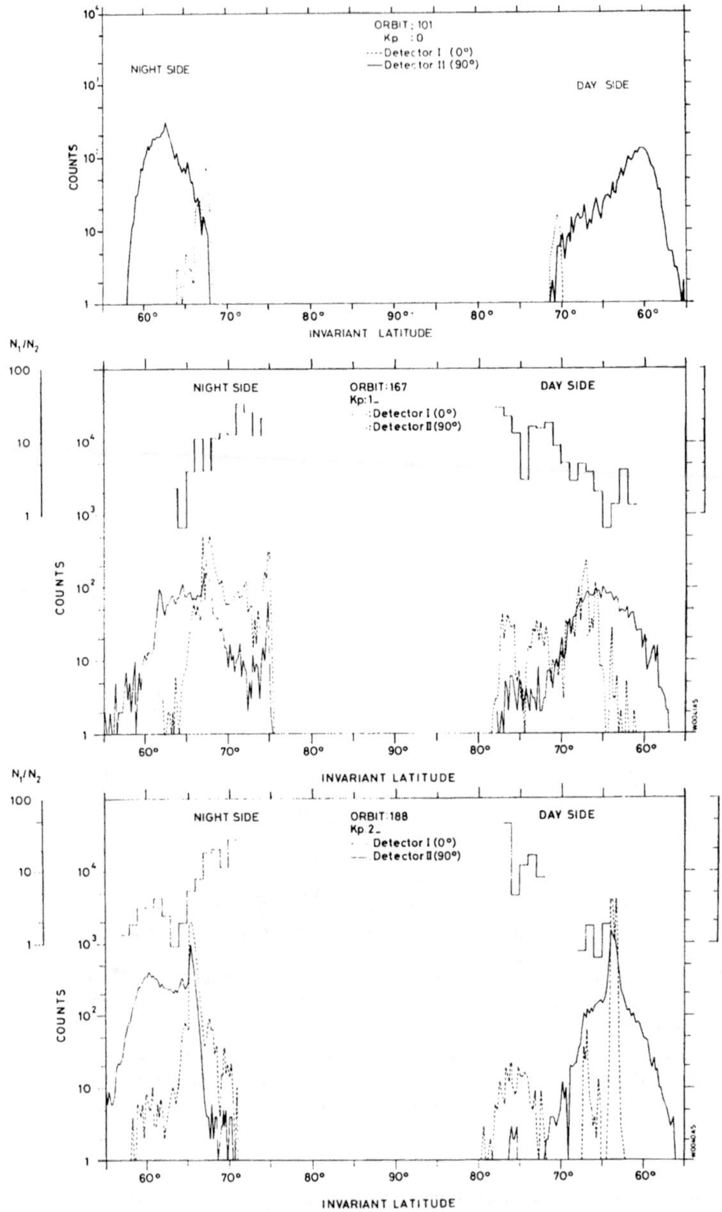

Fig. 5. Precipitated and trapped protons in the energy range 100 to 200 keV plotted vs. IN Lat for different K_p values.

line (for purposes of comparison) is <500 R. However, radio reflections probably exist there (Bates *et al.*, 1969) so it is reasonable to assume that for very quiet magnetic conditions the midnight sector lies near 70° IN Lat.

To show the distribution on the dayside also the observed count rates of trapped (150 to 215) keV and precipitated (115 to 180) keV protons measured under different geomagnetic conditions are plotted vs. IN Lat in Figure 5. The data presented at the top of the figures are representative for a magnetically 'quiet' day, $K_p = 0$. Passing north on the nightside, where the satellite is below the radiation belts, the satellite encounters the outer radiation belt somewhat below 60° IN Lat.

Near the northern trapping boundary there is a sudden change in the proton pitch angle distribution resulting in a region of precipitating protons. This nighttime region is about 2° wide and is located at 67° IN Lat. On the dayside the trapping boundary extends up to 71° IN Lat. The proton precipitation is located at this boundary.

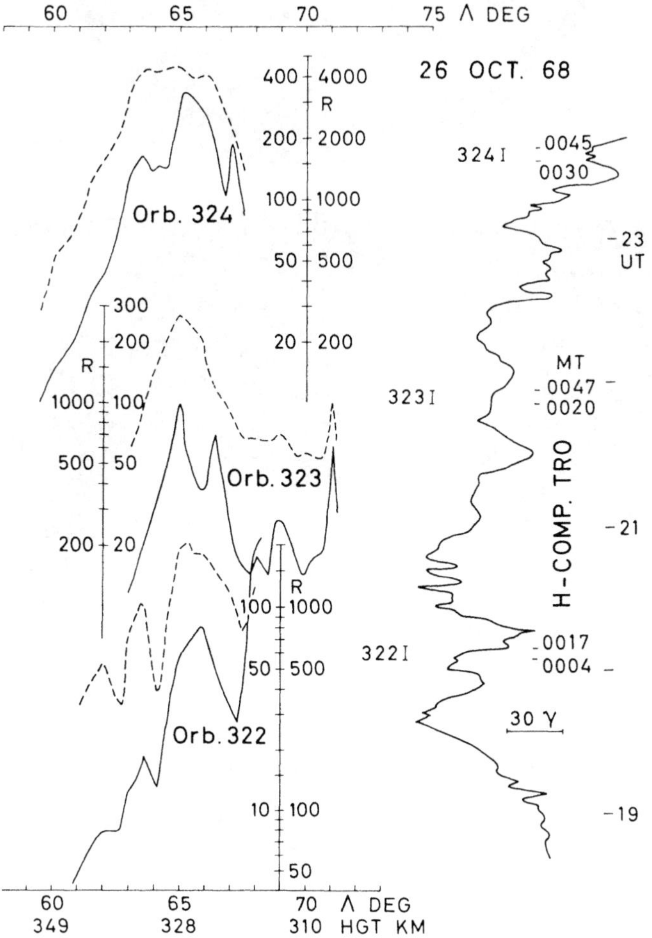

Fig. 6. Photometric data from orbits 322, 323, and 324 as a function of IN Lat (same as Figure 4).

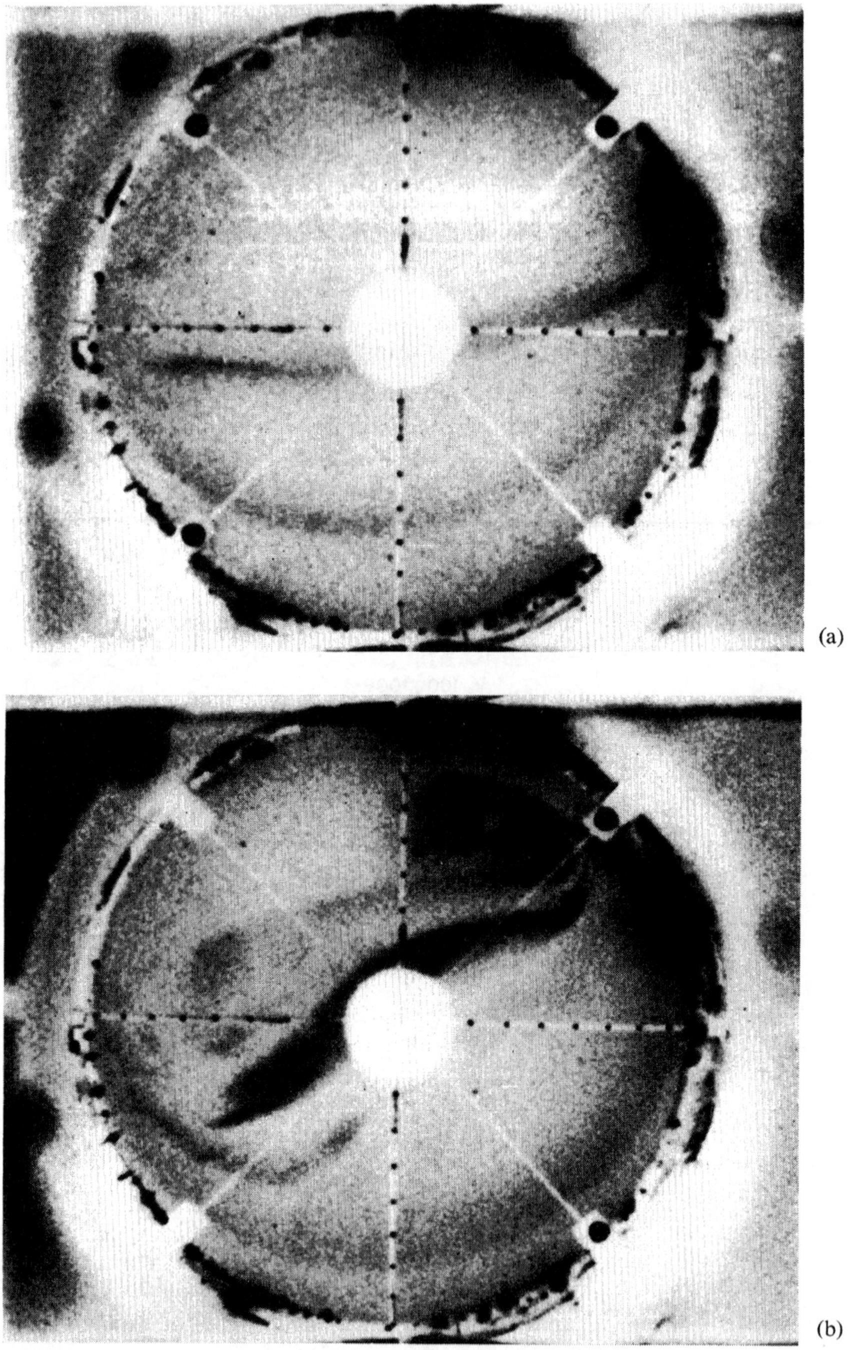

Fig. 7. All-sky camera picture from Tromsø, October 26, 1968. (a) 2152 UT. The satellite passed from south to north near the Tromsø zenith at this time on orbit 323. North is up and east is to the right. (b) 2345 UT corresponding to orbit 324 passing west of Tromsø just below the 80° zenith distance marker.

The other data presented in Figure 5 are from passes during 'disturbed' geomagnetic conditions. Here the ratio of the count rates in the two lowest energy channels for precipitated protons is given. This gives an indication of how the proton energy spectrum becomes softer with increasing IN Lat. The *e*-fold energy for the protons with energies above 100 keV is between 20 and 70 keV. The proton precipitation pattern for greater magnetic activity shows an expanded structured region on the nightside and the dayside. On the dayside the precipitation normally lies above 70° IN Lat. On many passes there is also a zone of proton precipitation at a lower latitude around 65°. This region can be absent or it can merge with the region at higher latitude.

F. INITIAL PHASE OF SUBSTORM

In Figure 6 orbit 323 illustrates the initial brightening of an arc just poleward of the 'stable H arc' which is usually the most equatorward evidence of the aurora. This pass was near the Tromsø zenith and the arcs are both visible in the Tromsø all-sky camera photograph (Figure 7a). The Tromsø magnetogram, *H* component, is shown to the right in Figure 6. Orbit 323 is seen to occur in the middle of a positive bay. The Tromsø Observatory reported relatively strong (400 R) Hβ emission during this pass (Pettersen, 1970) in good agreement with the observations reported here.

The passage of the westward traveling surge more than $1\frac{1}{2}$ hr later is shown in the Tromsø all-sky photograph in Figure 7b. This characterizes the major poleward expansion phase of the substorm. The passage is marked by the beginning of a negative bay in the *H* component at Tromsø (Akasofu, 1968b). Orbit 324 occurred westward of Tromsø after the initial poleward expansion in that region. It can be seen that the proton precipitation region has moved poleward into the main electron region.

It is evident from the data from orbit 323 that the initial brightening preparatory to the onset of the westward traveling surge involves only electrons. This supports the data of Akasofu *et al.* (1969) who found no H emission in the westward traveling surge. Analysis of ground based spectroscopic data by Montbriand (1971) showed that protons probably move poleward on the eastward side of the north-south oriented segments which occur after the initial passage of the surge.

The poleward expansion of the 100 keV protons precipitation region during a substorm is shown in Figure 8 as a poleward movement of the trapping boundary. The IN Lat of this boundary is plotted for both northern passes. It is seen that the proton precipitation region moves poleward with the onset of the negative bay. An equatorward expansion then occurs during the recovery period.

4. Average Relation of Electron and Proton Aurora to K_P and Dst

A. LATITUDE OF AURORAL ACTIVITY AS A FUNCTION OF MAGNETIC ACTIVITY

The relation of electron aurora to magnetic activity is well documented by all-sky camera surveys (Feldstein and Starkov, 1967). However, because the H emission is not easily separated from the electron aurora in all-sky camera data, the morphology

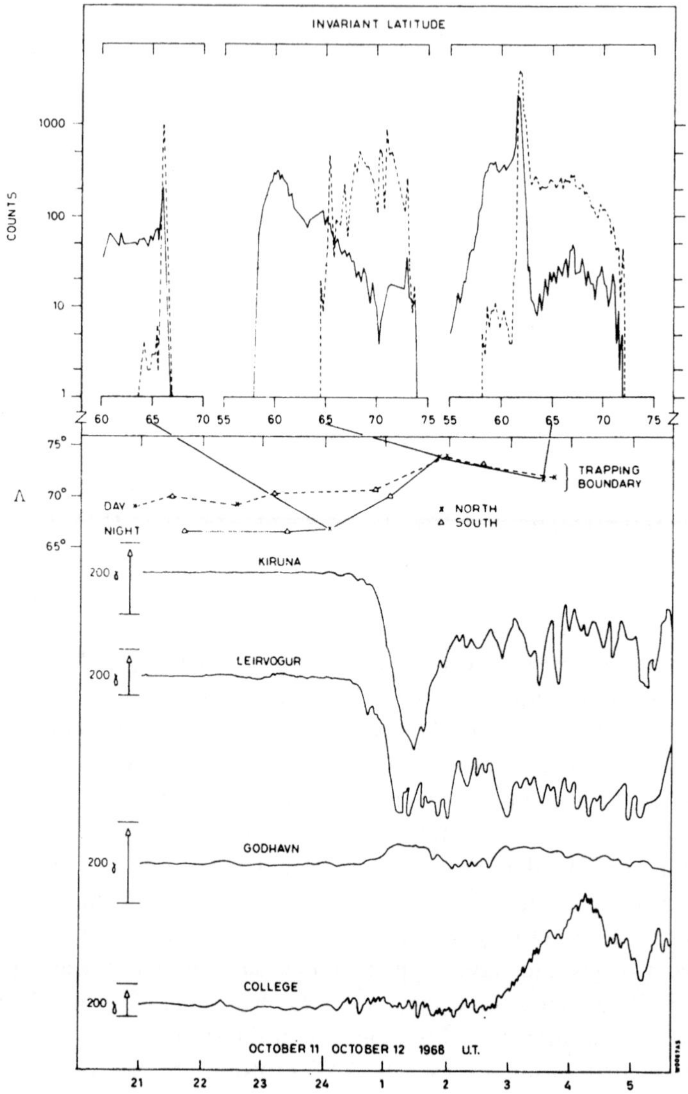

Fig. 8. The count rates of precipitated and trapped protons are shown as a function of IN Lat for three consecutive nightside passes. The location of the proton trapping boundary for both the nightside and the dayside, together with the magnetic records from Kiruna, Leirvogur, Godhavn and College are plotted vs. time.

of the proton precipitation zone has been sought through spectroscopic morphology studies. Wiens and Vallance Jones (1969), and Chubb and Hicks (1970) have found that the latitude of maximum luminosity due to proton precipitation varies with K_p in much the same way as the electron aurora, at least in the midnight-noon meridian.

The latitudinal distribution of proton precipitation can be plotted as a function of time and compared to magnetic indices. Figure 9 is a plot of contours of constant

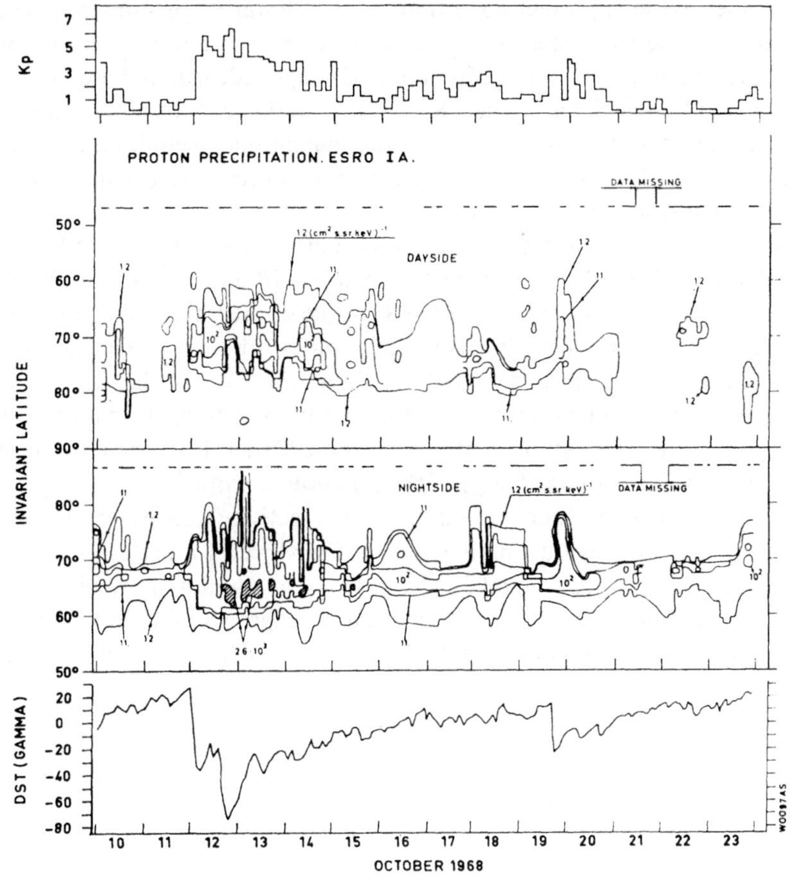

Fig. 9. Contours of constant intensities of precipitating protons within the energy bandpass 115 to 180 keV as a function of IN Lat and UT. Data for both day- and nightside together with K_p and Dst are given.

intensities of precipitating protons within the energy bandpass 115 to 180 keV as a function of IN Lat and UT for both the dayside and nightside. The LT coverage is the noon-midnight meridian ± 0.5 hr. The units of these intensities contours are $(cm^2 \text{ s sr keV})^{-1}$ average differential intensities within the energy band pass. The intensity contours are interpolated through periods of missing data indicated by the broken line, and can thus be somewhat uncertain. The 2 hr averages of Dst (h) in γ and the 3 hourly K_p index are included to indicate the magnetic activity.

An examination of this figure reveals the following general features. When magnetic activity as indicated by Dst and K_p increases there is an enhanced proton precipitation. The region of maximum proton precipitation moves to lower latitudes, the intensity increases, and there is an expansion of the whole region both poleward and equatorward boundary.

The intensity of proton precipitation is usually greater on the nightside than on the dayside. Under extremely quiet conditions no precipitating protons can be detected on the dayside, whereas some protons always seem to precipitate on the nightside. The low latitude boundary of precipitation on the nightside can under quiet conditions be found around 64° IN Lat. When K_p is high the lower latitude boundary on the nightside can move down to 56° IN Lat, the maximum precipitation flux occurs at 61 to 62° IN Lat. Thus, the extent of the proton precipitation is dependent on the strength of the magnetic disturbance.

In order to compare the electron and proton aurora using the ESRO-I photometric data, the IN Lat of the first equatorward peak of Hβ and the equatorward edge of 4278 Å, N_2^+ was listed as a function of the K_p value of the time of the pass. The results are plotted in Figure 10 where the average latitudes for the October and November 1968 lunations are shown as dots for Hβ and circles for N_2^+. The dashed line represents the plasmapause position as observed from OGO 5 by Chappell *et al.* (1970). These data here support the suggestion by Maehlum and O'Brien (1963) that the plasmapause boundary is at the equatorward edge of the luminous aurora.

Obviously, the low latitude boundary of the activity moves equatorward with increasing K_p. Not so obvious, perhaps, is the existence of a plateau in both the Hβ and N_2^+ curves between $K_p = 1$ and 3. This is also consistent with previous data if one considers the meaning of the 'percent occurrence' diagrams by Feldstein and Starkov (1967), Stringer and Belon (1967), and Sandford (1968). These 'occurrence' diagrams

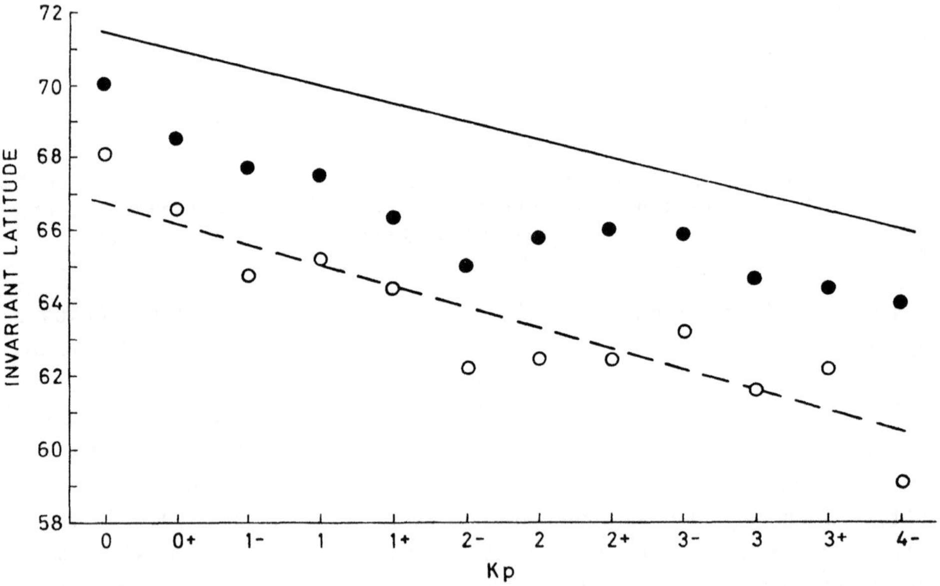

Fig. 10. The IN Lat. of the equatorward edge of N_2^+ emissions (circles) and the first equatorward peak of Hβ emission (dots) just after magnetic midnight, plotted as a function of the magnetic index, K_p. The rms deviation from these points is approximately 2° of latitude. The dotted line is the position of the plasmapause as a function of K_p from OGO 5 data (Chappell *et al.*, 1970). The solid line is the position of the equatorward peak of Lyα aurora from Chubb and Hicks (1970).

indicate that for K_p values between 1 and 3 the peak of percentage occurrence does not move equatorwards, but remains at 66.5° IN Lat. This corresponds to the plateau in Figure 10, which is also seen in the plots of intensity vs. IN Lat by Wiens and Vallance Jones (1969).

The solid line in Figure 10 is the position of the equatorward peak of the H Lyα aurora found by Chubb and Hicks (1970). The reason for the poleward displacement of that line is that the invariant given is at the satellite and the photometers were pointed towards the sub-satellite point. If this correction is made the curve compares favorably with the Hβ curve represented by dots. The Lyα data did not resolve the plateau, probably because integral values of K_p were used.

The third interesting effect to be seen in Figure 10 is the least obvious, but evidence for its existence can also be found in the literature. The slopes of the curves indicating the position of the electron and proton aurora differ by a significant amount. The average slope of the N_2^+ aurora curve and the line indicating the position of the plasma-pause is 0.6. The slope of the H aurora line is 0.5. This indicates that the average equatorward position of the electron aurora proceeds farther equatorward than the H aurora for the same increase in K_p. These data refer to geomagnetic midnight or shortly after.

It is generally agreed that the H aurora is observed equatorward of the electron aurora before midnight. However, observations in the Northern Hemisphere indicate a crossover in the midnight region (Stoffregen and Derblom, 1962; Derblom, 1967; Wiens and Vallance Jones, 1969). The H emission in the Southern Hemisphere is observed to remain equatorward of the electron aurora during the entire night (Eather and Sanford, 1966; Rees and Benedict, 1970). An exception are the early morning observations of poleward Lyα emission of Chubb and Hicks (1970) in the Southern Hemisphere. The latter observation was made during the southern summer (December 1967), but a seasonal effect which is opposite in the two hemispheres is difficult to visualize. If, however, the spatial relation of the electron and proton aurora is dependent on the magnitude of the magnetic storm, as the data given here seem to indicate, observations under different magnetic storm conditions could yield conflicting results.

Some support for the above view comes from the data of Wiens and Vallance Jones (1969) who plotted the latitudinal distribution of 5577 Å O I and 6563 Å (Hα) intensities as a function of latitude for three ranges of K_p. Table I below shows that the position of the luminosity associated with electron precipitation at 0300 geomagnetic time moves farther equatorward than that for proton auroras for the same increase in K_p.

TABLE I

Geomagnetic latitude of maximum luminosity at 0300 and (2100) geomagnetic time (Wiens and Vallance Jones, 1969).

	$K_p < 2_0$	$2_0 \leq K_p < 4_0$	$K_p \geq 4_0$
Hα (6563 Å)	68 (67)	67 (65)	66 (63)
O I (5577) Å)	67 (70)	64 (68)	62 (67)

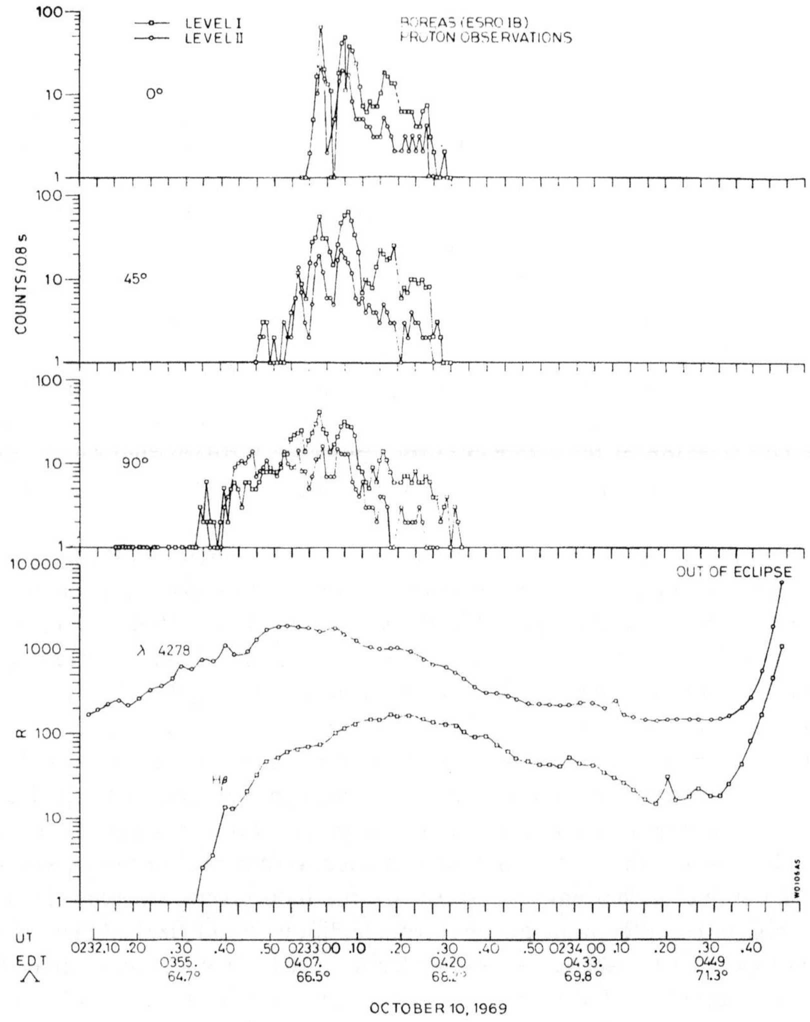

Fig. 11. Data from a northbound pass of ESRO I (Boreas) which show simultaneous detection of auroral light (N_2^+ and Hβ and of protons) in two energy channels at 0°, 45° and 90° with the magnetic field. The energy channels for the three detectors have nearly the same sensitivity. The 90° protons are recorded farther to the south than the 45° and 0° particles. The proton pitch angle distribution thus is peaked normal to the magnetic field. At 0232.58 UT ($\Lambda = 66.5°$), the 0° proton flux increases. This increase is almost 2 orders of magnitude in 3 sec. After this rapid increase, the proton pitch angle distribution becomes isotropic and remains so, except for a short time interval around 0233.03 UT when the 0° protons drop to a low value (depleted loss cone anisotropy). This holds for both the (120 to 190) keV level and the (200 to 350) keV level.

The data for 2100 hr are shown in parentheses and indicate the opposite trend. That is, before midnight, the proton aurora moves farther equatorward than the electron aurora for the same increase in K_p.

Thus, it seems that under certain conditions if the oval moves equatorward before midnight, the proton aurora moves ahead. An expansion after midnight puts the electron aurora equatorward. Figure 11 shows an example of an uncommonly large separation between the electron and proton emission regions. The all-sky camera at Kiruna showed the aurora to be expanding equatorward at this time. The magnetometer registered a simple negative bay of approximately 60 γ. K_p was 4+ and it was roughly 0400 geomagnetic time. It may be said, however, that even though the electron and proton precipitation regions seem to be separated here, this is not a 'typical' event and as a rule the main electron and proton precipitation regions coincide on the nightside.

B. EFFECT OF DR AND DP CURRENTS ON LATITUDINAL DISTRIBUTION OF AURORA

The reaction of the auroral oval to magnetic activity represented by K_p has been discussed in the preceding sections. Because this index mainly reflects the magnitude of the polar currents (DP), the influence of the ring current (DR) on the auroral oval is not observed. The magnetic field produced by the ring current is represented by the index Dst which commonly has a magnitude between +25 and $-125\ \gamma$.

Using all-sky camera data, Starkov and Feldstein (1967) plotted the IN Lat of the boundaries of the auroral oval as a function of Dst for two values of local Q, representing weak ($Q=3$) and moderate ($Q=5$) polar magnetic disturbances. The equatorward boundary moved poleward as Dst changed from $+25\ \gamma$ to 0. Then both the poleward and the equatorward boundaries moved equatorward as Dst changed from 0 to nearly $-120\ \gamma$. This is in agreement with the theoretical calculations of Akasofu (1966).

Although the width of the oval increased somewhat with the increase in ring current disturbance, DR, the spread was much greater (three times) as a function of Q for constant Dst. The main feature is that the center of the oval (half of the difference of the boundaries) moved monotonically equatorward with increasing DR.

The photometers aboard ESRO-I furnish a cross section of the distribution of luminosity of the instantaneous auroral oval at approximately 1.5 hr intervals. In order to separate the effects of DR and DP currents on the latitude distribution of auroral luminosity, an average cross section was determined for different values of Dst and K_p. That is, the average luminosity in each 1° IN Lat interval was assigned the value of Dst and K_p existing at the time of the reading. The values of K_p and Dst were grouped according to Table II.

TABLE II

Grouping of magnetic indices

Dst group 1	$\text{Dst} \geqslant +10\ \gamma$	K_p group 1	$0 \leqslant K_p \leqslant 1+$
Dst group 2	$+10\ \gamma > \text{Dst} \geqslant -15\ \gamma$	K_p group 2	$2 \leqslant K_p \leqslant 3-$
Dst group 3	$-15\ \gamma > \text{Dst} \geqslant -40\ \gamma$	K_p group 3	$3 \leqslant K_p$
Dst group 4	$-40\ \gamma > \text{Dst}$		

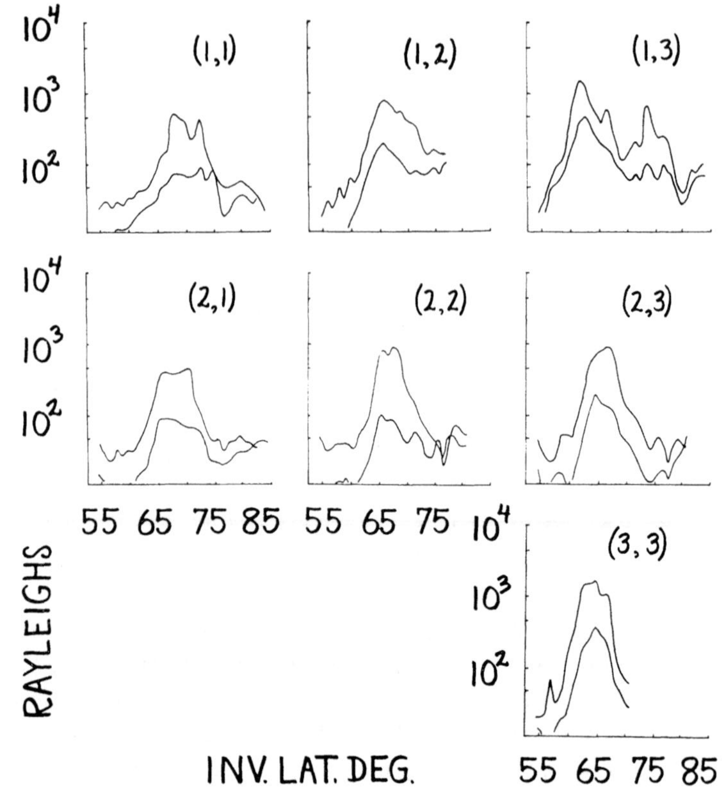

Fig. 12. A matrix (Dst, K_p) of averaged luminosity vs. IN Lat plots. The passes were taken near the midnight meridian and arranged according to the value of Dst and K_p at the time of the pass. The row and column numbers refer to the Dst and K_p grouping listed in Table III. The top trace in each case is 4278 Å N_2^+ and the bottom trace is 4861 Å Hβ.

Thus, each data point representing the luminosity observed in a 1° latitude interval was assigned a position in a 3 × 4 matrix according to the (Dst, K_p) value at the time of the observation. The average luminosity was calculated for each latitude interval in each magnetic group. The data were limited between 55° and 75° N IN Lat and to the magnetic time period between 2300 and 0100 hr.

The resulting matrix of 12 latitudinal luminosity distributions can be expected to separate the relative effects of DR and DP currents. The Figure 12 shows the preliminary plots obtained from the October and November lunations. Both electron and proton aurora are plotted but there is no large difference between them.

Despite the limitations imposed by the relatively small number of data points (100 to 1000 per 1° lat. interval) some trends are clearly visible. Table III shows the reversal of direction of motion of the main luminosity region as Dst values pass through 0 (for constant K_p), which was found by Starkov and Feldstein (1967).

For Dst values greater than 0 (group 1), an increase in K_p is accompanied by a distinct broadening of the emission zone. Both the poleward and the equatorward

TABLE III

The width of the main emitting region in degrees of latitude followed by the latitude and the intensity of the equatorward and poleward maxima 4278 Å N_2^+.

K_p	1	2	3
Dst 1	7° 68–73° 500–400 R	10° 66–74° 800–500 R	18° 62–75° 1500–700 R
2	7° 67–71° 500–500 R	7° 66–70° 1000–1000 R	7° 65–69° 1000–1000 R
3	–	–	6° 63–67° 1500–1200 R

edges move apart forming two distinct maxima. The equatorward maximum is brighter and has a greater rate of equatorward expansion.

At Dst values near 0 (group 2), the width of the emission zone remains quite constant for all K_p. There is equatorward motion but at a slower rate than for group 1.

For negative values of Dst (group 3 and 4), there are fewer data points available, but the width does not change very much with K_p. The whole emission region moves equatorward at a greater rate than for group 2 above.

At all levels of Dst, the brightness of the aurora increases with increasing K_p. Over all the K_p groups, however, increasing DR currents are associated with little change in brightness. The emission region narrows and moves equatorwards under these conditions, however, and the rate of equatorward motion is relatively greater on the polewared side.

Although the effects of DR and DP as shown here are not independent, some characteristics have been separated. Dependency between the two is most obvious in that the effects associated with K_p depend strongly on the level of Dst. The use of K_p as representative of DP currents is somewhat unfortunate because this index can include some effects of DR currents. The only independent effects seem to be that the emission region narrows with increasing DR currents and brightens with increasing K_p.

These results, however, must be regarded with caution. The relatively small number of data points resulted in relatively large intervals of magnetic indices. In view of the strong relation between the instantaneous aurora and the magnetospheric activity, however, it is suggested that this type of analysis should be used with satellite auroral data. Obvious improvements are the use of data which are not limited by sun and moon, and the use of magnetic indices which are more compatable with the temporal and spatial resolution of the data frame (AE, for example). Appropriate indices for introducing the effect of the neutral sheet current (Akasofu, 1966) would also be of interest.

References

Aarsnes, K., Amundsen, R., Lindalen, H. R., and Søraas, F.: 1970, *Physica Norvegica*, in press.
Akasofu, S.-I.: 1966, *Planetary Space Sci.* **14**, 587.
Akasofu, S.-I.: 1968a, *Polar and Magnetospheric Substorms*, D. Reidel Publishing Company, Dordrecht-Holland.
Akasofu, S.-I.: 1968b, *Planetary Space Sci.* **16**, 1365.
Akasofu, S.-I.: 1970a, private communication.
Akasofu, S.-I.: 1970b, Paper delivered, Spring Meeting Amer. Geophys. Union. Washington, D.C.
Akasofu, S.-I., Eather, R. H., and Bradbury, J.: 1969, *Planetary Space Sci.* **17**, 1409.
Andriyenko, D. A.: 1965, *Geomag. and Aeron.* **5**, 684.
Bates, H. F., Sharp, R. D., Belon, A. E., and Boyd, J. S.: 1969, *Planetary Space Sci.* **17**, 83.
Boyd, J.: 1969, Master's Thesis University of Alaska, College, Alaska.
Buchau, J., Whalen, J. A., and Akasofu, S.-I.: 1969, *J. Atmospheric Terrest. Phys.* **26**, 1155.
Buchau, J.: 1970, Paper delivered, Spring Meeting, Amer. Geophys. Union, Washington, D.C.
Chappell, C. R., Harris, K. K., and Sharp, R. W.: 1970, *J. Geophys. Res.* **75**, 50.
Chubb, T. A. A. and Hicks, G. T.: 1970, *J. Geophys. Res.* **75**, 1290.
Davidson, G. T.: 1965, *J. Geophys. Res.* **70**, 1061.
Davis, T. N., Deehr, C. S., and Leinbach, H.: 1959 *Sci. Rep. No. 5, Y/22.6/327* Geophysical Inst., University of Alaska.
Davis, T. N., Stenbaek-Nielsen, H., and Hallinan, T.: 1971, this volume, p. 160.
Deehr, C. S., Sten, T. A., Egeland, A., and Omholt, A.: 1970a, *ESRO/ELDO Scientific and Technical Review* **2**, 233.
Deehr, C. S., Gustafsson, G., Omholt, A., Anderson, L., Egeland, A., and Borg, H.: 1970b, *Physica Norvegica*, in press.
Derblom, H.: 1967, *Ann. Geophys.* **24**, 163.
Eather, R. H.: 1966, *J. Geophys. Res.* **71**, 4133.
Eather, R. H.: 1969, in *Atmospheric Emissions* (ed. by B. M. McCormac and A. Omholt), Van Nostrand Reinhold Company, New York, p. 233.
Eather, R. H. and Akasofu, S.-I.: 1969, *J. Geophys. Res.* **74**, 4794.
Eather, R. H. and Mende, S. B.: 1971, this volume, p. 255.
Eather, R. A. and Sandford, B. P.: 1966, *Australian J. Phys.* **19**, 25.
Egeland, A. and Omholt, A.: 1966, *Geofys. Pub.* **26**, No. 6.
Evans, J. E., Newkirk, L. L. and McCormac, B. M.: 1970, North Polar, South Polar, World Maps and Tables of Invariant Magnetic Coordinates for 6 Altitudes: 100, 300, 600, 1000 and 3000 km, Lockheed Palo Alto Research Laboratory.
Feldstein, Y. I.: 1969, *Revs. Geophys.* **7**, 179.
Feldstein, Y. I.: 1970, Paper delivered, May meeting COSPAR, Leningrad.
Feldstein, Y. I. and Starkov, G. V.: 1967, *Planetary Space Sci.* **15**, 209.
Fukushima, N. and Nagata, T.: 1967, *The Birkeland Symposium* (ed. by J. Holtet and A. Egeland) p. 195.
Galperin, Y. I.: 1963, *Planetary Space Sci.* **10**, 187.
Gustafsson, G.: 1970a, *Tellus* **21**, 6.
Gustafsson, G.: 1970b, *Physica Norvegica*, in press.
Hartz, T. R.: 1971, this volume, p. 225.
Hartz, T. R. and Brice, N. M.: 1967 *Planetary Space Sci.* **15**, 301.
Hartz, T. R., Montbriand, L. E., and Vogan, E. L.: 1963, *Can. J. Phys.* **41**, 581.
Hoffman, R. A. and Evans, D. S.: 1968, *J. Geophys. Res.* **73**, 6201.
Hultquist, B., Borg, H., Riedler, W., and Christophersen, P.: 1970, *Planetary Space Sci.*, submitted.
Johansen, O. E. and Omholt, A.: 1963, *Planetary Space Sci.* **11**, 1223.
Lassen, K.: 1969, in *Atmospheric Emissions* (ed. by B. McCormac and A. Omholt), Van Nostrand Reinhold Company, New York, p. 63.
Lassen, K.: 1970, *Physica Norvegica*, in press.
Maehlum, B. and O'Brien, B. J.: 1963, *J. Geophys. Res.* **70**, 557.
McDiarmid, I. B. and Burrows, J. R.: 1964, *Can. J. Phys.* **42**, 1135.
Mishin, V. M., Saifudinova, T. I., and Zhulin, I. A.: 1970, *J. Geophys. Res.* **75**, 797.
Montalbetti, R.: 1959, *J. Atmospheric Terrest. Phys.* **14**, 200.

Montbriand, L. E.: 1971, this volume, p. 366.
Nadubovich, Yu. A.: 1963, *Geomag. and Aeron.* **3**, 409.
Omholt, A.: 1956, *J. Atmospheric Terrest. Phys.* **9**, 18.
Omholt, A.: 1971, *Optical Aurora*, Springer-Verlag, in press.
Pettersen, H.: 1970, private communication.
Rees, M. H. and Benedict, P. C.: 1970, *J. Geophys. Res.*, **75**, 4763.
Riedler, W. and Borg, H.: 1970, *Physica Norvegica*, in press.
Romick, G. J.: 1970, Paper delivered, Spring meeting, Amer. Geophys. Union, Washington, D.C.
Sandford, B. P.: 1968, *J. Atmospheric Terrest. Phys.* **30**, 1921.
Sharp, R. D. and Johnson, R. G.: 1968, *J. Geophys. Res.* **73**, 969.
Sharp, R. D., Carr, D. L., and Johnson, R. G.: 1969, *J. Geophys. Res.* **74**, 4618.
Søraas, F., Aarsnes, K., Lindalen, H. R., and Møhl Madsen, M.: 1969, University of Bergen, Dept. of Physics, Report No. 8.
Starkov, G. V.: 1968, *Geomag. and Aeron.* **7**, 28.
Starkov, G. V. and Feldstein, Y. I.: 1967, *Geomagn. and Aeron.* **7**, 854.
Stoffregen, W. and Derblom, H.: 1962, *Planetary Space Sci.* **9**, 711.
Stringer, W. and Belon, A. E.: 1967, *J. Geophys. Res.* **72**, 245.
Whalen, J. A.: 1970, Paper delivered, Spring meeting, Amer. Geophys. Union, Washington, D.C.
Wiens, R. H. and Vallance Jones, A.: 1969, *Can. J. Phys.* **47**, 1493.

INSTITUTE DISCUSSION

The precipitation and auroral arcs are continuous around the auroral oval. During disturbed times the oval may tend toward a circle. R. Johnson pointed out that although the auroral observations may appear to be continuous on the basis of widely separated observations, there may well be two or more separate processes involved.

AURORAL EXCITATION AND IONIZATION INTENSITY AND TEMPERATURE VARIATIONS

G. G. SHEPHERD
York University, Toronto, Canada

Abstract. A description is given of the methods employed in the study of auroral excitation and ionization, with particular emphasis on emission rate profiles. A summary of these data so far obtained is presented. Recent measurements of vertical profiles for both rotational and Doppler temperatures are shown, and temporal variations of auroral intensity and Doppler temperature are discussed.

1. Introduction

This paper is intended as an introduction to the observational study of auroral excitation and its consequences. Although this information can in part be obtained from direct measurements, such as ion concentrations and soft electron spectra, most of the information so far gained has come from the emitted photons and only optical data are considered here. Vallance Jones (1969) presented a thorough review of auroral spectroscopy at the last symposium, and Swider (1969) described the atomic and molecular reactions; and for those aspects of auroral spectra and microscopic processes the reader is referred to those papers. Here, emphasis is placed on the spatiotemporal characteristics of the emission, the various techniques used to study its distribution, and some of the results that have been obtained. Particular emphasis is placed on vertical emission rate profiles, and a summary of those obtained by various techniques is given. Some recent data on vertical N_2^+ rotational temperature profiles obtained by Bunn (1970) are described and some new O I Doppler measurements made by Zwick (1970) with a wide angle Michelson interferometer are shown. The latter show the complex interrelationship between auroral intensity and temperature and their temporal variations.

2. Auroral Excitation and Ionization

A. MEASUREMENTS FROM THE GROUND

It is interesting that one of the first (1910) scientific auroral measurements was that of height (Størmer, 1965), and that it is still one of the most important and most difficult measurements one can make. Størmer's monumental parallactic photographic studies are a tribute to his perserverance, and experimental and mathematical skill. In addition to the experimental problems involved in obtaining simultaneous auroral photographs over long baselines, the data reduction involved laborious transformations between three coordinate systems, incorporating significant amounts of lens distortion into the calculations. Later, workers in Canada (Currie, 1934) collected data under exacting conditions that moderns experimenters can scarcely visualize.

The reduced data consisted mainly of the heights of well defined edges of auroral

B. M. McCormac (ed.), The Radiating Atmosphere, 150–159. All Rights Reserved.

forms, and while this explains much about the excitation, it is more valuable to have the vertical distribution of the emission intensity. The importance of 'luminosity profiles', or volume emission rate profiles, as they are called here, was recognized very early by Harang (1951). Currie and Weaver (1955) made a statistical determination of the vertical extent of auroral forms; the most frequently observed extent was about 30 km. In his Figure 85, Harang (1951) shows photographs isolating the N_2^+ 3914 Å and O I 5577 Å emissions. There are distinct differences between the two distributions (an observation of current interest). To convert photographic data to vertical profiles, one assumes that the auroral form is in the form of a thin sheet perpendicular to the line of sight. In fact, from one station one cannot distinguish a vertical sheet from a horizontal one, or any other configuration.

Belon *et al.* (1966) demonstrated a practical and realistic way of employing scanning photometers to obtain emission rate profiles. They selected aurora which adequately

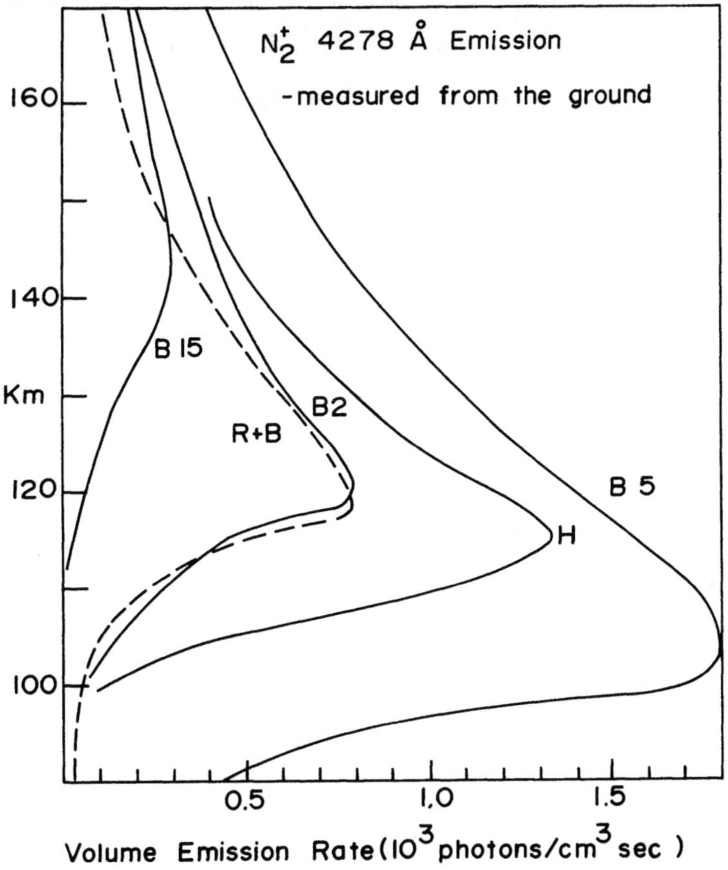

Fig. 1. Volume emission rate profiles of N_2^+ measured by ground based methods. The numbers in B2, B5 and B 15 indicate the order of entry in the table of Belon *et al.* (1966); the profiles were chosen to be the highest, lowest and average of their data. R + B indicates Romick and Belon (1967). All data are converted to 4278 Å equivalent except Harang's (H), which are in arbitrary units.

satisfied an ideal geometrical situation located at the magnetic zenith for one station and the magnetic horizon for the other station. From this selection, only aurora observed to satisfy the single sheet approximation was used. Romick and Belon (1967) extended this method to aurora which did not satisfy the thin sheet approximation, and used iterative procedures to calculate both horizontal and vertical distributions. It was necessary to employ analytical distribution equations with adjustable parameters; this constraint may be important because a unique solution may not otherwise exist.

In Figure 1, these ground based emission rate profile measurements are compared. The N_2^+ 3914 Å profiles of Belon *et al.* (1966), and Romick and Belon (1967) have been converted to 4278 Å emission by dividing by three. The data of Harang (1951) are in arbitrary units. The general features are similar but there are significant variations in the heights of peak emission.

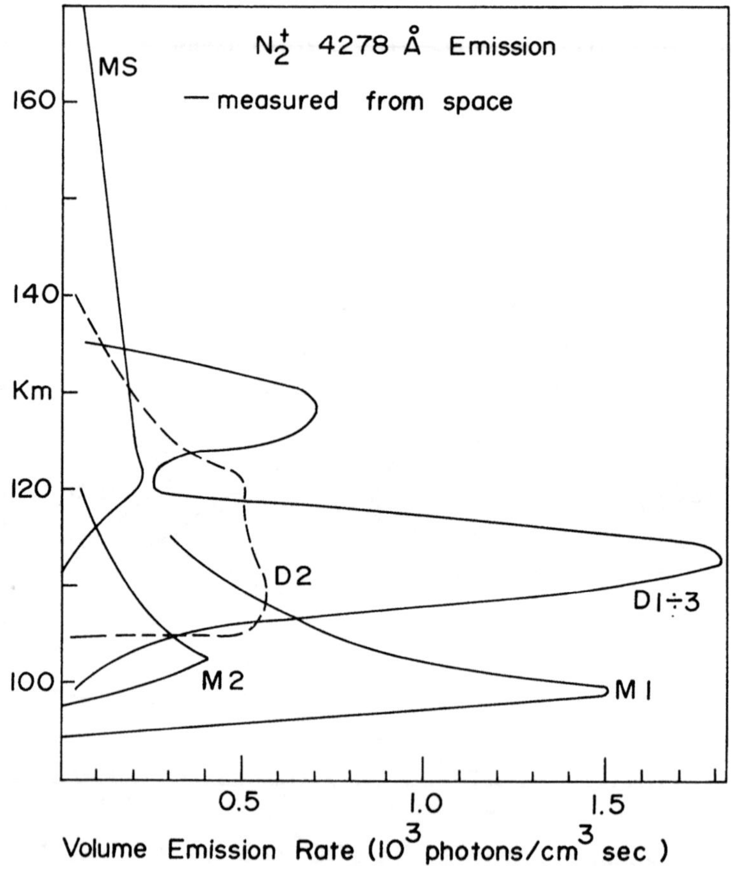

Fig. 2. Volume emission rate profiles of N_2^+ measured by rocket techniques. M1 and M2 refer to Maseide (1967), MS to Miller and Shepherd (1968), and D1 and D2 to Donahue *et al.* (1968). All data are converted to 4278 Å equivalent and the D1 curve was additionally divided by 3 to bring it on scale.

B. MEASUREMENTS FROM SPACE

Rocket flights offer a promising way to explore the volume distribution of auroral emission, and a number of optical devices have been flown through auroral forms. Some have been viewed perpendicular to the spin axis (Murcray, 1967; Baker, 1968) but generally data obtained in this way are difficult to interpret. However, a particularly interesting conclusion of Murcray's (1967) measurements is that the *horizontal* distribution of N_2^+ 3914 Å and O I 5577 Å emissions are different, a curious result that was also obtained by Romick and Belon (1967). Baker (1968), on the other hand, found the horizontal distributions of the two emissions to be similar.

Photometers viewing parallel to the rocket axis are better suited for deriving vertical distributions, so long as the rocket coning angle is small. This is because if the vehicle is at 130 km, say, then one can be sure that all the emission arises from above 130 km. However, horizontal spatial variations and temporal variations are still hazards with which to contend.

Volume emission rate profiles derived from forward viewing photometers by Donahue *et al.* (1968), Maseide (1967), and Miller and Shepherd (1968) are shown in Figure 2, where the data have been reduced to correspond to the N_2^+ 4278 Å emission. The magnitudes of the emission rates are similar to the ground based ones, which is encouraging, except for the Donahue *et al.* (1968) curve D1, which was an unusually strong aurora, particularly so for one with a peak emission at 115 km. In general, the rocket obtained distributions are much sharper than the ground based ones, which suggests a layered structure not evident from the ground based work.

The results so far obtained from proton aurora are collected in Figure 3; the data are from Miller (1969), Miller and Shepherd (1969) and Murcray (1966). The Hβ emission is about 50 times weaker than that for 4278 Å in the electron auroras shown. Murcray's (1966) data are in arbitrary units. Ground based heights obtained by Rees (1961) and Wiens (1968) are also shown. Although one thinks of proton aurora as a more stable and simply characterized phenomenon, the variation in emission rate profiles is rather dramatic. Direct measurements of protons from a number of rocket flights over Fort Churchill (Whalen, 1970), suggest that the M I curve is typical of most proton precipitation. Proton generated emission at the lower altitudes shown appears to be less common.

3. Vertical Temperature Distribution

The measurement of spectroscopic temperature in airglow and aurora has been reviewed by Shepherd (1969). Rotational temperature profiles from the N_2^+ emission have recently been deduced by Bunn (1970) using forward looking two-channel photometers (Miller and Shepherd, 1968) flown in four Black Brant II Rockets. The photometers are characterized by two spectral channels of about 5 Å bandwidth and separated by about 20 Å. Temperatures are determined from the ratio of the emission seen in the two channels, employing synthetic spectra and an accurate knowledge of the filter passbands and their temperature characteristic (filter temperatures are

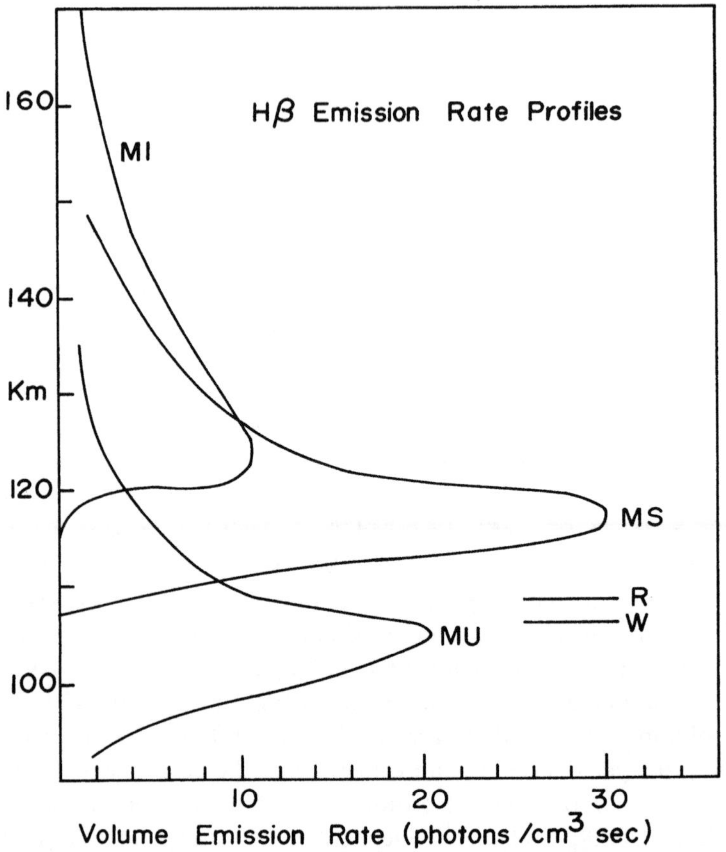

Fig. 3. Rocket measurements of volume emission rate profiles for Hβ in proton aurora. MI denotes Miller (1969), MS denotes Miller and Shepherd (1969) and MU is for Murcray (1968). R and W indicate ground based heights measured by Rees (1961) and Wiens (1968).

monitored throughout the flight). Volume emission rate profiles are derived for each channel separately, and from these the 'true' temperature profiles (corresponding to 5 km altitude intervals), and the total band emission rate are computed. Since each temperature value results from the ratio of two differentiated signal curves, each of which is subject to spatiotemporal effects, some data reduction problems are expected. On the other hand, the reduction involves a ratio, so electron flux variations should tend to cancel and the temperature profile should be smoother than the individual emission rate profiles. However, variations in the electron energy spectrum will perturb the derived temperature profile.

A detailed presentation of these data will be given elsewhere by Bunn *et al.* (1971); only some preliminary data from one flight are shown here in Figure 4. This flight, which was in a post-breakup auroral glow gave the emission rate profiles least complicated by spatio-temporal effects. This was confirmed by the particle data of Whalen and McDiarmid (1970), who observed nearly constant electron fluxes throughout the

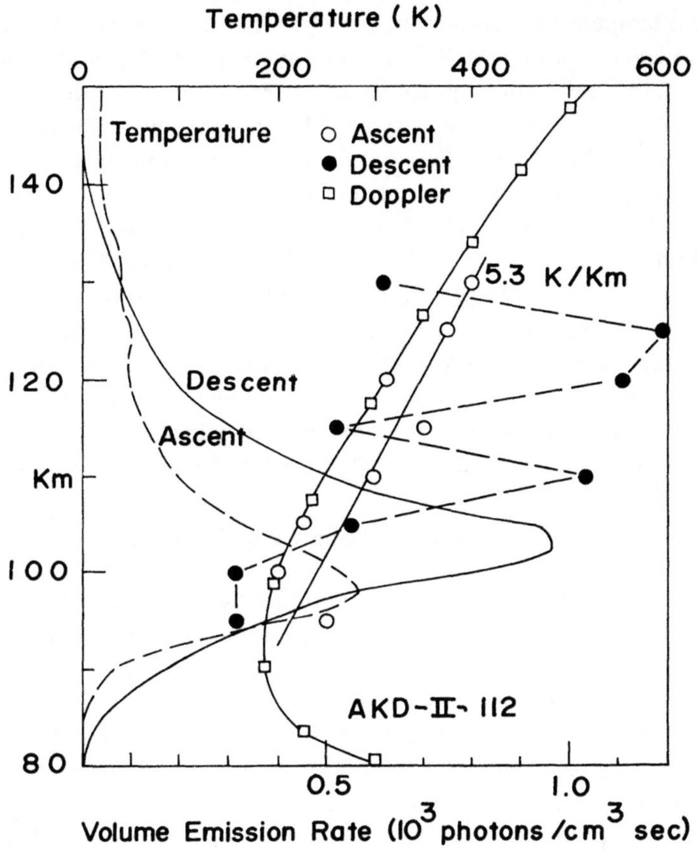

Fig. 4. Volume emission rate profiles and differential rotational temperatures measured from N_2^+ in a post-auroral glow with Black Brant Rocket AKD-II-112. The data have been provided by Bunn (1970). The Doppler temperature profile shown was measured with a ground based WAMI by Zwick (1970).

flight. Yet this 'ideal' flight displays complex behavior. The ascent and descent profiles are noticeably different. The ascent curve peaks at about 98 km and shows some temporal oscillations at 120 to 135 km and the emission continues above 150 km. This high altitude emission died away while the vehicle was at apogee but at intermediate altitudes strengthened and moved the peak up to 102 km.

The electron detectors (Whalen and Miller, 1970) showed the same integrated flux above 20 keV for ascent and descent, although there was a slight reduction in the higher energy flux during descent. Whalen and McDiarmid (1970) considered this spectrum to be soft, much softer than observed in a flight into the breakup which preceded this glow. Yet the altitude of the emission peak for ascent is as low as any shown in this paper. The similarity to Maseide's (1967) (M I, Figure 2) profile is striking (it was also for an auroral glow). All this confirms how strongly the behavior of the electrons with $E > 20$ keV is uncoupled from the main optical aurora.

The ascent temperature profile (open circles) follows a linear gradient of 5.3 K/km with a noise level of about 15 K. The descent profile (solid circles) is very erratic. The temperature minima do suggest a reasonable gradient but the high temperature values suggest that during descent a variable emission, high above the rocket, was being viewed. In other words, there were marked variations in the energy spectrum even though the total flux was relatively constant.

A Doppler temperature profile measured with a wide angle Michelson interferometer (Hilliard and Shepherd, 1966) by Zwick (1970) of a 200 kR type B aurora is shown in Figure 4 also. It agrees closely with the rotational temperature profile. An unusual feature of this profile is the inversion at low altitude. Although this resembles the known temperature inversion in this region, it does not resemble it closely enough, and further, it was observed in all auroras independently of their (assumed) altitude. This could be some peculiar observational effect (such as a low altitude wind shear) but it seems more likely to be horizontally extended aurora, invalidating the thin sheet assumption.

4. Intensity and Temperature Variations

Following a number of earlier measurements (Shepherd, 1969), Turgeon and Shepherd (1962) made some Doppler temperature measurements from the OI 5577 Å line in aurora, using a Fabry-Perot spectrometer, and found that "Repeated observations made on the same forms over periods of a few minutes have shown that the temperatures can exhibit large changes." (150 K in 1 min.) Hilliard and Shepherd (1966) greatly increased the time resolution of these measurements with a wide angle Michelson interferometer (WAMI) and revealed a frequently occurring systematic inverse relationship between auroral brightness and temperature. This relationship sometimes held for tens of minutes at one location in the sky and also for single pulsations.

Following on this, and the current interest in auroral pulsations, Zwick (1970) made some WAMI observations on some bright early morning pulsations. High resolution time histories of some of these pulsations are shown in Figure 5a, where fringe visibility is plotted against emission rate. A simple inverse temperature relationship would generate roughly linear plots of a positive slope in this representation; the behaviour is seen to be more complex than that. Many of the pulses generate open loops, and of those which do, the clockwise direction predominates. Of 41 histories studied, there were 14 clockwise, 4 counter clockwise, and 23 indefinite patterns.

Long duration runs on individual forms allowed Zwick (1970) to conclude that neutral heating is 'indetectable' in the sense that it is dominated by other effects. The other dominant effect is electron energy spectrum changes. In these histories we thus associate emission rate with electron flux, and visibility with electron energy. A clockwise pattern can be generated by an electron stream whose energy and flux vary as shown in Figure 5(b). In effect, it means that a low flux of high energy electrons is followed by a larger flux of low energy electrons. The time delay between the high and low energy electrons is consistent with dispersion over a travel distance of about 10^5 km.

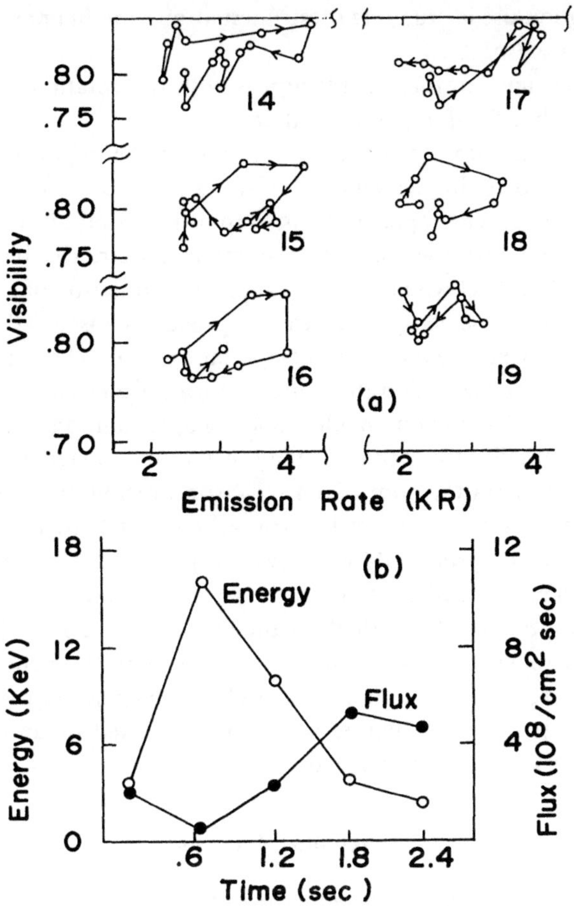

Fig. 5a. Time histories of some auroral pulsations observed by Zwick (1970) with a WAMI. 'Visibility' means Michelson's fringe visibility, which is interpreted as increasing with increasing electron energy.

Fig. 5b. Electron energy (the energy parameter in a power exponential energy spectrum) and flux variations that would generate a clockwise pulsation history similar to 15 and 18.

5. Discussion

Some definite points emerge from the data presented.

(a) The rocket measured emission rate profiles are distinctly different from the ground based ones. They are much thinner and may exhibit a layered structure. Each of the two methods has its own limitations but the rocket measured ones are more correct likely. If so, complex processes such as local acceleration of secondaries may be required to fit the data.

(b) The vertical rotational temperature profile is very sensitive to auroral energy spectrum fluctuations. The knowledge that the temperature profile is well behaved might be put to good use in studying these fluctuations and in improving the reliability of the emission rate profiles.

(c) The Doppler temperature profile agrees with the rotational one and also with current models within the limits of these data.

(d) The WAMI pulsation measurements reinforce the importance of energy spectrum fluctuations noted in the emission rate profiles and rotational temperature measurements. This process appears to be important in all aurora and to dominate some types. This demonstrates that electron energy spectra can be effectively studied from the ground. The clockwise pulsations suggest modulation of a distance source.

(e) The intermediate scale spatiotemporal processes discussed here must be resolved from microscopic one if correct interpretations are to be made. As an example, consider the puzzling observation that auroral forms are broader in the 3914 Å emission than in the 5577 Å emission. While a microscopic solution to this problem would be the most intellectually stimulating, it may in fact be a macroscopic one. Suppose, for example, that around the core of soft electron precipitation causing most of the optical aurora there is an extensive horizontal layer of hard precipitation, so hard that the 5577 Å emission produced by it is heavily quenched. This tentative explanation has not been studied; it is offered primarily as an example but it would also satisfy the point made by Evans (Walt, 1969) at the conference review of the last Institute. It was stated that rocket measurements of particles, with their enormous spatial and temporal resolution, normally find a structureless precipitation region above auroral forms which appear to be fluctuating wildly when viewed by an optical device that integrates if over a vast volume of space.

Acknowledgments

The work reported here was largely carried out at the University of Saskatchewan. I would like to thank Dean B. W. Currie for his support and encouragement during my years spent there. I am also indebted to my colleagues who executed much of the work described here, in particular, F. E. Bunn, J. R. Miller and H. H. Zwick. Financial support was provided by the National Research Council of Canada.

References

Baker, K. D.: 1968, *Ann. Geophys.* **24**, 285.
Belon, A. E., Romick, G. J., and Rees, M. H.: 1966, *Planetary Space Sci.* **14**, 597.
Bunn, F. E.: 1970, M. A. Thesis, York University, Toronto.
Bunn, F. E., Miller, J. R., and Shepherd, G. G.: 1971, to be published.
Currie, B. W.: 1934, *Terr. Mag.* **39**, 293.
Currie, B. W. and Weaver, J. T.: 1955, *Can. J. Phys.* **33**, 611.
Dohanue, T. M., Parkinson, T. D., Zipf, E. C., Doering, J. P., Fastie, W. G., and Miller, R. E.: 1968, *Planetary Space Sci.* **16**, 737.
Harang, L.: 1951, *The Aurorae*, Chapman and Hall Ltd., London.
Hilliard, R. L. and Shepherd, G. G.: 1966, *Planetary Space Sci.* **14**, 383.
Maseide, K.: 1967, *Planetary Space Sci.* **15**, 899.
Miller, J. R.: 1969, Ph.D. Thesis, University of Saskatchewan, Saskatoon.

Miller, J. R. and Shepherd, G. G.: 1968, *Ann. Geophys.* **24**, 305.
Miller, J. R. and Shepherd, G. G.: 1969, *J. Geophys. Res.* **74**, 4987.
Murcray, W. B.: 1966, *Planetary Space Sci.* **14**, 812.
Murcray, W. B.: 1967, *J. Geophys. Res.* **72**, 1047.
Rees, M. H.: 1961, *Mem. Soc. Roy. Sci. Liège, Ser. 5*, **4**, 609.
Romick, G. J. and Belon, A. E.: 1967, *Planetary Space Sci.* **15**, 1695.
Shepherd, G. G.: 1969, *Ann. Geophys.* **25**, 841.
Stormer, C.: 1965, *The Polar Aurora*, Clarendon Press, Oxford.
Swider, W.: 1969, *Atmospheric Emissions* (ed. by B. M. McCormac and A. Omholt), Van Nostrand Reinhold Company, New York, p. 367.
Turgeon, E. C. and Shepherd, G. G.: 1962, *Planetary Space Sci.* **9**, 295.
Vallance Jones, A.: 1969, *Atmospheric Emissions*, (ed. by B. M. McCormac and A. Omholt), Van Nostrand Reinhold Company, New York p. 47.
Walt, M.: 1969, in *Atmospheric Emissions* (ed. by B. M. McCormac and A. Omholt), Van Nostrand Reinhold Company, New York, p. 547.
Whalen, B. A.: 1970, private communication.
Whalen, B. A. and McDiarmid, I. B.: 1970, *J. Geophys. Res.* **75**, 123.
Whalen, B. A. and Miller, J. R.: 1970, private communication.
Wiens, R. H.: 1968, Ph.D. Thesis, University of Saskatchewan, Saskatoon.
Zwick, H. H.: 1970, Ph.D. Thesis, University of Saskatchewan, Saskatoon.

INSTITUTE DISCUSSION

S. Mende described his recent experimental effort in Alaska to measure the spatial extent of 3914 Å relative to that for 5577 Å. He could not find any difference; he also explained previous interpretations of a broader 3914 Å region on the basis of improper correction for scattering. G. Shepherd stated that he was unable to find a widening of 3914 Å over 5577 Å also.

Determining the rotational temperature from a rocket does not have any background subtraction problems as the background is negligible. D. Bryant stated that he had observed electron spectra change from 10 keV at the southern edge of an aurora to 2 keV at the northern edge of a fairly broad aurora. This could give a rotational temperature of a complex distribution or at least distort things.

The dependence of rotational temperature on electron energy spectra is uncertain, but G. Shepherd stated that this should be clarified. G. Culp stated that A. Stair's group had measured the dependence of rotational temperature on electron energy and found that there was a positive enhancement of rotational temperature near the threshold energy for production in 3914 Å.

AURORAL CONJUGACY AND TIME-DEPENDENT GEOMETRY OF AURORAS

T. N. DAVIS, T. J. HALLINAN, and H. C. STENBAEK-NIELSEN
Geophysical Institute
University of Alaska, College, Alaska, U.S.A.

Abstract. All-sky cameras and image orthicon TV carried onboard conjugate-flying aircraft have allowed studies of the degree of conjugacy and of time dependent auroral geometries. Auroras during quiet times and at low auroral zone latitudes during disturbed times are found to be conjugate; during disturbed times Southern Hemisphere auroras at higher latitudes are displaced eastward and equatorward of the Northern Hemisphere counterparts. Examination of small and large scale convolutions of auroras suggests that these are produced by the Kelvin-Helmholtz shear flow instability.

1. Introduction

Measurements of the degree of conjugacy of auroras using jet aircraft equipped with all-sky cameras were performed during the March equinoxes of 1967 and 1968. A third series of observations is planned for the Fall equinox in 1970. In addition to all-sky cameras, the 1968 expedition made use of image orthicon TV systems to obtain detailed observations in a small region over each aircraft. These observations and similar TV observations at various locations in the Northern Hemisphere have enabled investigation of temporally quasi-periodic phenomena and rapidly evolving small scale structure of auroras that may be indicative of the nature of electric fields and other phenomena associated with and perhaps controlling auroral morphology. Here we describe the results obtained thus far in three areas: auroral conjugacy, pulsations, and time dependent micromorphology.

2. Auroral Conjugacy

All conjugate aircraft flights have been made along the 256° dp meridian with the aircraft being maintained conjugate relative to the GSFC 12/66 time-dependent model of the internal geomagnetic field (Hendricks and Cain, 1966). The three flights during March 1967 occurred during magnetically undisturbed epochs ($K_p=0$ to 2 and College $K_p=0$ to 1). On each of these flights auroras were observed between dp latitudes 66° and 71°. Qualitative analysis of all data, and quantitative analysis of a limited portion, indicate that within the limits of error all auroras were conjugate relative to the GSFC 12/66 field model. The auroras exhibited similar intensity variation and remarkable similarity in detailed geometry, including identical individual ray structures in the two hemispheres. A slight tendency for the southern auroras to be eastward of the northern auroras was noted. A detailed description of the 1967 results and references to other conjugate studies are given by Belon *et al.* (1969). Whereas the March 1967 flights reached only to dp lat 67°, the aircraft range was extended so as to reach

B. M. McCormac (ed.), The Radiating Atmosphere, 160–169.

to dp lat 72° during the March 1968 flights. All-sky camera frame exposures were shortened from 8 to 3 sec, and image orthicon TVs with 40° by 53° field of view were mounted to view the zenith region. However, the most significant difference was that the 1968 flights (5 in all) occurred during disturbed times ($K_p = 2$ to 4 and College $K = 2$ to 6).

Incomplete analyses of the 1968 conjugate flight data together with the 1967 result reveal a major difference in auroral conjugacy during quiet and disturbed periods. During disturbed periods the normal auroral display exhibits a latitudinal demarcation line across which there is a change in the degree of conjugacy. Moving poleward in the evening hours on each flight, the conjugate flying aircraft simultaneously encountered the diffuse H rich arc that normally appears on the equatorward boundary of a display. Primarily homogeneous, this diffuse arc usually displays some structure near its poleward boundary where it would be classified as a multiple arc in the terminology specified in the *International Auroral Atlas* (1963). This arc system usually does not exceed 200 km in horizontal width. Poleward of it is a gap in the aurora of width near 100 km. Auroras appearing poleward of this gap are the normal electron produced forms lacking in H emission. The demarcation line referred to above lies in this gap in the visible aurora. Equatorward of the demarcation line during disturbed periods the auroras exhibited the same degree of conjugacy observed within all auroras measured previously during undisturbed periods. Intensity and form were similar in the two hemispheres, and the auroras were strictly conjugate relative to the GSFC 12/66 geomagnetic field.

Poleward of the demarcation line, corresponding auroras in the two hemispheres normally exhibit a pronounced displacement relative to the main geomagnetic field and do not necessarily vary together in intensity. Both the relative displacements and the noncoherent intensity variations present a considerable difficulty to detailed analysis. Two essentially identical auroras of different intensity may not be recognized as being identical. Even more critical is the problem of aspect; the appearance of any aurora depends to a remarkable extent upon the angle between the view direction and the direction of the local magnetic field. Nonconjugate but identical auroras viewed in their respective hemispheres from conjugate locations do not look the same. Therefore, it becomes difficult to determine whether corresponding but conjugately displaced auroras are or are not exactly identical in detail. At this time we are unable to make a firm statement on this point.

In general, the relative displacement of corresponding auroras poleward of the demarcation line is such that the Southern Hemisphere auroras are eastward and equatorward of their northern counterparts (Figure 1). The displacement typically amounts to 100 to 350 km. Its eastward component appears not to depend as much upon the degree of magnetic disturbance as does the meridional displacement, which evidently tends to increase with increasing activity. Poleward of the demarcation line the relative displacement of auroras shows no sign of latitudinal dependence.

With detailed analysis of the conjugate all-sky data approximately half complete there was only one exception to the relative eastward displacement of the Southern

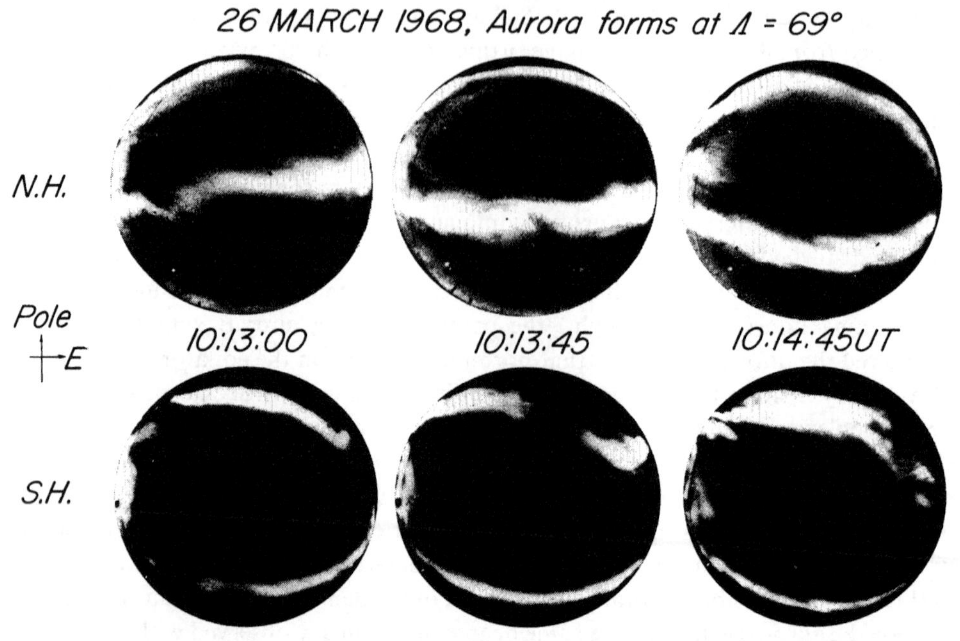

(a)

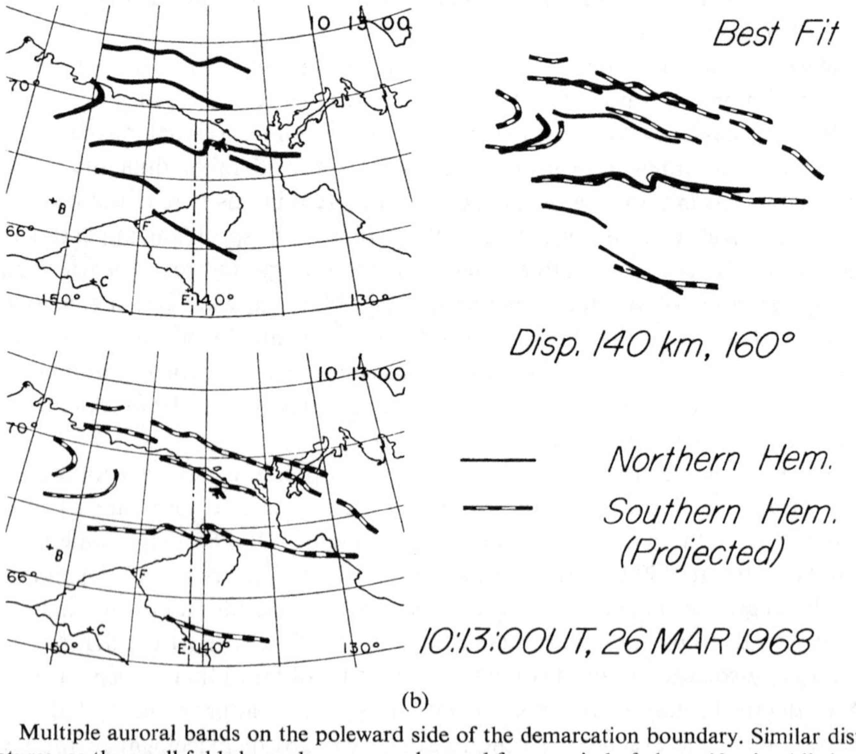

(b)

Fig. 1. Multiple auroral bands on the poleward side of the demarcation boundary. Similar distinct features, as the small fold shown here, were observed for a period of about 10 min. All-sky photographs are 3 sec exposures.

Hemisphere aurora. At about 0930 on March 26, 1969, a westward moving loop structure (westward surge in the nomenclature of Akasofu, 1964) observed in both hemispheres was relatively displaced with the Southern Hemisphere loop 210 to 260 km *west* of the Northern Hemisphere counterpart.

The relative intensities of corresponding auroras may at times vary similarly, but it is also common to observe a decrease of intensity in one hemisphere and an increase in the other. At any one time the auroras in either hemisphere may be more intense than in the other. However, the time averaged intensity is the greater in the Northern Hemisphere.

In at least one instance there appears to be a complete breakdown in conjugacy in the region dp lat 69 to 72°. A moderately bright rayed arc system occurred in the Northern Hemisphere, but for a period of 20 min no aurora appeared in the field of view of the Southern Hemisphere all-sky camera.

The occurrence of auroras which, in all respects, displays a high degree of dynamic conjugacy during magnetically undisturbed times and at relatively low dp lat during active times implies that these auroras are occurring on essentially undistorted dipole-like field lines. Auroras occurring poleward of the 'demarcation line' observed during disturbed periods, in addition to showing displacements in conjugacy, exhibit non-conjugate intensity variations. Clearly, these auroras are occurring on highly distorted field lines and, furthermore, the precipitated flux is asymmetric with respect to the geomagnetic equatorial plane. In the extreme case there is a complete breakdown in conjugacy and then the precipitation occurs on a given field line in one hemisphere only. In essence, the conjugate flight data show that auroras occur on both distorted and undistorted closed field lines and on field lines that are 'open' in the sense that they do not guide auroral particles from one hemisphere to the other.

As yet there does not seem to be a model for the geomagnetic field that explains even qualitatively the prevalent eastward and equatorward relative displacement of the Sounthern Hemisphere aurora. A goal of the Fall 1970 conjugate flight series is to attempt to determine if this displacement is seasonally dependent and if the observed greater average intensity of auroras in the Northern Hemisphere is also a function of season.

3. Quasi-Periodic Phenomena

One instance of conjugate *flickering aurora* was observed near dp lat 63° on March 26, 1968; this was near the time of the westward displacement of a Southern Hemisphere loop sturcture as described above. Flickering is the term applied by Beach *et al.* (1968) to a 10 ± 3 cps intensity variation observed in bright homogeneous auroras just prior to and during the breakup. Evidently, flickering is a phenomenon associated with all or most auroral breakups and consequently may be a significant consideration in trying to understand the nature of the breakup process.

A completely different type of quasi-periodic variation, pulsating aurora, is a commonly observed phenomenon at low auroral zone latitudes especially during the post-breakup phase of the display. The period of the pulsations ranges from 0.1 to

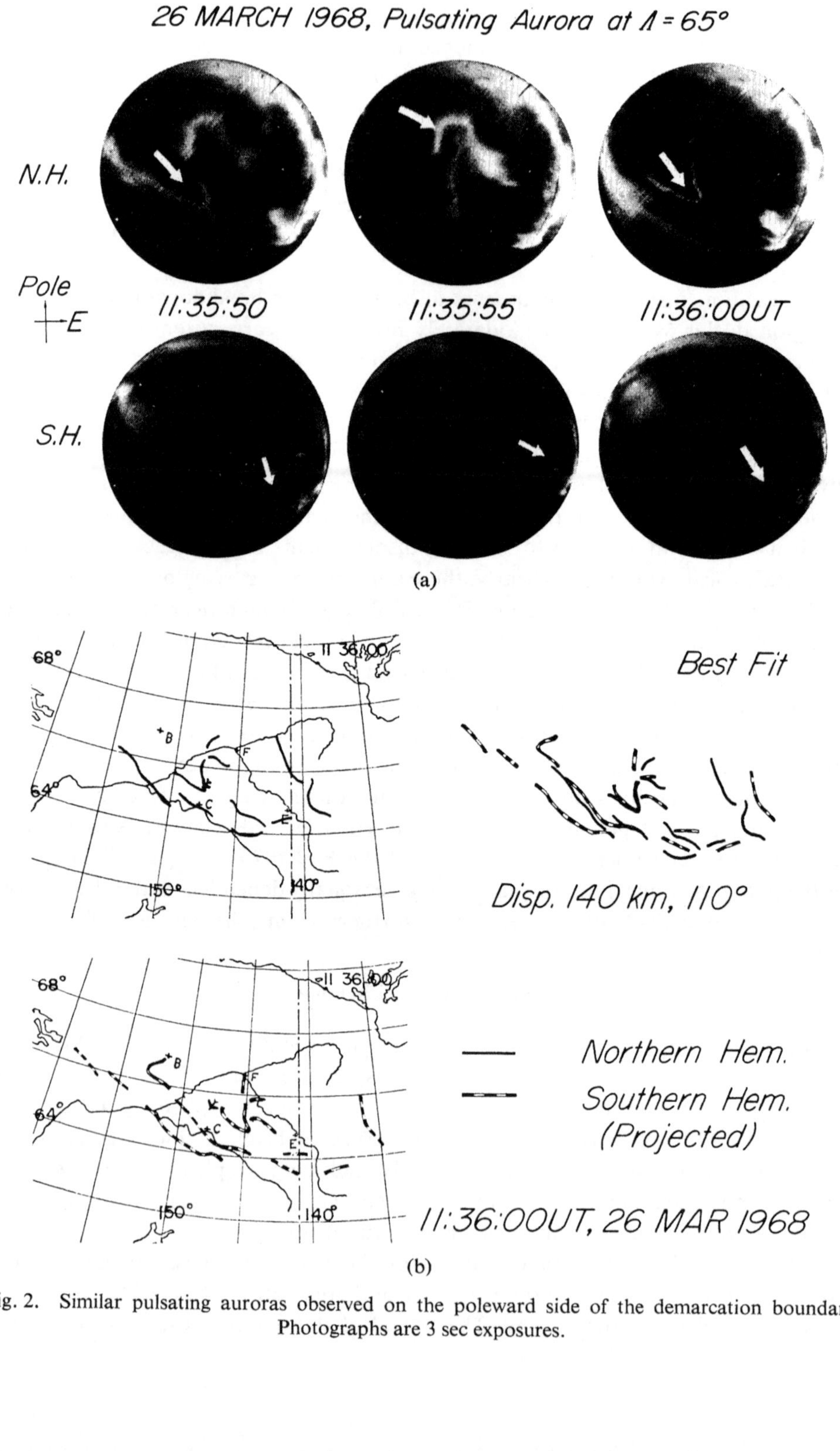

Fig. 2. Similar pulsating auroras observed on the poleward side of the demarcation boundary. Photographs are 3 sec exposures.

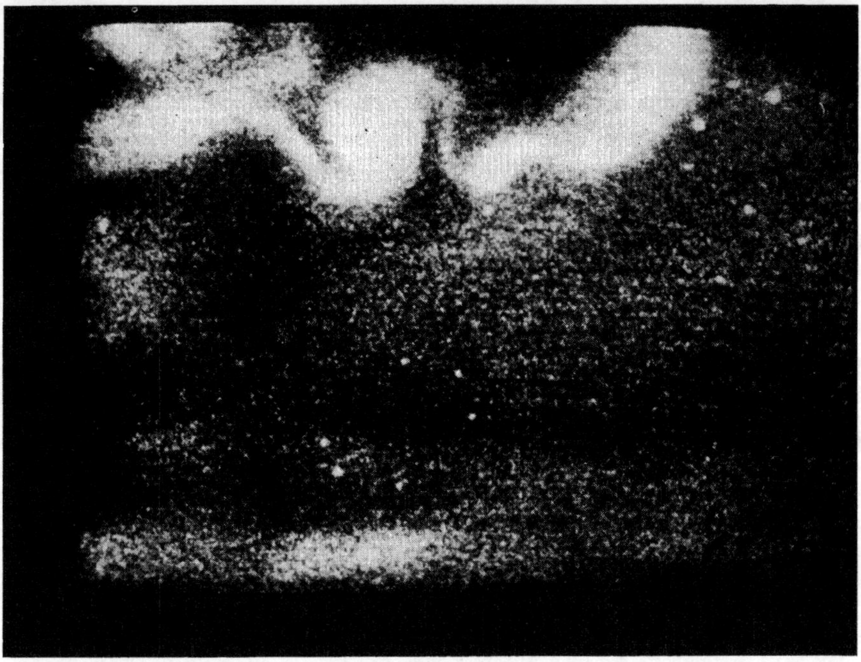

(A)

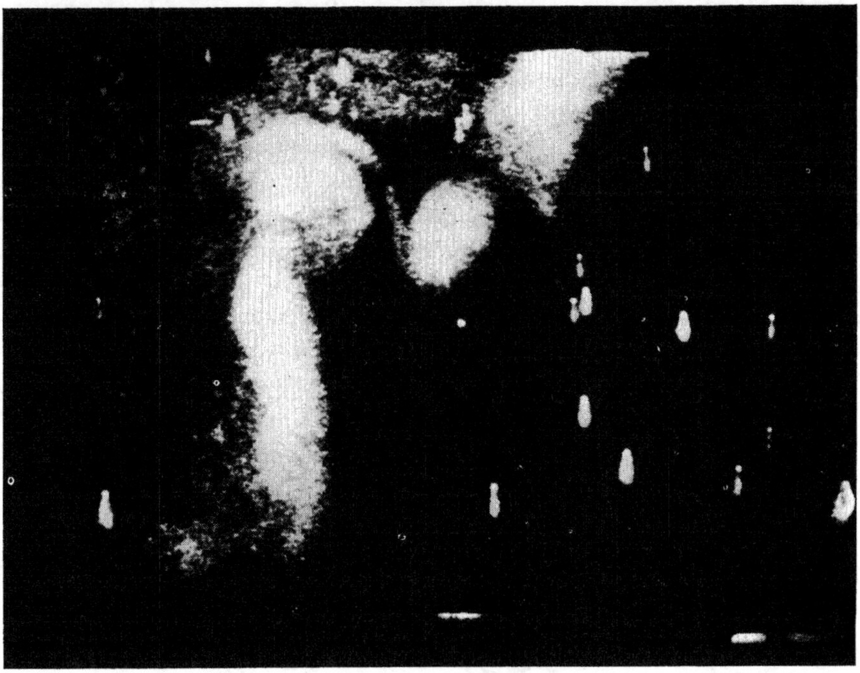

(B)

Fig. 3. Two examples of curl structures observed with TV systems. The exposure time is 1/30 sec and the field of view is approximately 15°. Elongated star images in Part B are due to camera motion (Hallinan and Davis, 1970).

over 10 sec, and various geometrical configurations (e.g., arcs, arc segments, patches) show pulsations. (For a review, see Omholt *et al.*, 1969). Pulsating auroras were observed during each of the 1968 conjugate flights. At dp lat 65° on March 26, 1968, relatively displaced pulsating forms were observed; eastward and equatorward displacements of the Southern Hemisphere were 140 to 220 km (Figure 2). Owing to the displacement these corresponding forms could be observed only with the all-sky cameras. Within the timing resolution of those instruments (several seconds) the observed pulsations pulsed on and off simultaneously. On the same night and on other nights the TV systems detected corresponding forms that pulsated but did not show relative displacement. Those observations were at dp lat 63 to 64°, near to or equatorward of the demarcation line described in the previous section. Approximately 50 pairs of individual pulses of 1 to 6 sec have been definitely identified in the data from both hemispheres. In all cases the pulsations are synchronous; some pulses with clearly defined onsets are obviously synchronous to within 0.1 sec.

4. Time Dependent Micromorphology

The conjugate aircraft data have proven useful in the study of the development and motions of both large and small scale auroral configurations. For example, with observations in both hemispheres it is possible to specify motions of geometrical configurations with reference to the direction of **B**; whereas, with observations in only one hemisphere one does not know if an observed configuration is mapable to the other hemisphere without changing the sign of a coordinate axis.

Examination of rayed auroral arcs viewed near the magnetic zenith (Hallinan and Davis, 1970) has shown that to a large extent the ray structure results from the formation of arrays of vortex like forms (Figure 3) to which the term *curls* has been attached. Observations in both hemispheres show that the curls are always wound in a counter clockwise sense as viewed antiparallel to **B**. An auroral arc or band exhibiting the curl structures often actually is in the form of thin closely spaced *arc elements*. When several curls result from one of these the whole assemblage has been referred to as a

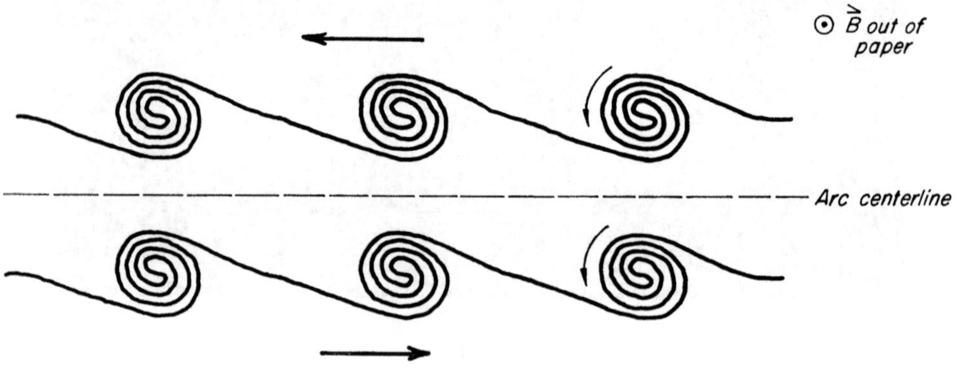

Fig. 4. Observed geometry and motions of curl structures (auroral rays seen end on) in an auroral arc.

curl system. Two or more arc elements in an arc can develop into curl systems, and when this happens the curl systems drift horizontally such that the drift direction reverses across the center of the arc. Relative to a point on the arc center the sense of drift is counter clockwise viewed antiparallel to **B** (Figure 4).

From observation of the development and shape of the curls, Hallinan and Davis (1970) concluded that the responsible mechanism is the Kelvin-Helmholtz shear flow instability. The required shear is thought to be produced by the $\mathbf{E} \times \mathbf{B}$ drift resulting from a negative space charge within a field aligned sheet. From the curl geometry and the observed growth rate (4.2 sec^{-1}) the required electric field must have a strength of 0.25 to 1 Vm^{-1}. A field of this magnitude is compatible also with the observed drift velocities of the curls. This **E** is highly transient and of limited spatial extent but may contribute to the excitation of certain auroral emissions.

Carlqvist and Boström (1970) independently reached similar conclusions regarding

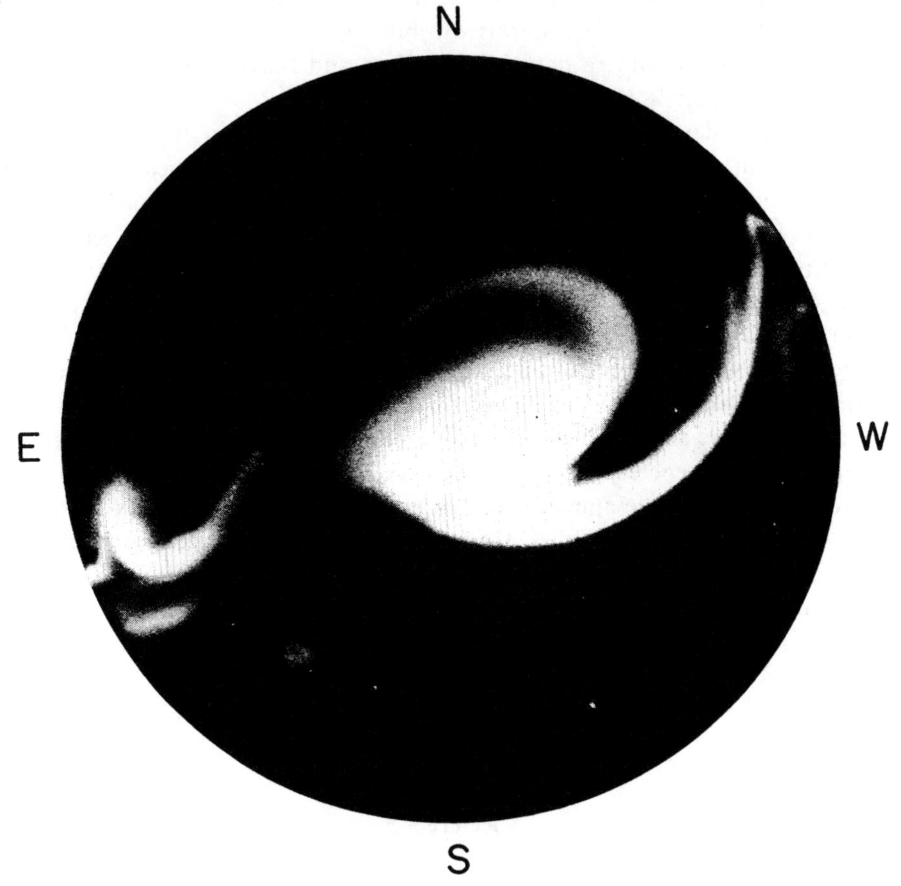

Fig. 5. All-sky photograph of an aurora containing small scale curl structures and larger spiral deformations. The exposure is 4 sec. Taken from Fort Yukon, Alaska at 0924 March 24, 1968 UT (Hallinan and Davis, 1970).

small scale irregularities in auroras on the basis of their own observations of relative drift rates and on theoretical implications of field aligned electric currents. They suggest that if field aligned currents exceed a critical value, negative space charge regions will develop within the current sheet at some point above the ionosphere.

The curls are a relatively small scale feature of the aurora; they are typically less than 1 km in cross section and the separation (λ) between adjacent curls is typically $\sim$5 km. A qualitatively similar but much larger feature (λ $\sim$100 km) has been noted by various observers (Webster, 1957; Gartlein, 1959; Akasofu and Kimball, 1964; Hallinan and Davis, 1970). These features, which we call *spirals*, differ from curls not only in scale size, but also in rotational sense. An all-sky photograph showing both curls and spirals appears as Figure 5. Whereas the curls are always counter clockwise viewed antiparallel to **B**, the spirals with occasional exception are clockwise. Approximately 2% of the spirals violate this rotation convention. Also the spirals occasionally are observed to unwind but the curls never exhibit this reversibility.

The statistical occurrence of the two forms differs also. While the curls seem to occur in any active aurora the spirals exhibit a striking latitudinal dependence. A comparison of all-sky camera data from College and Barrow indicated that the occurrence rate at dp lat 68.6° is over twice that at 64.6°.

The shape and development of the spirals suggests that, like the curls, the spirals must be the result of the Kelvin-Helmholtz shear instability. The shear, however, must be clockwise viewed antiparallel to **B**. An interesting question then arises: how is it possible to have both clockwise and counterclockwise shear in the same region? That both senses of shear are simultaneously possible can be seen by considering absolute and convective geometries along **B**. (Absolute geometry is independent of position along **B** whereas convective geometry changes with position; also, time dependent configurations, if absolute, are irreversible and if they are convective the time dependent configurations are reversible.) Based upon the characteristics of the curls and spirals we suggest that the curls correspond to absolute counter clockwise shear and the spirals correspond to convective clockwise shear. A consequence of this correspondence is that the apparent velocities of the curls are the same as the actual particle transverse drift velocities. However, the rotational motions of the spirals reflect the time rate of change of the convective shear pattern rather than the magnitude of the shear.

A mechanism for a convective Kelvin-Helmholtz instability based upon field aligned currents is discussed by Hallinan *et al.* (1970). Field-aligned currents do produce convective shear with the correct rotation sense, and the current density calculated with the Kelvin-Helmholtz dispersion relation (2.5×10^{-6} amp/m^2) is reasonable.

References

Akasofu, S.-I.: 1964, *Planetary Space Sci.* **12**, 273.

Akasofu, S.-I. and Kimball, D. S.: 1964, *J. Atmospheric Terrest. Phys.* **26**, 205.

Beach, R. F., Cresswell, G. R., Davis, T. N., Hallinan, T. J., and Sweet, L. R.: 1968, *Planetary Space Sci.* **16**, 1525.

Belon, A. E., Maggs, J.E., Davis, T. N., Mather, K. B., Glass, N. W., and Hughes, G. F.: 1969, *J. Geophys. Res.* **74**, 1.
Carlqvist, P. and Boström, R.: 1970, *Rept. No. 70-18*, Dept. Plasma Phys., Royal Inst. Technology, Stockholm, Sweden.
Gartlein, C. W.: 1959, *Ann. Geophys.* **15**, 31.
Hallinan, T. J. and Davis, T. N.: 1970, *Planetary Space Sci.*, **18**, 1735.
Hallinan, T. J., Davis, T. N., and Webster, H. F.: 1970, in preparation.
Hendricks, S. J. and Cain, J. C.: 1966, *J. Geophys. Res.* **71**, 346.
International Auroral Atlas: 1963, Edinburgh at the University Press, Edinburgh, Scotland.
Omholt, A., Kvifte, G. J., and Pettersen, H.: 1969, in *Atmospheric Emissions* (ed. by B. M. McCormac and A. Omholt), Van Nostrand Reinhold Company, New York, p. 131.
Webster, H. F.: 1957, *J. Appl. Phys.* **28**, 1397.

INSTITUTE DISCUSSION

Height information on curls is lacking. They may be short and stubby, but the observations described by N. Davis were looking almost directly up the field lines. In the future Davis plans to have 2 TV's linked by computer with an appropriate base line in the meridian plane. The instabilities leading to curls may be at about 1000 km altitude but for spirals they are out on the field lines, perhaps at about 6 R_E.

EQUATORIAL MODULATION IN A PULSATING AURORA

D. A. BRYANT, G. M. COURTIER, and G. BENNETT
S.R.C., Radio and Space Research Station, Ditton Park
Slough, SL3 9JX, Bucks., U.K.

1. Introduction

From the results of a rocket flight into a post-breakup aurora at Andøya, Norway on March 3, 1967, we concluded (Bryant *et al.*, 1969) that the mechanism causing a series of pulsations in electron intensity acted at or near the geomagnetic equator on the $L = 6$ magnetic shell. This conclusion was subject to an assumption that the pulsations observed at the rocket were temporal and not due to the motion of the rocket through a spatial structure in the electron precipitation. A similar result has now been obtained on a rocket flight into a pulsating aurora at Kiruna, Sweden, where $L = 5.4$, on October 17, 1969. On this occasion optical measurements made from the ground helped to establish that the pulsations in electron intensity were temporal, and thus removed the uncertainty associated with the earlier measurements. This report is based on a preliminary analysis of the recent flight.

2. Electron Detectors

The rocket, ESRO Skylark S29/2, carried channel multipliers and electrostatic analyzers to record electron intensities at 4 keV, 6 keV, and 10 keV continuously, and between 0.2 and 14 keV in successive steps of 2 keV in a 5 sec cycle. It also carried two geiger counters with 50% transmission thresholds of 22, and 40 keV for electrons. The detectors were all inclined at 45° to the rocket axis with the same azimuth and scanned pitch angles from 20 to 70° during the flight. There was unfortunately a failure in the 10 keV detector so we have selected the measurements taken at 4 keV, 6 keV, and >22 keV for the preliminary analysis.

3. Flight Conditions

The rocket was launched at 2316 LT (0055 GLT) on October 17, 1969 at the beginning of the recovery phase of a 400 γ magnetic bay while a broad pulsating band of light lay under the flight path. The emission at 3914 Å was recorded at the launch site and is shown in Figure 1. The photometer, whose total opening angle was 5°, was facing northwards in the plane of the rocket trajectory at an elevation of 66°. A scale in Figure 1 shows the distance between the magnetic field line at the centre of the region viewed by the photometer and the magnetic field line passing through the rocket at various times, assuming the light to have been emitted at an altitude of 120 km. This scale shows approximately when the photometer and the electron detectors were sampling the same region and illustrates the nonlinear progress of the

B. M. McCormac (ed.), The Radiating Atmosphere, 170–175. *All Rights Reserved.*

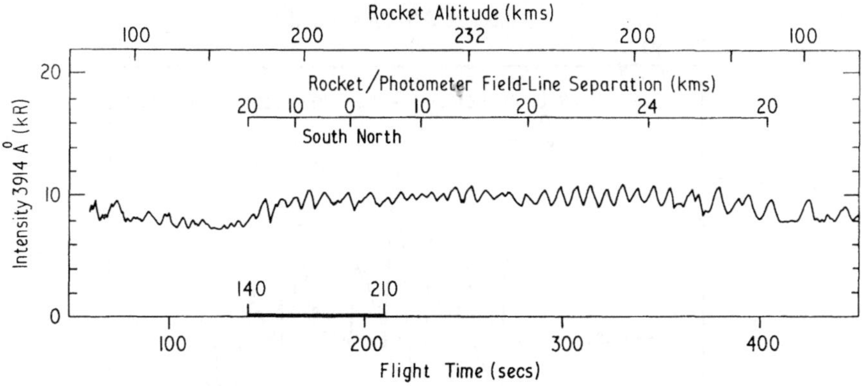

Fig. 1. The optical intensity at 3914 Å recorded by a photometer at the ground during the flight of the ESRO rocket S29/2. A scale shows the distance between the magnetic field line at the center of the region viewed by the photometer and the magnetic field line passing through the rocket at various times. It is indicated whether the rocket was south of north of the region viewed by the photometer. The pulsations in electron and optical intensities from 140 to 210 sec are discussed in the text.

rocket across magnetic field lines. It also indicates whether the rocket was south or north of the region viewed by the photometer.

For the greater part of the flight the mean intensity of the emission at 3914 Å was 10 kR. There are pulsations superimposed having a peak-to-peak amplitude of typically about 20% and a period of 8 sec. In this report we concentrate on the measurements taken during the pulsations from 140 to 210 sec when the photometer and the electron detectors were sampling approximately the same magnetic field lines.

4. Electron and Optical Pulsations

The electron and optical intensities measured from 140 to 210 sec are shown in Figure 2. The same logarithmic scale is used throughout but the normalization is arbitrary. The mean electron intensities were approximately 7×10^7 cm^{-2} sec^{-1} sr^{-1} keV^{-1} at 4 keV and 6 keV, and 8×10^6 cm^{-2} sec^{-1} sr^{-1} greater than 22 keV. The 4 keV and 6 keV electrons were almost isotropic over the range of pitch angles sampled (20 to 70°) but the higher energy electrons were anisotropic and showed a modulation at the spin period of the rocket (0.4 sec). The spin modulation has been removed for present purposes by averaging the counting rates of the >22 keV geiger counter over successive half spin cycles of 0.2 sec.

The electron intensities show pulsations that are remarkably similar to the optical pulsations, even to some of the fine details. At 4 keV and 6 keV the peak-to-peak amplitudes are typically about 20%, as are those of the optical pulsations; but for the >22 keV electrons the amplitude is greater thus producing a hardening of the electron energy spectrum at the pulsation maximum. The similarity of the measurements made from the rocket and the ground shows that the observed changes of electron intensity

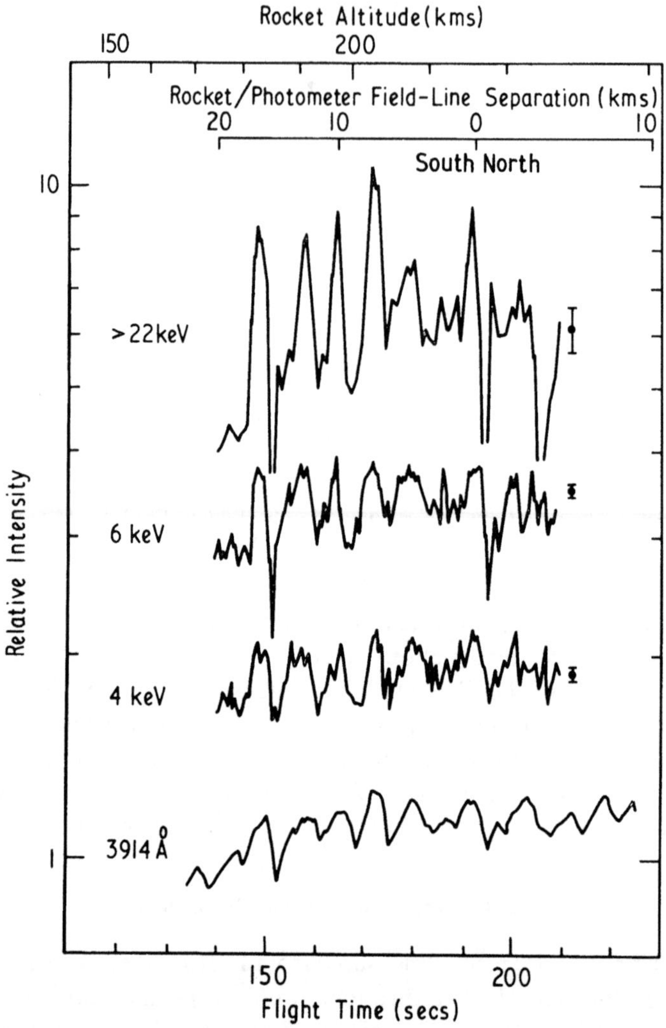

Fig. 2. Electron intensities recorded at the rocket and optical intensities recorded at the ground from 140 to 210 sec. The same logarithmic scale is used throughout but the normalization is arbitrary. The close similarity of the electron and optical pulsations shows that the electron pulsations were temporal and not spatial in origin.

were temporal rather than spatial in origin. Therefore, if we assume that the modulation was simultaneous at all energies at its source, we can find the distance of the source of the modulation from the dispersion in time between the pulsations at the different energies. To measure the time dispersion we plot in Figure 3 the correlation coefficients between the 6 keV electrons and the 4 and >22 keV electrons, also, the 6 keV auto-correlation, for various time shifts introduced between the measurements. The time shifts range from 2 sec in the sense to compensate for the 6 keV pulsations occurring in advance of those of the other component, to 2 sec in the opposite sense. The increment

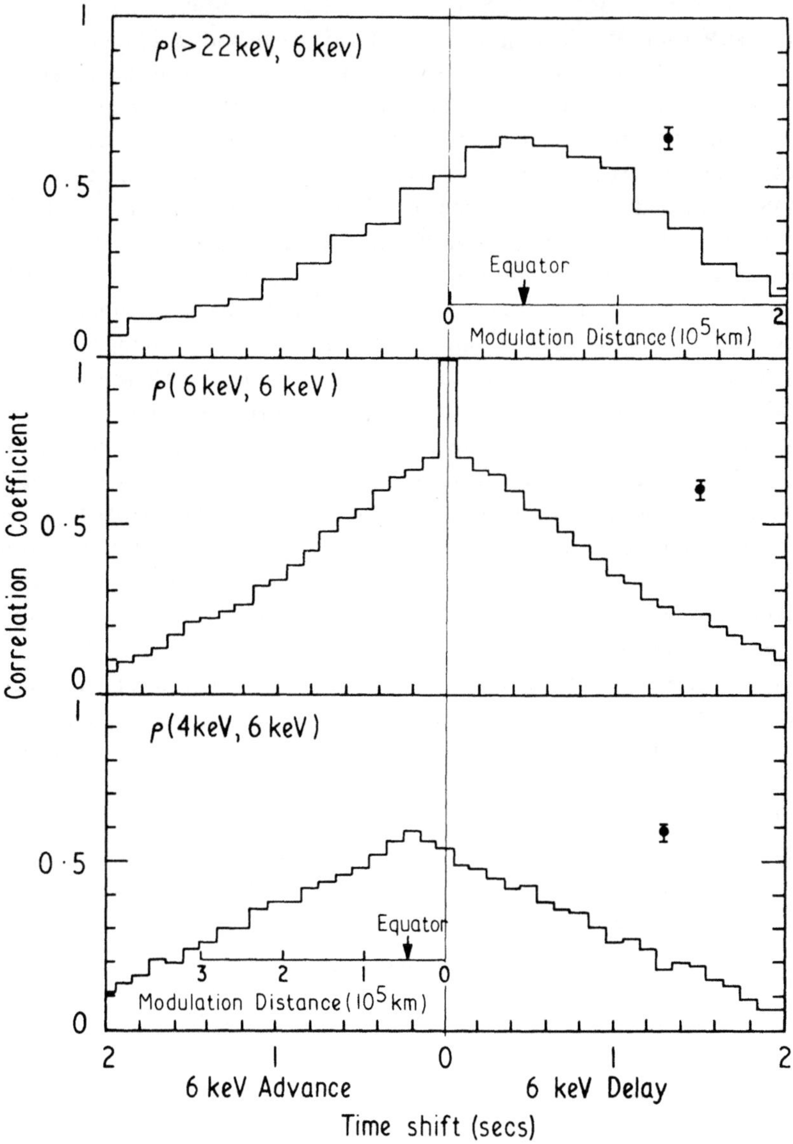

Fig. 3. Correlation coefficients ϱ between the 6 keV electrons and the 4 keV and > 22 keV electrons, and the 6 keV autocorrelation for various time shifts. The scales for the distance of the source of the modulation are drawn under the assumption that the observed dispersion arises from simultaneous modulation at a remote source.

of time shift is 0.1 sec except where the spin modulation in the >22 keV electrons limits the resolution to 0.2 sec. Any spin modulation in the 4 keV and 6 keV measurements would produce peaks in ϱ(4 keV, 6 keV) and the 6 keV autocorrelation at time shifts of 0, ± 0.4, ± 0.8 sec, etc., but there is no evidence for structure of this kind.

The correlation coefficients indicate that the pulsations occurred 0.2 ± 0.05 secs earlier at 6 keV than at 4 keV, and 0.4 ± 0.1 secs earlier at >22 keV than at 6 keV. Both of these results are in the right sense to be explained in terms of remote source where the modulation was simultaneous. The distance of the source of the modulation, measured along the electron trajectory, for various time delays is shown by the distance scales in Figure 3 which are derived simply from the electron speeds and the delay times. The speed used here for the >22 keV electrons is that of a 22 keV electron which we estimate is within 10% of the effective speed for this component. Both of the observed delays are consistent with the source of the modulation being $(4 \pm 1) \times 10^4$ km away along the electron trajectory. The distance of the geomagnetic equator, computed under a dipole approximation with allowance for changes of pitch angle along the trajectory, is 4.5×10^4 km as indicated in Figure 3. Therefore the source of the modulation is found to be within about 10^4 km of the geomagnetic equator on the magnetic shell $L = 5.4$. This result agrees with the earlier one (Bryant *et al.*, 1969), obtained under similar auroral conditions but without the knowledge that the observed intensity changes were temporal. As pointed out in the earlier paper there is no discrepancy between these results and the results obtained by Evans (1967), Mozer (1968), and Lampton (1967), which indicated sources of modulation within or near the ionosphere, since the latter were obtained under auroral conditions significantly different to the ones we have investigated.

An estimate of the size of the region over which the modulation process acts can be obtained with further reference to Figure 2. The pulsations at the ground and at the rocket were coherent to within about 1 sec for the whole period 140–210 sec while the rocket moved about 25 km across magnetic field lines (i.e. 20 km south of the region viewed by the photometers, to 5 km north). This distance traces along dipole field lines to become a radial distance of about 500 km at the equator so, by implication, the modulation process was also coherent to the same degree over this larger distance near the equator.

5. Conclusion

From the time dispersion between pulsations of 4 keV, 6 keV, and >22 keV electrons we conclude that the process causing the pulsations was located $(4 \pm 1) \times 10^4$ km away along the electron trajectory, i.e. within 10^4 km of the geomagnetic equator on the $L = 5.4$ magnetic shell. The modulation process is coherent to within 1 sec over a radial distance of at least 500 km.

Acknowledgments

We are greatly indebted to Dr. J. M. Vreux of the University of Liège for the loan of the photometer which played a vital part in this analysis, and to Drs. D. Köhn and D. E. Page of ESTEC for making their geiger counter data available before publication. The work described in this paper was carried out at the Radio and Space Research Station of the Science Research Council and is published with the permission of the Director.

References

Bryant, D. A., Courtier, G. M., and Johnston, A. D.: 1969, *J. Atmospheric Terrest. Phys.* **31**, 579.
Evans, D. S.: 1967, *J. Geophys. Res.* **72**, 4281.
Lampton, M.: 1967, *J. Geophys. Res.* **72**, 5817.
Mozer, F. S.: 1968, *J. Geophys. Res.* **73**, 999.

INFRARED SPECTRUM OF AURORA

A. VALLANCE JONES and R. L. GATTINGER

Astrophysics Branch, National Research Council
Ottawa, Canada

Abstract. An account is given of the history and methods of attempts to study the spectrum of the aurora in the photographic and near IR regions. Observed spectra are compared with synthetic spectra of which the relative intensities are based on the best overall estimates of the total intensities of the principal band systems.

1. Introduction

The observed IR spectrum of aurora is, for the most part, a continuation of the spectrum of the visible region. The region from 7000 to 9000 Å was first investigated by Meinel (1951) by means of a grating spectrograph with photographic plates. With the help of the techniques described below spectra have now been obtained as far as 1.7 μ. The spectrum between 1.1 and 2.0 μ has not yet been completely explored because of the insensitivity of the detectors available and the water vapor absorption bands centered at 1.13 μ, 1.38 μ and 1.88 μ. Beyond 2.0 μ thermal emission from the lower atmosphere begins to become important and observations must be conducted from a high altitude. Very little is known of auroral emissions in this region.

2. Experimental Methods

Meinel (1948) and Petrie (1950) obtained some of the first photographic IR spectra of aurora. A definitive spectrum was published by Meinel in 1951. A combination of a 5 in. aperture $f/0.8$ Schmidt camera, blazed 4 in. × 5 in. diffraction grating and a sensitive IR emulsion made it possible to obtain well exposed plates with spectral slit widths of 5 to 10 Å. Fan and Schulte (1954) and Omholt (1957) subsequently obtained spectra in as little as 60 min with an $f/0.65$ solid Schmidt spectrograph. Beyond 9000 Å plate sensitivities are lower so that even the strongest auroral displays show auroral lines very weakly.

An alternative to direct photography is to use an electron image converter tube ahead of the plate in a grating spectrograph. This was first done by Bagariaçki and Fedorova (1955) who obtained spectra with a resolution of 10 Å beyond 10500 Å. Earlier Bagariaçki and Mordukhovich (1952) had obtaind spectra to 11000 Å with an image converter and a prism spectrograph. Later Fedorova (1961, 1963) and Mironov *et al.* (1959) obtained even higher resolution spectra using a $3\frac{1}{2}$ in. $f/1.5$ spectrograph with an image converter. The image converter in this work (Krassovsky, 1949) had a S-1 (CsOAg) photocathode cooled by dry ice.

A different approach is to use a grating monochromator capable of scanning in wavelength and to detect the radiation by means of a photomultiplier. Very soon after CsOAg photomultipliers became available, spectra of aurora to 10600 Å were

B. M. McCormac (ed.), The Radiating Atmosphere, 176–184.

obtained with a 20 Å resolution by Lytle and Hunten (Hunten, 1958) by this method. Dufay (1959) and recently Harrison (1969) have obtained auroral spectra with spectral slit widths of 15 Å and 27 Å. In the latter study pulse counting and digital signal averaging were used to enhance the signal-to-noise ratio and to subtract the background OH bands. Vaisberg (1963) has described a system in which an CsOAg image converter tube is used between the exit slit of the monochromator and a photomultiplier having a visible region photocathode. This system could scan the 9000 to 12500 Å region with a spectral slit width of 20 Å in 30 to 50 min and had a minimum detectable brightness of about 100 R. Such a system does not offer any advantage over an ideal photomultiplier (or possibly over currently available tubes) but it would appear that earlier IR photomultipliers suffered from large nonthermal dark currents which could not be reduced sufficiently by cooling.

Beyond a limit which appears to between 1.15 and 1.25 μ, photoconductive detectors must be used. As at shorter wavelengths such detectors have been used almost exclusively with grating monochromators. As was first demonstrated by Connes and Gush (1960) a very considerable gain can be obtained by using a Michelson interferometer for nightglow spectroscopy beyond 1 μ. Similar advantages should be obtainable in auroral studies. Unfortunately, as is noted in Section 3, auroral emissions beyond 1.2 μ at ground level are weak compared to the nightglow background except for very intense aurora. However, intense aurora generally undergoes rapid motions and intensity variations. At least two attempts (Harang, 1963; Fjarlie, 1965) have been made unsuccessfully to obtain auroral spectra using the Michelson technique. No doubt these attempts failed because of the problems of low auroral intensity, auroral brightness variations and night airglow background; however, it should be possible to obtain auroral spectra.

The grating spectrometer with photoconductive detector has been shown by Gush and Vallance Jones (1955), Harrison and Vallance Jones (1957), and Shemansky (1960) to be capable of obtaining auroral spectra out to nearly 1.7 μ with resolutions of 50 to 200 Å. However, the necessity of waiting for many months for suitable auroral and meteorological conditions has made progress slow.

3. Observed Spectra

Very satisfactory spectra of the photographic region to 8800 Å are widely available (i.e., Chamberlain, 1961). The spectrum beyond this limit (that of the relatively sensitive N-plates) is less well known. A selection of the available results has been assembled in Figure 1. The 8500 to 9600 Å spectrum is taken from the work of Fedorova (1963). This spectrum was obtained by the image converter method with a 2 hr exposure to IBC II-III aurora on December 13, 1961. Similar spectra to 9150 Å have been obtained recently by Brown and Belon (1969) with a spectrometer with S-1 photomultyplier and memory system. In the region to 10200 Å, no strong auroral emissions seem to be present in the spectra published by Hunten (1958), Dufay (1959) or Fedorova (1961). There are relatively strong OH bands in this region however which could

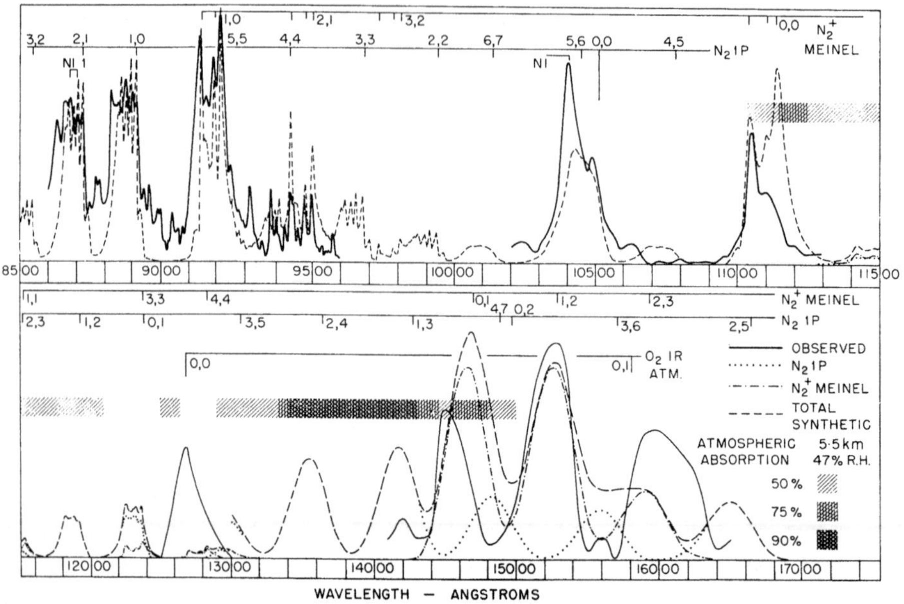

Fig. 1. Composite diagram of observed and synthetic spectra 8500 to 17000 Å as described in text (Sections 3 and 4). The synthetic spectra were calculated for a ground state temperature of 300 K assuming to change in rotational quantum number in excitation.

mask weaker auroral emissions. Brown and Belon (1969) have found enhancements of the OH bands in very intense low level aurora. Stair (1971) has reported enhancements associated with the auroral oval. In general, OH emissions do not appear in close association with aurora (Harrison, 1970) with the possible exception of very intense low level aurora as observed by Brown and Belon (1969).

Between 10200 Å and 11300 Å, the spectrum published by Harrison (1969) has been incorporated into Figure 1. This was obtained using a S-1 photomultiplier digital spectrometer system from IBC II-III aurora. A substantial OH background contribution (~20% of the peak auroral band brightness) was substracted from the observed spectrum to yield the auroral component of the spectrum. Fedorova (1963) has also published excellent spectra in this region but has not subtracted the OH contribution. Beyond 1.13 μ there is a region of good atmospheric transmission between 1.21 μ and 1.30 μ which has not been well explored. The 1.27 μ (0,0) O_2 IRatmospheric band falls in this region. Its spectrum has been recorded with an aircraft borne spectrometer by Noxon (1970); however, this band is strongly self absorbed by at mospheric O so that only about 6.5% is transmitted to the ground for a zenith path (Lowe, 1969; Evans *et al.*, 1970). The 1.27 μ band as given by Noxon (1970) has been incorporated into Figure 1; the asymmetry is due to the long time constant of the spectrometer.

Finally, in the transmission window between 1.44 μ and 1.80 μ, the observations of Harrison and Vallance Jones (1959) seem to be the only ones to have been published, except for the thesis reports of Reid (1958) and Shemansky (1960). These data were re-examined by Vallance Jones (1964) to give the averaged spectrum (with OH night-glow substracted) which is incorporated into Figure 1.

At longer wavelengths the spectral information is fragmentary. Noxon and Vallance Jones (1960) have reported the transitory appearance of emission between 2.1 μ and 2.2 μ on a single occasion during very intense aurora while Murcray (1957) noted 20% intensification of the 9.6 μ O_3 band during an auroral display. Stair (1971) has reported a strong signal at this wavelength above 90 km during a rocket flight. The surprisingly high intensity reported, (about 50 MR) makes it important to establish whether or not the signal originates in the undisturbed atmosphere above the rocket.

4. Interpretation of Spectrum

Most features of the IR spectrum belong to electronic band systems which either extend into the visible region or are closely related to the bands of the visible region. Some important atomic transitions also fall in the IR region. The wavelengths of observed and predicted features of the IR spectrum to 2.5 μ are listed in Table I, together with observed and predicted intensities.

TABLE I

Wavelength	Identification	Intensity IBC III (kR)
7037/81	N_2^+ M (4,1)	7
7059	N_2 1P (8,6)	(0.4)
7165	N_2 1P (7,5)	4
7240/82	N_2^+ M (5,2)	4
7274	N_2 1P (6,4)	15
7319	O II (2F)	
7330	O II (2F)	
7350	N_2 1P (11,10)	
7387	N_2 1P (5,3)	39
7479	N_2 1P (10,9)	(0.8)
7505	N_2 1P (4,2)	71
7613	N_2 1P (9,8)	1.7
7627	N_2 1P (3,1)	85
7684	O_2 (atm) (1,1) R branch	
7752	N_2 1P (8,7)	4
7754	N_2 1P (2,0)	54
7774	O I, 1	
7826/75	N_2^+ M (2,0)	34
7897	N_2 1P (7,6)	5
7987	O I, 19	
7995	O I, 19	
8048	N_2 1P (6,5)	5.5
8054/105	N_2^+ M (3,1)	27
8185/88	N I, 2 ($1\frac{1}{2}$, $2\frac{1}{2}$; $\frac{1}{2}$, $1\frac{1}{2}$)	
8205	N_2 1P (5,4)	1.7

Table I (Continued)

Wavelength	Identification				Intensity IBC III (kR)
8216/23	N I,	2	(2½, 2½)		
8242	N I,	2			
8293/348	N_2^+	M	(4,2)		10
8446	O I,	4			
8542	N_2	1P	(3,2)		27
8643	O_2	(atm)	(0,1)		
8629	N I,	8			
8680/4	N I,	1			
8703	N I,	1			
8712	N I,	1			
8719	N I,	1			
8723	N_2	1P	(2,1)		81
8912	N_2	1P	(1,0)		94
9145/212	N_2^+	M	(1,0)		95
9204	N_2	1P	(5,5)		12.5
9436	N_2	1P	(4,4)		28
9431/502	N_2^+	M	(2,1)		45
9680	N_2	1P	(3,3)		36
9733/809	N_2^+	M	(3,2)	N/O	8.3
9940	N_2	1P	(2,2)		17
10133	N_2	1P	(6,7)		7
10395	(N I)				
10404	(N I)				
10448	N_2	1P	(5,6)		7
10510	N_2	1P	(0,0)		41
10781	N_2	1P	(4,5)		8
10830	He I,	1			
11036/134	N_2^+	M	(0,0)		81
11520	N_2	1P	(2,3)	N/O	4
11416/521	N_2^+	M	(1,1)	N/O	6.3
11931	N_2	1P	(1,2)	N/O	23
12249/369	N_2^+	M	(3,3)	N/O	7.5
12374	N_2	1P	(0,1)	N/O	25
12867/13001	N_2^+	M	(4,4)	N/O	5
13061	N_2	1P	(3,5)	N/O	6
13635	N_2	1P	(2,4)	N/O	17
14269	N_2	1P	(1,3)	N/O	20
14523/692	N_2^+	M	(0,1)		38
14890	N_2	1P	(4,7)	N/O	4.2
14977	N_2	1P	(0,2)	N/O	8.3
15114/297	N_2^+	M	(1,2)		40
15717	N_2	1P	(3,6)		10
15748/947	N_2^+	M	(2,3)		15
16643	N_2	1P	(2,5)		12.5
17687	N_2	1P	(1,4)		8
18252	N_2	1P	(4,8)		4.6
19619	N_2	1P	(3,7)		6.2
21210	N_2	1P	(2,6)		5.0
21281	N_2^+	M	(0,2)		5.2
22411	N_2^+	M	(1,3)		12
23652	N_2^+	M	(2,4)		10

Beyond 1.0 μ the intensities may be compared with an earlier list (Vallance Jones, 1964). The latter are generally similar except in the case of 3,4 and 2,3 1P N_2 bands for which the more recent Franck-Condon factors (Benesch *et al.*, 1966) differ from those on which the earlier calculation was based and for the 2,2 Meinel N_2^+ band for which an error led to a value ten times too large appearing in the previous table. As discussed by Vallance Jones (1971) the best estimates for the 0,0 N_2 1P and N_2^+ Meinel bands should be reduced to about 40 and 80 kR, respectively, for IBC III aurora. Indeed, the spectrum of Harrison (1969 reproduced in Figure 1 would indicate an intensity of 23 kR for the 0.0 Meinel band; however, this value was uncorrected for water vapor absorption which reaches values well over 50% for continuum at 1.11 μ (Taylor and Yates, 1957). Water vapor absorption of a molecular band of complex structure is also likely to be severe. The fact that the observed contour of the 0,0 Meinel band is so different from the synthetic spectrum indicates that the band is

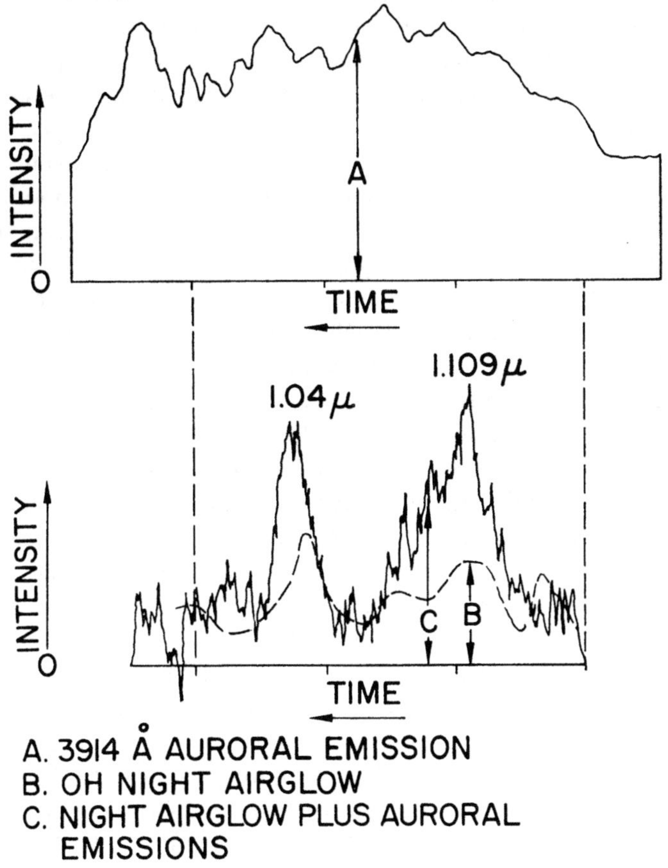

Fig. 2. Observed spectrum of aurora from 10000 to 11500 Å. 1 min scan at spectral slit width of 100 Å. The zenith distance was 60° and the aurora was a band of intensity IBC II to III. The upper tracing is the output of a 3914 Å photometer having the same field of view as the spectrometer.

indeed strongly absorbed. This conclusion is supported by a spectrum obtained by Gattinger (1961) by means of a scanning spectrometer provided with a 3914 Å intensity monitor. This spectrum, together with a nightglow background and the monitor trace, is reproduced in Figure 2. In this spectrum, obtained at a high elevation angle, the water vapor absorption is less severe; the ratio I (0,0 Meinel)/I(0,0 1P)$\simeq$2.0 in this observation.

The observed spectrum beyond 1.2 μ is of low resolution and the expected bands of N_2 and N_2^+ are considerably overlapped so that it is difficult to determine individual band intensities from the results. In Figure 1 the positions of the P_1 heads of the N_2 1P bands and the R_2 and Q_1 heads of the N_2^+ Meinel bands are indicated. Particularly in the case of the 1P bands, the head wavelength provides little idea of the intensity distribution and consequently synthetic spectra were calculated for the individual bands and plotted on the figure for both band systems. The method of calculation and the spectroscopic constants assumed are described by Gattinger (1971).

Finally, the synthetic spectra for the bands, normalized to the semi-empirical intensity values of Table I, were added to produce the dotted line plotted on Figure 1, between 1.0 and 1.70 μ. The fit between the observed average spectrum and the synthetic spectrum is fairly good for the stronger bands but there are some discrepancies especially beyond 1.55 μ.

In particular there is an unexplained region of emission between 1.61 μ and 1.63 μ where the synthetic spectra have a minimum. It would be tempting to ascribe this emission to an incomplete subtraction of the OH background, since there is a peak in the OH spectrum at about 1.625 μ at this resolution. On the other hand Harrison and Vallance Jones (1959) report definite peaks at 1.614 μ in two separate strong spectra while Reid (1958) and Shemansky (1960) obtained spectra with unidentified peaks at 1.61 μ, 1.62 μ, and 1.63 μ. These observations suggest that there may indeed be an auroral emission in this wavelength range. In the review by Vallance Jones (1964) this emission was ascribed to the 2,5 N_2 1P band on the basis of the wavelength of 1.6163 μ (Turner and Nicholls, (1954) for the band center. It appears that this value is in error since the band center is closer to 1.65 μ from the synthetic spectra and also in the laboratory spectra of Saum and Benesch (1970a). Likewise the band centers of the 1.4 and 3.6 bands seem closer to 1.74 μ and 1.56 μ than the values quoted by Turner and Nicholls (1954) and tabulated in the earlier review.

Until further spectra have been obtained it does not appear that it is valuable to attempt further to analyze the older observations. Laboratory spectra obtained by Saum and Benesch (1970a, b) provide some guidance as to what further features might be expected. They suggest that OI lines at 1.589 μ, 1.687 μ, 1.823 μ, 2.650 μ, 2.764 μ, 3.307 μ, 3.661 μ, and 4.559 μ should certainly be present since they share a common upper state with the observed visible region features 5330 Å (Mult. 12), 5436 Å (Mult. 11), 6046 Å (Mult. 22), 6157 Å (Mult. 10), 3947 Å (Mult. 3), 6455 Å (Mult. 9), 7254 Å (Mult. 20) and 4368 Å (Mult. 5), respectively. None of these are strong features so the IR emissions may be quite weak. Saum and Benesch (1970b) suggest that the 1.129 μ ($3^5P-4^5S^0$; $3^3P-3^3D^0$) transition and the 1.317 μ ($3^3P-4^3S^0$) multiplet which are observed to be strong in the laboratory spectrum may be the strongest IR auroral lines beyond 1 μ. One might expect these lines to be comparable to the 8466 Å and 7774 Å multiplets which have an IBC III intensity around 10 kR. Saum and Benesch (1970a) show that the $W^3\Delta_u - B^3\Pi_g$ and $B^3\Pi_g - W^3\Delta_u$ N_2 bands are prominent in the laboratory spectrum from 2.4 to 4 μ (the $\Delta v=2$ and 3 sequences of the former and the $\Delta v=-2$ sequence of the latter). There is no reason why these bands should not be present in aurora although observation may be difficult.

The IR atmospheric bands, of which the 0,0 band occurs at 1.27 μ, appear to occur at times in association with aurora (Megill *et al.*, 1970; Noxon, 1970). Noxon (1970) has reported the results of a series of observations in the auroral zone and these show that the band occurs with zenith intensities typically about 400 kR with the ratio I(1.27 μ)/I(3914 Å) attaining values typically of several hundred in weak aurora and on occasion as high as 10^3. The intensity did not rise proportionately in brighter aurora and Noxon (1970) suggested that there might indeed be a limiting value of the order of 1000 kR. Megill *et al.* (1970) observed peak intensities as high as 6 MR in a horizontally looking rocket borne photometer on a passage up through the height range of an aurora for which the corresponding 3914 Å horizontal intensity was about 90 kR. The zenith intensity, however, may have been no higher than 600 kR and possibly less depending on the geometry of the emitting layer. There are, moreover, other series of observations by Wood (1969) and Evans (1969) carried out by means of a sensitive filter photometer. The former study detected a 200 kR enhancement of the 1.27 μ band associated with 20 kR of 5577 Å but only on one occasion whereas the latter showed enhancements in the auroral oval region of 200 to 300 kR but not associated with aurora. Noxon (1970) has suggested that electric field heating of aurorally produced electrons may be the most attractive explanation of the $O_2(^1\Delta)$ emission.

5. Conclusions

There is still much experimental work to be done in the 1.1 to 2.0 μ region before the spectrum is fully defined. Beyond 2 μ little is known of the spectrum so there may be some surprises in store. However, experimental studies are very difficult because of enormous fluxes of thermal radiation from the lower atmosphere. Consequently, experimental difficulties akin to those encountered in the dayglow are added to those of the IR.

References

Bagariaçki, B. A. and Mordukhovich, M. I.: 1952, *Dokl. Akad. Nauk.* **82**, 45.
Bagariaçki, B. A. and Fedorova, N. I.: 1955, *Dokl. Akad. Nauk.* **103**, 1009.
Benesch, W., Vanderslice, J. T., Tilford, S. G., and Wilkinson, P. G.: 1966, *Astrophys. J.* **144**, 408.
Brown, N. B. and Belon, A. E.: 1969, *Trans. Am. Geophys. Union*, **50**, 258.
Chamberlain, J. W.: 1961, in *Physics of the Aurora and Airglow*, Academic Press, New York.
Connes, J. and Gush, H. P.: 1960, *J. Phys. Radium* **21**, 645.
Dufay, M.: 1959, *Ann. Geophys.* **15**, 134.
Evans, W. F. J., 1969, private communication.
Evans, W. F. J., Wood, H. C., and Llewellyn, E. J.: 1970, *Can. J. Phys.* **48**, 748.

Fan, C. Y. and Schulte, D. H.: 1954, *Astrophys. J.* **120**, 563.
Fedorova, N. I.: 1961, *Geomag. Aeron.* **1**, 702.
Fedorova, N. I.: 1963, *Aurorae and Airglow, Results of the International Geophysical Years, Researches, Pub. House Acad. Sci., U.S.S.R.* **10**, 80.
Fjarlie, E. J.: 1965, University of Saskatchewan, Ph.D. Thesis.
Gattinger, R. L.: 1961, University of Saskatchewan, Ph.D. Thesis.
Gattinger, R. L.: 1971, to be published.
Gush, H. P. and Vallance Jones, A.: 1955, *J. Atmospheric Terrest. Phys.* **7**, 285.
Harang, L.: 1963, private communication.
Harrison, A. W.: 1969, *Can. J. Phys.* **47**, 599.
Harrison, A. W.: 1970, *J. Geophys. Res.* **75**, 1330.
Harrison, A. W. and Vallance Jones, A.: 1957, *J. Atmospheric Terrest. Phys.* **11**, 192.
Harrison, A. W. and Vallance Jones, A.: 1959, *J. Atmospheric Terrest. Phys.* **13**, 291.
Hunten, D. M.: 1958, *Ann. Geophys.* **14**, 167.
Krassovsky, V. I.: 1949, *Dokl. Akad. Nauk.* **66**, 53.
Lowe, R. P.: 1969, *Phil Trans. Roy. Soc., London* **A264**, 163.
Megill, L. R., Despain, A. M. , Baker, D. J., and Baker, R. D.: 1970, *J. Geophys. Res.* **75**, 4775.
Meinel, A. B.: 1948, *Publ. Astron. Soc. Pacific* **60**, 373.
Meinel, A. B.: 1951, *Astrophys. J.* **114**, 583.
Mironov, A. V., Prokudina, V. S., and Shefov, N. N.: 1959, *Spectral, Electrophotometrical and Radar Researches of Aurorae and Airglow, IV Section of IGY Program, (Aurorae and Airglow)* **1**, 20.
Murcray, W. B.: 1957, *Nature* **180**, 139.
Noxon, J. F.: 1970, *J. Geophys. Res.* **75**, 1875.
Noxon, J. F. and Vallance Jones, A.: 1960, *J. Atmospheric Terrest. Phys.* **18**, 342.
Omholt, A.: 1957, *J. Atmospheric Terrest. Phys.* **10**, 320.
Petrie, W.: 1950, *J. Geophys. Res.* **55**, 143.
Reid, J. H.: 1958, University of Saskatchewan, M.A. Thesis.
Saum, K. A. and Benesch, W. M.: 1970a, *Appl. Opt.* **9**, 195.
Saum, K. A. and Benesch, W. M., 1970b *Appl. Opt.*, **9** 1419.
Shemansky, D. E.: 1960, University of Saskatchewan, M.Sc. Thesis.
Stair, A. T., Jr., Huppi, E. R., Sandford, B. P., Murphy, R. E., O'Neil, R. R., Hart, A. M., Huppi, R. J., and Pendleton, W. R.: 1971, this volume p. 185.
Taylor, J. H. and Yates, H. W.: 1957, *Opt. Soc. Am.* **47**, 223.
Turner, R. G. and Nicholls, R.: 1954, *Can. J. Phys.* **32**, 468.
Vaisberg, O. L.: 1963, *Aurorae and Airglow, Results of the International Geophysical Year Researches, Pub. House Acad. Sci., U.S.S.R.* **10**, 54.
Vallance Jones, A.: 1964, *Mem. Soc. Roy. Liège*, Series 5, **9**, 289.
Vallance Jones, A.: 1971, *Space Sci. Rev.*, in press.
Wood, H. C.: 1969, University of Saskatchewan, M.Sc. Thesis.

INSTITUTE DISCUSSION

W. Williams reported on some daytime high latitude (Fairbanks) balloon observations at 10.5 to 12.5 μ at an elevation angle of 45°. Fluctuations of up to a factor of 10 in 1 sec were observed. Such an effect has never been observed at low latitudes. The radiance was as high as 10 MR. J. Noxon commented that hot electrons can create much N_2 vibrational energy which can transfer efficiently to CO_2. The 10.4 μ CO_2 laser band and the 11.3 μ emission from HNO_3 are in that region.

The theoretical spectrum of the N_2^+ Meinel 1-2 band was shown by R. Gattinger, who stated that the branch at 1.51 μ could be used for measuring the Meinel intensity without interference from OH emissions.

INFRARED INVESTIGATIONS OF AURORA AND AIRGLOW*

A. T. STAIR, Jr., E. RAY HUPPI,[1] BRIAN P. SANDFORD,[1] RANDALL E. MURPHY,[2] ROBERT R. O'NEIL,[3] and ALAN M. HART[3]

Radiation Effects Branch
Optical Physics Laboratory
Air Force Cambridge Research Laboratories
Bedford, Mass., U.S.A.

and

RONALD J. HUPPI[1] and W. R. PENDLETON[3]

Electro-Dynamics Laboratories
Utah State University
Logan, Utah, U.S.A.

Abstract. This paper presents some recent results of the continuing IR auroral/airglow studies.

1. Auroral Region Enhancements of OH Emission

Infrared measurements of the OH emission of the atmosphere have been made using the AFCRL optical IR KC-135 jet aircraft. A number of flights through the auroral regions have shown definite enhancements of the OH emission, particularly in and near active aurora; but there is no spatial correlation with the visible auroral forms.

Using a sensitive IR PbS radiometer viewing vertically with a 9° full angle field of view, atmospheric OH measurements of various transitions of the first overtone, $\Delta v=2$, have been obtained from the aircraft at altitudes on the order of 12 km. The transmission of the two filters used in the radiometer is shown in Figure 1.

The narrow band filter (Figure 1a) was chosen to measure primarily the 5-3 OH band and to stay clear of known N_2 and O_2 band systems. The wide band filter (Figure 1b), often employed because it gives a larger integrated signal than the narrow band filter, covers the 2-0 to 5-3 OH bands in the 1.35 to 1.70 μ region. The region covered by the wide band filter also includes the possible emissions from the Meinel system of N_2^+ $(A^2\Pi_u \rightarrow X^2\Sigma_g^+)$, the first positive bands of N_2 $(B^3\Pi_g \rightarrow A^3\Sigma_u^+)$ and the $O_2(^1\Delta)$ 0–1 band; but none of the observed signal in the data presented in this paper is attributable to these sources. This has been confirmed by checking the correlation between the signals obtained with the broad band filter and the narrow band filter. In all cases the signal levels obtained with the narrow band filter correlated closely with the broad band levels. A further indication that the Meinel system and 1P bands do not contribute any significant emission in the 1.35 to 1.70 μ region is found in the absence of any detailed correlation between visible aurora, which is recorded by an

* The superscripts behind the various authors' names identify the particular section for which he is primarily responsible.

B. M. McCormac (ed.), The Radiating Atmosphere, 185–204.

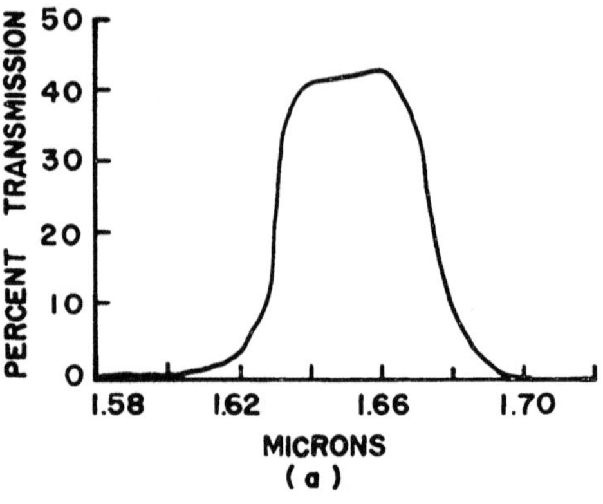

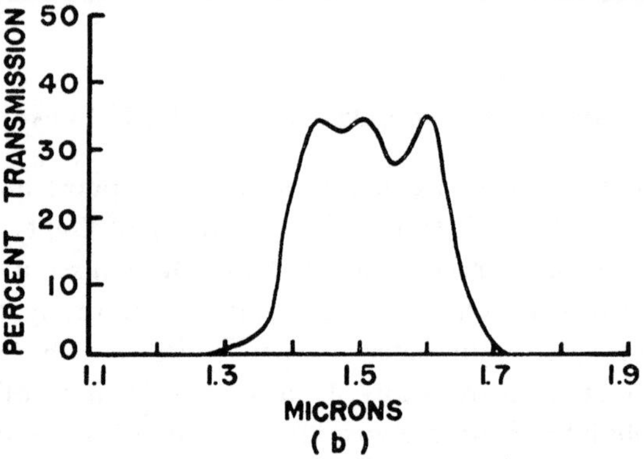

Fig. 1. Radiometer filter transmission (a) 1.655 μ used to measure primarily the (5-3) OH band, (b) 1.5 μ includes the (2-0), (3-1), (4-2), (5-3) bands of OH emission.

all-sky camera and a visible-region photometer, and the OH radiometer signals. Also, data taken in twilight conditions have been excluded from this report to avoid effects, such as the enhancement of the $O_2\,(^1\Delta_g)0\text{–}1$ emission and the normal variations of OH, which often occur under these conditions.

The visible region photometer (operated by Photometrics Inc.) also views vertically with a 5.5° full angle field of view. It is used for auroral correlation measurements of the 5577 Å $O(^1S-{}^3P)$ and the 4278 Å N_2^+ (0–1) emissions.

In the observations that follow, the 5577 Å and 4278 Å intensities may have absolute errors because continuum emissions are included; however, the intensity changes do give an accurate indication of auroral activity. The peak-to-peak noise of the PbS radiometer system is indicated in the figures for comparison with the signal variations.

The OH (1.65 μ), 5577 Å and 4278 Å emission observed on the night of November 2 to 3, 1969, is illustrated in Figures 2a and 2b. This flight proceeded from Boston

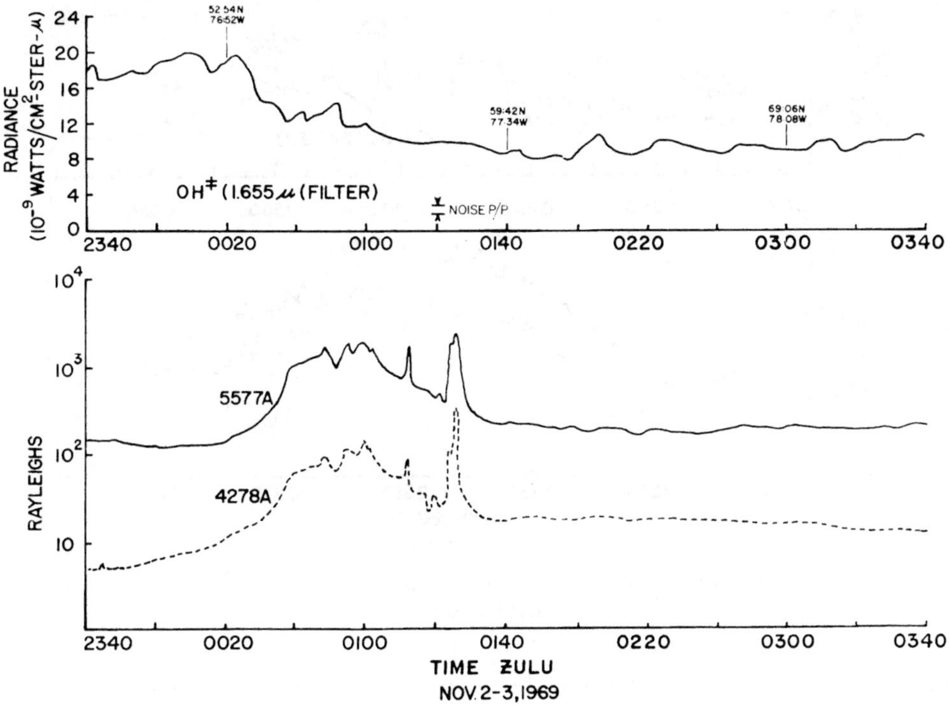

Fig. 2a.

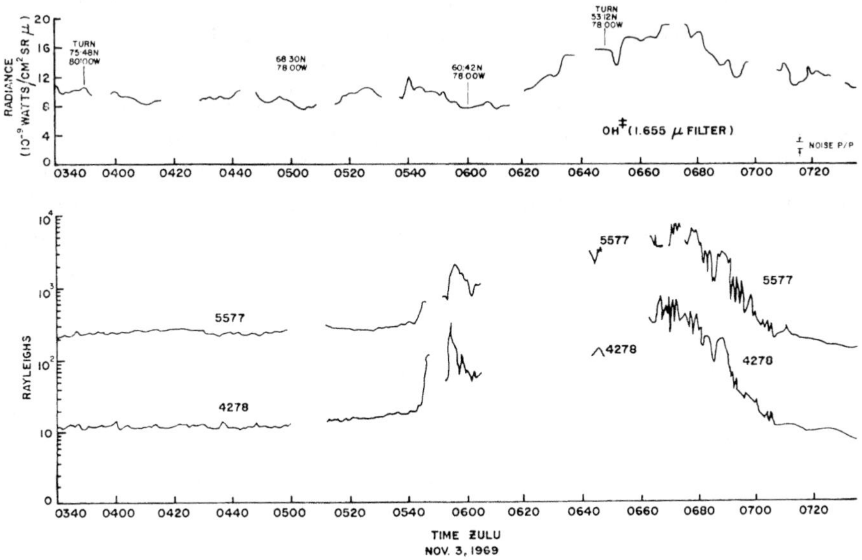

Fig. 2b.

Fig. 2a–b. Visible and near IR measurements. (a) Outbound leg, November 2 to 3, 1969, approaching auroral region. (b) Inbound leg, November 3, 1969, under a homogenous band of aurora.

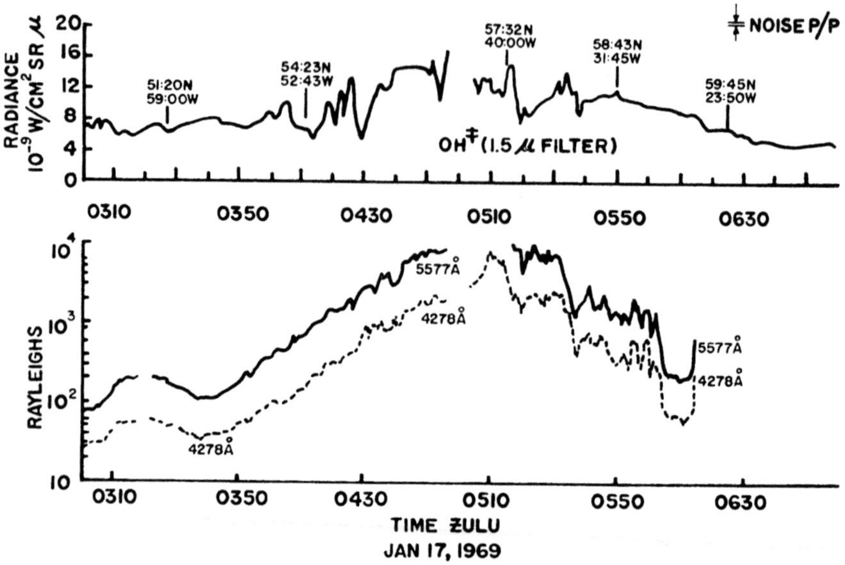

Fig. 3. Visible and near IR measurements taken January 17, 1969, during a flight from Pease AFB, New Hampshire, to Oslo, Norway.

through the auroral zone and into the polar cap at 76°N (geographic) (near Thule, Greenland) and returned to Boston on the same track. A large PCA event was in progress on this night but had decreased to 0.9 dB of 30 MHz absorption at Thule at 0400 UT. Hydroxyl enhancements, which are typical of those seen in the auroral region, were observed in and near the auroral oval, both on the outbound leg and the inbound leg, but enhancement was not observed in the polar cap region. An increase in emission in the 1.655 μ channel was observed in the outbound leg (Figure 2a) immediately south of the auroral oval. This enhancement occurred just prior to entering an area of visible aurora. During the inbound leg (Figure 2b) an OH enhancement occurred in the same region as the visible aurora and coincided quite closely with a broad region of pulsating patch aurora, but there was no detailed correlation between visible and OH fluctuations. The particles which produced the PCA event did not appear to alter the OH levels significantly from those typically observed in the polar cap, while a careful analysis of the 4278 Å emission shows about 15 R of emission over the polar cap which can be attributed to the PCA particles.

Measurements of OH using the 1.5 μ filter, 4278 Å and 5577 Å emissions made on the night of January 17, 1969, on a flight between Boston and Oslo are given in Figure 3. Here the OH intensity increased by a factor of 3 in the same general region of the rather stable homogeneous and patch aurora, but the large fluctuations in OH, like the ones observed between 0425 to 0430 hr, have no counterpart in the visible photometer's data or in the all-sky camera photos of this aurora.

In Figures 4a and 4b the OH (1.3 to 1.7 μ), 5577 Å and 4278 Å emissions are recorded on a flight on December 12, 1968, that proceeded from Boston into the auroral oval, thence east and west within the oval for 4 hr and then returned to Boston. Here the OH signal is generally enhanced in the auroral oval, but again there is no evidence of any detailed correlation between the OH and the visible aurora of the photometric intensities. The visible aurora on this occasion was generally very active and bright.

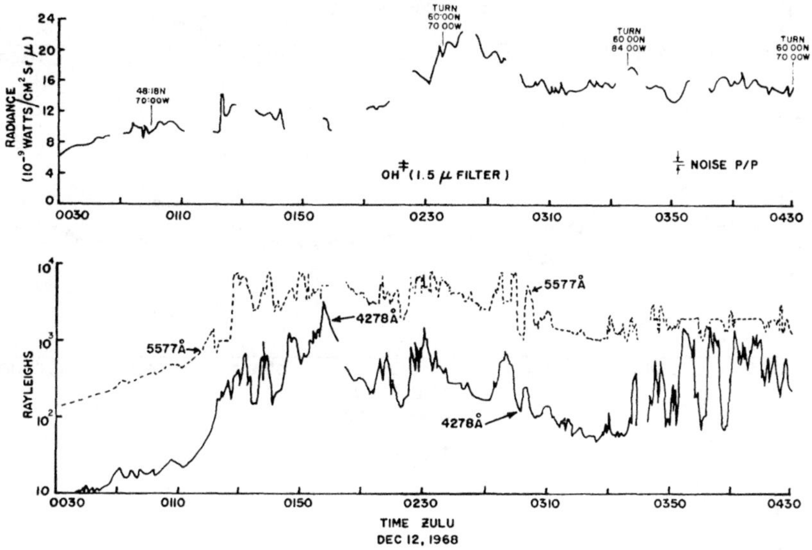

Fig. 4a.

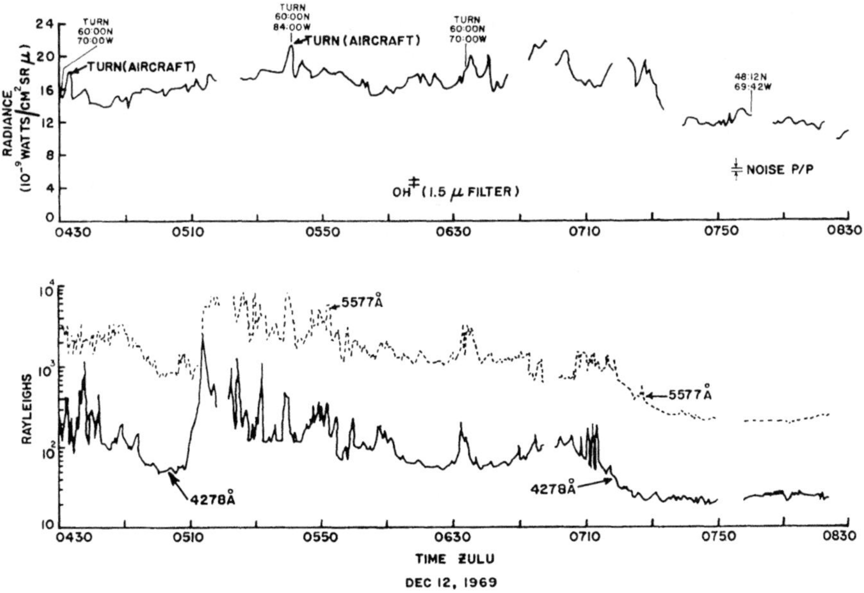

Fig. 4b.

Fig. 4a–b. Visible and near IR measurements taken December 12, 1968, from Boston into the auroral oval and return. (a) 0030 to 0430 hr (b) 0430 to 0830 hr.

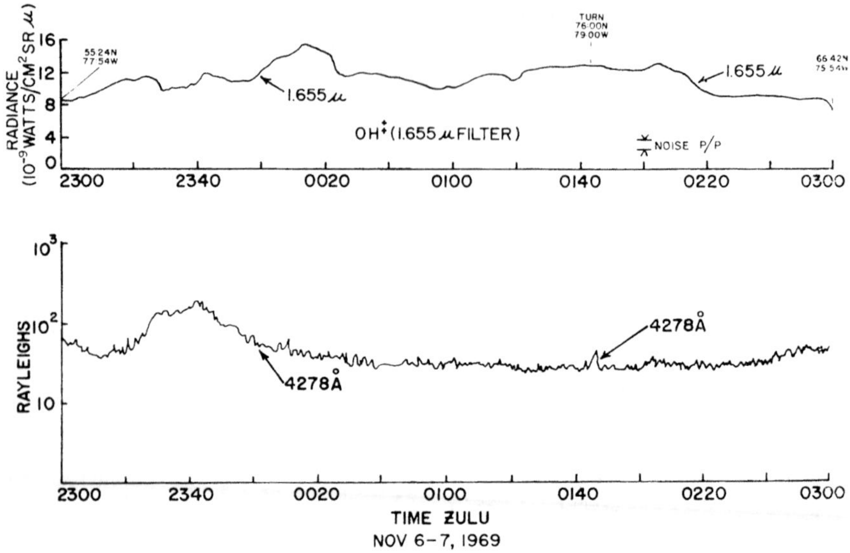

Fig. 5a.

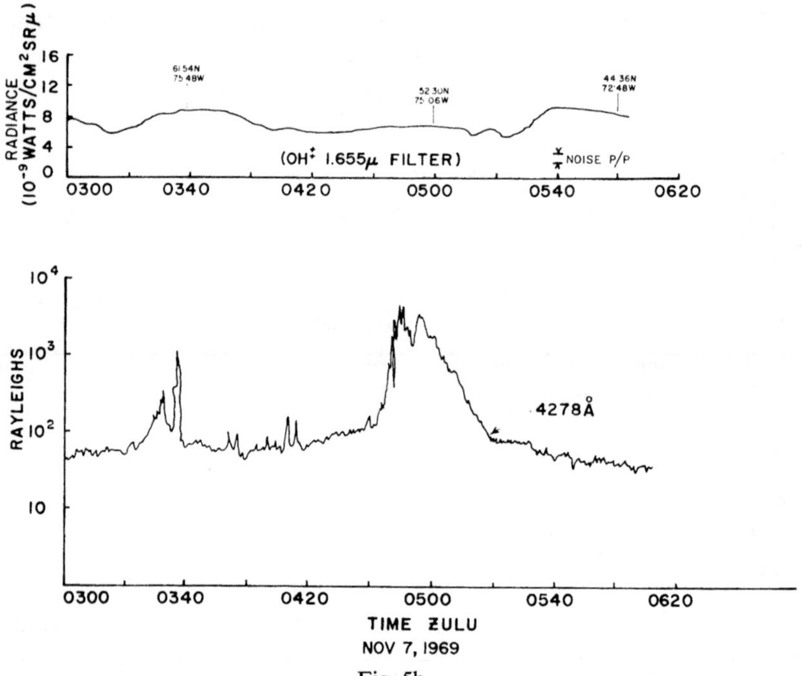

Fig. 5b.

Fig. 5a–b. Visible and near IR measurements taken November 6 to 7, 1969, during a flight north from Boston to 76:00 north and return. (a) 2300 to 0300 hr. (b) 0300 to 0620 hr.

A more complex situation occurred on November 6 to 7, 1969, (Figures 5a and 5b) during a flight with the same route as on November 2 to 3. The levels of OH (1.65 μ) are not as high as those observed on the occasions discussed this far, but there appears to be no relationship between OH enhancements and the regions of auroral occurrence and the level of OH is quite high in the polar cap up to 76 °N.

In contrast to these observations no comparable OH enhancements have been seen in many flights across the Atlantic, USA, and Pacific between latitudes of 15 to 45 °N. At these latitudes the OH signals are typically $(9 \pm 3) \times 10^{-9}$ Wcm^{-2} sr^{-1} μ^{-1} an hour or so after sunset and fall steadily to a value of $(6 \pm 2) \times 10^{-9}$ Wcm^{-2} sr^{-1} μ^{-1} just before sunrise. Although there is no evidence of any one-to-one association between the OH and the visible aurora, there does seem to be a strong tendency for the OH enhancement of the type reported here to occur within or close to the auroral oval. We cannot as yet produce any firm evidence to indicate the origin of these high levels of OH emission.

2. Laboratory Determination of the Dipole Moment Function of OH

Since the first atmospheric observations (Meinel, 1950) of the vibration-rotation bands of the OH radical, investigators have been performing spectral measurements of OH to obtain a better understanding of the physics and chemistry of the earth's atmosphere. From both laboratory and atmospheric measurements the mechanical and electrical properties of the OH molecule have been determined. While the mechanical parameters used to describe OH have been determined accurately, Ferguson and Parkinson (1962) have shown the electrical character, namely, the dipole moment function, is arbitrary without a knowledge of the intensity of the fundamental, $\Delta v = 1$ sequence. Only qualitative estimates of the intensity of the fundamental sequence have been available previously. Recent intensity measurements of the fundamental and first overtone sequences of OH are described briefly and a unique dipole moment function given.

Vibrationally exited OH was produced by the reaction of

$$H + O_3 \rightarrow OH^{\ddagger} + O_2 \quad v \leqslant 9 \tag{1}$$

which is exothermic by 3.3 eV and capable of populating the 9th vibrational level. Figure 6 shows a schematic of the experimental system which has been described by Stair *et al.* (1965). Spectra were obtained with an asymmetrical 0.8 m Czerny-Turner monochromator having 2 in. Schaffer corrected curved slits. An image reduction system at the exit slit focussed the radiation onto a 3 by 3 mm PbS detector cooled with liquid N. The monochromator was equipped with either of two 4 by 4 in. gratings blazed at 1.6 μ and 3.0 μ, having 600 and 300 lines mm^{-1}, respectively. The total pressure in the reaction cell was a few Torr including the He carrier for the O_3. By varying the reactant pressure and the optical path, it was found that the spectra indicated no self-absorption. The spectra of the fundamental vibration-rotation bands from 3000 to 3800 cm^{-1} and the bands of the overtone from 5700 to 6900 cm^{-1} are

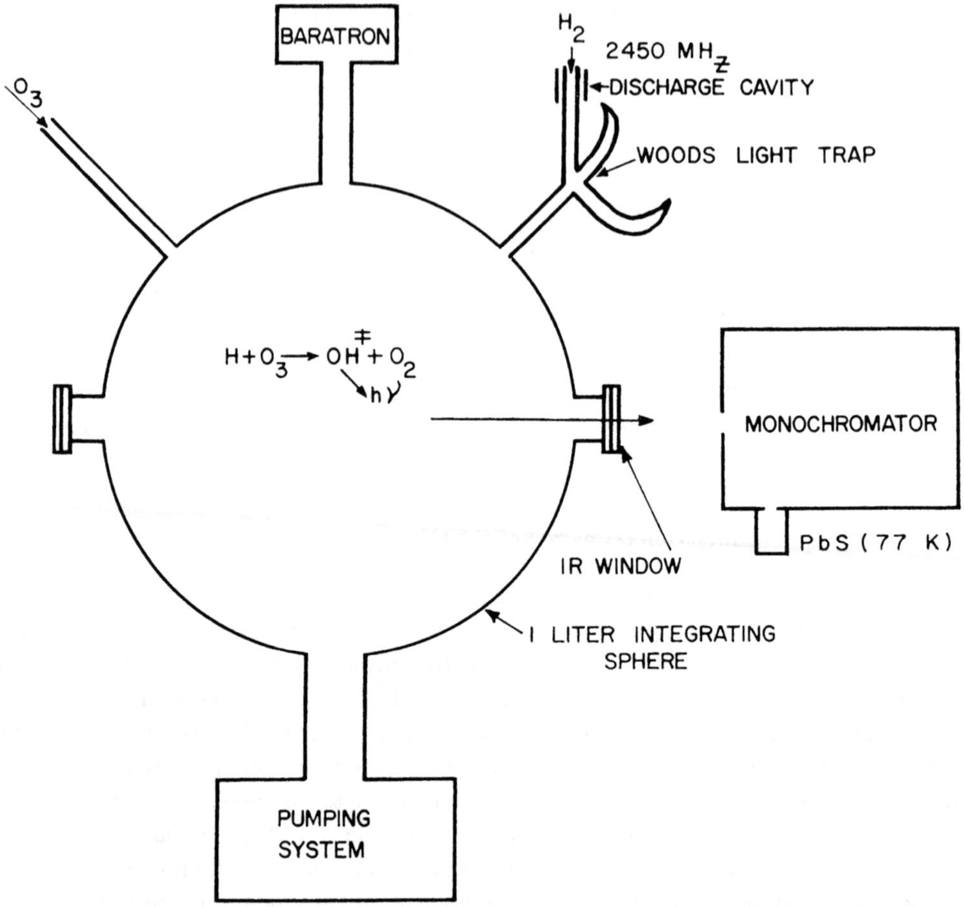

Fig. 6. Schematic of the Apparatus.

shown in Figures 7 and 8. The spectra have been compressed on the wave number scale for the purpose of display. The resolution is sufficient to resolve the spin doublet components of each rotational line but not sufficient to resolve the lambda doubling. Line identification was easily established from the close agreement between the measured values and the reported line positions given by Dieke and Crosswhite (1948) from their UV studies of OH.

The intensity of emission produced by transitions between vibrational quantum state v' to v (for vibrational temperature T_{vib}) and rotational quantum state K' to K (for rotational temperature T_{rot}) is proportional to

$$\frac{N}{Q(K')}\,(v_{vK}^{v'K'})^4\, S_{vK}^{v'K'} \exp(-hcF(K')/kT_{rot}) \exp(-hcG(v')/kT_{vib}) \qquad (2)$$

where N is the number of molecules, $v_{vK}^{v'K'}$ is the wave number of the resulting radiation, $Q(K')$ is the rotational partition function, $F(K')$ is the rotational term value, $G(v')$

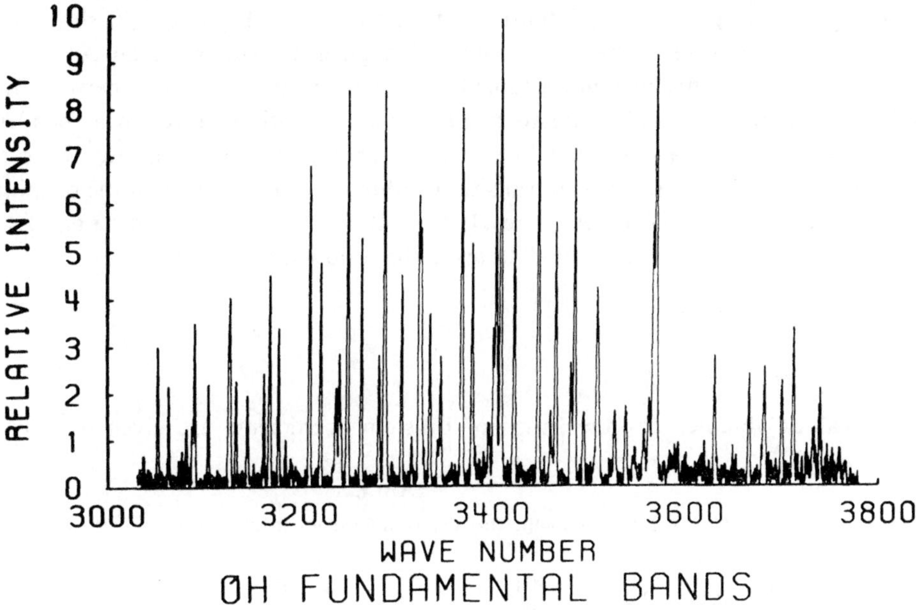

Fig. 7. Corrected emission spectrum of OH from the H + O_3 reaction, fundamental bands.

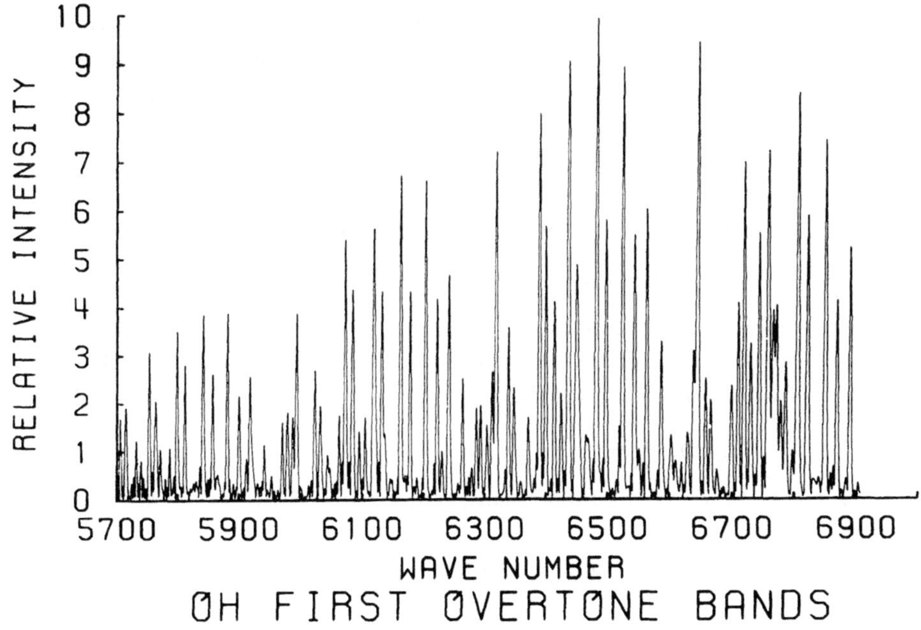

Fig. 8. Corrected emission spectrum of OH from the H + O_3 reaction, first overtone bands.

is the vibrational term value, and $S_{vK}^{v'K'}$ is the line strength (Bruce *et al.*, 1967). The usual logarithmic plots of the intensity data according to Equation (2) showed the rotational temperature to be a few hundred degrees Kelvin while an equilibrium vibrational temperature was not attained. By forming intensity ratios between lines in different sequences which originate from the same vibrational level, the dependence on the population of a particular level was eliminated. In addition, intensity ratios were taken for the same rotational transition within the sequence so that the rotational dependence was minimized. As a result the ratio of Einstein coefficients, $A_v^{v'}/A_{v''}^{v'}$ were calculated from the intensity data using the following relation:

$$\frac{I_v^{v'}}{I_{v''}^{v'}} = \frac{A_v^{v'} \nu_{vK}^{v'K'}}{A_{v''}^{v'} \nu_{v''K}^{v'K'}} \tag{3}$$

where $v' > v > v''$.

The values obtained for the ratios of Einstein coefficients are given in Table I.

TABLE I

Ratio of Einstein coefficients determined from spectroscopic measurements

$\dfrac{A^2{}_0}{A^2{}_1} = 0.44 \pm 0.03$	$\dfrac{A^3{}_1}{A^3{}_2} = 1.15 \pm 0.05$	$\dfrac{A^4{}_2}{A^4{}_3} = 3.3 \pm 0.7$

The ratios are higher than had been anticipated from theoretical models of either Heaps and Herzberg (1952) or Cashion (1963). Both of these models are based on the Morse oscillator for which it has been shown

$$\int \Psi_{v'}(r) \{e^{\beta(r-r_e)}\}^l \Psi_v(r) \, dr = 0 \quad \text{for} \quad |v' - v| > l. \tag{4}$$

The dipole moment function is usually expressed as a Taylor series expansion

$$M(r) = M_i (r - r_e)^i$$

which according to Equation (4) is seen to be arbitrary unless the intensity of the fundamental sequence has been determined. Since the Einstein coefficients depend upon the square of the matrix elements, i.e.,

$$\langle \Psi_{v'} | M(r) | \Psi_v \rangle^2$$

the discrepancy between the theoretical predictions and the measured values is not surprising.

Ferguson and Parkinson (1962) have fitted a polynomial of degree five to the dipole moment function based on intensity measurements in the overtones. They give the following expansion for the dipole moment function:

$$M_{FP}(r) = M_1 (r - r_e) + M_2 (r - r_e)^2 + M_3 (r - r_e)^3 + M_4 (r - r_e)^4 + M_5 (r - r_e)^5 \tag{5}$$

where r_e is the equilibrium internuclear separation, $M_1=1$, $M_2=0.036$ Å^{-1}, $M_3=-0.129$ Å^{-2}, $M_4=-0.543$ Å^{-3}, and $M_5=0.441$ Å^{-4}. Intensities predicted for the overtone bands using this expression for the dipole moment function are reported to agree with the experimental data to within 15%. Combining Equations (4) and (5) shows that the dipole moment function given by

$$M(r) = M_{FP}(r) + P \exp(\beta(r - r_e)) \tag{6}$$

reproduces the overtone intensity predictions and explicitly contains the arbitrary exponential factor where P is an undetermined constant and β is the Morse constant appropriate to OH. Each of the ratios listed in Table I were used to calculate the two roots of P. The required integrations were carried out numerically and the values -0.472 and -0.246 obtained for P. From intensity arguments involving the higher vibrational levels the later value has been determined to be correct, and hence the dipole moment function is uniquely defined.

The integrated intensity S_0^1 of the 0-1 fundamental band may be approximated from the relation

$$\frac{S_0^2}{S_0^1} = \frac{A_0^2}{A_1^2}\left(\frac{v_0^1}{v_0^2}\right)^2 = \frac{\langle 2\,|M(\mathrm{r})|\,0\rangle^2\, v_0^2}{\langle 1\,|M(\mathrm{R})|\,0\rangle^2\, v_0^1} \tag{7}$$

where $M(r)$ is given by Equation (6) with -0.246 taken as the value of P. Using the value quoted by Benedict and Plyler (1954) for S_0^2 as 4.0 atm^{-1} cm^{-2} one finds

$$S_0^1 = 34 \text{ atm}^{-1} \text{ cm}^{-2}. \tag{8}$$

The integrated intensity for the 1-0 band determined above is lower than the order of magnitude estimate of 100 atm^{-1} cm^{-2} given by Benedict and Plyler (1954). Recent absolute measurements of the $\Delta v=1$ region reported by Huppi and Stair (1969) from the aircraft indicate that the intensity of the OH fundamental bands are considerably weaker than predicted using the earlier estimates for either S_0^1 or A_0^1. These atmospheric observations, however, are consistent with the determination of the integrated intensity of the 1-0 fundamental band given here. This measurement is useful for determining the highest altitudes where OH rotational temperature measurements are representative of the kinetic temperature. The radiative lifetime for the 1-0 transition (converting $S_0^1 \cong 34$ atm^{-1} cm^{-2}) is ~0.1 sec. Transition probability calculations for higher levels are comparable; consequently, if 10 collisions are sufficient to relax the rotational states, rotational-kinetic equilibrium is a valid assumption to altitudes on the order of 130 km.

3. Measurement of the Vibrational Temperature of Atmospheric Nitrogen by Means of Electron Beam Induced Luminescence

On March 13, 1970, as part of the ARPA/AFCRL atmospheric IR measurements program, an Aerobee 150 was launched from Fort Churchill to measure the vibrational temperature of atmospheric N_2. The rocket and all payload instrumentation functioned successfully; measurement of N_2 vibrational temperature began at 140 km on rocket ascent, continued through apogee (200 km), and terminated with loss of telemetry at 80 km. Preliminary analysis of the data indicates signals are of sufficient magnitude to yield vibrational temperatures at altitudes of 140 km and less. The experimental technique used a four-channel interference filter photometer to monitor luminescence induced by a rocket based electron beam with an extended boom collector. This technique has been previously proposed as a method of measuring N_2 atmospheric density (also measured in the present experiment) and has been successfully used to measure N_2 rotational temperature (DeLeeuw and Davies, 1969).

In the past several years a series of theoretical papers, principally by Walker (1968) and colleagues, have estimated the vibrational temperature of N_2 at altitudes of 100 km and higher with vibrational temperatures in excess of the kinetic temperature. In a recent paper Walker *et al.* (1971) calculate a vibrational temperature at 150 km in the range of 1200 to 2400 K depending on efficiency assumed for quenching of $O(^1D)$ by N_2. The vibrational temperature of N_2 is relevant to mechanisms that determine the atmospheric IR background in two ways. Vibraluminescence, the collisional transfer of vibrational energy from N_2 to IR active molecules, primarily CO_2, and the subsequent radiative decay of CO_2 is considered the principal loss process for N_2 vibrational energy in the atmosphere. The (001) vibrational mode of CO_2 is very nearly resonant with the $N_2(v=1)$ vibrational level resulting in a large cross section for vibrational energy transfer. The CO_2 (001) mode relaxes by radiative decay resulting in IR emission at 4.3 μ, 10.4 μ, 13.9 μ, and 15.0 μ. In addition some chemical processes involving N_2 are very dependent on the N_2 vibrational state. For example, the rate for the reaction

$$O^+ + N_2 \rightarrow NO^+ + N$$

is critically dependent on the N_2 vibrational state (Schmeltekoff *et al.*, 1967). Thus, the N_2 vibrational population is an essential factor in determining the magnitude of reaction rates of chemical processes instrumental in determining the composition and IR radiance of both a normal and an auroral disturbed atmosphere.

A. EXPERIMENTAL TECHNIQUE

Electron induced luminescence was used as a remote diagnostic technique to measure the vibrational population of ground state N_2. Figure 9 illustrates potential energy curves for a limited number of the electronic states of N_2 and N_2^+. Electrons (2.5 keV) were used to simultaneously ionize and excite ground state $N_2(X^1\Sigma_g^+)$ to the $B^2\Sigma_u^+$ state of N_2^+. The N_2^+ molecules in turn spontaneously decay to the N_2^+ ground state, $X^2\Sigma_g^+$, with the emission of the N_2^+ 1 N bands. Monitoring the intensity of any band with a given upper vibrational may be used as an index of the population of that vibrational level. The population of the N_2^+ $(B^2\Sigma_u^+)$ vibrational levels excited by electron impact (Figure 10) may be used in turn as an index of the ground state $N_2(X^1\Sigma_g^+)$ vibrational population. It has been assumed in Figure 10 that the electron excitation cross sections of the $N_2(X^1\Sigma_g^+)$ vibrational levels to the N_2^+ $(B^2\Sigma_u^+)$ vibrational

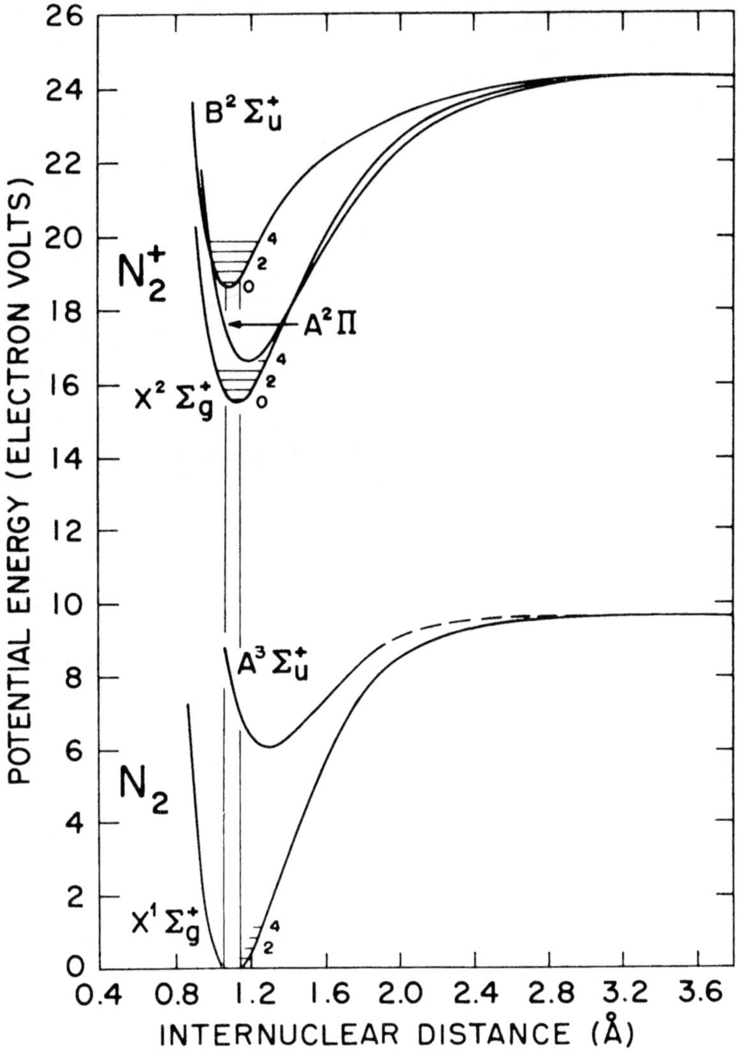

Fig. 9. Potential energy curves for N_2 and N^+_2.

levels are proportional to the Franck-Condon factors of Nicholls (1962). Figure 10 indicates that N_2 vibrational temperatures of approximately 1000°K and greater are reflected significantly in the N_2^+ 1 N system $v=0$, 1 and 2 vibrational population. The N_2^+ 1 N system is an appropriate choice for measuring N_2 vibrational temperature for the following considerations:

1. The corresponding vibrational levels of the $N_2(X^1\Sigma_g^+)$ and $N_2^+(B^2\Sigma_u^+)$ states conveniently have similar internuclear separations (Figure 9) and thus a change in ground state vibrational population is reflected in a large relative change in N_2^+ 1 N optical emission.

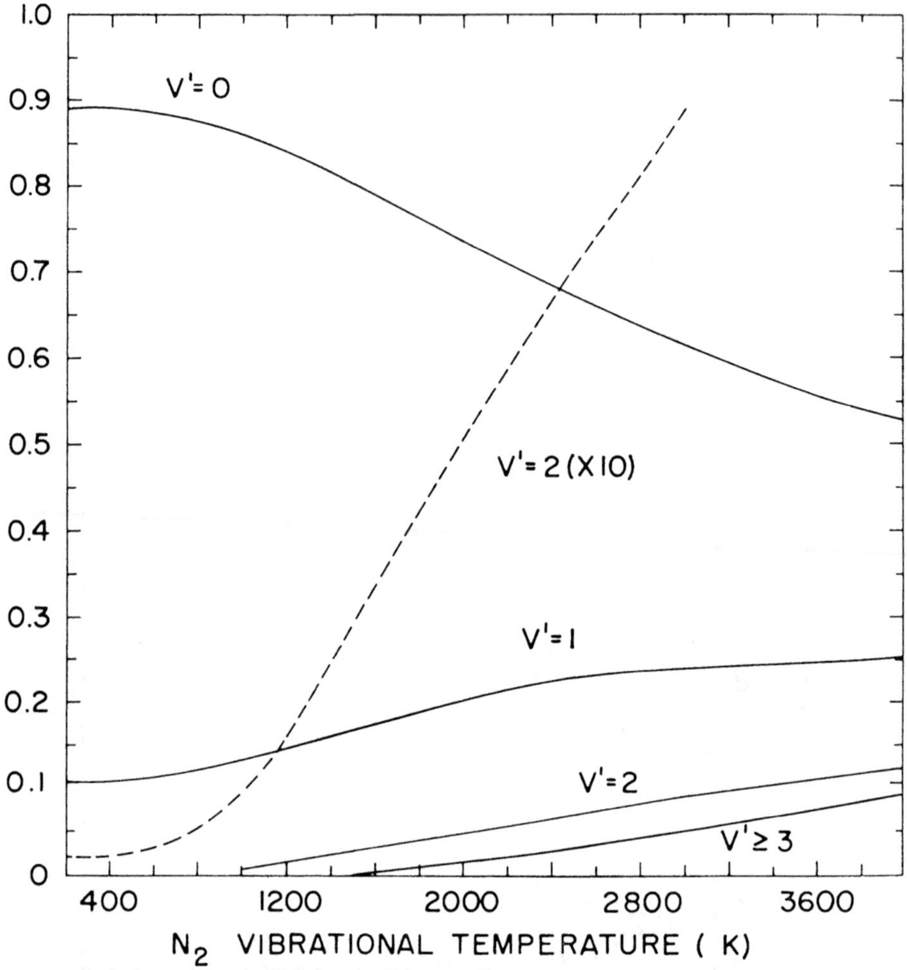

Fig. 10. Fractional population of N_2^+ 1N system vibrational level.

2. The electron excitation cross section of the 1 N system is of sufficient magnitude to provide adequate photometer signal-to-noise ratios.

3. The N_2^+ 1 N system conveniently radiates at wavelengths on the order of 4000 Å appropriate for the use of interference filters and low dark current photomultipliers in rocket instrumentation.

B. ROCKET INSTRUMENTATION

The payload (Figure 11) consisted of four photometers monitoring three N_2^+ 1 N vibrational levels, boom mounted Faraday cup and optical backstop, and a series of instruments designed to monitor the auroral electron energy and flux. The electron beam was collected by a segmented Faraday cup to define electron beam position and to avoid vehicle charge buildup. The electron beam was modulated at 350 Hz and

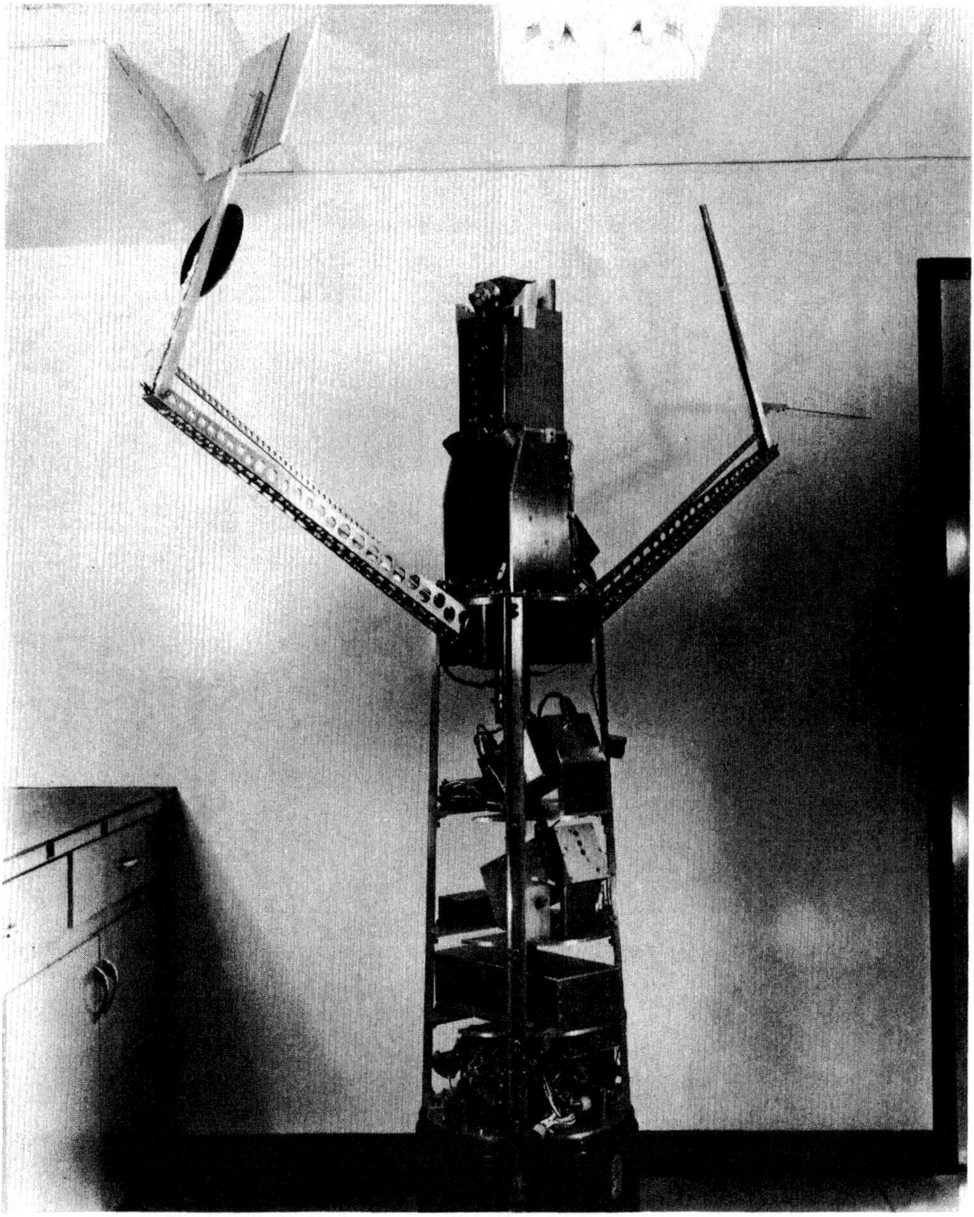

Fig. 11. Rocket instrumentation with booms elevated. The primary boom supports an optical backstop, to limit stray light from entering photometers, and a Faraday cup to collect electron beam. The secondary boom supports the Langmuir probe and maintains symmetrical distribution of mass.

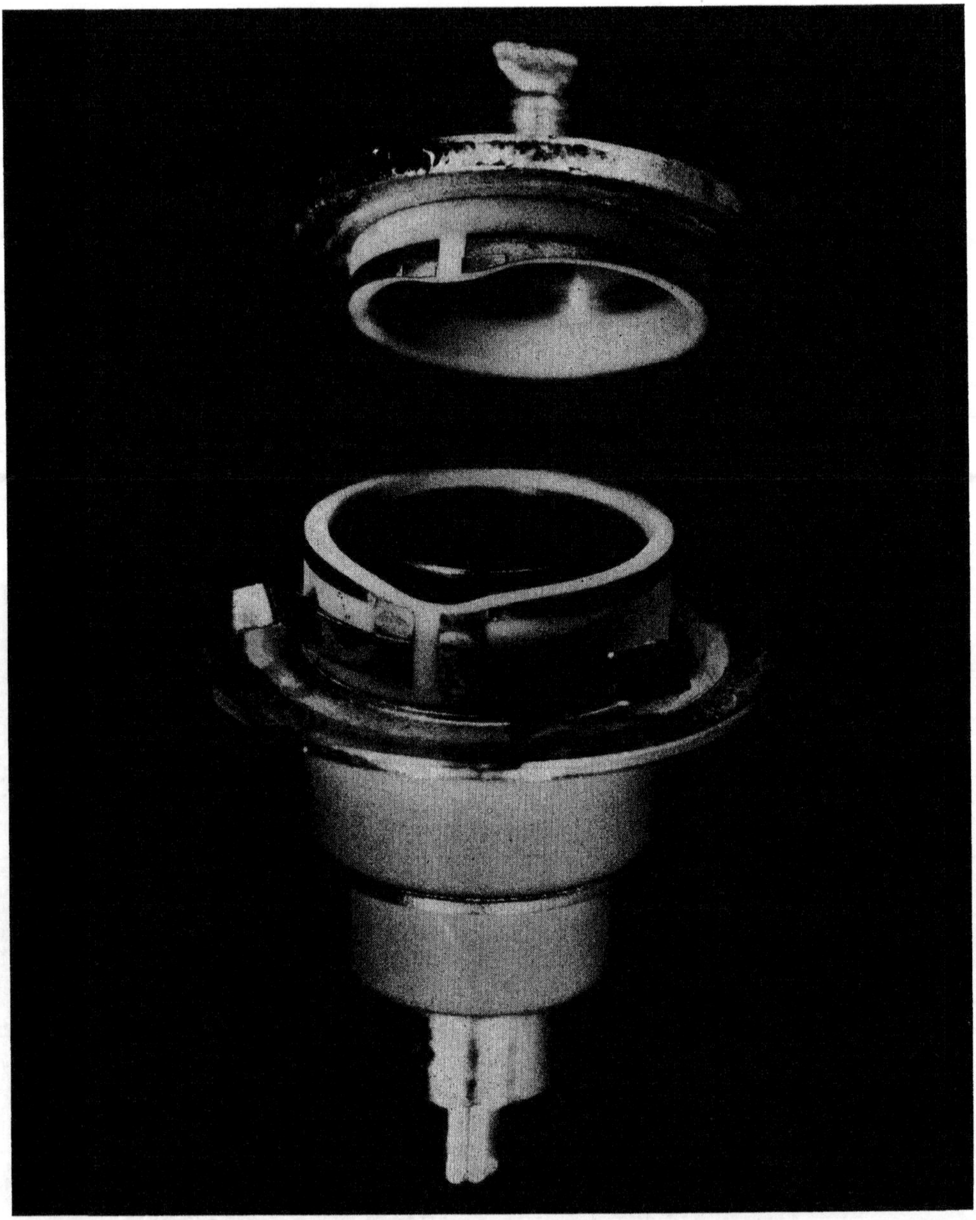

Fig. 12. Electron gun manufactured by Machlett Laboratories, Inc. The gun is opened by application of approximately 15 V and 40 amp to the break seal which fractures the ceramic end cap by thermal shock.

phase sensitive detection techniques were used. The electron gun was manufactured by Machlett Laboratory, Inc., who had previously developed the design and break seal opening technique jointly with Ion Physics Corp. The gun, operated at 2.5 keV with 10 mA peak current, used thermal shock to crack a ceramic seal at an altitude appropriate to avoid poisoning of the oxide coated cathode (see Figure 12). The optical axis of the four photometers was tilted by means of thin prisms such that a common length (approximately 5 cm) of the electron beam luminescence was observed. The photometer fields of view, coincident at the observed path length, allowed for some deflection of the beam by the rocket rotating in the earth's magnetic field.

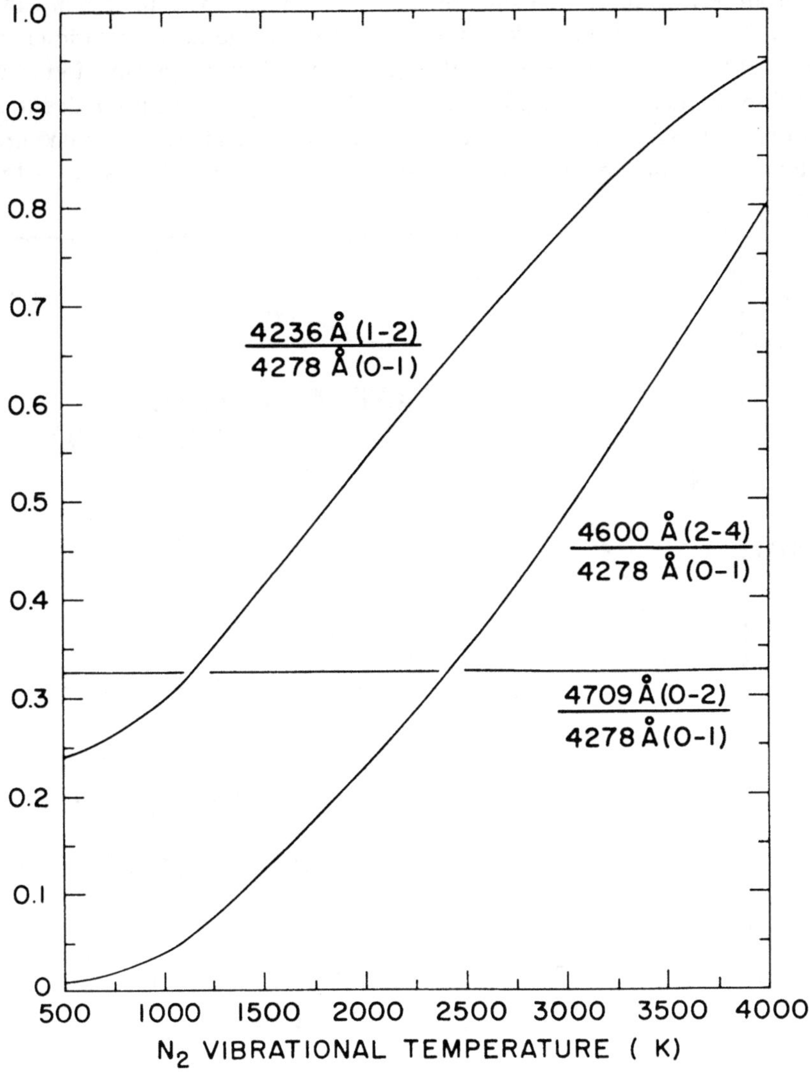

Fig. 13. Ratio of N_2^+ 1N transitions as measured by 4 channel photometer.

C. RESULTS

Figure 13 indicates the anticipated ratios of the four photometers as a function of N_2 vibrational temperature. The ratios of Figure 13 assume an interference filter transmission efficiency based on a N_2^+ rotational temperature of 300 K. Preliminary analysis of these data (based on the ratios of Figure 13) in the altitude range of 100 to 135 km indicates a maximum vibrational temperature on the order of 1200 K at 135 km.

Figure 14 compares preliminary results of the present measurement with the vibrational temperature model of Walker *et al.* (1969). The temperatures given in Figure 14 are presently in the form of upper limits for vibrational temperature due to uncertainties in the photometer signal ratios. When the kinetic temperature approaches 500 K or more (135 km) the associated change in the rotational structure of the N_2^+ 1N system may significantly change the effective transmission of an interference filter to its design transition and also the leakage contribution due to adjacent transitions in the sequence. This effect has been considered for rotational temperatures up to 600 K and contributes in part the upper limit nature of the results of Figure 14.

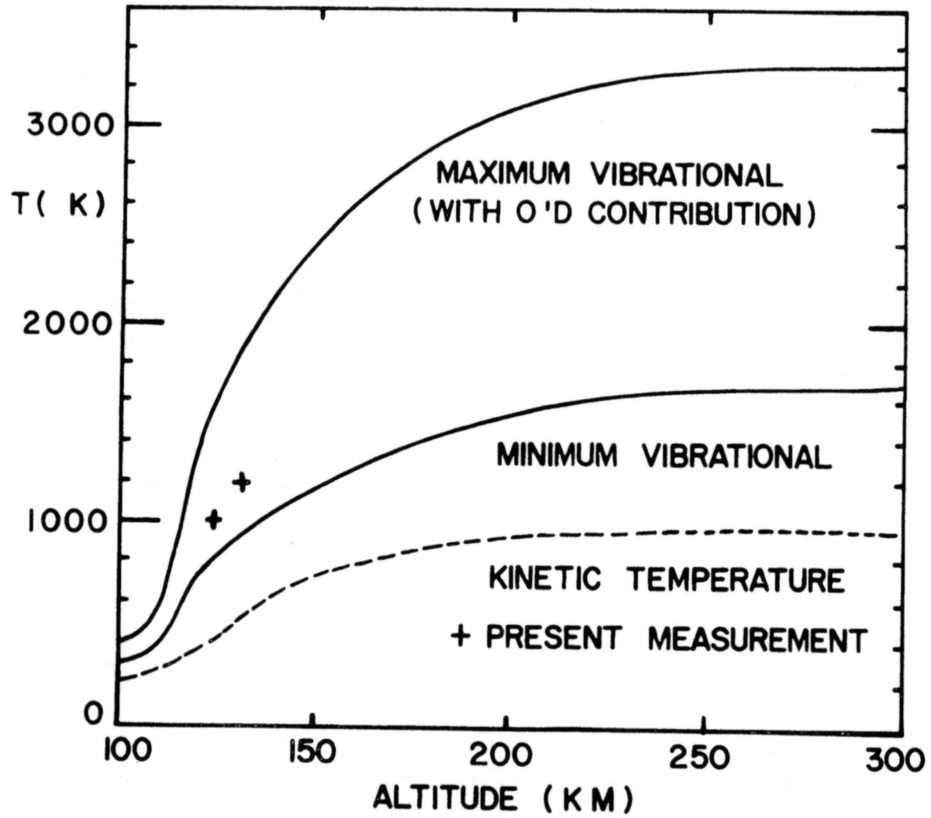

Fig. 14. Vibrational temperature results of present experiment and the theoretical model of Walker, *et al.* (1969). The temperatures of the present measurement are in the form of upper limits.

The relative N_2 number density profile in this altitude range is in excellent agreement with the CIRA 1965 mean model atmosphere.

Two members of the N_2^+ 1N $v=0$ progression, the (0-1) and the (0-2) bands, were monitored to ensure against misinterpretation of spurious signal sources and also of an index of the location and geometry of the electron beam. Data analysis is continuing and additional results will be reported at a later date.

Acknowledgments

In field efforts, such as aircraft and rocket borne experiments, there are many other valuable supporting people, such as the aircraft crews from L. G. Hanscom Field, integration and launch support from the rocket group (CRE) of AFCRL. In particular, we are indebted to Mr. H. P. Gauvin, Chief of the Optical Physics Laboratory, and to Dr. Thomas Walter, Major John Reed, and Mr. John W. La Spina of this branch (CROR) and Mr. Earl Pound and Mr. Gary Frodsham of Utah State University for their support and participation. This research in Section 1 has been partially sponsored by the Defense Atomic Support Agency. The Research in Section 3 has been sponsored by the Advanced Research Projects Agency of the Department of Defense.

References

Benedict, W. S. and Plyler, E. K.: 1954, *Energy Transfer in Hot Gases*, NBS circular **523**, 57.
Bruce, M. H., Stair, A. T., and Kennealy, J. P.: 1967, *J. Chimie Physique* **64**, 36.
Cashion, J. K.: 1963, *J. Mol. Spec.* **10**, 182.
DeLeeuw, J. H. and Davies, W. E. R.: 1969, *Small Rocket Intrumentation, Techniques*, North-Holland Publishing Company, Amsterdam, Holland, p. 144.
Dieke, G. H. and Crosswhite, H. M.: 1948, *The Johns Hopkins Bumblebee Report*, No. 87.
Ferguson, A. F. and Parkinson, D.: 1962, *Planetary Space Sci.* **11**, 149.
Heaps, H. S. and Herzberg, G.: 1952, *Z. Phys.* **133**, 48.
Huppi, E. R. and Stair, A. T., Jr.: 1969, in *Atmospheric Emissions* (ed. by B. M. McCormac and A. Omholt), Van Nostrand Reinhold Company, New York, p. 471.
Meinel, A. B.: 1950, *Astrophys. J.* **111**, 555.
Nicholls, R. W.: 1962, *J. Quant. Spectrosc. Radiative Transfer* **2**, 433.
Penner, S. S.: 1959, *Quantitative Molecular Spectroscopy and Gas Emissivities*, Addison Wesley Publications, Reading, Mass., p. 23.
Schmeltekoff, A. L., Fehsenfeld, F. C., Gilman, G. I., and Ferguson, E. E.: 1967, *Planetary Space Sci.* **15**, 401.
Stair, A. T. Jr., Kennealy, J. P., and Stewart, S. P.: 1965, *Planetary Space Sci.* **13**, 1005.
Walker, J. C. G.: 1968, *Planetary Space Sci.* **16**, 321.
Walker, J. C. G., Stolarski, R. S., and Nagy, A. F.: 1969, *Ann. Geophys.* **25**, 831.

INSTITUTE DISCUSSION

Absorption spectra of OH are very difficult to measure.

The upper altitude for which the rotational temperature of OH is equal to the kinetic temperature was discussed. E. Llewellyn believes that temperature errors will occur for data taken above 125 km.

The correlation of OH intensities with magnetic activity has not been made yet. The enhancement of OH as the aircraft in the NASA auroral expedition approaches the auroral oval from the polar cap has not been checked.

K. Baker reported on rocket observations of 1.27 μ from $O_2(^1\Delta_g)$ and 3914 Å from $N_2{}^+$ in an aurora. The 1.27 μ emission was a maximum at 85 km and there was a lack of correlation between the 3914 Å and 1.27 μ emission region. The energy in 1.27 μ was essentially equal to the total energy in precipitating electrons; therefore, some other excitation mechanism is needed. J. Noxon suggested that the temperature of the *E* region could be raised significantly by conduction downward from a high altitude heat source. It was suggested by W. Evans that the rocket observations could be interpreted as pure night glow; however, K. Baker feels that it then becomes difficult to explain the observed 3914 Å radiance.

J. Noxon discussed $O_2(^1\Delta_g)$ observations. The emission region has to be below 130 km altitude. R. Johnson pointed out that the electron energy to reach 130 km is 2 keV. The nonstructured $O_2(^1\Delta_g)$ emission cannot be directly attributed to its long lifetime as it probably does not live out its lifetime. The time history of $O_2(^1\Delta_g)$ emission from an aurora for over 1 hr is needed.

PART IV

AURORAL INTERPRETATIONS

IONOSPHERIC EFFECTS OF AURORAL ELECTRIC FIELDS

JAMES C. G. WALKER
Dept. of Geology and Geophysics, Yale University, New Haven, Conn., U.S.A.

1. Introduction

Large electric fields, directed perpendicular to the geomagnetic field, are present in the auroral ionosphere. The magnitudes of these fields can be deduced from motions of auroral forms and from geomagnetic perturbations associated with the auroral electrojet. These phenomena suggest electric field magnitudes of about 50×10^{-3} V m^{-1} (cf., Boström, 1967). Electrostatic measurements of auroral electric fields ranging up to 120×10^{-3} V m^{-1} have been made by Potter and Cahill (1969). Ion cloud releases have also been used to study auroral zone electric fields. In a series of rocket flights conducted in Sweden, Föppl *et al.* (1968) detected electric fields between 2×10^{-3} V m^{-1} and 20×10^{-3} V m^{-1}. Similar measurements in Norway by Wescott *et al.* (1969) detected electric fields which were generally about 50×10^{-3} V m^{-1}, but which ranged as high as 180×10^{-3} V m^{-1}. The rocket measurements all indicate that auroral electric fields are highly variable.

Auroral electric fields result from convection in the magnetosphere. Theories of their production have been discussed by Boström (1968) and Schield *et al.* (1969) among others. It is not our purpose to review these theories in this paper; rather, we wish to describe some of the effects of electric fields on the auroral ionosphere. We shall not discuss the role of the neutral wind (Swift, 1963) although winds at auroral latitudes may be large enough to be important.

2. Motion of Ions and Electrons

At E region altitudes the motions of ions and electrons are limited by collisions with neutral molecules. Under these conditions the drift velocity of a charged particle, in the binary approximation of Chapman and Cowling (1952), is

$$\mathbf{U} = \frac{1}{B}\left\{\frac{\alpha}{1+\alpha^2}\mathbf{E}_\perp + \frac{\alpha^2}{1+\alpha^2}\left(\frac{\mathbf{E}_\perp \times \mathbf{B}}{B}\right) + \alpha\mathbf{E}_\parallel\right\} \text{ m sec}^{-1} \tag{1}$$

(cf., Chapman, 1956). In this expression $\mathbf{E}_\perp$ V m^{-1} is the perpendicular component of the electric field and $\mathbf{E}_\parallel$ V m^{-1} is the parallel component; B weber m^{-2} is the magnetic field; $\alpha = \omega/v$ where v sec^{-1} is the collision frequency and $\omega = eB/m$ sec^{-1} is the cyclotron frequency. In the expression for ω, e coulombs is the charge on the particle and m kg is the mass.

In auroras α is of the order of one for the ions at about 130 km (cf., Boström, 1964; Rees and Walker, 1968). At the same altitude α is of order -1000 for the electrons.

B. M. McCormac (ed.), The Radiating Atmosphere, 207–216.

The ratio of α_i to α_e varies little with altitude. Both quantities increase approximately exponentially with a scale height of about 14 km (cf., Walker, 1970). We conclude that at auroral altitudes (above 100 km) the motion of electrons parallel to $\mathbf{E}_\perp$ can be neglected compared to the motion of ions in this direction. In the $\mathbf{E} \times \mathbf{B}$ direction the electrons drift with a velocity equal to E/B m sec^{-1}, while the ions drift less rapidly in the same direction. Because α_e is so much larger than α_i the motion of ions parallel to the magnetic field can be neglected compared to the motion of electrons.

We find, therefore, that electric current in the ionospheric plasma parallel to the magnetic field is carried almost entirely by electrons. The parallel current is

$$J_{\parallel} = \frac{|e|\,N}{B}\,\alpha_e E_{\parallel}\ \text{amp m}^{-2}, \tag{2}$$

where N m^{-3} is the electron number density. The Pedersen current, parallel to $\mathbf{E}_\perp$, is carried almost entirely by the ions, and is

$$J_{\mathrm{P}} = \frac{|e|\,N}{B}\left(\frac{\alpha_i}{1+\alpha_i^2}\right) E_\perp\ \text{amp m}^{-2}. \tag{3}$$

Hall current flows in the $-\mathbf{E} \times \mathbf{B}$ direction because of the relative motion of ions and electrons. At altitudes above 100 km where $\alpha_e \gg 1$ the Hall current is

$$J_{\mathrm{H}} = \frac{|e|\,N}{B(1+\alpha_i^2)}\left(\frac{\mathbf{B} \times \mathbf{E}_\perp}{B}\right)\text{amp m}^{-2}. \tag{4}$$

3. Influence of Field Induced Drifts on Electron Density

The ease with which electrons can move parallel to the magnetic field inhibits the formation of large parallel electric fields. Parallel fields calculated by Rees *et al.* (1971) for an auroral arc are of the order of 10^{-6} V m^{-1}, too small to be detected experimentally (Mende, 1968; Potter and Cahill, 1969). In the auroral oval the magnetic field B is approximately 0.5×10^{-4} weber m^{-2} so we see, from Equation (1), that the parallel component of the ion drift velocity is only about 2×10^{-2} m sec^{-1}. This velocity is too small to affect the ion density at E region altitudes (cf., Walker and McElroy, 1966).

It appears that the influence of perpendicular drift, however, may be significant. The important quantity is the perpendicular drift of the ions because, as we have just shown, parallel motion of the ions is negligible. If the drift of the ions leads to an excess or deficiency of ions relative to electrons at any point in the ionosphere, electrons flow parallel to the magnetic field to remove the discrepancy. The steady state continuity equation for auroral ionization therefore is

$$\nabla \cdot (N\mathbf{U}_i) - q + \beta N^2 = 0 \tag{5}$$

where q m^{-3} sec^{-1} is the rate of production of ionization and β m^3 sec^{-1} is the recombination coefficient.

In the neighborhood of 130 km $U_i \sim 10^4$ E m sec^{-1}, and $\beta \sim 10^{-13}$ m^3 sec^{-1} (cf., Walker and Rees, 1968). We can assess the importance of perpendicular drift by equating the divergence term in Equation (5) to the recombination term. We find that drift is important if

$$\frac{1}{NE_\perp} \nabla_\perp \cdot (N\mathbf{E}_\perp) \gtrsim \frac{10^{-17} N}{E_\perp}. \tag{6}$$

In the auroral oval outside the region of a strong arc $N \sim 4 \times 10^{10}$ m^{-3} (Walker *et al.*, 1968; cf., Föppl *et al.*, 1968) and we take $E_\perp \sim 0.05$ V m^{-1}. We find that drift is important if either the density or the perpendicular field vary on a horizontal scale smaller than 100 km. The electric field measurements cited in the introduction show ample evidence for small scale variations, and Walker *et al.* (1968) have presented evidence for a horizontal variation of the electron density on a scale of about 20 km. We conclude that E region electron densities in the auroral oval outside auroral arcs are substantially influenced by field induced drifts of ionization.

As an illustration of this phenomenon let us consider a simple model in which the ionization rate, q, is enhanced in a relatively narrow band. We assume that $q = q_b = 1.6 \times 10^8$ m^{-3} sec^{-1} for $x < 0$ and $x > L = 10$ km. For $0 < x < L$ we assume that $q = q_e = 1.6 \times 10^9$ m^{-3} sec^{-1}. We assume that the x component of the ion drift velocity is $U = 500$ m sec^{-1} for all x, corresponding to a constant electric field of about 0.05 V m^{-1}. The continuity equation (Equation (5)) becomes

$$U\, \mathrm{d}N/\mathrm{d}x - q(x) + \beta N^2 = 0 \tag{7}$$

with

$$N = N_0 = \sqrt{q_b/\beta}\ \mathrm{m}^{-3} \quad \text{at} \quad x = -\infty .$$

The solution is

$$N(x) = \begin{cases} N_0, & x < 0 \\ \sqrt{\dfrac{q_e}{\beta}} \tanh\left[\dfrac{\sqrt{\beta q_e}}{U} x + \tanh^{-1}\left\{\sqrt{\dfrac{\beta}{q_e}} N_0\right\}\right], & 0 < x < L \\ \sqrt{\dfrac{q_b}{\beta}} \tanh\left[\dfrac{\sqrt{\beta q_b}}{U} (x - L) + \tanh^{-1}\left\{\sqrt{\dfrac{\beta}{q_b}} N(L)\right\}\right], & x > L. \end{cases} \tag{8}$$

For the model we have adopted the background density, $N_0 = 4 \times 10^{10}$ m^{-3}. If there were no electric field, the density in the region of enhanced ionization would be $N_e = \sqrt{q_e/\beta} = 1.3 \times 10^{11}$ m^{-3}. With the assumed field induced drift, however, the maximum density is $N(L) = 6.6 \times 10^{10}$ m^{-3}. We see, therefore, that field induced drift of ionization can substantially reduce the amplitude of variations in background auroral densities caused by variations in the ionization rate with a horizontal scale on the order of 10 km.

Horizontal variations in the perpendicular electric field can produce substantial variations in the E region electron densities without any change in the rate of production of ionization. Imagine, for example, a situation in which the continuity equation (Equation (5)) is dominated by the transport term. In this situation, the equation becomes

$$1/N\,\nabla N = -\,1/U_{\mathrm{i}}\,\nabla U_{\mathrm{i}}\,. \tag{9}$$

If we assume that the spike in the electric field observed by Potter and Cahill (1969) corresponded to a horizontal variation rather than a temporal variation we find that this spike could have caused a reduction in the electron density by about a factor of 2 over a region only 5 km wide, without any change in the rate of production of ionization.

In strong auroral arcs the effect of field induced drifts on ionization density is less important. In part this is because the rate of recombination of ions and electrons is proportional to the square of the electron density, which means that the chemical lifetime of the ions is small in regions where the electron density is large. In addition, there is evidence that the electric field is reduced in strong arcs (Aggson, 1969; Potter and Cahill, 1969; Wescott *et al.*, 1969), presumably because the electric currents that flow in strong arcs are large enough to load the magnetospheric generator. If, in Equation (6), we assume that $N = 10^{12}\ \mathrm{m}^{-3}$ in an arc (Walker and Rees, 1968) and that $E_{\perp} = 0.01\ \mathrm{V\ m}^{-1}$ we find that the flux transport term in the continuity equation is important only if the density or the electric field vary on a horizontal scale of about 1 km. In regions where ionospheric currents are large enough to load the magnetospheric generator there will be a tendency for the field to decrease when the density increases and vice versa, so the influence of the field gradient and the density gradient terms in the continuity equation will tend to cancel.

4. Field Aligned Currents

A. REVERSE CURRENTS

Energetic charged particle precipitation associated with auroras seldom consists of an electrically neutral stream. Most of the time electron fluxes are larger than proton fluxes by many powers of ten. The ionosphere, however, cannot depart appreciably from electrical neutrality in the steady state, so a reverse current must flow parallel to the magnetic field lines on which auroral electron precipitation occurs (Cole, 1967). We wish to demonstrate that this reverse current develops essentially instantaneously, so that there can never be any significant difference between the current carried by energetic bombarding electrons and the reverse current carried by thermal ionospheric electrons.

Let $F\ \mathrm{m}^{-2}\ \mathrm{sec}^{-1}$ be the downward flux of energetic electrons and let ϱ coulombs m^{-3} be the net charge density in the ionosphere. If we assume that all quantities vary only in the vertical direction, and if we neglect vertical advection of charge, the equation of charge conservation is

$$\partial\varrho/\partial t = |e|\;\partial F/\partial z - \partial J_{\|}/\partial z \tag{10}$$

where $J_{\parallel}$ amp m^{-2} is the vertical current density, given by Equation (2). For present purposes we can neglect the variation of N and α_e. The equation therefore becomes

$$\partial \varrho / \partial t = |e| \, \partial F / \partial z - \sigma_0 \, \partial E_{\parallel} / \partial z \tag{11}$$

where

$$\sigma_0 = \frac{|e| \, N}{B} \alpha_e \text{ mho m}^{-1} .$$

From Coulomb's law we have

$$\partial E_{\parallel} / \partial z = \varrho / \varepsilon_0 \tag{12}$$

where

$$\varepsilon_0 = 10^7 / 4\pi c^2$$

and c m $\sec^{-1}$ is the speed of light; therefore,

$$\partial \varrho / \partial t = |e| \, \partial F / \partial z - (\sigma_0 / \varepsilon_0) \, \varrho . \tag{13}$$

The first term in this equation does not depend on ϱ, so ϱ approaches a steady value in a time of order ε_0/σ_0. At 100 km $\alpha_e \sim 300$ and $N > 10^{10}$ m^{-3} (Walker *et al.*, 1968). We find, therefore, that $\sigma_0 > 10^{-2}$ mho m^{-1} and the time constant for the buildup of ϱ and hence of $E_{\parallel}$ and the associated reverse current is less than 10^{-9} sec. Evidently the parallel conductivity of the ionosphere above 100 km is so large that no significant accumulation of charge can occur, even when the bombarding flux varies rapidly. In the absense of horizontal variations charge cannot escape horizontally from a region of bombardment, and it cannot escape downwards because the parallel conductivity becomes negligible at low altitudes. Therefore, the reverse current must always exactly balance the current carried by the bombarding flux. The magnitude of the reverse current in the topside ionosphere may be about 2×10^{-7} amp m^{-2} in the absence of strong auroral bombardment (Walker *et al.*, 1968) and about 2×10^{-5} amp m^{-2} in a bright auroral arc (Belon *et al.*, 1966; Cloutier *et al.*, 1970). The electric field required to drive currents of this magnitude is too small to affect directly the density distribution of minor ions in the topside ionosphere (Schunk and Walker, 1970), but Rees *et al.* (1971) have shown that these currents are large enough to affect electron temperatures.

These considerations concerning the reverse current enable us to assess the conditions that would be required to produce a parallel electric field of the order of 50×10^{-3} V m^{-1} at an altitude of 110 km. The presence of such a field was inferred from the results of a rocket experiment by Donahue *et al.* (1970). From the data presented by Donahue *et al.* we can deduce that the current carried by the bombarding flux in the aurora in question was less than 10^{-5} amp m^{-2}. At 110 km the parallel conductivity, $\sigma_0 > 10^{-12} N$ mho m^{-1}. Therefore, a parallel field of 50×10^{-3} V m^{-1} would be required to drive the reverse current only if the ambient electron density, $N < 2 \times 10^8$ m^{-3}. The ionization rate in this aurora was about 5×10^9 m^{-3} $\sec^{-1}$, so N would increase to values larger than 2×10^8 m^{-3} within 0.1 sec of the onset of auroral bombardment. We conclude that charge carried into the auroral ionosphere by precipitating electrons cannot cause the large parallel field inferred by Donahue *et al.* (1970) unless, as these authors suggest, there is some mechanism, presumably an instability, that reduces the parallel conductivity by several orders of magnitude.

B. SHEET CURRENTS

Since auroral bombardment is always accompanied by an equal and opposite reverse current there is no detectable magnetic perturbation that can be directly associated with the current carried by the bombarding particles. Field aligned currents, however, have been detected by magnetic measurements in a number of rocket and satellite experiments (Zmuda *et al.*, 1966, 1967; Findlay *et al.*, 1969; Cloutier *et al.*, 1970). Evidently these currents are not reverse currents. Instead they must be associated with variations in the horizontal current flowing in the E region of the ionosphere (Boström, 1964, 1968).

Let us consider a model of a strong auroral arc, following the treatment of Boström (1964). We assume that the arc has a width of L m and that electron densities within the arc are of the order of 10^{12} m^{-3} (Walker and Rees, 1968). We assume that horizontal variations of electron density and electric field within the arc are small. The arc is bounded by relatively narrow transition regions in which the density and the field vary rapidly from their values within the arc to their values in the surrounding ionosphere. We assume, further, that current densities in the background plasma are much smaller than current densities in the arc.

If $\mathbf{E}_\perp$ within the arc is directed across the arc, a Hall current flows along the arc in the $\mathbf{B}\times\mathbf{E}_\perp$ direction. Its magnitude is $\Sigma_H E_\perp L$ amp, where Σ_H mho is the height integrated Hall conductivity. A Pedersen current flows across the arc, in the direction of $\mathbf{E}_\perp$, with a magnitude of $\Sigma_p E_\perp$ amp m^{-1}, where Σ_p mho is the height integrated Pedersen conductivity. In order to conserve charge at the edges of the arc, where the Pedersen current drops to small values, there must be sheet currents in the transition regions of magnitude equal to the magnitude of the Pedersen current.

As an illustration, let us consider an arc in the Northern Hemisphere 10 km in width in which $\Sigma_p = 100$ mho and $\Sigma_H = 65$ mho (Rees and Walker, 1968). Let us assume that $E_\perp = 4\times10^{-3}$ V m^{-1} within the arc directed towards the north. Then the electrojet current flowing eastwards along the arc is 2.6×10^3 amp, and there are sheet currents of 0.4 amp m^{-1} flowing downwards on the southern edge of the arc and upwards on the northern edge.

If $\mathbf{E}_\perp$ within the arc is directed along the arc instead of across it, the electrojet is a Pedersen current of magnitude $\Sigma_p E_\perp L$ amp, parallel to $E_\perp$, and the sheet currents are maintained by the Hall current, of magnitude $\Sigma_H E$ amp m^{-1}, flowing in the $\mathbf{B}\times\mathbf{E}$ direction. Taking the model we have just described, but assuming, now, that $\mathbf{E}_\perp$ is directed toward the west, we find a westward electrojet of magnitude 4.0×10^3 amp flowing along the arc and sheet currents of magnitude 0.26 amp m^{-1} flowing downwards on the southern edge of the arc and upwards on the northern edge.

These currents correspond exactly to the observations of Cloutier *et al.* (1970) on a rocket flight at Fort Churchill that penetrated an auroral arc. We conclude that the electric field in this arc had a magnitude of about 4×10^{-3} V m^{-1} and was directed more nearly along the arc than across it. The significant point in this interpretation is the direction of the currents and of the field. The magnitudes of the currents depend on the relative magnitudes of Σ_p and Σ_H, and these are different for arcs at different altitudes (Rees and Walker, 1968). But the model in which $\mathbf{E}_\perp$ is directed across the arc produces an eastward electrojet; however, a westward electrojet was observed.

The electric field observations cited in the Introduction have shown that electric fields in the neighborhood of auroral arcs are directed nearly perpendicular to the arcs, and have magnitudes closer to 50×10^{-3} V m^{-1} than to 4×10^{-3} V m^{-1}. In order to reconcile these observations we may argue that the component of the field directed across the arc is short circuited by the sheet currents, so that all that remains within the arc is the small component of the field that is directed along the arc.

The current density in the sheet currents is governed by the equation of current continuity

$$\partial J_{\parallel}/\partial z + \nabla_\perp \cdot \mathbf{J}_\perp = 0\,. \tag{14}$$

In order to apply this equation to a strong auroral arc, however, we would need information, which we do not have, on the horizontal variations of $\mathbf{E}_\perp$ and N at the edges of an auroral arc. At present the best we can do is refer to the observations of Cloutier *et al.* (1970), which indicate sheet current densities of order 5×10^{-5} amp m^{-2}. Evidently sheet current densities can be a little larger than reverse current densities in strong auroral arcs.

In the background auroral ionosphere, nevertheless, it is possible for sheet current densities to be considerably larger than the corresponding reverse current densities. Consider the model we discussed earlier, in which the bombarding flux and the ionization rate in the background plasma are enhanced over a relatively narrow region. Assuming that the electric field did not vary horizontally, we showed that an enhancement in the ionization rate by as much as a factor of 10 over a region 10 km wide enhances the ionospheric electron density by a factor of only 1.6. In this model the electric field sweeps most of the ions that are produced out of the region of enhanced ionization rate. The electrons that are produced along with the ions are left behind and must escape by flowing upwards along the magnetic field. The excess ions recombine ultimately on the downstream side of the enhancement with electrons that flow down the magnetic field from higher altitudes.

We can estimate the order of magnitude of the sheet current associated with an enhancement in the background ionization rate by assuming that most of the ionization is produced at altitudes near 130 km, where α_i is close to 1, and horizontal transport is significant. Then almost all of the electrons that are produced in the enhancement must flow upwards in the sheet current, and the electron flux is approximately equal to the height integral of the rate of production of ionization. For bombarding electrons with energies of about 1 keV, the height integral of the ionization rate is about 30 times the flux of bombarding particles, so we find that the sheet current density is about 30 times the reverse current density. For our model of a background enhancement, the reverse current density is about 2×10^{-6} amp m^{-2} and the sheet current density is about 6×10^{-5} amp m^{-2}.

This current density is comparable to the sheet current densities flowing at the edges of strong auroral arcs, and we have argued that such currents may be large enough to load the magnetospheric generator, leading to a reduction in the perpendicular electric field. It is possible, therefore, that our assumption of constant electric field in a background enhancement is invalid. Quite modest increases in the auroral ionization rate may cause reductions in the auroral electric field.

5. Thermal Effects

The total rate of dissipation of electric field energy is given by $\mathbf{J} \cdot \mathbf{E} \sim \sigma_p E_\perp^2$ watt m^{-3} where σ_p mho m^{-1} is the Pedersen conductivity. If we integrate this expression over altitude in an auroral arc with $\Sigma_p = 100$ mho and $E_\perp = 4 \times 10^{-3}$ V m^{-1} we find that the integrated joule heating rate in the arc is 1.6×10^{-3} watt m^{-2} = 1.6 erg cm^{-2} sec^{-1}. The energy flux deposited by the bombarding particles, however, is of the order of 50 erg cm^{-2} sec^{-1} (Belon *et al.*, 1966). What fraction of this energy is converted into heat is not known, but it seems likely that joule heating makes a minor contribution in auroral arcs.

In the background ionosphere, on the other hand, where $\Sigma_p \sim 2$ mho (Rees and Walker, 1968) and $E_\perp \sim 50 \times 10^{-3}$ V m^{-1} we find an integrated joule heating rate of 5×10^{-3} watt m^{-2} = 5 erg cm^{-2} sec^{-1}. The energy flux carried by bombarding particles is of the order of 0.2 erg cm^{-2} sec^{-1} (Walker *et al.*, 1968), so we conclude that the energy budget of the auroral atmosphere outside auroral arcs is dominated by joule heating.

The joule heat is divided approximately evenly between the ions and the neutral molecules. The electrons are not heated significantly by perpendicular electric fields because they drift at right angles to the electric field. Ion temperatures in the auroral oval have been studied by Rees and Walker (1968); Walker (1970) has derived the following formula for the ion temperature at E region altitudes

$$k(T_i - T_n) = \frac{M_n}{3} \frac{E_\perp^2}{B^2} \left(\frac{\alpha_i^2}{1 + \alpha_i^2} \right) \text{ joule} \tag{15}$$

where k joule deg^{-1} is Boltzmann's constant and M_n kg is the mass of the neutral molecules. For $B = 0.5 \times 10^{-4}$ weber m^{-2} this becomes

$$T_i - T_n = 4.5 \times 10^5 \frac{E_\perp^2 \alpha_i^2}{1 + \alpha_i^2} \text{ K}. \tag{16}$$

At altitudes above about 150 km, where $\alpha_i \gg 1$, we have $T_i - T_n = 7$ K in an auroral arc, where $E_\perp = 4 \times 10^{-3}$ V m^{-1}, and $T_i - T_n = 1100$ K in the surrounding ionosphere, where $E_\perp = 50 \times 10^{-3}$ V m^{-1}. We predict, therefore, high ion temperatures in the auroral ionosphere outside auroral arcs, and low ion temperatures in auroral arcs. Ion temperature measurements may constitute one of the easiest ways of determining the magnitude of auroral electric fields.

Walker (1970) has discussed the conditions under which high ion temperatures could lead to significant excitation of the infrared band of molecular oxygen through the charge-exchange process

$$O_2^+ (X^2\pi_g) + O_2 (X^3\Sigma_g^-) \rightarrow O_2 (a^1\Delta_g) + O_2^+ (X^2\pi_g) - 0.98 \text{ eV}. \tag{17}$$

In addition to a large cross section, the mechanism requires electric fields of the order of 150×10^{-3} V m^{-1} in regions where the electron density is as high as 3×10^{11} m^{-3}. If, as we have argued above, high ionospheric densities effectively short circuit the auroral electric field it seems unlikely that the required conditions can be met.

Field aligned currents do not directly affect ion temperatures in the auroral ionosphere because the electrons carry the current. Electron temperatures in the topside ionosphere, however, are significantly affected by field aligned currents with densities of the order of 10^{-5} amp m^{-2} or larger. The problem has been studied by Schunk and Walker (1970) and by Rees *et al.* (1971), who find that joule heating is negligible, but that two other terms are important. One is the thermoelectric heat flow, which is directed parallel to the electron drift velocity. The other is the expansion term that corresponds to the divergence of the electron drift velocity. Both of these terms depend on the direction of the current. Reverse currents caused by electron bombardment, for example, are directed downwards and lead to a reduction in topside electron temperatures. Sheet currents, on the other hand, are directed downwards on one side of an auroral density perturbation and upwards on the other side. We expect, therefore, that field aligned currents can produce the fine structure in topside electron temperatures and, through the temperature dependence of the density scale height, in topside densities that have been observed in the auroral oval (Calvert, 1966; Findlay *et al.*, 1969; Dyson and Zmuda, 1970).

There is no evidence for field aligned currents large enough to cause direct excitation of optical emissions by electron impact (cf., Walker, 1970).

6. Conclusion

Auroral electric fields affect both densities and temperatures in the auroral ionosphere. It appears that the coupling between the ionosphere and the magnetospheric source of electric fields is close. Horizontal variations in the perpendicular component of the electric field produce changes in ionospheric electron density. Electron density variations, however, cause field aligned currents which may significantly affect the electric field.

Acknowledgments

On various aspects of this research I have enjoyed the collaboration of E. B. Hooper, Jr.,

R. A. Jones, M. H. Rees, and R. W. Schunk. The work has been supported, in part, by the Atmospheric Sciences Section of the National Science Foundation, under Grant GA-10601.

References

Aggson, T. L.: 1969, in *Atmospheric Emissions* (ed. by B. M. McCormac and A. Omholt), Van Nostrand Reinhold Company, New York, p. 305.

Belon, A. E., Romick, G. J., and Rees, M. H.: 1966, *Planetary Space Sci.* **14**, 597.

Boström, R.: 1964, *J. Geophys. Res.* **69**, 4983.

Boström, R.: 1967, in *Aurora and Airglow* (ed. by B. M. McCormac), Reinhold Publishing Corporation, New York, p. 293.

Boström, R.: 1968, *Ann. Geophys.* **24**, 681.

Calvert, W.: 1966, *J. Geophys. Res.* **71**, 3665.

Chapman, S.: 1956, *Nuovo Cimento Suppl.* **4**, 1385.

Chapman, S. and Cowling, T. G.: 1952, *The Mathematical Theory of Non-Uniform Gases* (2nd ed.), Cambridge University Press, Cambridge.

Cloutier, P. A., Anderson, H. R., Park, R. J., Vondrak, R. R., Spiger, R. J., and Sandel, B. R.: 1970, *J. Geophys. Res.* **75**, 2595.

Cole, K. D.: 1967, *Planetary Space Sci.* **15**, 873.

Donahue, T. M., Zipf, E. C., and Parkinson, T. D.: 1970, *Planetary Space Sci.* **18**, 171.

Dyson, P. L. and Zmuda, A. J.: 1970, *J. Geophys. Res.* **75**, 1893.

Findlay, J. A., Dyson, P. L., Brace, L. H., Zmuda, A. J., and Radford, W. E.: 1969, *J. Geophys. Res.* **74**, 3705.

Föppl, H., Haerendel, G., Haser, L., Lüst, R., Melzner, F., Meyer, B., Neuss, H., Rabben, H.-H., Rieger, E., Stöcker, J., and Stoffregen, W.: 1968, *J. Geophys. Res.* **73**, 21.

Mende, S. B.: 1968, *J. Geophys. Res.* **73**, 991.

Potter, W. E. and Cahill, L. J.: 1969, *J. Geophys. Res.* **74**, 5159.

Rees, M. H., Jones, R. A., and Walker, J. C. G.: 1971, *Planetary Space Sci.*, in press.

Rees, M. H. and Walker, J. C. G.: 1968, *Ann. Geophys.* **24**, 193.

Schield, M. A., Freeman, J. W., and Dessler, A. J.: 1969, *J. Geophys. Res.* **74**, 247.

Schunk, R. W. and Walker, J. C. G.: 1970, *Planetary Space Sci.* **18**, 1535.

Swift, D. W.: 1963, *J. Geophys. Res.* **68**, 2131.

Walker, J. C. G.: 1970, *Planetary Space Sci.* **18**, 1043.

Walker, J. C. G., Brace, L. H., and Rees, M. H.: 1968, *J. Geophys. Res.* **73**, 7285.

Walker, J. C. G. and McElroy, M. B.: 1966, *J. Geophys. Res.* **71**, 3779.

Walker, J. C. G. and Rees, M. H.: 1968, *Planetary Space Sci.* **16**, 459.

Wescott, E. M., Stolarik, J. D., and Heppner, J. P.: 1969, *J. Geophys. Res.* **74**, 3469.

Zmuda, A. J., Heuring, F. T., and Martin, J. H.: 1967, *J. Geophys. Res.* **72**, 1115.

Zmuda, A. J., Martin, J. H., and Heuring, F. T.: 1966, *J. Geophys. Res.* **71**, 5033.

BASIC PROCESSES AND CROSS SECTIONS IN PROTON AURORAS

R. J. McNEAL and J. H. BIRELY

Space Physics Laboratory, The Aerospace Corporation, El Segundo, Calif., U.S.A.

1. Introduction

A proton aurora is a weak diffuse emission readily identified by Doppler-shifted H lines in its spectrum (Eather, 1967). It is caused by precipitation of H^+ with a broad energy distribution that peaks near 1 keV (Chase, 1968). For theoretical analysis of this aurora, cross sections for ionization and excitation of atmospheric constituents by energetic H^+ and H are needed over a wide energy range. Recent measurements of these cross sections are reviewed in this paper, and preliminary results from this laboratory for excitation of some of the H emissions are reported.

2. Basic Processes

Above about 50 keV the principal interactions of H^+ with the atmosphere are ionization,

$$H^+ + M \rightarrow H^+ + M^{+(*)} + e, \tag{1}$$

and excitation,

$$H^+ + M \rightarrow H^+ + M^*, \tag{2}$$

where M is any atmospheric constituent, and an asterisk denotes the possibility of product excitation. At lower energies the charge-changing collisions,

$$H^+ + M \rightarrow H^{(*)} + M^{+(*)}, \tag{3}$$

$$H + M \rightarrow H^+ + M^{(*)} + e, \tag{4}$$

and

$$H + M \rightarrow H^- + M^{+(*)}, \tag{5}$$

become important. These processes lead to conversion of an initially incident H^+ flux into a mixed flux of H^+, H, and H^-. In the low keV energy range H is the dominant flux constituent, and H^- is a very minor component (Allison, 1958). The mixture flux can be followed through the atmosphere by Monte-Carlo techniques (Davidson, 1965; Prag *et al.*, 1966), and cross sections for Equations (3) through (5) are basic to such calculations.

In the proton aurora excitation of atmospheric constituents occurs in Equations (1) and (2), in their analogs for H projectiles, and in charge-changing collisions. Hydrogen atoms are excited in Equation (3) and in

$$H + M \rightarrow H^* + M. \tag{6}$$

B. M. McCormac (ed.), The Radiating Atmosphere, 217–222.

Emissions from H* are the Doppler shifted lines that are the characteristic signatures of proton auroras.

3. Laboratory Measurements

Cross sections for Equations (1) through (6) are usually obtained by passing a beam of H^+ or H through a low pressure gas and measuring the collision products along the beam track or the collision induced changes in beam composition as functions of target pressure.

The cross section for electron capture by H^+ in Equation (3), σ_{10}, has been measured over a broad energy range with a variety of experimental techniques. There is generally good agreement among the various experimenters, except at high energies (above about 500 keV), where σ_{10} is small and of little auroral interest. The cross section for stripping of H (Equation (4)), σ_{01}, is also available over a broad range, except that there are no data for N_2 and O_2 targets below 4 keV. For electron capture by H (Equation (5)), $\sigma_{0\bar{1}}$, the data cover a rather narrow range, but this cross section is relatively unimportant in auroral analysis.

In Figure 1 are collected some recent results for σ_{10}, σ_{01}, and $\sigma_{0\bar{1}}$ for H^+ and H incident on N_2. A similar plot for O_2 and most other atmospheric constituents,

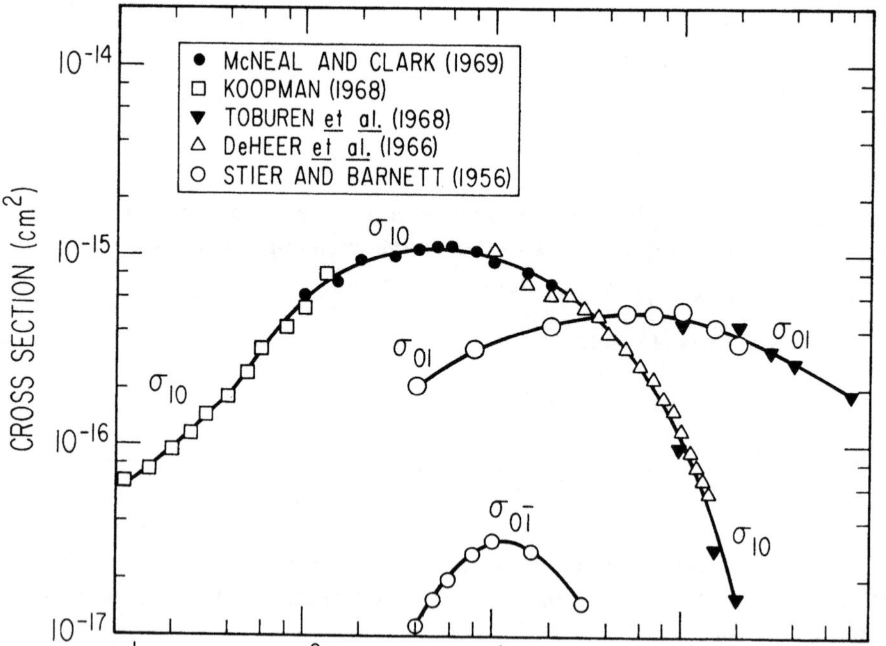

Fig. 1. Charge-changing cross sections for H^+ and H incident on N_2. Representative recent data are plotted and references to previous measurements can be found in the works cited.

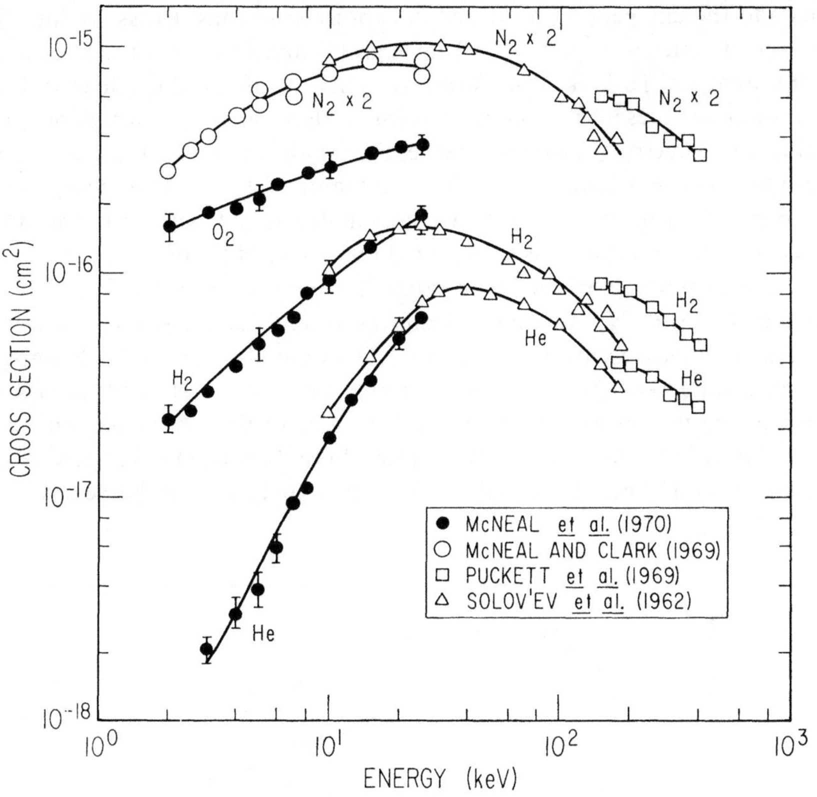

Fig. 2. Cross sections for ionization of He, H_2, O_2, and N_2 by energetic H. The data for N_2 have been multiplied by 2 for convenience in plotting.

except O, can be constructed from the sources for the figure. There are in general very few data for O for any cross section of auroral interest, because its reactivity virtually precludes its use in single beam, thin target experiments. Stebbings *et al.* (1964) have reported measurements of σ_{10} for O obtained in a crossed beam experiment.

Ionization cross sections for H^+ and H impact are obtained by collecting the charge produced along the beam track. For H^+ DeHeer *et al.* (1966) have reviewed recent results. Fairly large discrepancies (up to 50%) are often found in data from different laboratories, apparently because of the difficulty in avoiding systematic errors in the electron collection required in the H^+ experiments. Fewer data are available for H impact ionization. Where more than one set of results exist, agreement for atmospheric targets is fairly good at low energies, but poorer at high energies.

Some recent results for H impact ionization cross sections are shown in Figure 2. At 10 keV and above, H^+ and H cross sections for Equation (1) are similar, but at 1 keV cross sections for H impact are significantly larger than for H^+ impact on many important targets (McNeal, 1970).

In a proton aurora light emission results from excitation of atmospheric constituents and the energetic H produced in capture collisions. Cross sections for producing excited states in Equations (1) through (6) are, therefore, the key parameters in auroral analysis. In laboratory measurements all of the difficulties in beam experimentation are present in excitation studies, along with the formidable problems of absolute photometry. As a consequence, the excitation studies that have been done have resulted in cross sections that differ by a factor of 2 or more in many cases, even for H^+ bombardment. Very few data are available for H bombardment, and most H impact cross sections have been measured relative to H^+ cross sections.

One of the most widely studied processes is the emission of the N_2^+ 3914 Å band in collisions of H^+ with N_2. In Figure 3 are the recent data. The large scatter of the various absolute measurements is apparent. Data are also shown for H impact with N_2. The latter data were obtained by measuring the ratio of the H^+ and H cross sections and normalizing to a measurement of the H^+ cross section. A choice must be made in the normalization, and both sets of data for H bombardment have been normalized in Figure 3 to the H^+ results of Dufay *et al.* (1966) at 15 keV. Relative cross sections

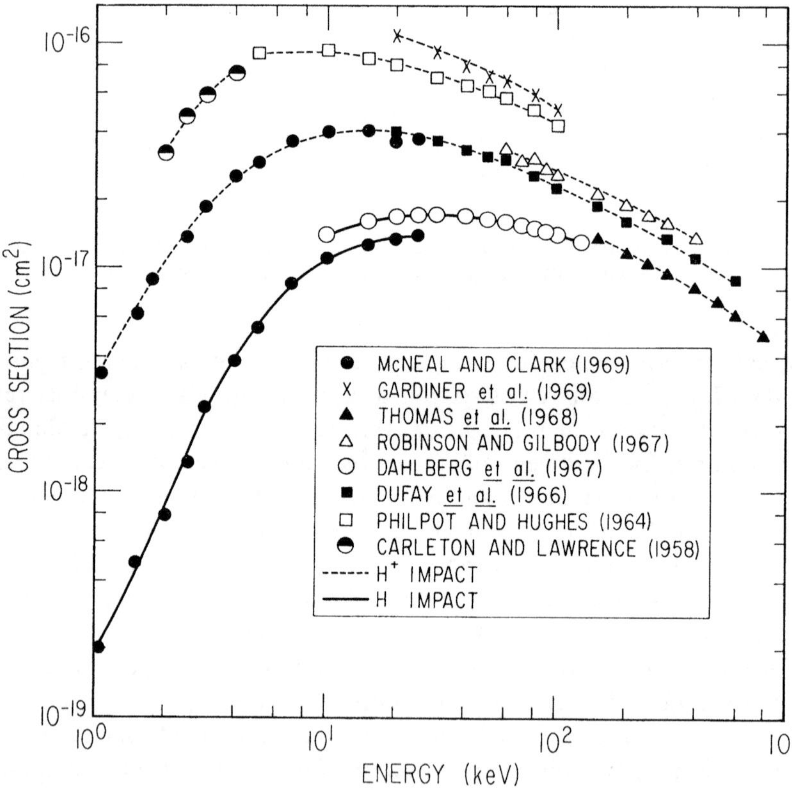

Fig. 3. Cross sections for emission of N_2^+ 3914 Å radiation in collisions of H^+ and H with N_2. Dotted curves connect data obtained with H^+ projectiles, and solid curves connect results for H impact, normalized (McNeal and Clark, 1969) to the H^+ results of Dufay *et al.* (1966) at 15 keV.

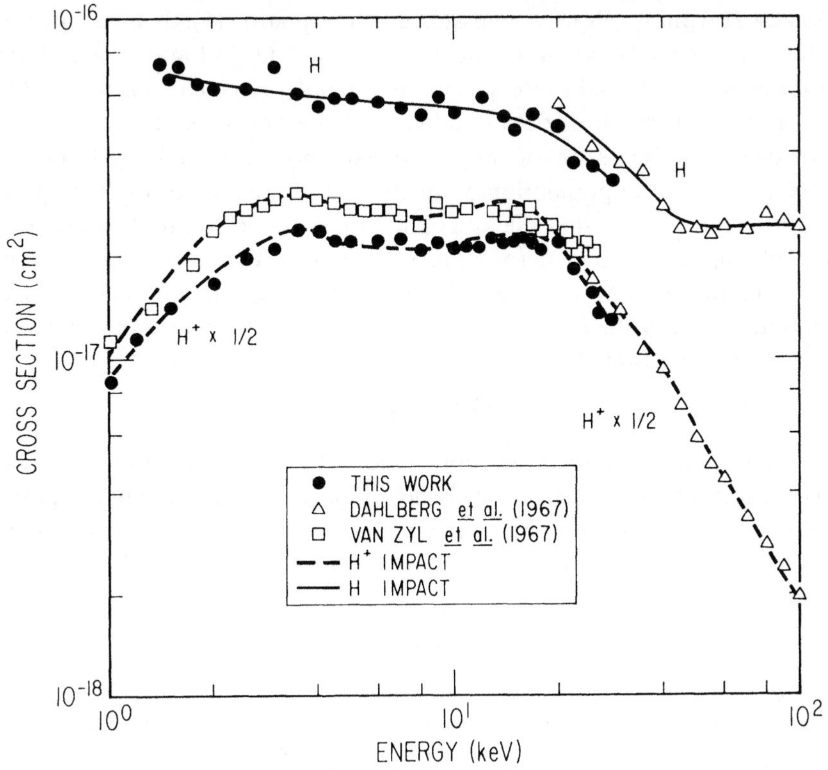

Fig. 4. Cross sections for emission of Lyα radiation in collisions of H^+ and H with N_2. The photometer used in this laboratory has been calibrated by reference to previous results for H^+ on Ar (Andreev *et al.*, 1966; Gaily *et al.*, 1968), as described by Birely and McNeal (1971). The H^+ data have been reduced by one-half for clarity in plotting.

for 1 to 25 keV H^+ impact obtained in this laboratory are also shown, normalized in the same way at 15 keV.

At present the scatter in these data must be considered to represent the magnitude of systematic errors in excitation experiments with H^+, primarily in photometer calibration. For other auroral spectral features there are fewer data, and extension and improvement of these experiments remain as major challenges to laboratory investigators. Data are particularly lacking for H impact excitation of atmospheric targets at low energies.

Hydrogen atom emissions that result from Equations (3) and (6) are of special interest in proton auroras. Many more data are available at energies of auroral interest for Equation (3) than for (6). Absolute capture cross sections into specific states of H have recently become available, for example H^+ on N_2 and O_2 to yield H in the 3*s*, 3*p*, and 3*d* states (Hughes *et al.*, 1970a).

For atmospheric gas targets there are very few data for Equation (6). Hughes *et al.* (1970b) have reported cross sections for population of the 3*s* level in collisions of

7.5 to 35 keV H with N_2. For Lyα emission due to Equation (6), there are no published data below 20 keV for N_2 and no data at all for O_2 or O. In Figure 4 are plotted preliminary results from this laboratory for emission of Lyα in collisions of 1 to 30 keV H with N_2, along with previous data obtained at higher energies. Data for H^+ impact are also shown. The results are for emission due to population of the 2*p* level. Measurements are in progress on population of the 2*s* level; there are no results at present for H on N_2. It is noteworthy that the 2*p* population cross section for H impact is larger than for H^+ impact. At low energies process (6) must, therefore, contribute most of the Lyα radiation in an aurora, because the H atoms are the dominant constituents of the mixed flux, and the cross section for excitation in Equation (6) is larger than for excitation in Equation (3).

Acknowledgments

It is a pleasure to acknowledge the contribution of Dr G. A. Paulikas whose interest and encouragement have been extremely helpful in the measurements program at the Space Physics Laboratory. Financial support by the U.S. Air Force Space and Missile Systems Organization (SAMSO) under contract F04701-70-C-0059 is also acknowledged.

References

Allison, S. K.: 1958, *Rev. Mod. Phys.* **30**, 1137.
Andreev, E. P., Ankudinov, V. A., and Bobashev, S. V.: 1966, *Sov. Phys. JETP* **23**, 375.
Birely, J. H. and McNeal, R. J.: 1971, *J. Geophys. Res.*, in press.
Carleton, N. P. and Lawrence, T. R.: 1958, *Phys. Rev.* **109**, 1159.
Chase, L. M.: 1968, *J. Geophys. Res.* **73**, 3469.
Dahlberg, D. A., Anderson, D. K., and Dayton, I. E.: 1967, *Phys. Rev.* **164**, 20.
Davidson, G. T.: 1965, *J. Geophys. Res.* **70**, 1061.
DeHeer, G. J., Schutten, J., and Moustafa, H.: 1966, *Physica* **32**, 1766.
Dufay, M., Desesquelles, J., Druetta, M., and Eidelsberg, M.: 1966, *Ann. Geophys.* **22**, 614.
Eather, R. H.: 1967, *Rev. Geophys.* **5**, 207.
Gaily, T. D., Jaecks, D. H., and Geballe, R.: 1968, *Phys. Rev.* **167**, 81.
Gardiner, H. A. B., Pendelton, W. R., Jr., Merrill, J. J., and Baker, D. J.: 1969, *Phys. Rev.* **188**, 257.
Hughes, R. H., Stigers, C. A., Doughty, B. M., and Stokes, E. D.: 1970a, *Phys. Rev.* **A1**, 1424.
Hughes, R. H., Filippelli, A. R., and Petefish, H. M.: 1970b, *Phys. Rev.* **A1**, 21.
Koopman, D. W.: 1968, *Phys. Rev.* **166**, 57.
McNeal, R. J.: 1970, *J. Chem. Phys.* **53**, 4308.
McNeal, R. J., Clark, D. C., and Klingberg, R. A.: 1970, *Phys. Rev.* **A2**, 131.
Philpot, J. L. and Hughes, R. H.: 1964, *Phys. Rev.* **133**, A107.
Prag, A. B., Morse, F. A., and McNeal, R. J.: 1966, *J. Geophys. Res.* **71**, 3141.
Puckett, L. J., Taylor, G. O., and Martin, D. W.: 1969, *Phys. Rev.* **178**, 271.
Robinson, J. M. and Gilbody, H. B.: 1967, *Proc. Phys. Soc. (London)* **92**, 589.
Solov'ev, E. S., Il'in, R. N., Oparin, V. A., and Fedorenko, N. V.: 1962, *Soviet. Phys. JETP*, **15**, 459.
Stebbings, R. F., Smith, A. C. H., and Ehrhardt, H.: 1964, *J. Geophys. Res.* **69**, 2349.
Stier, P. M. and Barnett, C. F.: 1956, *Phys. Rev.* **103**, 896.
Thomas, E. W., Bent, G. D., and Edwards, J. L.: 1968, *Phys. Rev.* **165**, 32.
Toburen, L. H., Nakai, M. Y., and Langley, R. A.: 1968, *Phys. Rev.* **171**, 114.
Van Zyl, B., Jaecks, D., Pretzer, D., and Geballe, R.: 1967, *Phys. Rev.* **158**, 29.

PART V

PARTICLE PRECIPITATION

PARTICLE PRECIPITATION PATTERNS

T. R. HARTZ

Communications Research Centre, Dept. of Communications, Ottawa, Canada

Abstract. The general pattern of auroral particle precipitation identified by Hartz and Brice (1967) has been re-examined in the light of more recent observations. It is shown to be an adequate description of almost all of the observations, with the exception of phenomena in the midday sector of the auroral oval. A consideration of such daytime observations, and notably of LF noise emiss ions that appear related to the low energy electron flux through the Cerenkov process, has led to the conclusion that a three zoned pattern might better describe the precipitation pattern. This paper gives a description of the suggested third zone and relates it to the other two zones which are already well established, and also to the cusp region on the magnetospheric surface.

1. Introduction

The past decade has seen a tremendous pile up of data relevant to the questions of where and when auroral particles precipitate. A number of statistical studies have been made of the various observations, one such being that of Hartz and Brice (1967) who identified a general precipitation pattern that was two zoned in nature. That particular study was based mainly on observations made with ground based instrumentation, and since then there have been many additional data from the satellite programs. Accordingly, while that pattern still appears to be a valid description of the particle influx, it is instructive to re-examine it in the light of the more recent, and in most cases, the much more detailed observations that now exist to see if some modifications might be made to the earlier conclusions.

It will be obvious that the process of arriving at a particle precipitation pattern from the detailed observations involves generalizations of several kinds. The satellite data include an ambiguity between the time and spatial coordinates which must be overcome by some form of statistical averaging. The ground data can give the temporal variation at several fixed locations from which a general pattern may be obtained by extrapolation and interpolation procedures. In each case the fine details in the data tend to be sacrificed at the expense of the overall picture. There is thus a constant danger that the real physical situation might be obscured and replaced by a pattern forced by the processing used, as one finds, for example, in the case of the 'spirals' in some of the older literature (cf., Meek, 1955). Moreover, to get complete diurnal coverage a variety of data needs to be considered; the observations referred to here are those with which the adjective 'auroral' has been associated, and include parameters whose relationship with the aurora may sometimes be questioned. Nevertheless, it is worth noting that the generalizations are not intended to provide a faithful description of the aurora, but rather to reveal a gross pattern or patterns for the particle precipitation which may be meaningful in terms of the magnetospheric processes.

B. M. McCormac (ed.), The Radiating Atmosphere, 225–238.

2. The 'Splash' and 'Drizzle' Precipitation Zones

The study carried out by Hartz and Brice (1967) tried to look closely at the distinguishing characteristics of the different phenomena. They examined riometer data on auroral absorption, all-sky camera data, results on magnetic variations, micropulsations, mantle aurora, various types of radio aurora, VLF radio noise and auroral hiss data, observations of spread *F* and sporadic *E* ionosonde echoes, balloon X-ray data, and satellite observations of the 10 keV and 40 keV electron flux. In almost every case they found indications of two general classes of characteristics which could be distinguished into two different diurnal patterns. These were identified, on the one hand, as discrete events which they associated with a 'splash' type of particle precipitation, and on the other hand, as diffuse events associated with a 'drizzle' type of precipitation. The former tended to maximize just prior to geomagnetic midnight, while the latter peaked in the late morning or forenoon hours (Figure 1).

It should be noted that this is an idealized diagram; it represents the latitude and time regions in which a number of auroral parameters occur on the average, and in

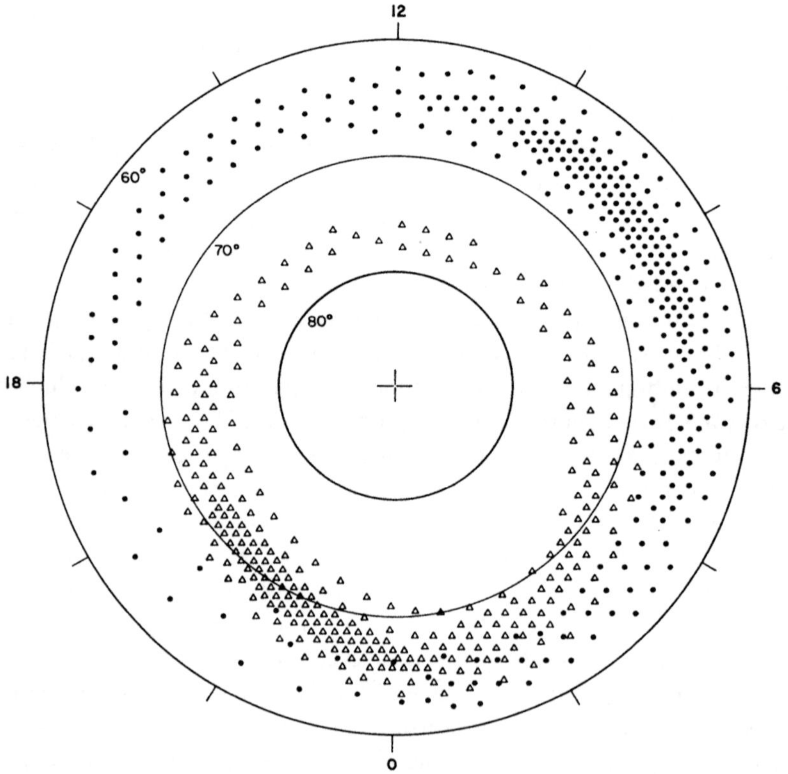

Fig. 1. An idealized representation of the two zoned auroral particle precipitation pattern in the Northern Hemisphere where the average flux is indicated approximately by the density of symbols and the coordinates are geomagnetic latitude and geomagnetic time (Hartz and Brice, 1967).

which regions they show similar characteristics. The Hartz and Brice analysis tended to force a set of observations into one of only two categories. Consequently, they smoothed out a lot of detailed features and fine structure, played down the degree of overlap of the two regions, and generalized the particular observational parameters. They also portrayed the two zones as rather narrower than warranted by the observations in order to accentuate the distinction between the two; probably a more realistic representation would have shown some overlapping at all hours of the day.

It will be obvious that the inner or splash zone corresponds with what has become generally known as the auroral oval. Feldstein's (1966) results on percentage occurrence, of course, do not show a nighttime maximum, but it is well known that the most active auroral features appear at this time (Akasofu, 1964). It is also apparent that the outer or drizzle zone is not represented in the conventional auroral statistics, such as the all-sky camera data of Feldstein (1966). The precipitation here is at about a constant geomagnetic latitude and undoubtedly is of more energetic electrons that can lead to *D* region phenomena such as auroral absorption (cf., Hartz *et al.*, 1963) and to the mantle aurora identified by Sandford (1964). The diurnal maximum falls in the forenoon hours at a latitude that is associated with magnetic field lines which lie deep in the trapped particle zones. Accordingly, Hartz and Brice (1967) suggested that this morning peak must constitute the principal sink for the outer Van Allen radiation belt. The splash precipitation, by contrast, occurs at the feet of field lines that are at or very close to the boundary between the open and closed families of lines in the magnetosphere.

Regarding the interrelation of events in the two zones, one can find many instances in the literature that strongly suggest a sequential association (cf., Collins and Forsyth, 1959), but this was first demonstrated conclusively by the work of Jelly and Brice (1967). Following the commencement of the spectacular phenomena that mark the substorm in the night sector, the drizzle zones show a build up and decay of activity on a somewhat longer time scale which apparently occurs at all longitudes but is most noticeable in the late morning sector (Jelly, 1970). This paper will not dwell on this aspect, however; instead it will look more closely at the characteristics of the phenomena in the inner zone or auroral oval

There are philosophical arguments which lead one to expect that phenomena on the day-and nightsides of this oval should differ. In the one case the boundary between the open and closed lines lies deep in the magnetospheric tail in association with the plasma sheet, and in the other case it is linked to the neutral point or neutral line on the surface of the magnetosphere and thus probably to the magnetosheath plasma. Observationally, there are substantial data that indicate the auroral phenomena near noon have quite different properties from those at night. Instead of aurora that is primarily substorm related, of limited duration, localized, rather sporadic, and at a height near 105 km, the daytime phenomenon apparently occurs virtually 100% of the time at heights in excess of about 140 km (Starkov and Feldstein, 1967; Starkov, 1968; Lassen, 1969). Recently, Akasofu (1970) and Romick (1970) have reported midday auroras with intense red (6300 Å) color, characteristic of the greater heights and low energy particle precipitation; similar findings have been reported by Eather (1969) and Eather and Mende (1971). These auroras tend to be of the glow type with some arcs and ray structure that exhibit a green (5577 Å) lower border. They occur in the higher latitude portion of the oval, above about 75° geomagnetic, whereas the lower latitude side of the oval shows aurora that appears typical of a more energetic particle influx. The magnetometer data also show very significant day-to-night differences in the oval, as described by Mishin *et al.* (1970), and somewhat similar conclusions have been expressed in regard to a number of other parameters (cf., the 'zone of confusion' referred to by Axford and Hines (1961)). Accordingly, the general pattern for particle precipitation presented by Hartz and Brice (1967), and outlined briefly above should probably be modified somewhat. In the remainder of this paper further experimental justification for a modification will be presented and a more appropriate precipitation pattern will be suggested.

3. LF Radio Noise and Associated Electron Flux

The ISIS 1 satellite was launched in January 1969 into a 90° inclination orbit with apogee and perigee heights of about 3500 and 600 km (Florida, 1969). Onboard are several experiments that have given results on the particle influx; a soft particle spectrometer to measure the electron and proton flux at energies less than about 10 keV and a complex of detectors to measure the particle flux at energies greater than this limit. There are also two radio noise experiments that can measure noise emissions generated by the particle influx and some of the low frequency results will be described here. In addition, analogous LF results have been obtained from a similar noise experiment on the Alouette 2 satellite, and a statistical study of those data has been made from which the average particle flux may be inferred (Hartz, 1969, 1970b; Barrington *et al.*, 1971).

The noise experiment of immediate concern involves a swept frequency receiver covering the frequency range from about 100 kHz to 15 MHz. In this range it is possible to observe a number of characteristic bands of noise emissions, some of which can be identified as having an ionospheric or magnetospheric origin (Hartz, 1969). This paper will discuss one such noise band which exists at the lower end of the observing range of frequencies for the satellite receiver, below the lower of the local electron plasma frequency, f_N, and the local electron cyclotron frequency, f_H. This region of the spectrum supports a mode of radio propagation that is generally known as the ordinary or whistler mode. For this mode the phase refractive index is greater than unity and under suitable circumstances may take very large values. In describing the noise emissions that are observed by the satellite receiver this large refractive index must be kept in mind since it can influence both the generation and propagation of emissions. Note that the emissions referred to here are not whistlers, but rather the broadband and quasi-continuous noise that exists in the so-called whistler mode of radio propagation.

A sample recording from the Alouette 2 receiver is shown in the upper portion of Figure 2. The receiver sweeps through its full range once each 30 sec, and only the

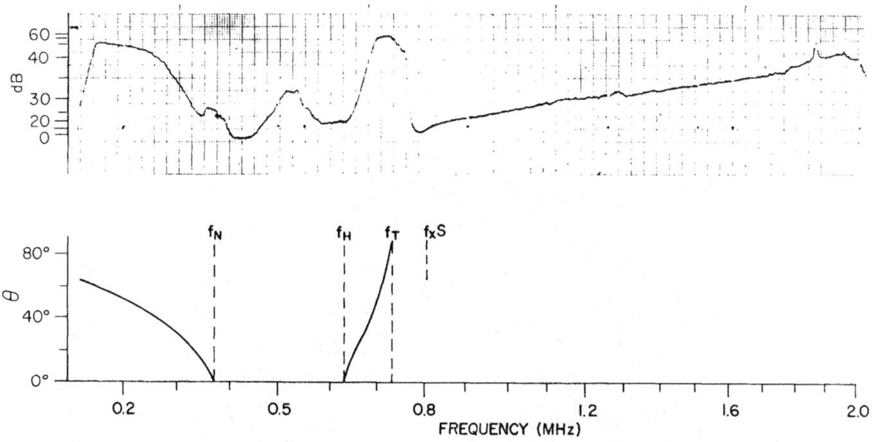

Fig. 2. (a) Sample recording of the background radio noise intensity from the swept frequency receiver on Alouette 2 obtained at 2143:49 UT on February 3, 1969. The indicated relative intensity scale is applicable at 0.2 MHz and the frequency scale is given below.
(b) A graphical representation describing the above sample record and indicating the local characteristic frequencies of the ionosphere and the spectral regions where, as a function of θ, the refractive index may take very large values in this instance. f_N is the plasma frequency, f_H the electron gyrofrequency, f_T the upper hybrid frequency, and f_xS the limiting frequency for the extraordinary mode.

low frequency portion of a sweep is shown here. Also shown in the lower portion of the diagram are the local characteristic frequencies at the satellite at the time. These mark discontinuities in the local propagation modes and indicate the frequency regions in which characteristic noise bands are expected. On the assumption that the noise emissions are generated by the Cerenkov process (Jørgensen, 1968, 1971), the following relationship obtains between V_{11}, the component of the electron velocity parallel to the field, and the wave phase velocity, V_p, where θ is the angle between the wave normal and the magnetic field direction;

$$V_p = V_{11} \cos \theta .$$

Thus the requirement for Cerenkov noise generation is that the wave phase velocity should be less than that of light, or that the refractive index, n, should exceed unity. For low energy electrons n would have to be quite large; the lower portion of Figure 2 shows the frequency ranges as a function of θ for which very large values of n are obtained in this example. Clearly the two intense noise bands observed in the sample record are consistent with this particular generation mechanism (cf., Jørgensen, 1971). Accordingly, it may be possible to take a measure of the noise intensity in the whistler mode or lower frequency band, at say 0.2 MHz, as a measure of the low energy electron influx. This is of some importance since there are available statistical data for a number of years of satellite noise measurements in the whistler mode, but it is only very recently that soft electron data in any quantity have been obtained (Barrington *et al.*, 1971; Heikkila, 1970).

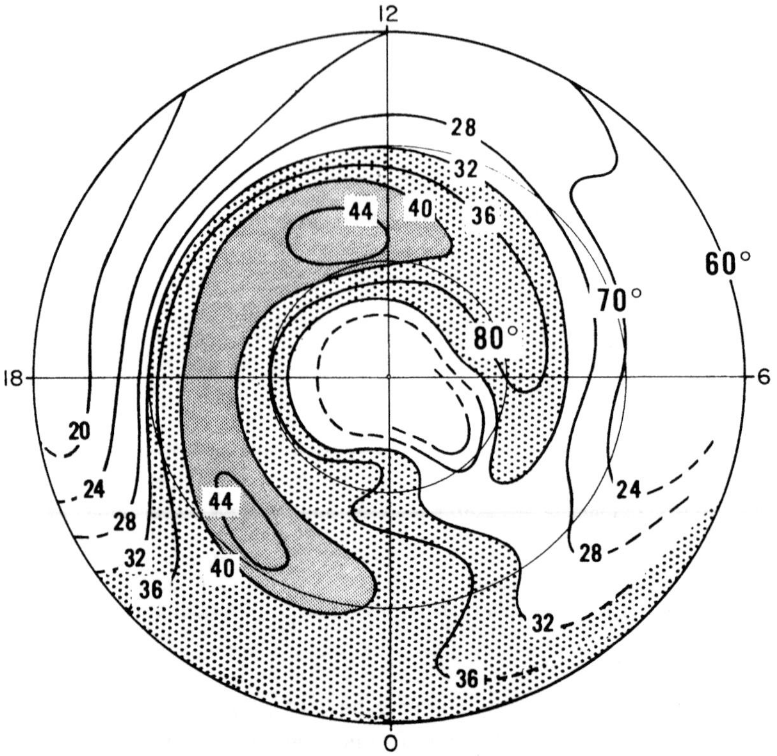

Fig. 3. Contour diagram of the average whistler mode noise intensity at 0.2 MHz as a function of IN Lat and LMT. The contours are labeled in dB relative to the receiver threshold.

A statistical survey of Alouette 2 noise recordings has been carried out to determine the latitudinal and diurnal variation of the intensity of the whistler mode emissions at a fixed frequency of 200 kHz (Barrington *et al.*, 1971; Hartz, 1970b). The data used were obtained at the Ottawa and Resolute Bay telemetry stations and covered two 6 mo periods in 1966 and 1967. The results for the higher latitudes are shown in Figure 3. Here is plotted in contour form the average intensity of the noise in dB relative to the receiver threshold, as a function of IN Lat and local mean time. It may be seen that the maximum intensity is associated with the auroral oval latitudes but that the day and afternoon side of the oval is favored over the midnight to dawn sector which is characterized by quite low values (cf., Hughes and Kaiser, 1971).

If this interpretation for the noise generation mechanism can be substantiated then the maximum regions in Figure 3 are, statistically, where the peak influx of low energy electrons should occur. From the refractive index considerations the indicated energies are of the order of, or less than, about 1 keV. There is now considerable supporting evidence for this interpretation in the form of direct particle measurements from satellites, and also from electron density and auroral observations on the dayside of the oval, as noted above. Specifically, the ISIS 1 data obtained with the soft particle spectro-

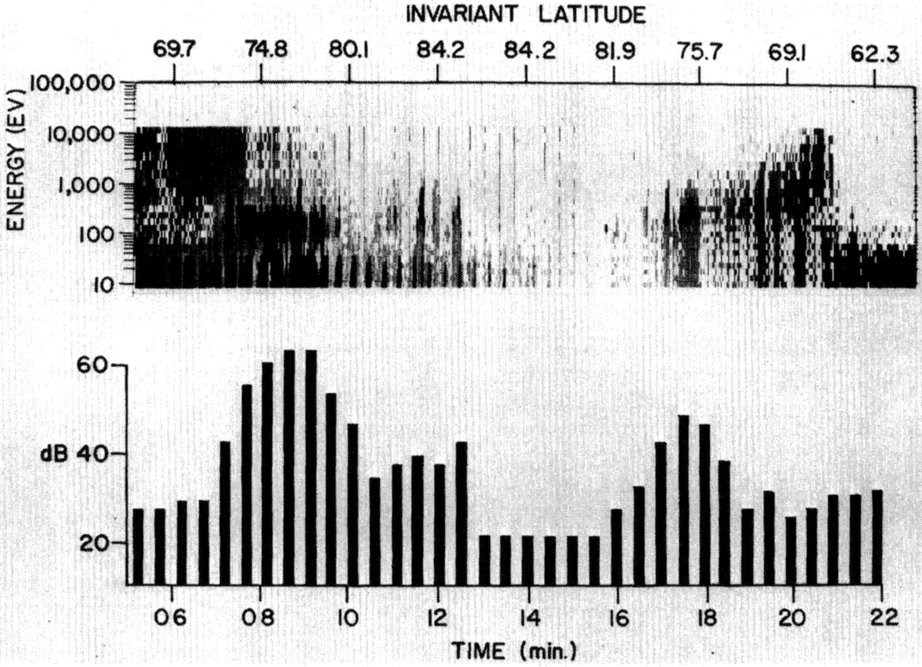

Fig. 4. (a) Electron data from the soft particle spectrometer on ISIS 1 obtained on February 4, 1969, between 0205 and 0223 UT on a trans-polar pass, with the density of the trace for each spectral sweep modulated by the counting rate (Courtesy of W. J. Heikkila and J. D. Winningham.) (b) The corresponding radio noise intensity at 0.2 MHz in dB relative to the receiver threshold.

meter by Heikkila and Winningham (1970) and from the radio noise experiment by Hartz (1970a) show a very significant correlation. This is illustrated in Figure 4 for a sample pass over the polar regions. The particle spectrometer measures the electron flux in the energy range from 10 eV to 10 keV each sec, and these data are portrayed in the upper portion of the diagram using a grey scale to represent the flux. The radio receiver sweeps through the spectrum once every 30 sec, so that the noise intensity at 200 kHz, shown by the histogram in the lower portion, cannot be compared in the same detail. Nevertheless, the general correlation is quite evident. In this example these data cover an entire pass across the polar cap, starting on the dayside near 65° IN Lat and 10 hr LMT and ending near 60° IN Lat and 22 hr LMT. It is apparent that the regions of most intense noise correspond with those for which the satellite encounters a high electron flux at energies of a few hundred eV. A detailed examination of the particle data shows both a satellite roll modulation and some additional fine structure which is rather more prominent in the nighttime than the daytime data. For the midday interval one finds typically a fairly uniform intense electron flux of energy in the several hundred eV range which correlates very well with the whistler mode noise intensity. This is taken as justification for interpreting Figure 3 in terms of a soft electron influx.

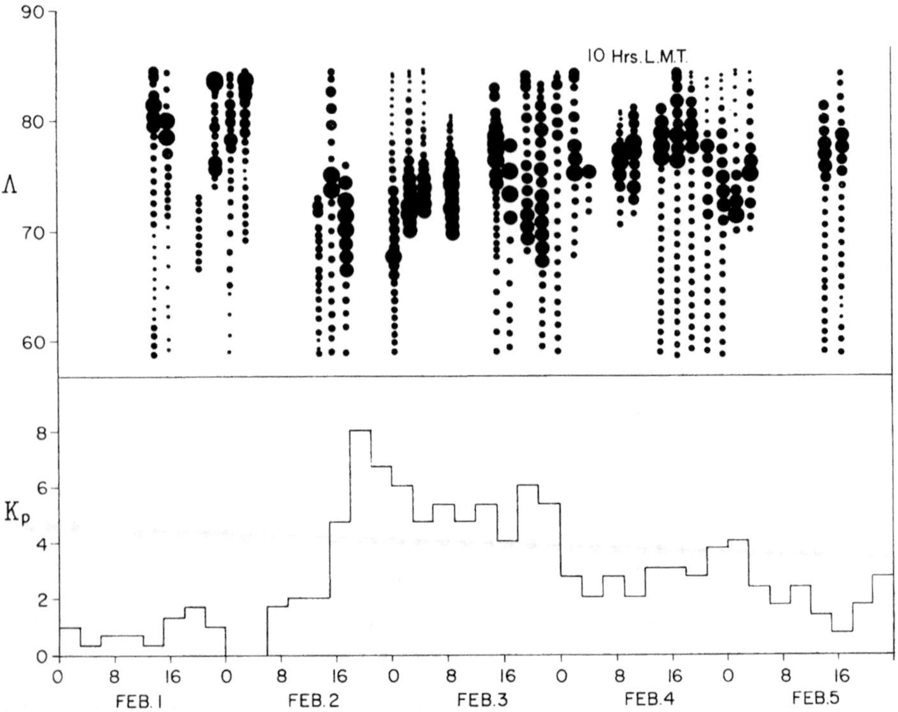

Fig. 5. (a) The noise intensity at 0.2 MHz on the high latitude ISIS 1 satellite passes during the interval, February 1 to 5, 1969, plotted as a function of IN Lat and UT. These data are all for the same LT, namely 10 hr LMT. The intensity is represented by the size of the data points, which range in 9 dB steps from < 20 dB for the smallest to > 56 dB for the largest.
(b) The corresponding K_p indices for the 5 day interval.

The dependence of the electron influx on geomagnetic disturbances was investigated by studying the noise observations from ISIS 1 and Alouette 2, taken over the northern auroral zone for a 5 day interval (February 1 to 5, 1969) that included a major storm. A summary of these data for the time sector near 10 hr LMT is shown in Figure 5. The upper portion of the diagram represents the noise intensity at 200 kHz for a series of south-to-north passes during the 5 day interval, all at about the same LT. The lower portion of the figure shows the K_p index during the period. On the first day, while K_p was low, the noise is seen to have been most intense (i.e., the represented data points have maximum size) at a latitude near 80°. Later, during the high K_p conditions the latitude at which the noise is most intense is seen to be significantly lower though the noise intensity does not appear to have changed significantly. Thus, a good correlation can be seen between K_p and the colatitude of the midday noise zone, but the peak intensity of the noise and hence the particle flux does not seem to be disturbance dependent.

On the other hand, for the night hours the corresponding noise data for this storm show a much more erratic pattern, as is shown in Figure 6 for 02 hr LMT. Here

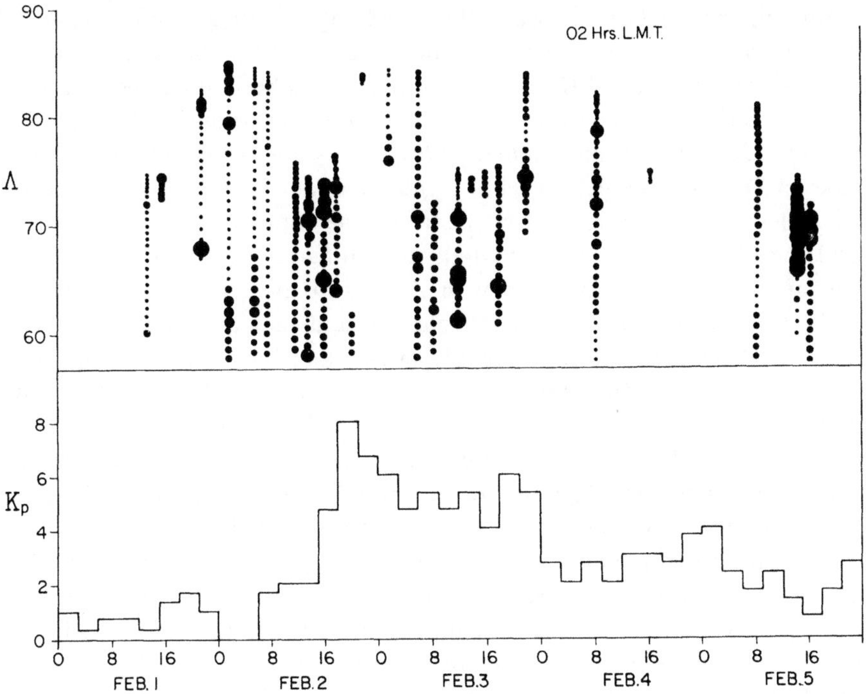

Fig. 6. (a) The noise intensity at 0.2 MHz on the high latitude Alouette 2 satellite passes during the February 1 to 5, 1969, interval plotted as a function of IN Lat. and UT. These data are all for about the same local time, namely 02 hr LMT. The intensity on each receiver sweep is represented by the size of the data point, which range in 8 dB steps from < 20 dB for the smallest to > 52 dB for the largest. (b) The corresponding K_p indices for the interval.

very large changes can be seen to have occurred from one receiver sweep to the next, implying a fine scale structure that is consistent with the nighttime 'splashes' or substorm phenomena. Furthermore, the average noise level is much reduced, indicating that the particle flux in the hundred eV range is generally lower as already suggested by the example in Figure 4 and the statistical plot in Figure 3.

4. Particle Precipitation from the Magnetopause Cusp Regions

A comparison of Figure 3 with the Feldstein (1966) auroral oval, the Akasofu (1964) substorm picture, or the Hartz-Brice (1967) general precipitation pattern (Figure 1) reveals some significant differences. The midday portion of the oval above 75° IN Lat., shows a characteristic low energy precipitation that does not show the short term substorm properties nor does it appear to be disturbance related except in latitude. The energy spectrum is quite narrow and the most logical interpretation is that these are magnetosheath particles that have entered the magnetosphere through the cusp region associated with the neutral point or neutral line (Hartz, 1970b; Heikkila and Winningham, 1971). Considerable theoretical and experimental support can be found for this explanation (Lebeau, 1965; Piddington, 1965; Pletnev *et al.*, 1965; Spreiter and Summers, 1967; Andrews and Thomas, 1969; Sato and Colin, 1969; Willis, 1969; Mishin *et al.*, 1970). The cusp region has been described as a region of trapped gas bounded by the geomagnetic field on the inside and by a free streamline on the outer side (Spreiter and Summers, 1967; Willis, 1969). Leakage of such particles may take place down the field lines at a rate determined by the solar wind particle flux and velocity, with a further dependence on the relative orientation of the dipole axis and the direction of the solar wind. Accordingly, one might expect a midday particle precipitation in association with the cusp region, that the particle energies should be those of the magnetosheath plasma, and that this influx should not be primarily K_p dependent (except probably in latitude) but rather dependent on the solar wind parameters. The present observations appear to be consistent with this explanation.

The statistical results shown in Figure 3 would indicate that the low energy electron flux extends well into the afternoon and evening hours. The latitude of the evening maximum appears to be significantly higher than that of the auroral oval, more so than appears to be the case on the day side (cf., Burch, 1968; Hoffman, 1969; Johnson and Sharp, 1969; Hughes and Kaiser, 1971). While it is recognized that the cusp region must have a finite east-west extent, it is not expected that all of the low energy particle flux should be considered in this same category. It has been emphasized earlier (Hartz and Brice, 1967) that different zones in the precipitation pattern are categorized according to properties other than only energy, so that a critical examination of the general characteristics of this flux is required. It has already been shown that the nighttime observations exhibit much more structure than do those from the dayside. Furthermore, there are indications at night of a somewhat harder spectrum (Burch, 1968) and an apparent association with localized auroral or substorm features (Sharp and Johnson, 1971). Although evident to some extent near midday, such features appear to accompany a more general background of softer particles (cf., Figure 4). It seems likely that the low energy flux on the nightside can be considered to belong to the splash precipitation. If so the energies are apparently those of the plasma sheet or of the tail region at slightly higher latitudes (Bame *et al.*, 1967; Sharp and Johnson, 1971) without having been accelerated in the localized precipitation process. The time of night at which such an influx should maximize on the average, presumably, is the same as for the other splash phenomena (Hartz and Brice, 1967), i.e., just prior to geomagnetic midnight, and this is seen to be borne out by Figure 3.

5. The General Pattern of Auroral Particle Precipitation

On the basis of the foregoing discussion regarding the midday maximum at auroral oval latitudes, there seems to be ample justification for suggesting a modification to the pattern depicted in Figure 1. This modification would distinguish phenomena on the dayside of the oval from those at night on the basis of general characteristics rather than on simply an energy basis. The revised pattern is presented in Figure 7,

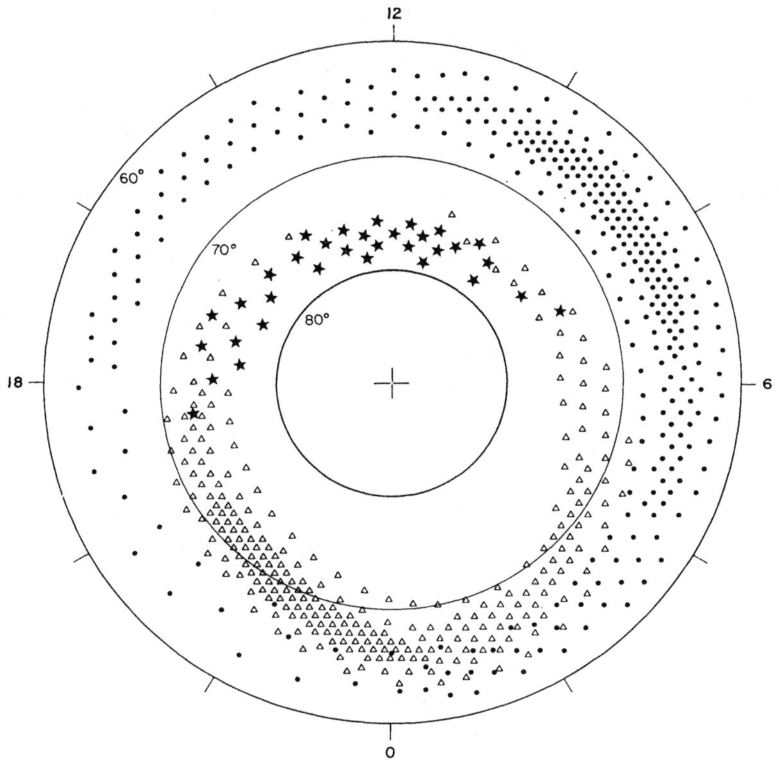

Fig. 7. An idealized representation of a three zoned auroral particle precipitation pattern. The coordinates are again geomagnetic latitude and geomagnetic time, and the average flux is indicated approximately by the density of symbols. The splash type of precipitation is represented by the triangles, the drizzle type by the dots, and the quasi-continuous inflow at midday by the stars.

with the stars representing the distinctive midday influx and the rest of the diagram essentially corresponding to the earlier version. Again this is an idealized diagram to represent the average situation; actually the observations of particular phenomena suggest rather more overlapping of the three principal zones with a more gradual transition from one to the other than is shown here. Furthermore, the nighttime splash zone is depicted on the basis of the more energetic flux and does not adequately represent the low energy splash events which tend to lie at slightly higher latitudes.

In each of the three principal regions identified in Figure 7 one finds, in general, a rather distinctive type of precipitation. The distinguishing characteristics are summarized in Table I. Insofar as possible, these are generalized characteristics and are not limited to only one observable parameter. Two of the categories have already been described by Hartz and Brice (1967), while some aspects of the third must still be considered as rather tentative at this stage and are designated as such with parentheses. Obviously, these characteristics of the precipitation must be taken in a relative rather than an absolute sense.

TABLE I

Precipitation zone characteristics

Noon peak ★	Night peak △	Morning peak ●
Continuous or quasi-continuous	Sporadic, abrupt onset	Long duration, gradual onset
Widespread, some fine structure	Localized	Widespread, little structure
Stable, fine structure changes	Rapidly changing, impulsive	Quasi-stable, slow changes plus quasi-periodic modulation
Not substorm associated	Substorm associated	Substorm associated
Flux not K_p dependent	Flux K_p dependent	Flux K_p dependent
Latitude shift with K_p	Latitude shift with K_p	No latitude shift with K_p
(No time shift with K_p)	No time shift with K_p	Time shift with K_p
Peaked spectrum, principal flux ~ few tenths keV	Steep spectrum, principal flux ~ few keV	Fairly flat spectrum, principal flux ~ few tens keV
Mainly *F* region	Mainly *E* region	Mainly *D* region
(Maximum in summer)	Maximum at equinoxes	(Minimum in summer)

The foregoing discussion has concentrated on the electron influx and very little consideration has been given here to the role of the protons in the auroral phenomena. As a result this precipitation pattern is incomplete until this has been done.

6. Summary

This paper has re-examined the general pattern of auroral particle precipitation identified by Hartz and Brice (1967). Further statistical data have been studied on a variety of high latitude phenomena that apparently are associated with auroral electrons and the earlier two zoned pattern still appears to be an adequate description of almost all of the observations. In the midday region of the auroral oval, however, a particle influx that is characteristically different from the splash events has been recognized. The properties of some of the phenomena observed in this region have been examined and were found to fit into a separate category. As a consequence a modification is suggested to the Hartz-Brice general precipitation pattern to include this third zone. The resulting pattern would then consist of three distinctive zones within which, on the average, the observable phenomena show common characteristics. The splash zone is mainly associated with the nighttime hours; it involves short term processes that are generally described as substorms, is located in or near the boundary between the open and closed field lines in the magnetospheric tail, and presumably is fed from the socalled plasma sheet and/or the immediately adjacent tail regions. The drizzle zone is mainly a morning or forenoon feature, though it may extend to other hours of the day; it appears to involve a later stage in the substorm, is associated with the trapped zone of particles in the magnetosphere, and apparently constitutes the sink for such particles. The third zone is a daytime feature; it appears to be continuous or quasi-continuous rather than substorm related, and apparently represents the direct entry of solar wind (i.e., magnetosheath) plasma through the magnetospheric cusp regions.

It must be emphasized that the process of identifying these distinctive regions has smoothed out a lot of detail that was inherent in the original data. The emphasis has been placed on the dominant or most outstanding characteristics in each instance, so that the regions where the different zones overlap, or where two different phenomena occur simultaneously, are not represented reliably. Thus the final pattern is somewhat idealized; nevertheless, it should serve a useful purpose in studies of the gross behavior of the magnetosphere.

References

Akasofu, S.-I.: 1964, *Planetary Space Sci.* **12**, 273.
Akasofu, S.-I.: 1970 *Trans. Am. Geophys. Union*, **51**, 370.
Andrews, M. K. and Thomas, J. O.: 1969, *Nature* **221**, 223.
Axford, W. I. and Hines, C. O.: 1961, *Can. J. Phys.* **39**, 1433.
Bame, S. J., Asbridge, J. R., Felthauser, H. E., Hones, E. W., and Strong, I. B.: 1967, *J. Geophys. Res.* **72**, 113.
Barrington, R. E., Hartz, T. R., and Harvey, R. W.: 1971, *J. Geophys. Res.*, to be published.
Burch, J. L.: 1968, *J. Geophys. Res.* **72**, 3585.
Collins, W. C. and Forsyth, P. A.: 1959, *J. Atmospheric Terrest. Phys.* **13**, 315.
Eather, R. H.: 1969, *J. Geophys. Res.* **74**, 153.
Eather, R. H. and Mende, S. B.: 1971, this volume, p. 255.
Feldstein, Y. I.: 1966, *Planetary Space Sci.* **14**, 121.
Florida, C. D.: 1969, *Proc. IEEE* **57**, 867.
Hartz, T. R.: 1969, *Proc. IEEE* **57**, 1042.
Hartz, T. R.: 1970a, *Trans. Am. Geophys. Union* **51**, 376.
Hartz, T. R.: 1970b, in *The Polar Ionosphere and Magnetospheric Processes* (ed. by G. Skovli), Gordon and Breach Publishing Company, New York, in press.
Hartz, T. R. and Brice, N. M.: 1967, *Planetary Space Sci.* **15**, 301.
Hartz, T. R., Montbriand, L. E., and Vogan, E. L.: 1963, *Can. J. Phys.* **41**, 581.
Heikkila, W. J.: 1970, *Nature* **225**, 369.
Heikkila, W. J. and Winningham, J. D.: 1970, *Trans.Am. Geophys. Union* **51**, 375.
Heikkila, W. J. and Winningham, J. D.: 1971, *J. Geophys. Res.* **76**, 883.
Hoffman, R. A.: 1969, *J. Geophys. Res.* **74**, 2425.
Hughes, A. R. W. and Kaiser, T. R.: 1971, this volume, p. 336.
Jelly, D. H.: 1970, *Can. J. Phys.* **48**, 335.
Jelly, D. H. and Brice, N. M.: 1967, *J. Geophys. Res.* **72**, 5919.
Johnson, R. G. and Sharp, R. D.: 1969, in *Atmospheric Emissions* (ed. by B. M. McCormac and A. Omholt), Van Nostrand Reinhold Company, New York, p. 219.
Jørgensen, T. S.: 1968, *J. Geophys. Res.* **73**, 1055.
Jørgensen, T. S.: 1971, this volume, p. 327.
Lassen, K.: 1969, in *Atmospheric Emissions* (ed. by B. M. McCormac and A. Omholt), Van Nostrand Reinhold Company, New York, p. 63.
Lebeau, A.: 1965, *Ann. Geophys.* **21**, 167.
Meek, J. H.: 1955, *J. Atmospheric Terrest. Phys.* **6**, 313.
Mishin, V. M., Saifudinova, T. I., and Zhulin, I. A.: 1970, *J. Geophys. Res.* **75**, 797.
Piddington, J. H.: 1965, *Planetary Space Sci.* **13**, 363.
Pletnev, V. D., Skiridin, G. A., Shalimov, V. P., and Shvachunov, I. N.: 1965, *Geomag. Aeron.* **5**, 485.
Romick, G. J.: 1970, *Trans. Am. Geophys. Union.* **51**, 370.
Sandford, B. P.: 1964, *J. Atmospheric Terrest. Phys.* **26**, 749.
Sato, T. and Colin, L.: 1969, *J. Geophys. Res.* **74**, 2193.

Sharp, R. D. and Johnson, R. G.: 1971, this volume, p. 239.
Spreiter, J. R. and Summers, A. L.: 1967, *Planetary Space Sci.* **15**, 787.
Starkov, G. V.: 1968, *Geomag. Aeron.* **8**, 28.
Starkov, G. V. and Feldstein, Y. I.: 1967, *Geomag. Aeron.* **7**, 294.
Willis, D. M.: 1969, *Planetary Space Sci.* **17**, 339.

INSTITUTE DISCUSSION

L. Thomas discussed the potential for experimentally investigating the dayside entry of particles in the winter. It was suggested that the 2 to 3 erg cm^{-2} sec^{-1} of low energy precipitating particles reported by D. Winnigham for ISIS 1 should result in sufficient 3914 Å radiation in the polar cap in darkness to be observed.

Some discussion occurred on whether the noon 'rain' precipitation was really unassociated with substorms. L. Montbriand indicated the lack of correlation may be due to the presence of a succession of substorms.

LOW ENERGY AURORAL PARTICLE MEASUREMENTS FROM POLAR SATELLITES

R. D. SHARP and R. G. JOHNSON
Lockheed Palo Alto Research Laboratory, Palo Alto, Calif., U.S.A.

1. Introduction

The past few years have seen a tremendous increase in the available information on the details of the morphology of the low energy auroral particles from experiments on low altitude, polar orbiting satellites. For the purposes of this review, we will define low energy particles as those below the threshold of a conventional Geiger tube, that is $E \leqslant 40$ keV for electrons and $E \leqslant 500$ keV for protons. Table I lists some polar satellites which have carried experiments in this energy range. While not comprehensive, the table does include those experiments which have provided the principal information to date on the morphology of the particle precipitations and also lists the more recent successful experiments known to have resulted in large bodies of data which promise to provide the basis for detailed statistical studies.

The unique role of polar satellites in auroral research is in making the acquisition of these large bodies of data possible by providing global coverage with a single experiment, independent of land masses and the density of ground observatories, and thereby allowing for statistical studies of the detailed characteristics of the precipitating particles: their spatial, energy and angular distributions, and their variation between hemispheres, with season, and throughout the solar cycle. Due principally to the development of the channel multiplier (Evans, 1965), the instrumentation problems in this energy range have recently become more tractable, but they are still not completely routine. Added difficulties are encountered because of the highly variable nature of the auroral phenomena. For these reasons we will try, when possible, to focus on statistical studies rather than illustrative examples and give the greatest weight to those effects seen in more than one experiment.

2. Auroral Electron Precipitation

A. LOCATION

The recently published satellite data on the latitudinal and LT distributions of the auroral electron precipitations generally are consistent with the concept of the two zones of auroral precipitation first suggested by Piddington (1965) from a study primarily of ground based observations. Zone I in his model had a latitudinal extent between about 63° and 73°, with the strongest precipitation in the dawn hemisphere. Zone II overlapped Zone I at LT near midnight and extended poleward beyond 80° near local noon. A further study by Hartz and Brice (1967) led to the idealized re-

B.M. McCormac (ed.), The Radiating Atmosphere, 239–254.

TABLE I
Polar satellites

Satellite	Launch date	Particle and central or threshold energy or Energy range (keV)	References
INJUN 3	Dec. 12, 1962	e >10	Fritz and Gurnett (1965)
1963-42A	Oct. 29, 1963	e >0.08, >2, >21 p >4	Sharp *et al.* (1966); Sharp and Johnson (1968c); Sharp *et al.* (1969)
INJUN 4	Nov. 21, 1964	e >5	Craven (1970)
1965-90A	Nov. 8, 1965	e >0.015, >1, >2, >10, >25, >60 1, 2, 5, 12, 28, 70 p >10, >21, >38, >56	Johnson *et al.* (1967); Sharp *et al.* (1967); Johnson and Sharp (1969)
1967-65B (AURORA 1)	June 29, 1967	e 0.05–100 p 0.10–100	Maehlum (1968); Burch (1968, 1969)
OGO 4	July 28, 1967	e 0.7, 2.3, 7, 25	Hoffman and Evans (1968); Hoffman (1969a, b, 1970)
INJUN 5	Aug. 8, 1968	e 0.04–15 p 0.04–15	Frank and Ackerson (1970); Ackerson and Frank (1970)
ESRO 1A (AURORAE)	Oct. 3, 1968	e 1.3, 3, 6, 13 p 1, 6	Reidler and Hultqvist (1969); Liszka *et al.* (1970)
ISIS 1	Jan. 30, 1969	e 0.01–10 p 0.01–10	Heikkila and Winningham (1970); Heikkila (1970)
OV1 18	May 17, 1969	e >0.2, >1.2, >4, >7, >21 1, 2, 5, 12, 24 p >3, >9, >26, >38 1, 3, 8	Johnson *et al.* (1969); Paschmann *et al.* (1969)
OGO 6	June 5, 1969	e 1–20 p 1–20	Evans (1970)

presentation of these two zones, which is discussed in this volume by Hartz (1971) (see his Figure 7).

Published data on electron precipitation from five polar orbiting satellites have supported this two-zone concept. Johnson *et al.* (1966) first reported the existence of two distinct dayside zones of auroral electron precipitation based on observations made with satellites 1965-90A and 1963-42A (see also Sharp and Johnson, 1968a). Their 'soft' and 'hard' zones, the 'soft' and 'auroral' zones identified in the data from the Aurora 1 satellite by Burch (1968), and the 'burst' and 'band' regions studied by Hoffman (1969a) with the data from the OGO 4 satellite, appear to be caused by the precipitation of electrons whose properties are grossly consistent with those derived from the ground based studies. Heikkila (1970) has also reported that the preliminary results from the ISIS 1 satellite confirm these precipitation patterns.

Some important recent results on the spatial and LT distributions of these separate zones of electron precipitation have been derived from statistical studies based on the data from some of these satellites (Hoffman, 1969b; Sharp *et al.*, 1969). They show that the soft highly structured electron precipitations extend poleward of the auroral oval at all LT.

An important question still to be resolved is whether the positions of the LT maxima in the precipitation patterns derived from the ground based studies are consistent with statistical studies of the data from the long-lived satellites. Some results are now available from two such satellites which had extensive LT coverage. Hoffman (1969b) reports no indication of a LT maximum in the probability of occurrence of the 'burst' phenomena, but the average intensity of these precipitations as a function of LT was not reported. From a statistical study of the data from two threshold detectors on the INJUN 4 satellite Craven (1970) derived the percentage of occurrence contours for energy fluxes of $\geqslant 5$ keV electrons in excess of 0.4 erg cm^{-2} sec^{-1} sr^{-1} with pitch angles of about 40°. He finds a MLT maximum centered at about 04 hr and a latitudinal distribution diverging equatorward from the auroral oval at LT approaching noon. The auroral oval was determined by Feldstein (1966) from observations of discrete aurora using IGY all-sky camera photographs. At LT near noon the electrons causing these aurora typically have energies below the threshold of the INJUN 4 detector (Johnson and Sharp, 1969). Thus, the divergence in latitudinal distributions is not surprising. Whether the intermediate position of the LT maximum found by Craven (1970), with respect to the maxima of the two zones derived by Hartz and Brice (1967), is also due to the high detector threshold is a question which remains to be resolved.

B. ANGULAR DISTRIBUTIONS

It has been known for some time that the angular distributions of the integral number flux of energetic auroral electrons ($E > 40$ keV) are generally anisotropic, with maxima at high pitch angles, and approach isotropy with increasing intensity of the precipitation (O'Brien, 1964; Parthasarathy *et al.*, 1966). Similar results for the integral energy flux of electrons with energies greater than 80 eV were reported by Sharp and Johnson

(1968a, b, c) and have recently been confirmed for electrons with energies above 5 keV by Craven (1970). Hoffman (1969a) on the other hand, has reported different results for the differential number flux of electrons with energy of 2.3 keV. His angular distributions were generally isotropic (except for a small subgroup of events peaked along the field line which we shall discuss below). In the 'burst' region he reported no evidence of angular distributions peaked at high pitch angles. He did not state, however, over what intensity range of events this conclusion was based upon. The integral energy flux measurements discussed above were frequently dominated by higher energy electrons than those studied by Hoffman and so his results are not necessarily inconsistent with them. The possibility exists, for example, that the lower energy electrons approach isotropy at a lower intensity level than the more energetic electrons and that Hoffman's analysis did not include these low intensity events. Recent preliminary results on the angular distributions of differential number fluxes of low energy electrons from the OV1 18 satellite have shown that they indeed exhibit a significant anisotropy (peaking at high pitch angles) in low intensity events. The angular distribution results from this satellite promise to be more reliable than can be generally obtained from experiments utilizing separate detectors at different pitch angles. Due to a malfunction in the stabilization system the satellite is slowly tumbling with a 3-axis magnetometer providing orientation information. On occasion, it is oriented such that the various detectors can be intercalibrated against each other, leading to more precise angular distribution results than can be otherwise expected. Figure 1 shows some of these results at 5 keV derived by Paschmann (1970) from three sets of magnetic spectrometers oriented at three angles spanning a range of 90°. Results at the uniform pitch angles of 20° and 60° were obtained by passing smooth curves through the measured points at the particular angles assumed by the detectors at any one time and interpolating to the angles cited. The preliminary results indicate that a tendency toward anisotropy at low intensities occurs over the energy region from about 1 to 16 keV (Paschmann *et al.*, 1969). An intercomparison of these results with the integral energy flux and number flux measurements discussed above, in order to determine if the condition of isotropy is reached in lower intensity events for these softer particles than say for the >40 keV electrons analyzed by Parthasarathy *et al.* (1966) requires an assumption on the energy spectrum. The results are inconclusive; that is, one can interrelate the various results without resorting to this hypothesis using reasonable assumptions on the spectral shape. What is needed to resolve this question is an experiment covering the range from a few keV to >40 keV on one satellite.

Another interesting related result pertaining to the electron angular distributions has now been confirmed with measurements from two satellites. Sharp *et al.* (1969) reported a dramatic latitudinal dependence to the averaged pitch angle distributions of electron integral energy fluxes. They were almost isotropic near the center of the zone and became highly anisotropic in the high and low latitude wings. A similar result from the INJUN 4 experiment has been recently reported by Craven (1970).

As mentioned previously, Hoffman and Evans (1968) reported an interesting sub-

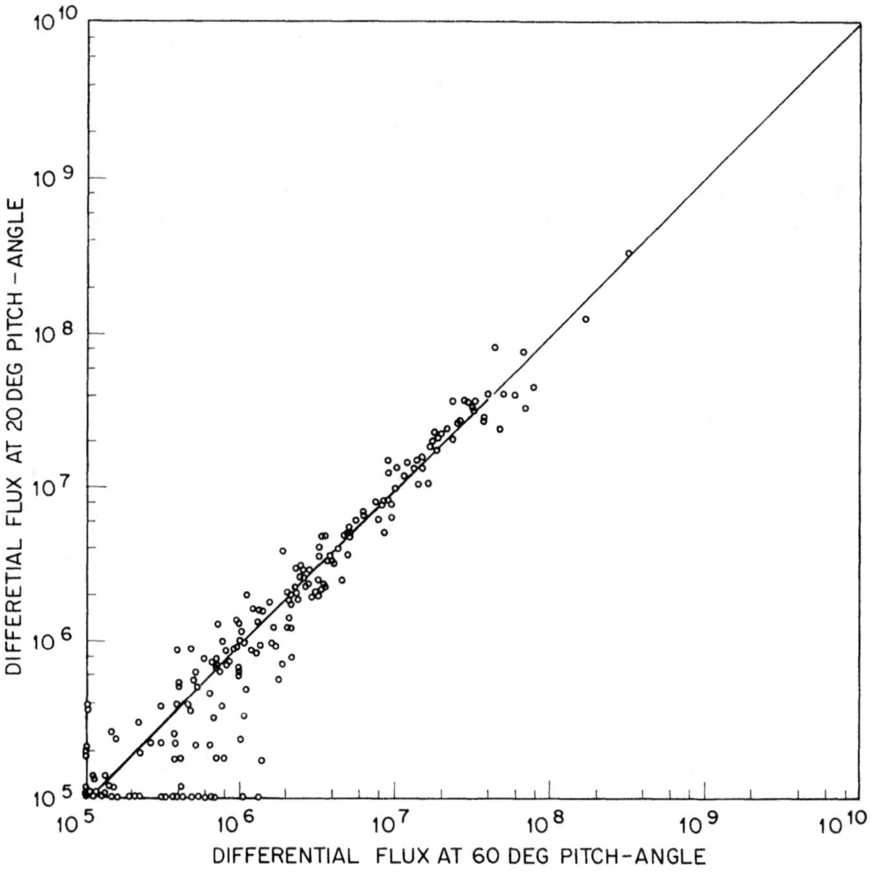

Fig. 1. OV1 18 satellite data for the differential fluxes of 5.4 keV electrons (cm^{-2} sec^{-1} sr^{-1} keV^{-1}) at 20° and 60° pitch angles. The solid line represents an isotropic angular distribution.

group of the data in the 'burst' region with pitch angle distributions maximizing along the magnetic field line. Their results were obtained at the single energy of 2.3 keV. This type of event has now been studied using the data from the OV1 18 satellite over a wide range of energies. In an early paper by Hoffman (1969b) it appears that events with anisotropies of up to a factor of 2 are relatively common (see his Figure 10). In his most recent paper, however, Hoffman (1970) states that departures from isotropy of this magnitude are not significant because of the difficulty in making these relative measurements with independent detectors. The significant events, therefore, form a relatively small subgroup of the data in better agreement with the results from the OV1 18 satellite, where the effect is measured with greater precision.

The most interesting new result to appear in the OV1 18 data is that this angular distribution effect is strongly energy dependent with the largest departures from isotropy observed in the lowest energy channel (Paschmann *et al.*, 1969). Figure 2 shows a typical event. The open circles are the data taken at about 30° pitch angle

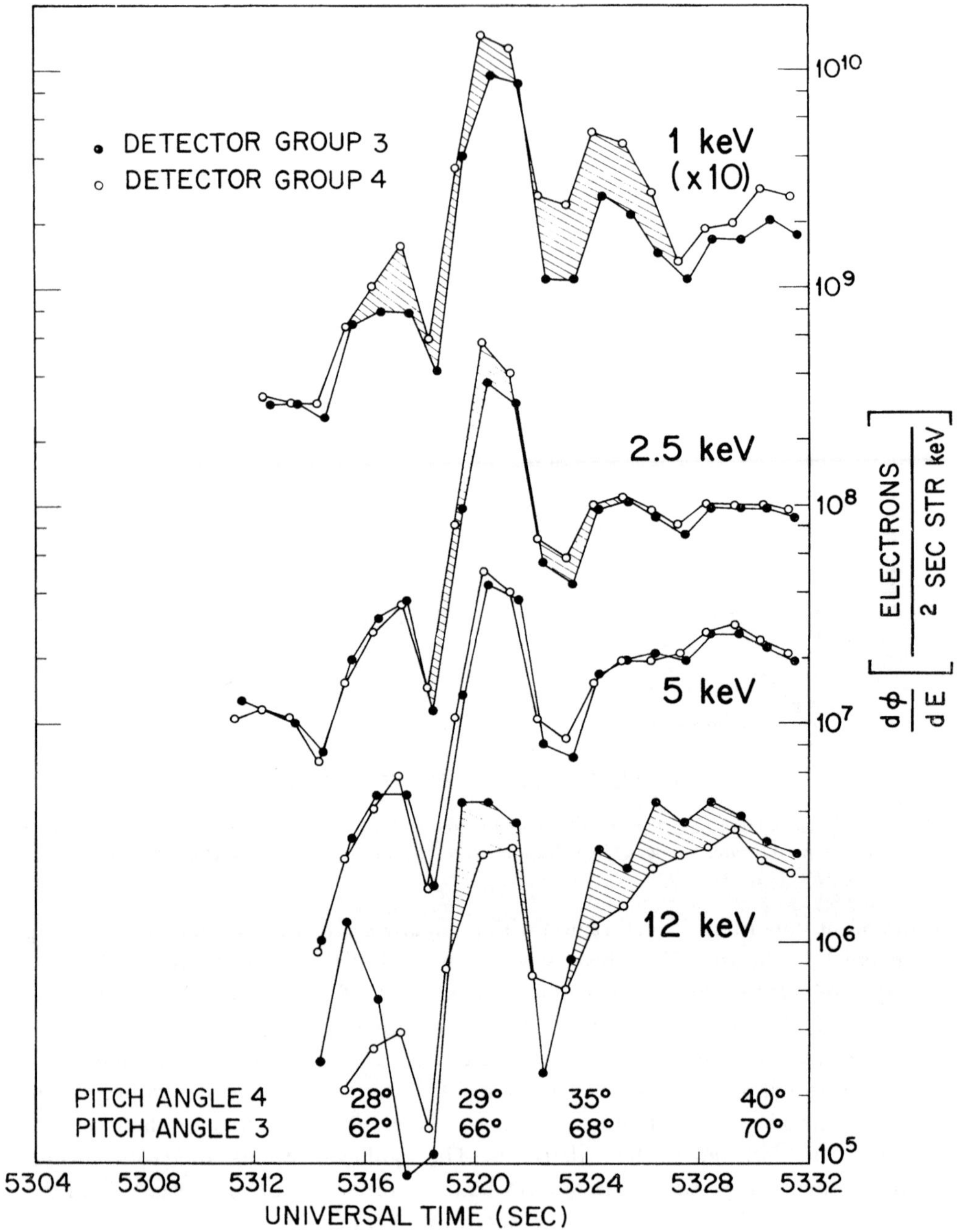

Fig. 2. An example of an anisotropic event measured on the OVI 18 satellite on May 13, 1969, at IN Lat 65°N, MLT = 02:40 (Paschmann *et al.*, 1969).

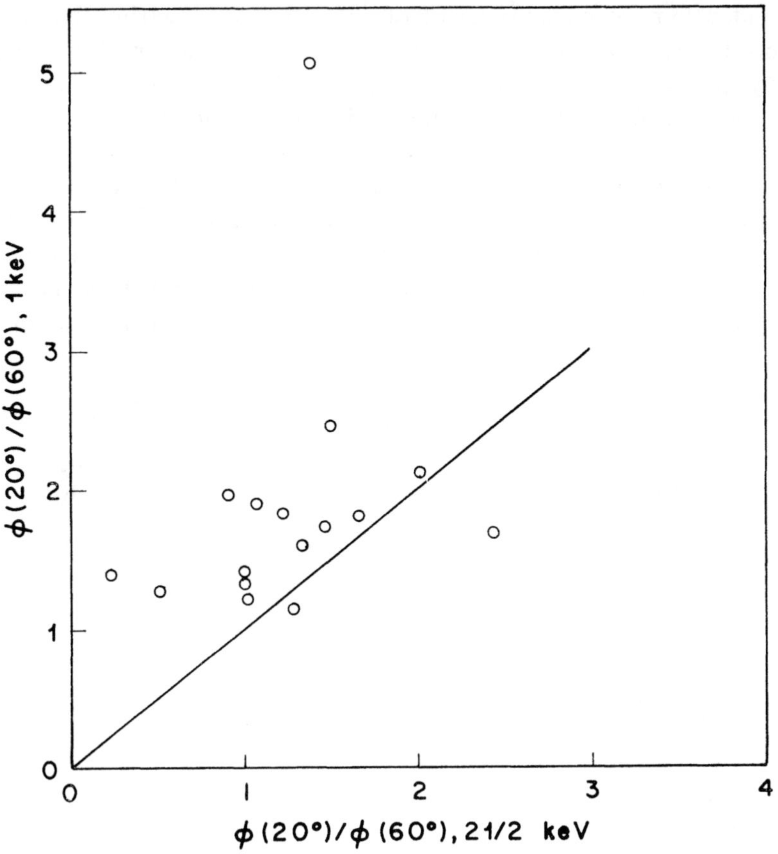

Fig. 3. OVI 18 satellite measurements of 16 anisotropic electron events. A scatter plot of the anisotropy parameter at 1 keV vs. that at 2.5 keV. The solid line represents equal anisotropy at the two energies.

and the filled circles are taken at about 65° pitch angle. One sees that angular distributions peaked along the field line are found at 1 and $2\frac{1}{2}$ keV. The 5 keV results are isotropic and the 12 keV angular distributions are peaked at the high pitch angles. A plot based on the data from 16 of these peaked events is shown in Figure 3 (Paschmann, 1970). The anisotropy parameter which is plotted is the ratio of the differential number flux at 20° pitch angle to that at 60° pitch angle obtained by interpolation as described earlier. The values of this parameter for the 1 keV electrons are plotted against those for the $2\frac{1}{2}$ keV electrons. The typical range of anisotropies and the strong energy dependence of the phenomena are illustrated.

C. ENERGY DISTRIBUTIONS

The most dramatic feature in the energy spectrums of the auroral electrons is the difference between the dayside 'soft' and 'hard' zones. Some typical spectrums for both these regions were shown by Johnson and Sharp (1969). Recently, Burch (1969) has computed some average properties of the 'soft' zone spectrums on both the day and nightsides. He finds that in both cases a substantial fraction of the energy flux is carried by electrons with energies of less than 1 keV. The average electron spectrum in the dayside 'hard' zone has been computed as a function of IN Lat by Sharp *et al.* (1969) and is shown in Figure 4. They found a dramatic spectral hardening with decreasing latitude in reasonably good agreement with the assumption that the electrons were precipitating from a quasi-trapped parent population that was drifting inward in L under conservation of the first two adiabatic invariants (the solid curve). This result is in agreement with the conclusions of Hartz and Brice (1967) that this precipitation region is associated with electrons from the outer radiation belt and

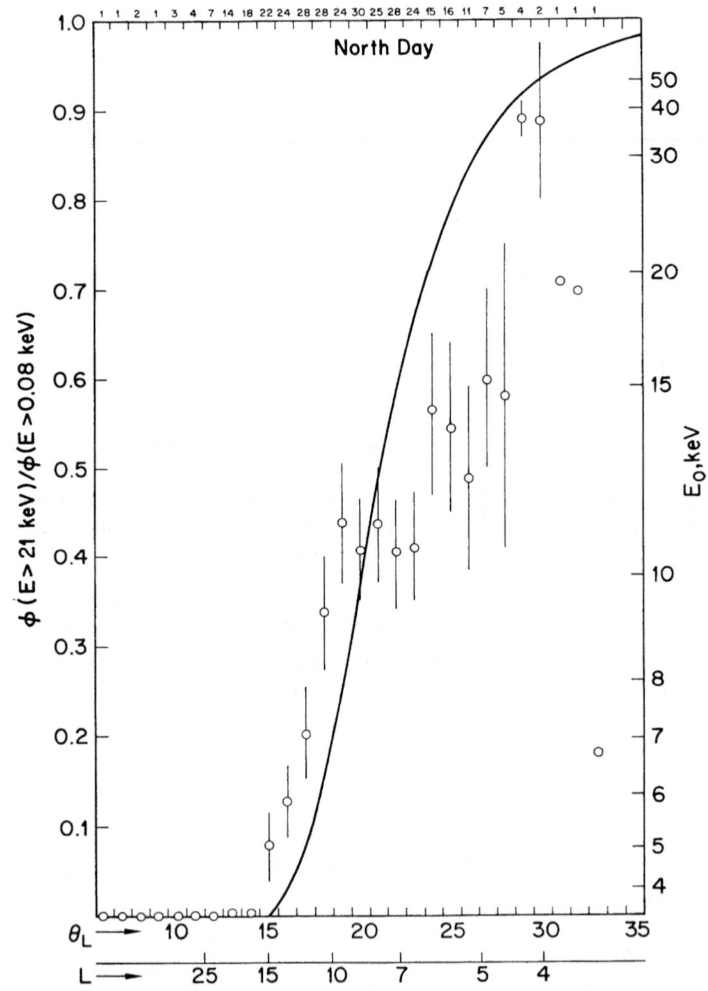

Fig. 4. Average values of an electron spectral parameter; the fraction of the energy flux in the form of electrons with energies greater than 21 keV from the dayside crossing of the auroral zone by the satellite 1963-42A (Sharp *et al.*, 1969).

supports the hypothesis that the inward drift and consequent acceleration of auroral electrons trapped on closed field lines may be the source of the energetic outer belt electrons.

A new result of great interest is the finding of a low latitude 'soft edge' to the night-side auroral electron precipitations. The effect was first reported by Sharp *et al.* (1969) from a study of the latitudinal dependence of the average electron spectrums. Their results (see Figure 6 of that work) were based on measurements with very limited spectral resolution. More detailed evidence of this phenomenon is now available in the data from the ESRO I (AURORAE) satellite. Figure 5 shows these data from two auroral zone traversals as reported by Liszka *et al.* (1970). The parameter S is the ratio of the differential flux at 1 keV to that at 6 keV. One sees a dramatic low latitude softening of the energy spectrum as well as the high latitude 'soft' zone. The frequency of occurrence, latitudinal extent, and the LT dependence of these events remain to be investigated.

Measurements with the OGO 3 satellite in the equatorial plane have shown that the inner edge of the plasma sheet is generally found at radial distances of 4 to 10 R_E

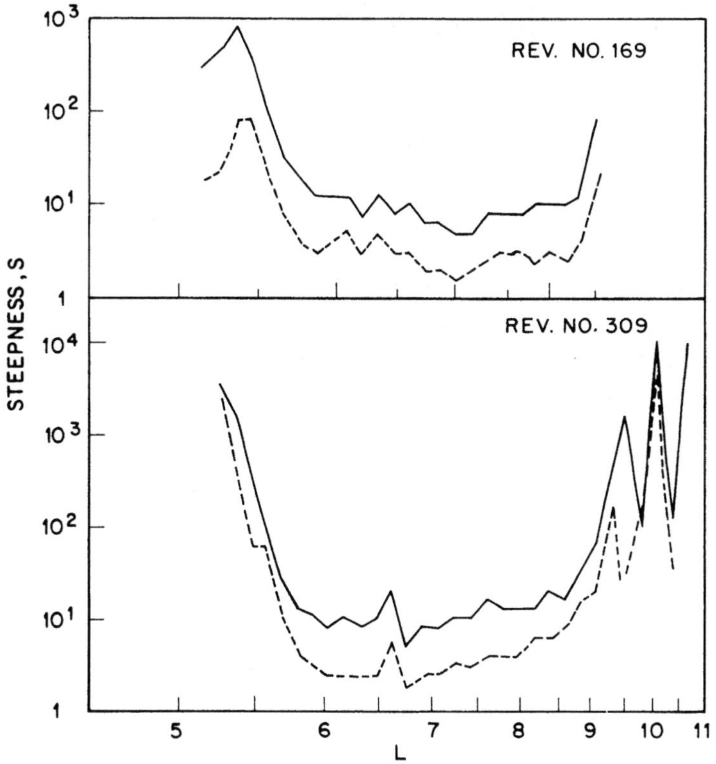

Fig. 5. Electron spectral data for two auroral zone traversals of the ESRO 1 satellite (Liszka *et al.*, 1970). The parameter S is the ratio of the differential flux at 1 keV to that at 6 keV. Solid curves correspond to precipitated electrons, broken to trapped electrons.

in the midnight sector (Schield and Frank, 1969). It is characterized by a dramatic softening of the electron spectrum with decreasing radical distance (Frank, 1968; Vasyliunas, 1968). Vasyliunas (1969) measured the location of this soft inner edge at relatively low altitudes using the data from several OGO 1 traversals of the high latitude horns of the plasma sheet, and made a comparison with ground based studies of the location of the auroral oval and of radar aurora. He concluded that the location of the inner edge of the plasma sheet corresponded to that of the equatorial boundary of the auroral oval. Figure 6 shows his suggested model. The low latitude spectral

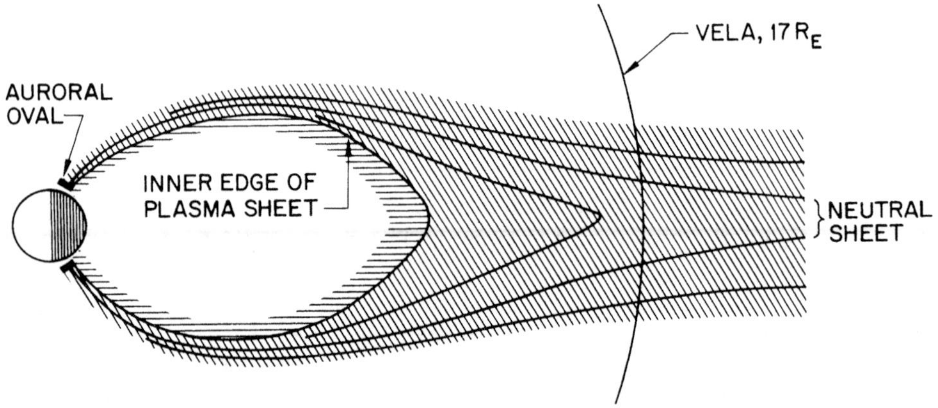

Fig. 6. Configuration of the magnetic field lines and the plasma sheet on the nightside of the magnetosphere and their relation to the auroral oval according to Vasyliunas (1969).

softening observed in the polar satellite data tends to support this general model. It is probably the low altitude signature of the soft inner edge of the plasma sheet. The latitudinal extent of the feature of about $\frac{1}{2}$ L value shown by the limited data in Figure 5 corresponds well with the typical radial extent of the comparable feature seen in the equatorial plane (Schield and Frank, 1969). Referring again to Figure 6, one sees that the magnetic field lines from the region of the high latitude soft zone on the nightside map into the vicinity of the outer edge of the plasma sheet at large radial distances in the tail. One can probably relate the spectral softening of the electrons at high latitudes in the nightside auroral zone with the spectral softening observed by the VELA satellites as they move in latitude from the center of the plasma sheet toward the magnetosheath at 17 R_E in the tail (Hones, 1968).

Turning now to the dayside, some recent results indicate that the magnetic field lines from the high latitude dayside soft zone of precipitation map into the magnetosheath region on the front of the magnetosphere. A comparison of the spectra of the magnetosheath particles with the auroral particles in the soft zone provides the evidence in support of this concept. Figure 7 shows a typical magnetosheath electron spectrum (Hones, 1968) and an average representation of the soft day zone electron spectrum given by Burch (1970) for the energy range 0.1 to 1.3 keV. The spectrums are seen to be quite comparable. A similar result is obtained for protons. Figure 7 shows a typical

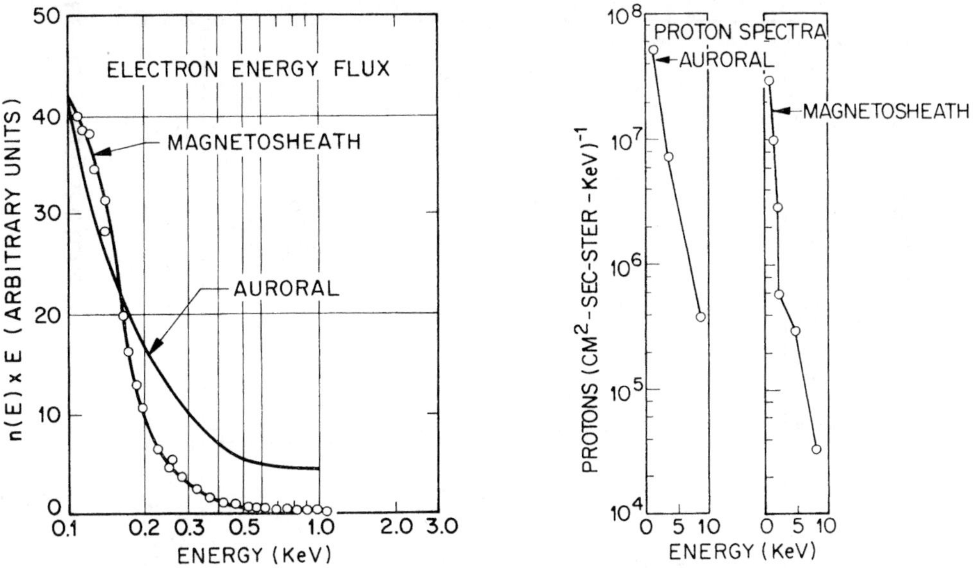

Fig. 7. Comparisons of proton and electron spectrums in the magnetosheath with those in the soft dayside auroral zone. The magnetosheath electron spectrum was reported by Hones (1968) from VELA 3A measurements. The auroral spectrum was computed from the average soft day zone results given by Burch (1969), i.e., a differential spectrum of the form $n(E) \sim E^{-2.26}$. The magnetosheath proton spectrum was measured on IMP 4 on June 11, 1967, at 19.8 R_E and MLT = 17:12 (Frank, 1970). The auroral proton spectrum was measured on OVI 18 on April 28, 1969, at IN Lat = 69°S and MLT = 11:49 (Johnson *et al.*, 1969).

magnetosheath proton spectrum (Frank, 1970a) compared to an auroral proton spectrum from a newly detected class of soft, high latitude dayside auroral proton precipitation events (Johnson *et al.*, 1969). These events shall be further discussed below.

3. Protons

A. LOCATION

The results for the Northern Hemisphere of a recent statistical study of the latitudinal dependence of the auroral proton precipitations at LT near midnight (Sharp *et al.*, 1969) are shown in Figure 8. The data were obtained with a single 4 keV threshold detector on satellite 1963-42A over a 5 day period. The study showed that during the period of the experiment the average zones of proton and electron precipitation were generally overlapping at this LT. The proton zone was displaced equatorward by about $1\frac{1}{2}°$ from the electron zone in the Northern Hemisphere but was not detectably displaced in the Southern Hemisphere. The width and location of the nightside proton precipitation zone was in good correspondence to that of the quiet time ring current reported by Frank (1970b) from his OGO 3 experiment at high altitudes near the equatorial plane. Figure 8 shows the inbound pass of OGO 3 on July 7, 1966, at about 2300 LT, which Frank indicates is a typical traversal of the quiet time ring current belt.

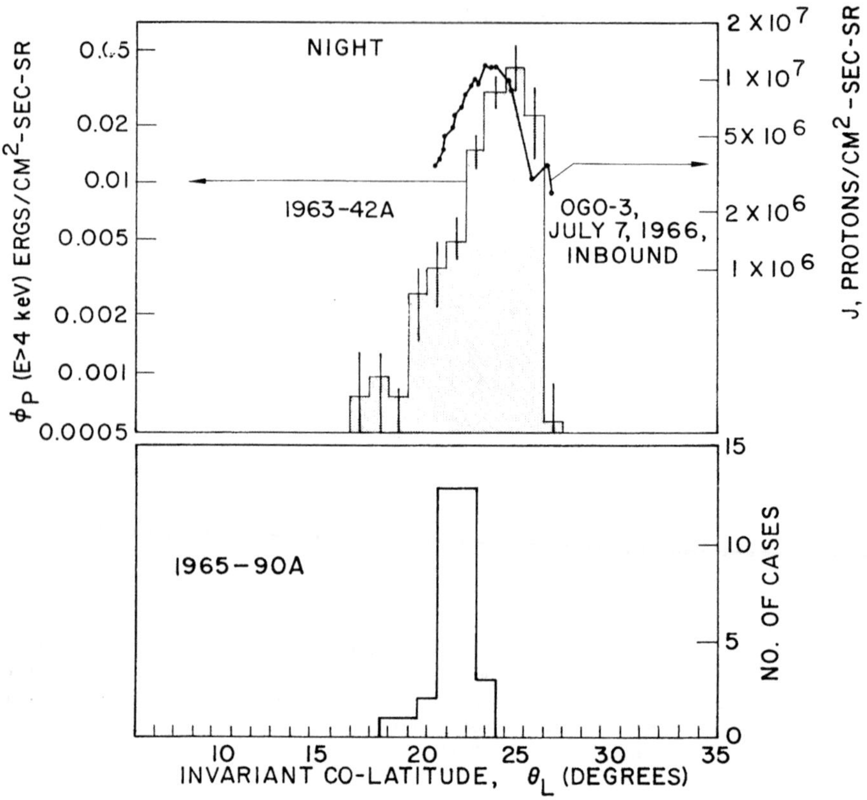

Fig. 8. A comparison of low energy ($E \lesssim 50$ keV) proton results from three satellites at LT near midnight.

The lower histogram in Figure 8 shows the location of the peaks of the auroral proton precipitations at LT near 2300 LT as obtained from the low altitude polar satellite 1965-90A over a $2\frac{1}{2}$ day period in 1965 (Sharp *et al.*, 1967). These data were obtained from detectors with energy thresholds spanning the range from about 10 to 60 keV. The general correspondence of these three sets of data from the nightside is suggestive that either the ring current is the source of these auroral protons (leaky bucket model), or conversely, the source of the auroral protons is also the source of the ring current (splash catcher model) (see O'Brien (1962) for a discussion of these two alternative models with respect to the auroral electron precipitation). Detailed arguments supporting the former of these two hypotheses have recently been presented by Eather and Carovillano (1970).

The dayside results from these same three satellite experiments are shown in Figure 9. Frank's data are for the outbound pass of OGO 3 on September 8, 1966, near local noon which he refers to as typical of the quiet time ring current at this LT. One sees that, in contrast to the nightside, the dayside results are not in good correspondence. The 1965-90A results suggest two separate precipitation zones while the 1963-42A

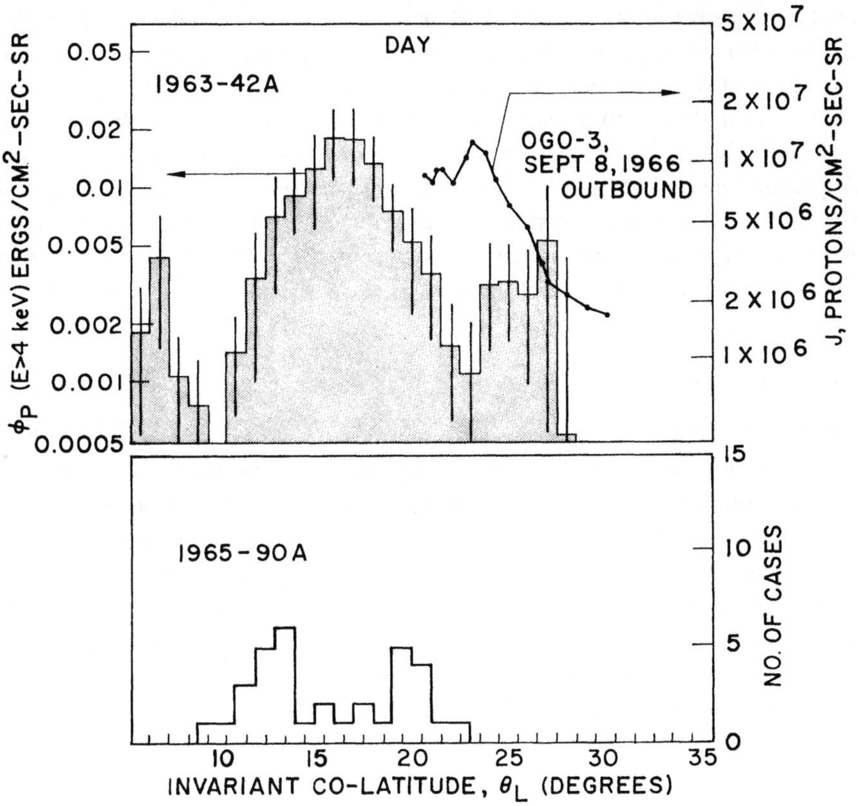

Fig. 9. Same as Figure 8 for LT near noon.

experiment resulted in a single dominant zone midway between them. The peak of the ring current appears to be at lower L values than the auroral precipitation zones, but the OGO 3 data corresponding to the higher latitudes have not been published. The 1963-42A results were obtained over a fairly active period following a magnetic storm and may not be comparable to the quiet time results of the other two experiments. Also, the 1963-42A proton detector was less sensitive than the detectors used in the 1965-90A flight (Sharp *et al.*, 1967, 1969). The ring current appears to be most closely associated in latitude with the low latitude dayside zone seen in 1965.

Recent preliminary results from the ESRO I (AURORAE) satellite confirm that the proton and electron precipitation zones are generally overlapping in the evening to midnight sector (Riedler and Hultqvist, 1969). These authors also report that they have seen instances of a second proton zone at high latitudes ($\Lambda_L \approx 85°$) in the early evening, and that the protons in this zone have a softer spectrum than those at the lower latitudes.

B. ANGULAR DISTRIBUTION

On the basis of a preliminary analysis of the data from 1965-90A at the positions of the peaks of the proton precipitations, Sharp *et al.* (1967) reported angular distributions consistent with isotropy over the loss cone within about a factor of 2. Riedler and Hultqvist (1969) find the same result for 6 keV protons in the heart of the zone and also report that the angular distributions tend to be peaked at high pitch angles at the low latitude edge of the zone. Riedler (1969) also reports instances where the 6 keV proton channel at 10° pitch angle has higher count rate than a similar channel at 80°. This is an important result since the precipitation of ring current protons by pitch angle diffusion (Eather and Carovillano, 1970) does not lead to such peaked distributions. It will be of great interest to see how frequently these distributions are found.

C. ENERGY DISTRIBUTIONS

The 1965-90A results for protons with energies above 10 keV at the peaks of the precipitation regions resulted in spectrums that were generally consistent with an exponential number flux shape with characteristic energies that were usually in the 10 to 20 keV range (Sharp *et al.*, 1967). The nightside spectra were often found to harden dramatically with decreasing latitude (Sharp and Johnson, 1968a). This latitudinal effect has now also been observed on the dayside and for protons with energies below 10 keV (Johnson *et al.*, 1969). The recent OV1 18 results also have revealed a new class of soft narrow proton precipitation events on the dayside (Johnson *et al.*, 1969) that were not seen in the earlier satellite data because of the high energy thresholds of the detectors. Figure 7 shows an example of the proton energy spectrum in one of these events. Their morphology has not yet been deduced but preliminary results indicate that they tend to occur at relatively high latitudes. As discussed previously, their spectrums are often comparable to those reported for protons in the magnetosheath.

4. Conclusions

We find that many detailed characteristics of the auroral particles are becoming available from recent polar satellite experiments. Coupled with the large quantity of new information being acquired from ground and rocket based observations and from the high altitude satellites, there is hope that the principal processes leading to the auroras will soon be understood. Individual experiments are generally still limited by shortcomings in energy, angular, and temporal resolution or by lack of sufficiently long term or global coverage, and much work remains to be done. We look forward to further results from the several newly acquired bodies of data which are currently being analyzed.

Acknowledgments

This work was supported partially by the Advanced Research Projects Agency through the Office of Naval Research and partially by the Lockheed Independent Research Program.

References

Ackerson, K. L. and Frank, L. A. 1970, *Trans. Am. Geophys. Union* **51**, 388.

Burch, J. L.: 1968, *J. Geophys. Res.* **73**, 3585.

Burch, J. L.: 1969, Paper presented at the NATO Advanced Study Institute – Production and Maintenance of the Polar Ionosphere, Tritten, Norway, April 9–18.

Craven, J. D.: 1970, *J. Geophys. Res.* **75**, 2468.

Eather, R. H. and Carovillano, R. L.: 1970, 'The Ring Current as the Source Region for Proton Auroras', Boston College preprint.

Evans, D. S.: 1970, private communication.

Evans, D. S.: 1965, *Rev. Sci. Instr.* **36**, 375.

Feldstein, Y. I.: 1966, *Planetary Space Sci.* **14**, 121.

Frank, L. A.: 1968, in *Earth's Particles and Fields* (ed. by B. M. McCormac), Reinhold Publishing Company, New York, p. 57.

Frank, L. A.: 1970a, in *Particles and Fields in the Magnetosphere* (ed. by B. M. McCormac), D. Reidel Publishing Company, Dordrecht-Holland, p. 319.

Frank, L. A.: 1970b, *J. Geophys. Res.* **75**, 1263.

Frank, L. A. and Ackerson, K. L.: 1970, *Trans. Am. Geophys. Union*, **51**, 388.

Fritz, T. A. and Gurnett, D. A.: 1965, *J. Geophys. Res.* **70**, 2485.

Hartz, T. R. and Brice, N. M.: 1967, *Planetary Space Sci.* **15**, 301.

Heikkila, W. J.: 1970, *Nature* **225**, 369.

Heikkila, W. J. and Winningham, J. D.: 1970, *Trans. Am. Geophys. Union* **51**, 375.

Hoffman, R. A.: 1969a, *J. Geophys. Res.* **74**, 2425.

Hoffman, R. A.: 1969b, 'OGO 4 Satellite Measurements of Low Energy High Latitude Electron Precipitation', GSFC preprint X-612-69-198.

Hoffman, R. A.: 1970, 'Auroral Electron Drift and Precipitation: Cause of the Mantle Aurora', GSFC preprint X-646-70-205.

Hoffman, R. A. and Evans, D. S.: 1968, *J. Geophys. Res.* **73**, 6201.

Hones, E. W., Jr.: 1968, in *Physics of the Magnetosphere* (ed. by R. Carovillano, J. F. McClay, and H. R. Radoski), D. Reidel Publishing Company, Dordrecht-Holland, p. 392.

Johnson, R. G. and Sharp, R. D.: 1969, in *Atmospheric Emissions* (ed. by B. M. McCormac and A. Omholt), Van Nostrand Reinhold Company, New York, p. 219.

Johnson, R. G., Sharp, R. D. Shea, M. F., and Shook, G. B.: 1966, *Trans. Am. Geophys.* **47**, 64.

Johnson, R. G., Meyerott, R. D., and Evans, J. E.: 1967, in *Aurora and Airglow* (ed. by B. M. McCormac), Reinhold Publishing Company, New York, p. 169.

Johnson, R. G., Sanders, T. C., Sharp, R. D., Shook, G. B.: 1969, *Trans. Am. Geophys. Union* **50**, 654.

Liszka, L., Borg, H., and Riedler, W.: 1970, 'Preliminary Results of ESRO 1A (AURORAE) Observations of Variations in the Low Energy Electron Spectrum, and Their Relation to Magnetic Activity', Kiruna Geophysical Observatory preprint.

Maehlum, B.: 1968, *J. Geophys. Res.* **73**, 3459.

O'Brien, B. J.: 1962, *J. Geophys. Res.* **67**, 3687.

O'Brien, B. J.: 1964, *J. Geophys. Res.* **69**, 13.

Parthasarathy, R., Berkey, F. T., and Venkatesan, D.: 1966, *Planetary Space Sci.* **14**, 65.

Paschmann, G., Carr, D. L., and Shelley, E. G.: 1969, *Trans. Am. Geophys. Union* **50**, 654 (abstract).

Paschmann, G.: 1970, private communication.

Piddington, J. H.: 1965, *Planetary Space Sci.* **13**, 565.

Riedler, W.: 1969, 'ESRO 1 Measurements of Low Energy Auroral Particles from February 23 to March 2, 1969', Kiruna Geophysical Observatory preprint.

Riedler, W. and Hultqvist, B.: 1969, 'First Results of 1 and 5 keV Proton Measurements from the ESRO 1 (AURORAE) Satellite', Kiruna Geophysical Observatory Report 696.

Schield, M. A. and Frank, L. A.: 1969, 'Electron Observations between the Inner Edge of the Plasma Sheet and the Plasmasphere', University of Iowa Technical Report 69-46.

Sharp, R. D. and Johnson, R. G.: 1968a, in *Earth's Particles and Fields* (ed. by B. M. McCormac), Reinhold Book Corporation, New York, p. 113.

Sharp, R. D. and Johnson, R. G.: 1968b, *Ann. Geophys.* **24**, 291.

Sharp, R. D. and Johnson, R. G.: 1968c, *J. Geophys. Res.* **73**, 969.

Sharp, R. D., Evans, J. E., and Johnson, R. G.: 1966, *Planetary Space Sci.* **14**, 85.
Sharp, R. D., Carr, D. L., and Johnson, R. G.: 1969, *J. Geophys. Res.* **74**, 4618.
Sharp, R. D., Johnson, R. G., Shea, M. F., and Shook, G. B.: 1967, *J. Geophys. Res.* **72**, 227.
Vasyliunas, V. M.: 1968, *J. Geophys. Res.* **73**, 2839.
Vasyliunas, V. M.: 1969, 'Low Energy Particle Fluxes in the Geomagnetic Tail', MIT Center for Space Research preprint CSR-P-69-17.

AURORAL PRECIPITATION PATTERNS

R. H. EATHER* and S. B. MENDE

Lockheed Palo Alto Research Laboratory, Palo Alto, Calif., U.S.A.

Abstract. There are at least four distinct regions of auroral particle precipitation. The auroral oval and nightside proton aurora are well known; they are well separated in the evening sector and overlap in the morning sector. There is a zone of soft electron precipitation at latitudes above the auroral oval where ~0.5 keV electrons precipitate; no protons are observed in this region. Electron energies in this soft zone decrease systematically with latitude (for a given K_p). There is a zone of soft electron (~100–200 eV) and soft proton precipitation at latitudes above the auroral oval near the midday meridian, and these particles probably came directly from the magnetosheath. No higher energy protons (as in the nightside proton aurora) were observed near midday during quiet magnetic conditions.

1. Introduction

The measurements to be described in this paper were made during the NASA Airborne Auroral Expedition from November to December 1969. The aircraft completed a total of 14 research flights, 2 of which were in the region of the midday aurora. The average K_p value was 1+ with a maximum of 4+ during only one 3 hr period; thus, these results apply to quiet magnetic conditions. Lockheed Palo Alto Research Laboratories installed a vertical looking six channel tilting filter photometric system (Eather and Reasoner, 1969) on the aircraft for high sensitivity measurements of the following auroral and airglow emissions: 4278 Å N^+, Hβ, 5200 Å N I, 5577 Å O I, 5876 Å He I, 5995 Å O_2^+ and 6300 Å O I. All channels were used with a full included field of view of 3° and regular photometric calibrations were carried out with a C^{14} radioactive light source.

2. Analysis

The data were scaled every 2 min of UT throughout all flights; if intensities varied appreciably during a 2 min period then a 2 min average was estimated. Each data point was then grouped into its appropriate IN Lat magnetic time interval.

Such averages include both proton and electron excitation as well as airglow. A second program subtracted the proton excited component (using spectral ratios) (Eather, 1968) and also the airglow contribution to the O I and N I emissions.

Similar programs to the above were also run to order the data in terms of the number of degrees of IN Lat from the auroral oval (Feldstein and Starkov, 1968). It was expected that this would reduce the latitudinal smearing of the results by allowing for the systematic movements of the auroral oval with time, and for the expansion of the oval with magnetic activity. Starkov (1969) has given the following expression for the geomagnetic colatitude of the equatorward boundary of the oval

$$\theta = 18.0 + 0.9\, Q + 5.1 \cos (t - 12°)$$

* Present Address: Physics Department, Boston College, Chestnut Hill, Mass., U.S.A.

B. M. McCormac (ed.), The Radiating Atmosphere, 255–266.

where t is the magnetic local time (MLT) in degrees, and the magnetic index, Q, is characteristic for the midnight meridian at IN Lat 65°.

3. Results

A. HELIUM EMISSION

A search for 5876 Å He I emission on the first half of the expedition yielded negative results. If He emission does occur sporadically during active aurora (Stoffregen, 1969), then it must certainly be a relatively rare phenomenon. Consequently, the search for He emission was discontinued after eight flights and channel 6 of the system was then used to measure 5200 Å N I for the remaining flights.

B. HYDROGEN EMISSION

Figure 1 shows averaged Hβ emission for three magnetic time periods: dusk to midnight, midnight to dawn, and midday. The nighttime results agree well with the findings of the 1st Convair Expedition (Eather, 1969) both in the location and absolute intensities of the Hβ emission. The daytime results indicate that the zone of Hβ emission on the dayside is some 10° higher in latitude, perhaps somewhat narrower in latitudinal extent, and only about 30 to 50% of the nighttime intensity.

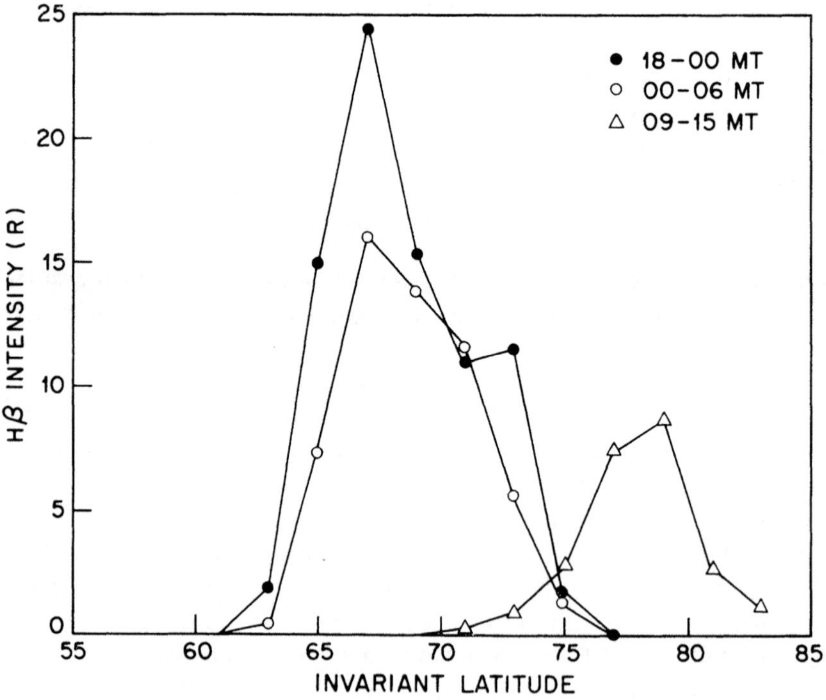

Fig. 1. Averaged Hβ emission as a function of IN Lat for three magnetic time intervals.

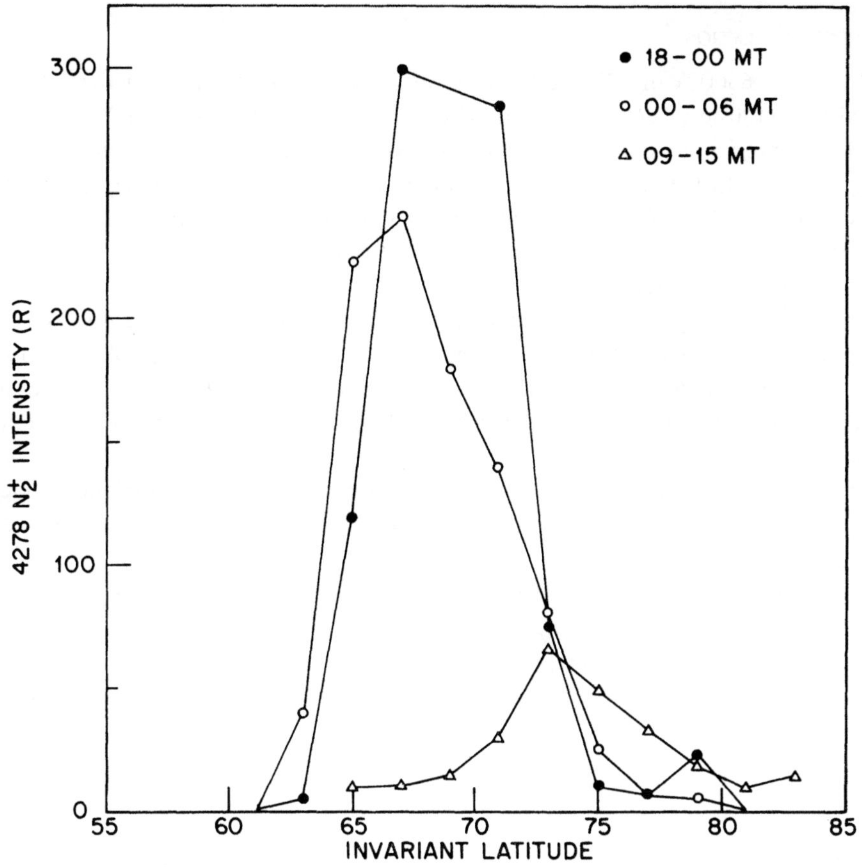

Fig. 2. Averaged 4278 Å N_2^+ emission (corrected for proton excitation) as a function of IN Lat for three magnetic time intervals.

Comparison with the 4278 Å N_2^+ distributions in Figure 2 shows that the daytime Hβ emission locates in the poleward side of a much broader region of electron precipitation.

c. 4278 Å N_2^+ emission

Figure 2 shows averaged 4278 Å N_2^+ emission (with the proton excited component subtracted) for the dusk to midnight, midnight to dawn, and midday time periods. Again, the nighttime results agree well with the results from the 1st Convair Expedition (Eather, 1969). The daytime results indicate a much weaker broader region of electron precipitation which peaks some 6° higher in latitude than at night and extends to higher latitudes. The nature of the aurora, giving rise to the 4278 Å N_2^+ emission on the high latitude edge of the emission zone, was similar to that described for the 1st Convair Expedition (Eather, 1969), viz., fairly short-lived, weak structured aurora. No uniform background glow or uniform 'mantle aurora' was detected during these flights.

D. SPECTRAL RATIOS

A plot of the 6300 Å or 4278 Å N_2^+ ratio as a function of IN Lat for the 1st Convair Expedition (Eather, 1969) led to the discovery of a nighttime zone of soft electron (~600 eV) precipitation located on the poleward edge of the normal auroral zone.

A similar plot for the present expedition is shown in Figure 3 and confirms the previous result. The 6300 Å/4278 Å nighttime ratio rises from normal values in the auroral zone to peak at a value of ~8.0 at 77° IN Lat (compared with a peak value of ~6.0 at 77° for the first Convair Expedition).

Also shown in Figure 3 is the 6300 Å/4278 Å plot for daytime aurora. The peak is shifted some 4° higher in IN Lat, and increased very markedly in amplitude to a peak value of 28.0. This is a very significant result as it shows that there is a zone of very soft electron precipitation on the dayside which locates on the poleward edge of normal dayside aurora.

We have concluded that the daytime soft zone is dominantly low energy electron excitation. We base this conclusion on the similar temporal behavior of the measured

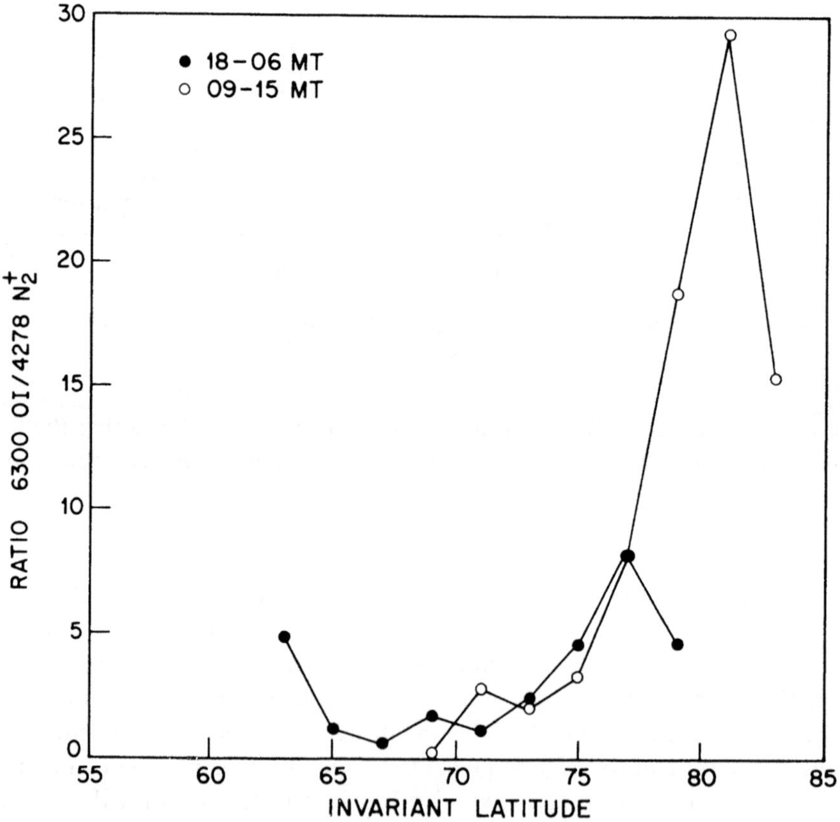

Fig. 3. The ratio 6300 Å OI/4278 Å N_2^+, corrected for proton excitation and airglow as a function of IN Lat for nighttime and daytime intervals.

4278 Å N_2^+ and 6300 Å O I signals. Alternative forms of selective 6300 Å O I excitation, such as electric fields, photoelectrons, or any form of heating would not result in such good correlation with N_2^+ emissions, unless such mechanisms were concommitant with low energy electron precipitation. Although the dayside protons precipitate in the same region as this soft zone they cannot be the particles exciting the high 6300 Å O I intensities as the spatial structure of the Hβ, and the 6300 Å and 4278 Å emissions is quite different (Figure 6). The high 6300 Å/4278 Å ratios correspond to very low electron energies.

This result is confirmed when we examine plots of the 5200 Å N I/4278 Å N_2^+ ratio (Figure 4). 5200 Å N I is a long lifetime forbidden transition, thus it is favored at high altitudes and is an indicator of soft particle precipitation. The peaking of the night-time curve at 77° IN Lat confirms that soft electrons precipitate poleward of normal aurora on the nightside; the ratio increases from ~0.05 in normal aurora to ~0.4 in the nightside soft zone. On the dayside the peak moves to an 81° IN Lat and increases in magnitude to ~0.8. This again is in agreement with the 6300 Å/4278 Å results.

We can estimate the soft electron energy on the dayside in the following way: The peak 6300 Å/4278 Å was ~30 and the corresponding 5577 Å/4278 Å ratio was ~6. This leads to a 6300 Å/5577 Å ratio of ~5. Now the ratio of excitation rates to the O('D) and O('S) levels is 15 (Rees, 1970) so this means in the absence of quenching

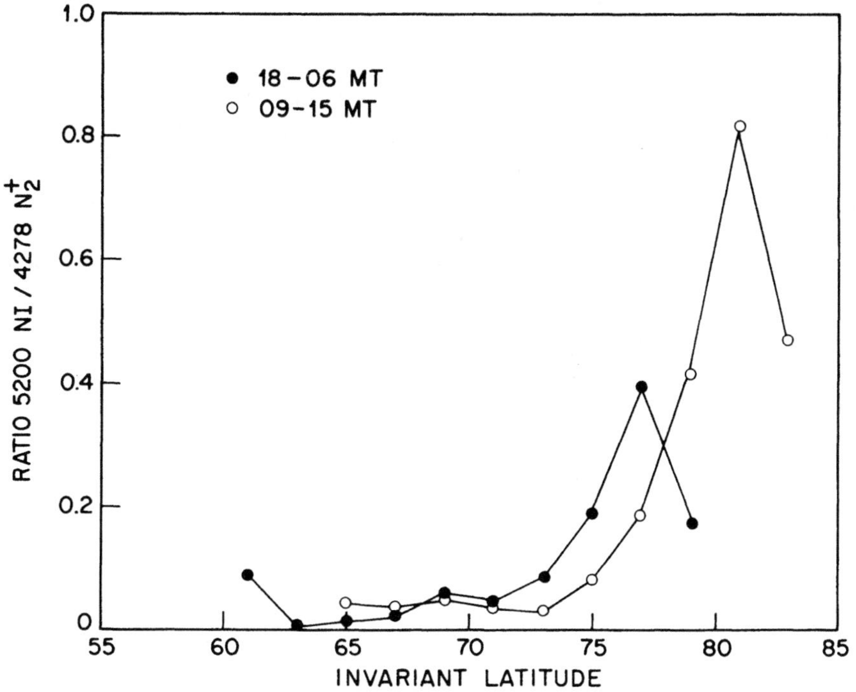

Fig. 4. Same as Figure 3 but for 5200 Å N I/4278 Å N_2^+.

we would expect a 6300 Å/5577 Å ratio of ~15 at equilibrium. We can write the observed ratio as

$$I/I_0 = \frac{1}{1 + \beta_i(X)\, n(X)/A} = \frac{1}{3}$$

where $\beta_i(X)$ is the quenching rate coefficient for excited state, i, by particle X with concentration $n(X)$, and A is the sum of radiation transition probabilities from state i. A quenching coefficient, $\beta 0_D(N_2) = 5 \times 10^{-11}$ cm^{-3} sec is adopted with N_2 the quenching species (Hunten and McElroy, 1966) and $1/A = 110$ sec. Hence

$$n(N_2) \simeq 3.6 \times 10^8 \text{ cm}^{-3}.$$

Jacchia's (1964) model atmosphere gives this N_2 density at a height of ~290 km and electrons will need energies ~125 eV (Rees, 1964) to penetrate to this height. The above estimation of electron energy in the soft dayside zone is approximate, of course, but we can conclude with certainty that the average electron energy will be between 100 and 200 eV.

E. PROTON-ELECTRON RELATIONS

There has been disagreement for many years on the question of proton-electron 'crossover.' It is firmly established that the proton aurora locates equatorward of electron aurora before midnight, but after midnight there is a disagreement as to whether the Axford and Hines (1961) prediction of a crossover with proton precipitation locating poleward of electron precipitation is correct. Stoffregen and Derblom (1962), Derblom (1968), Wiens (1968), and Yevlashin (1969) believe that there is a crossover, whereas Eather and Jacka (1966), Eather (1967), and Rees and Benedict (1970) maintain that the proton zone remains equatorward or intermixes with the electron zone. There is a danger in using statistical results to try to resolve this question as the relative morphology will be a function of substorm activity and overall magnetic activity (Montbriand, 1969). Montbriand (1969) and Yevlashin (1969) indicate the proton aurora may remain equatorward of electron aurora in the morning hours only during very quiet magnetic conditions.

During this expedition a total of 11 N-S passes was flown through the morning auroral zone to try to resolve this question. Relative locations of the proton (Hβ) and electron (4278 Å N_2^+ corrected for proton excitation) precipitation zones during these passes are shown in Figure 5. The only conclusion one can make is that there is certainly no systematic crossover in the morning hours; nor is there a definite separation of the zones with the protons remaining equatorward. There appears to be an overall mixing of the two zones of precipitation in the morning hours (and during quiet magnetic conditions).

Similar latitudinal plots for four passes through the midday aurora are shown in Figure 6. Again the regions of proton and electron precipitation are intermixed, though in this case the electron precipitation is more widespread to lower latitudes. In fact, the proton precipitation appears to locate on the poleward half of the electron region

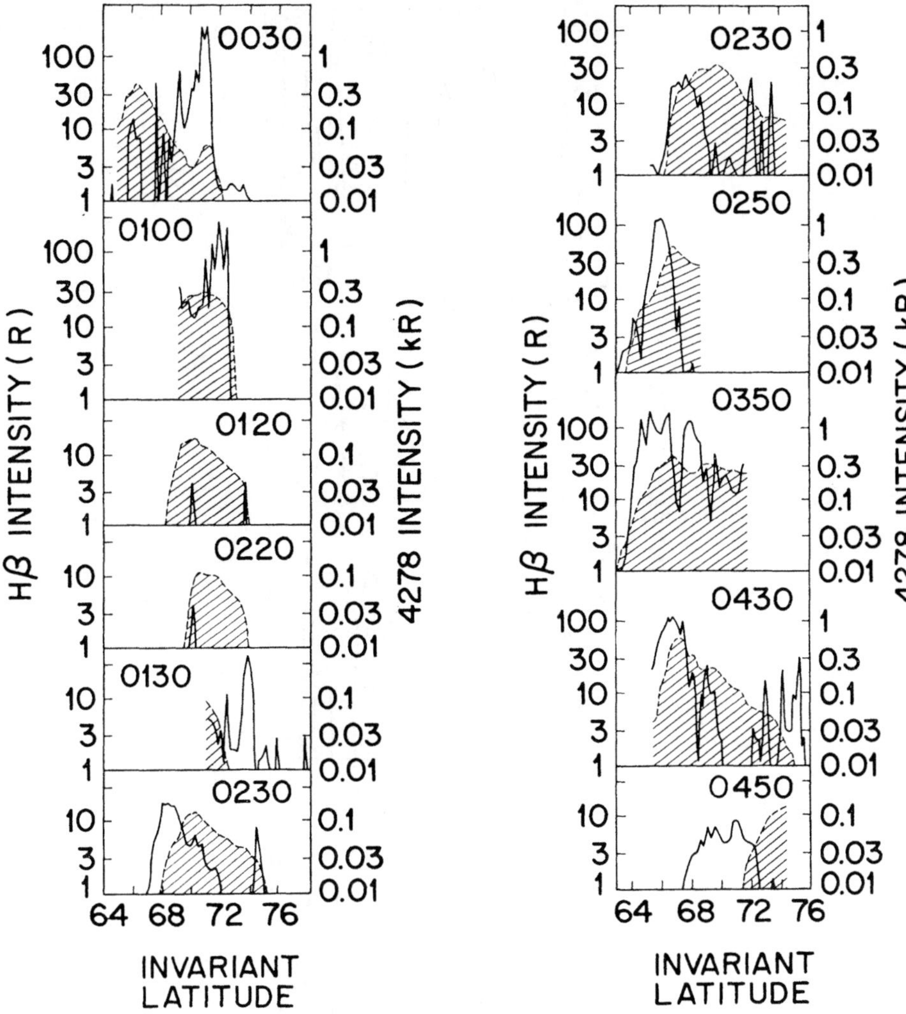

Fig. 5. Relative location of proton (Hβ) and electron (4278 Å N_2^+) corrected for proton excitation precipitation during the morning hours for 11 flights. Each flight through this region is labeled with the magnetic time near the center of the meridional pass.

(see also Figures 1 and 2), corresponding to the region we have shown to be the dayside soft zone.

This led us to suspect that the dayside proton at these high latitudes may also be 'soft', i.e., has an average energy much lower than nighttime proton energies. Tilting filter Hβ measurements confirm that the Hβ profile on the dayside is narrower and has a smaller Doppler shift than the nightside profile. The records are too noisy and the wavelength resolution too poor to make quantitative estimates of proton energies.

We conclude that both low energy electrons and low energy protons precipitate near midday in a zone that lies on the high latitude side of the normal auroral oval.

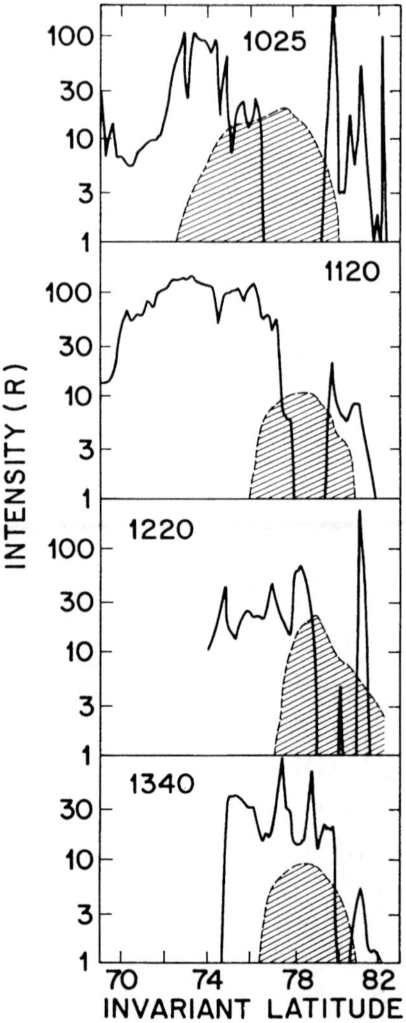

Fig. 6. Same as Figure 5 but for flights near midday. K_p was 0–1+.

We did not detect proton precipitation equatorward of the auroral oval during these flights. Unfortunately, we only have a few flights through this region so statistics are poor. A systematic investigation of the midday region would be of considerable interest.

F. REORDERING OF DATA IN 'OVAL COORDINATES'

The results were also ordered as functions of magnetic time and distance (in degrees) from the equatorial boundary of the auroral oval (Section 2).

1) This procedure did *not* result in a narrower distribution (in latitude) of the 4278 Å N_2^+ emission. The width at half intensity of the averaged 4278 Å N_2^+ was

7 to 8°, and the peak located 3 to 4° higher in latitude than the equatorial boundary of the oval.

Feldstein's oval (Feldstein and Starkov, 1967) for quiet moderate magnetic activity is typically shown as 3 to 5° in width, but this is based on a statistical averaging of stronger aurora than is recorded on all-sky cameras (4278 Å $N_2^+ \gtrsim 200$ R). Our results suggest that significant energy input is in the form of weaker auroras, and our sensitivity level of ~1 R of 4278 Å N_2^+ allows a better determination of the latitudinal level of magnetic activity characterized by moderate disturbance ($K_p < 4$); significant energy precipitates over about 12° IN Lat, with half-power points covering 7 to 8° IN Lat.

2) Plots of the 6300 Å/4278 Å ratio in this oval coordinate system showed the same soft zones as in Figure 3 with a strong peaking in the ratio at higher latitudes. There was no narrowing in the indicated width of the region of soft precipitation.

3) The position of the proton aurora relative to the auroral oval was examined by plotting the 4278 Å N_2^+ (electron excited) and Hβ distributions in oval coordinates, and also for three time sectors 1800 to 2400 LT, 0000 to 0600 LT, and 0900 to 1500 LT. In the evening hours the peak in the Hβ distribution was ≈3° equatorward of the peak in the 4278 Å N_2^+ distribution; in the morning hours the two peaks coincided within 1°; in the midday sector the Hβ peak was ~4° poleward of the 4278 Å N_2^+ peak.

These results in oval coordinates confirm the conclusions reached in Section 4.

4. Summary and Discussion

The results presented in this paper, together with results from the 1st Expedition (Eather, 1969), show that there are two zone like regions within the auroral oval where soft electrons precipitate. The average energy of electrons in the nightside soft zone is ~0.5 keV and the averaged precipitated energy is only about 0.1 erg cm^{-2} sec^{-1} (corresponding to 20 R 4278 Å N_2^+) (Dalgarno *et al.*, 1965). These compare to average energies of ~few keV in the nightside auroral oval and average precipitated energy of 0.7 to 2.2 erg cm^{-2} sec^{-1} (200 to 600 R of 4278 Å N_2^+).

The average energy of electrons in the dayside soft zone is only 100 to 200 eV, though average 4278 Å N_2^+ intensities are about the same as the nightside (corresponding to ~0.13 erg cm^{-2} sec^{-1}), implying precipitating fluxes are higher by a factor of 3 to 4. These soft zones extend to about 80° IN Lat at midday. At higher latitudes (the polar cap) precipitation seems to be a rare occurrence.

It should be emphasized that the zones of soft electron precipitation are a statistical averaging of contributions from discrete, often short lived auroral forms; they do not represent a diffuse continuous precipitation of low energy electrons. No evidence was found for any diffuse auroral continuum of 'mantle aurora' at these high latitudes, and we can set upper limits of 2×10^{-3} erg cm^{-2} sec^{-1} for any diffuse energetic particle precipitation across the polar cap.

There is a zone of soft proton precipitation associated with the dayside soft electron zone. The soft electron energy (100 to 200 eV) is similar to electron energies in the magnetosheath, and the presence of the soft protons in the same region leads us to suspect these particles come directly from the magnetosheath. (Magnetosheath proton energies are typically about 1 keV.)

There is no proton precipitation associated with the nightside soft electron zone in the dusk-midnight sector.

Normal (auroral energies ~10 keV) proton precipitation locates equatorward of normal electron precipitation in the dusk-midnight sector. No proton precipitation was observed equatorward or within the normal dayside zone of electron precipitation during these flights.

Eleven flights through the midnight-dawn sector showed that proton and electron

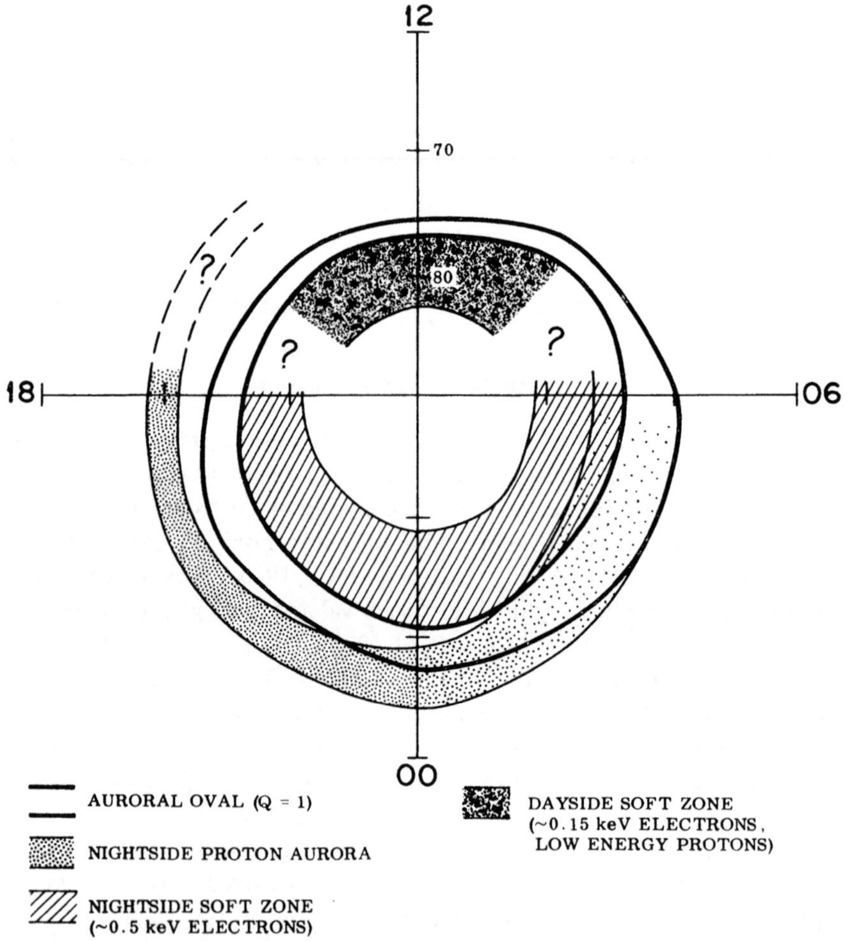

Fig. 7. The various zones of particle precipitation (as characterized by energy) at high latitudes. (a) The auroral oval; electron energy ~ 5–10 keV with the addition of some ~ 30 keV electrons in the morning and day sectors (enclosed by heavy lines). (b) Nightside proton aurora-proton energy ~ 10 keV, with some hardening toward the equator. (c) Nightside soft electron zone; electron energy ~ 500 eV. (d) Dayside soft electron and soft proton zone; electron energy ~ 150 eV, proton energy is unknown but less than nighttime energies.

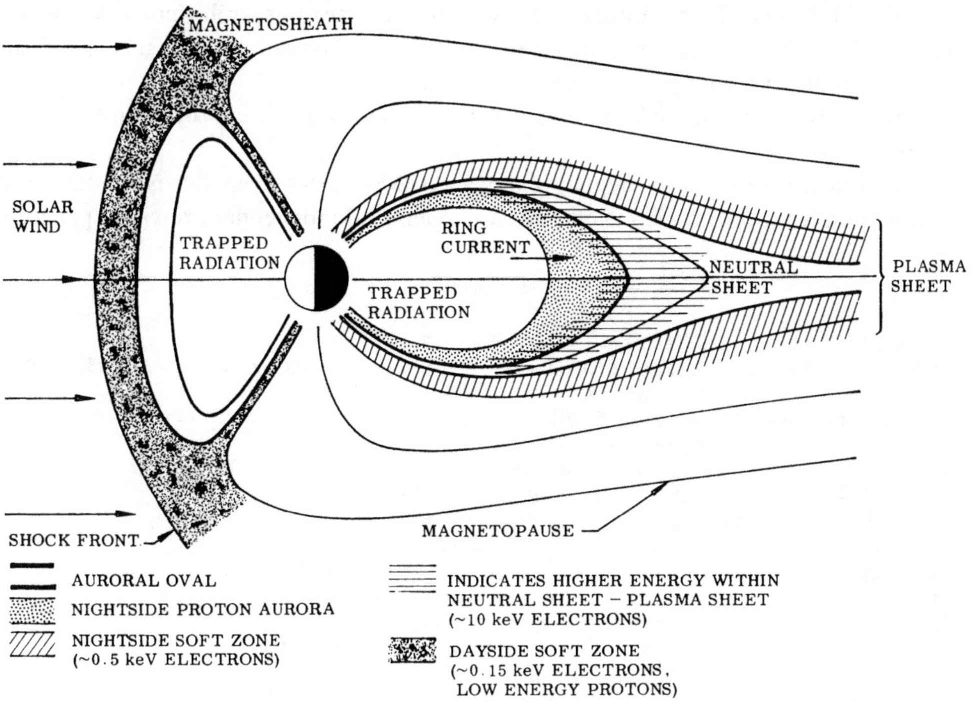

Fig. 8. A suggested magnetospheric model of the sources of precipitated particles as shown in Figure 7.

precipitation occur in the same region. No evidence was found for a systematic cross-over of the zones, at least during quiet magnetic conditions.

We conclude that there are at least three distinct energy divisions of precipitating electrons at high latitudes. These can be classed as 'normal' (the auroral oval), 'soft' (the nightside soft zone) and 'very soft' (the dayside soft zone). Nonphotometric data suggest even a fourth division of 'hard' or high energy region associated with the peaks in radio-noise absorption and auroral X-rays in the morning quadrant.

There are at least two distinct energy divisions of precipitating protons: these can be classed as 'normal' (the nightside proton aurora) and 'soft' (associated with the daytime soft electron zone).

These ideas are summarized schematically in Figure 7. Further observations are needed on the dayside in the morning (0600 to 1000 LT) and evening (1400 to 1800 LT) sector to complete this picture of particle precipitation patterns.

A suggested magnetospheric model showing suggested source regions for the precipitation regimes in Figure 7 is shown without comment in Figure 8. We are presently working on justifying various features of this model.

Acknowledgments

The 1969 NASA Airborne Auroral Expedition was managed by Mr. L. C. Haughney and Dr. M. Bader of the Airborne Sciences Office, Ames Research Center. We would like to thank them, and the pilots and crew of the Convair for the success of this expedition under trying polar winter conditions.

We are especially grateful to Dr. John E. Evans for his continued interest and support of this work.

This research was supported by Contract NASw 1997 from the National Aeronautics and Space Administration and by a Lockheed independent research program.

References

Axford, W. I. and Hines, C. O.: 1961, *Can. J. Phys.* **39**, 1433.
Dalgarno, A., Latimer, I. D., and McConkey, J. W.: 1965.: 1965, *Planetary Space Sci.* **13**, 1008.
Derblom, H.: 1968, *Ann. Geophys.* **24**, 163.
Eather, R. H.: 1967, *Rev. Geophys.* **5**, 207.
Eather, R. H.: 1968, *J. Geophys. Res.* **75**, 119.
Eather, R. H.: 1969, *J. Geophys. Res.* **74**, 153.
Eather, R. H. and Jacka, F.: 1966, *Australian J. Phys.* **19**, 241.
Eather, R. H. and Reasoner, D. L.: 1969, *Appl. Opt.* **8**, 227.
Feldstein, Y. I. and Starkov, G. V.: 1967, *Planetary Space Sci.* **15**, 209.
Feldstein, Y. I and Starkov, G. V.: 1968, *Planetary Space Sci.* **16**, 129.
Hunten, D. M. and McElroy, M. B.: 1966, *Rev. Geophys.* **4**, 303.
Jacchia, L. G.: 1964, *Smithsonian Astrophys. Obs. Spec. Rept.* **170**.
Montbriand, L. E.: 1969, Thesis, University of Saskatchewan, Saskatchewan.
Rees, M. H.: 1964, *Planetary Space Sci.* **12**, 722.
Rees, M. H.: 1970, private communication.
Rees, M. H. and Benedict, P. C.: 1970, Preprint, *L.A.S.P.*, University of Colorado.
Starkov, G. V.: 1969, *Geomag. Aeron.* **9**, 614.
Stoffregen, W.: 1969, *Planetary Space Sci.* **17**, 1927.
Stoffregen, W. and Derblom, H.: 1962, *Planetary Space Sci.* **9**, 711.
Wiens, R.: 1968, Ph.D. Thesis, University of Saskatchewan, Saskatchewan.
Yevlashin, L. S.: 1969, Preprint, *Polar Geophysical Institute*, Murmansk, U.S.S.R.

DIRECT OBSERVATION OF TEMPORAL AND SPATIAL STRUCTURE IN AURORAL ELECTRONS

DAVID S. EVANS*

Goddard Space Flight Center, Greenbelt, Md., U.S.A.

In the effort to deduce those processes which energize and precipitate the electrons which in turn produce the discrete aurora, the study of the temporal variations and spatial structure of these electron beams is assuming great importance. This is because such structures in particle beams are a direct reflection of processes that have governed the particles behavior, generally in the more recent than distant past.

Ground based optical observations using all-sky cameras, photometers, and TV image intensifiers have provided an excellent guide as to the range of spatial and temporal structure that may be expected in auroral electron precipitation. Maggs and Davis (1968) reported that individual auroral microstructures average only 200 to 300 m thickness while many are less than 100 m in thickness. All-sky photographs indicate the average width of the overall form is on the order of km (Kim and Volkman, 1963).

In the temporal realm such studies have exposed pulsations in auroral brightness ranging from the 10 Hz flicker aurora associated with the breakup phase of a display (Beach *et al.*, 1968) to beyond the 5 to 15 sec period range often seen in the post-midnight hours (Cresswell and Davis, 1966). Such optical studies of auroral pulsations are supplemented by investigations of auroral zone X-rays using balloon borne X-ray detectors. Pulsations observed with this technique (which is sensitive only to the higher energy electron precipitations) range upward in period from the very rapid 5 to 25 msec bursts reported by Parks *et al.* (1967). Bursts of X-rays on the order of 0.2 sec long (microbursts) are commonly observed in the morning hr (Anderson *et al.*, 1966) while X-ray pulsations in the 5 to 15 sec period range are typically associated with the postbreakup phase of the aurora (Barcus *et al.*, 1966). Occasionally X-ray pulsations in the 100 sec period range often exhibiting a very pure frequency are observed (Barcus and Christensen, 1965).

While the ground based and balloon studies thus provide a good impression of the scale of such temporal and spatial behavior in auroral electron precipitation, the indirect nature of the observations and the fact that the phenomena under study is removed some 100 km from the point of observation make the direct observation of the auroral electrons most essential in achieving a full understanding of this phenomena.

For the time and spatial scales of interest here – from a fraction of a sec to tens of sec and from less than 100 m to a few km, respectively – polar orbiting satellite ob-

* Now at ESSA Laboratories, Boulder, Colo., U.S.A.

B. M. McCormac (ed.), The Radiating Atmosphere, 267–278.

servations are of little value. The velocity of such a satellite is about 8 km sec^{-1} so that an experiment would be carried over an average patch of aurora in a second or so without any chance to examine temporal variations on the part of the auroral electrons of more than a few tenths sec time scale. Moreover, the telemetry rates allocated to satellite experiments are invariably too low to allow a full set of particle data to be obtained within this time period.

Sounding rockets, on the other hand, are much better suited as a vehicle for such direct studies of auroral electrons. Depending on the trajectory chosen, the velocity of the rocket perpendicular to magnetic field lines can range from, at times, almost zero to an extreme of about 3 km sec^{-1}. Moreover, data rates, particularly on a per experiment basis, can be more than ten times that available from the typical satellite. It should be pointed out that while the lower limits of spatial and temporal resolution presently available using sounding rocket techniques are about 1 m and 0.001 sec, respectively, flight time and impact range considerations normally prevent the study of pulsations of more than about 50 sec period or precipitation dimensions of more than some tens of km.

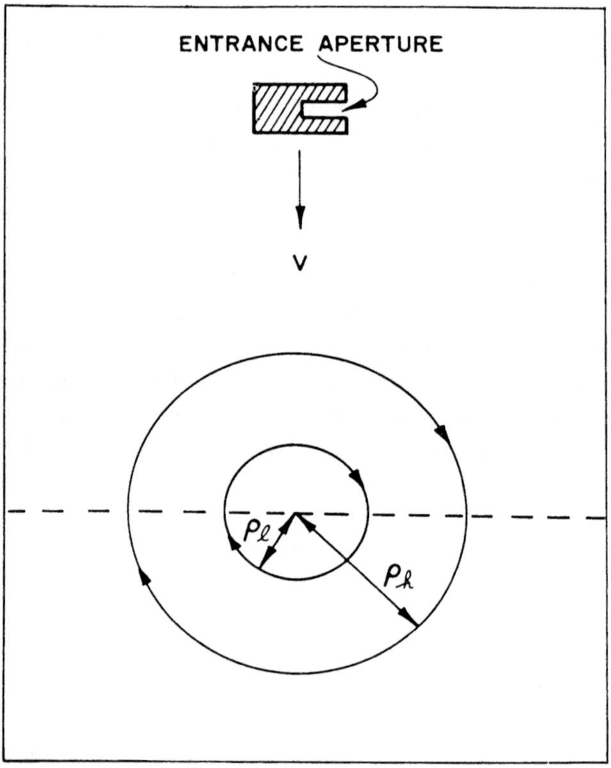

Fig. 1. A schematic illustration showing a rocket borne electron detector being carried into a region of enhanced electron precipitation. Note that in this aspect the detector responds to low energy electrons first (Mozer, 1968).

The most serious problem encountered in the direct observations of auroral electrons using the sounding rocket technique is the difficulty in removing the spatial-temporal-motion ambiguities in the interpretation of an observed electron flux variation. It is only in especially fortuitous circumstances that simultaneous ground based observations can aid the interpretation of the rocket data. Thus the choice as to whether an electron flux variation observed onboard a rocket is interpreted as a temporal or spatial effect almost always rests upon plausibility arguments.

Mozer (1968) observed many rapid variations in the flux of >65 keV electrons and performed a rather detailed analysis of these data in an effort to use the rocket observations themselves to separate spatial from temporal effects. For example,

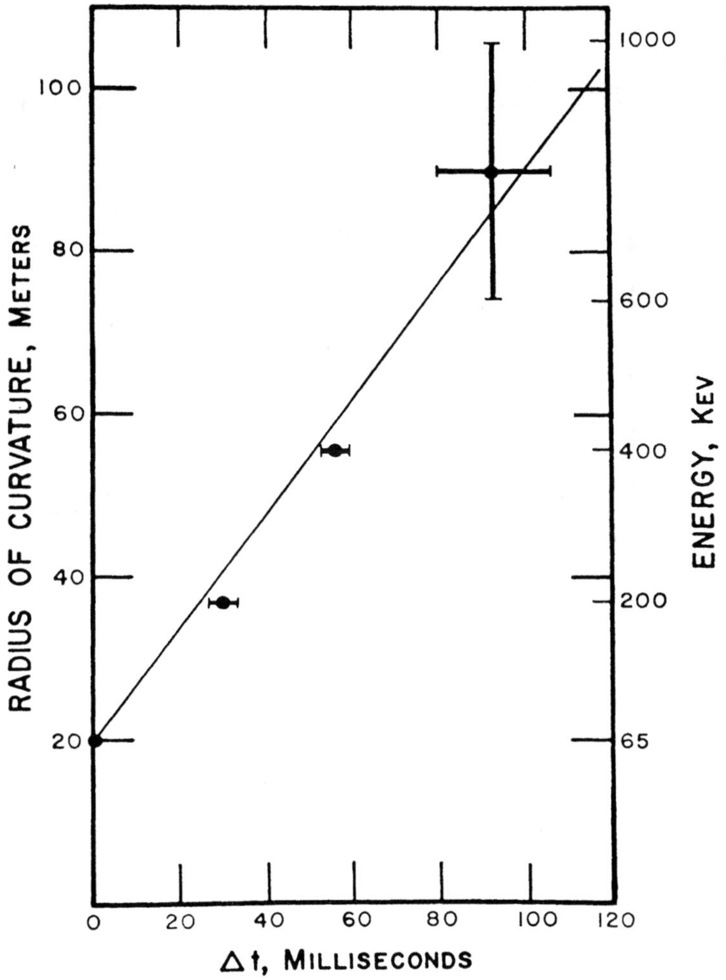

Fig. 2. The expected 'arrival times' of electrons of differing energies as a rocket instrument moving at 700 m sec^{-1} entered a region of enhanced electron precipitation with the aspect shown in Figure 1 (Mozer, 1968).

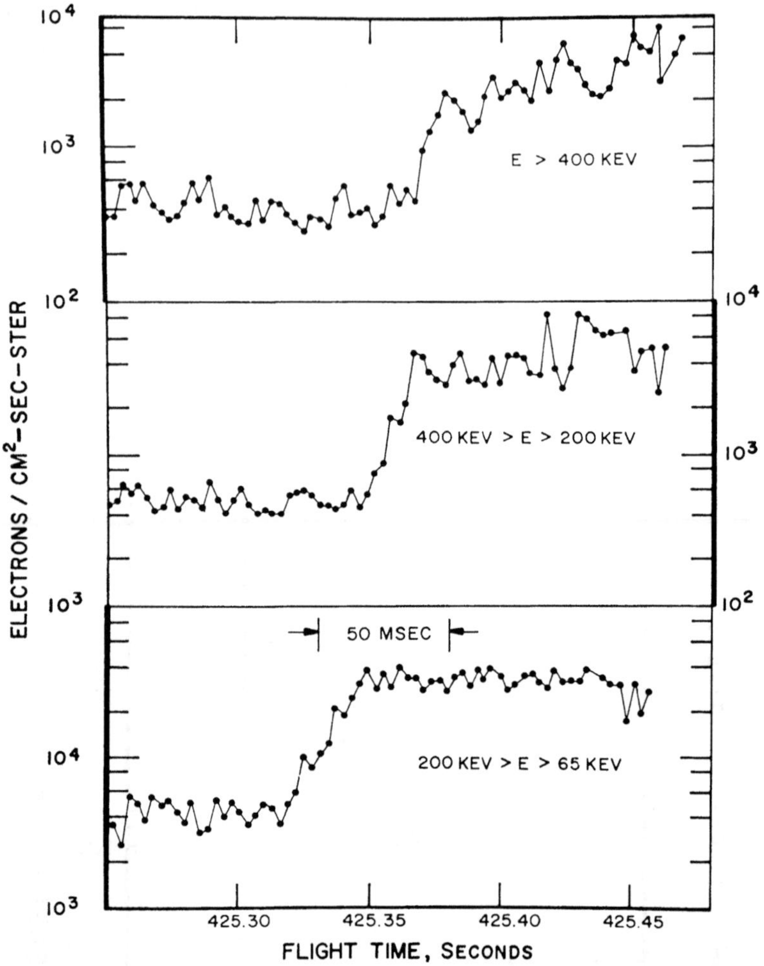

Fig. 3. A representative rapid fluctuation in electron flux observed by Mozer (1968) which was consistent with the 'spatial' model outlined in Figures 1 and 2.

Mozer visualized the situation in which the rocket borne detector moved into a region of greatly enhanced electron precipitation. For the geometry shown in Figure 1, as the rocket passed the sharp spatial gradient in particle flux, the counter would first respond to low energy electrons and then some time later the higher energies would be detected (note that, for this specific geometry, this is in spite of the larger Larmour radius of the energetic electrons). Figure 2 shows the expected 'arrival times' of particles at the mouth of the detector if the rocket had indeed passed through a sharp gradient with the aspect shown in Figure 1. The experimental points plotted in Figure 2, which are in good agreement with the model, are taken from the electron intensity variation, as observed in three energy intervals (Figure 3). This particular example, in which the intensity of energetic electrons increased by a factor of 10 within 50 msec, could then be interpreted as the rocket moving through a purely spatial gradient in electron flux. If this were indeed the case, it would imply a spatial scale for the order of magnitude intensity change of only 20 m, about the Larmour radius of the electrons involved, and a very remarkable spatial limiting of the electron beam.

Mozer also considered models where the observed intensity change was purely temporal in nature and, assuming that the modulation was introduced into the beam instantaneously over all electron energies, computed the differences in arrival times of various energy electrons as a function of distance to the source of modulation. Figure 4 plots the electron arrival times computed for Mozer's complement of solid state detectors for an equatorial source of modulation and gives some notion of the

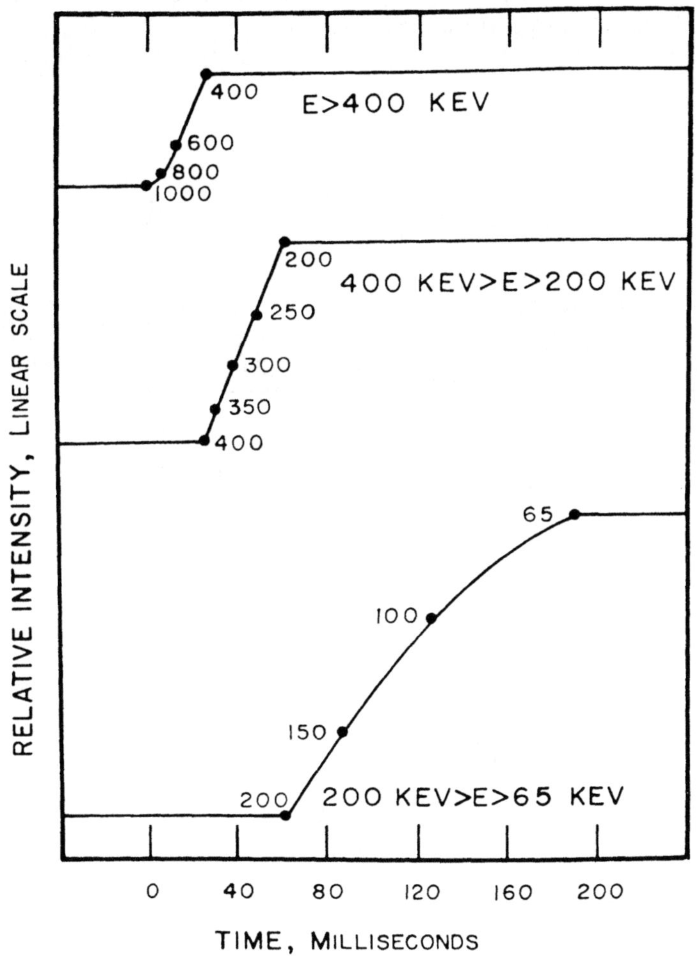

Fig. 4. The computed arrival times of electrons exciting Mozer's solid state detector assuming an equatorial modulation source (Mozer, 1968).

time scales involved. Figure 4 also illustrates the desirability of extending electron measurements to energies lower than 65 keV as the differences in arrival times would be greatly enhanced and thus allow velocity dispersion effects to be more easily exposed.

Variants used on a purely spatial or purely temporal model to explain particle intensity fluctuations, such as introducing a motion on the part of the precipitation region or allowing electrons of differing energies to be modulated at different instants in time, permits any intensity change to be explained, although in an ad hoc and undesirable manner.

Of the 75 instances of electron flux variations treated by Mozer (1968) only six were consistent with the spatial gradient model shown in Figure 1 and none had the proper time profile to be consistent with a temporal modulation of the beam at a large distance from the earth as in Figure 4. In view of this, Mozer could only conclude that

(a) If the variations were purely temporal, the modulation was introduced into the beam no more than 6000 km from the rocket, or

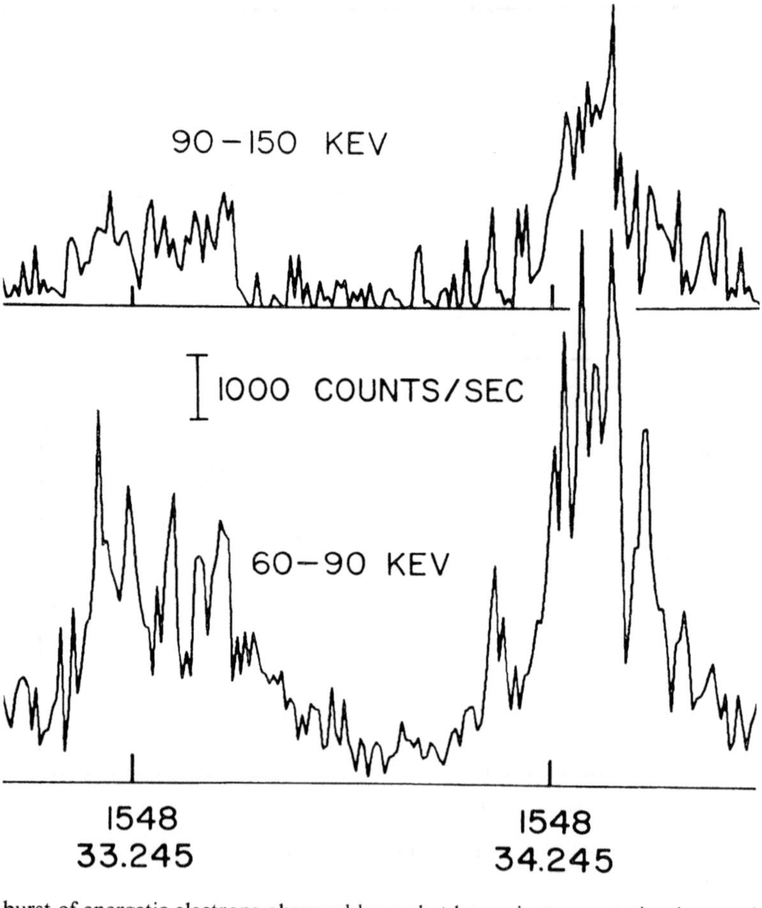

Fig. 5. A burst of energetic electrons observed by rocket borne instrumentation in association with X-ray microburst activity (Lampton, 1967).

(b) If the variations were purely spatial, then the gradient in electron intensity could be as high as a factor of 10 within a few Larmour radii.

Occasionally circumstances allow one to interpret a particle flux variation, observed directly by a rocket borne detector, as being a purely temporal phenomena. One of the best examples is that of Lampton (1967) who launched a rocket during an episode of X-ray microburst activity as determined by a coordinated balloon observation. During the flight several short lived bursts of energetic electrons were observed (Figure 5). The similarity in structure between these bursts and X-ray microbursts, coupled with the high level of microburst activity, served to identify these electrons as the parents of X-ray microbursts. In turn, the knowledge from balloon experiments that X-ray microbursts are purely temporal in nature completed the proof that these electron bursts were also temporal.

Having established this fact, Lampton (1967) was able to perform a velocity dispersion analysis to estimate the distance between the rocket and the source responsible for the electron pulse. Source distances obtained in this manner were grouped around two values: $2 \pm 2\ R_E$ (i.e., no dispersion within the available time resolution) and $22 \pm 6\ R_E$ (which could correspond to the conjugate ionosphere assuming the magnetic field line to be dipolar). Lampton thus suggested that the production of these electron bursts was located in or near the auroral ionosphere and, perhaps, was associated with a plasma instability.

Evans (1967) observed several episodes of 10 Hz pulsations in energetic electron fluxes during a single rocket flight into a breakup aurora (Figure 6). These fluctuations were interpreted as being temporal on plausibility grounds (the alternative required an ordered spatial structure in the electron precipitation coupled with a proper motion on the part of this bombardment which, taken together, appeared to be contrived). In view of the later reports of the flicker aurora (Beach *et al.*, 1968) which occurs under the same geophysical conditions, at the same frequency, and with roughly the same pulsation amplitude as was observed during the rocket flight, the assumption of temporal variation rather than spatial seems well founded.

Under this assumption a cross correlation was made between detectors which responded to electrons of differing velocities ranging from about 5×10^9 cm sec^{-1} to 1.5×10^{10} cm sec^{-1}. The result was that there were no significant differences in the arrival times of electrons of higher velocities and some indication that electrons of lower velocities arrived at the rocket somewhat earlier (about 10 msec) than the higher energy electrons. On the basis of this, the conclusion was made that the mechanism producing the 10 Hz variation in the beam was located in the auroral ionosphere not more than some 1200 km from the rocket. A beam driven plasma instability was suggested as the mechanism responsible for this phenomena.

Winiecki (1967) encountered rapid fluctuations in auroral electrons very similar to those observed by Evans (1967) and, under the same assumptions and analysis, also deduced a distance of about 1000 km between the rocket and the source of modulation.

The latest example of rapid pulsations of auroral electrons is reported by Arnoldy (1970) who observed such pulsations during a rocket flight over a breakup aurora.

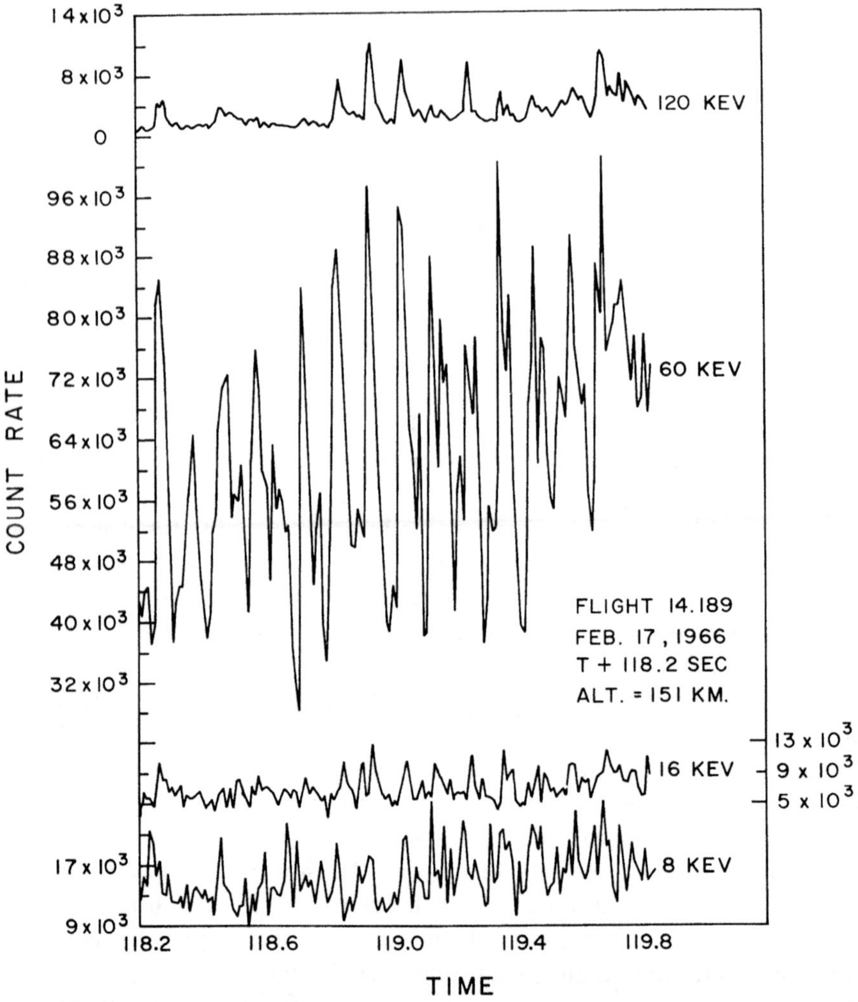

Fig. 6. An example of 10 Hz fluctuations observed in energetic electrons and probably associated with a flicker aurora (Evans, 1967).

The phase of the aurora and the temporal behavior of the particles themselves indicate that this observation also is to be associated with a flicker aurora. Arnoldy did not attempt a cross correlation analysis to obtain a source distance but he did use pitch angle distribution arguments (the distribution appeared to peak along the field line) to suggest that some process had affected these electrons relatively close to the earth.

Observations of particle fluctuations in a somewhat longer period range than discussed above were made by Bryant *et al.* (1967). On a rocket flight through a post-breakup visual aurora they encountered 10 sec period quasi-periodic variations in the flux of 1 to 10 keV electrons. The rocket data alone could not establish whether the fluctuations were temporal or due to vehicle motion through a striated electron precipitation. However, because a spatial model would have required a great deal of order on the part of such a spatial structure and, more importantly, because optical pulsations in this period range are most common on the part of the post-breakup aurora, the entirely justifiable assumption was made that the observed particle pulsations were purely temporal. Having established this point, Bryant *et al.* proceeded with the now standard velocity dispersion analysis. Unlike the results for the very rapid electron flux variations that were discussed above, the source distances obtained in this case were consistent with the electron modulation mechanism being at great distance from the rocket, i.e., about 10 R_E. If the relevant magnetic field line (Andenes, $L=6.0$) was near dipolar, this distance would correspond, in fact, to a position on the field line near the equatorial plane.

Very recently Bryant (1971) reported a similar pulsation in the flux of primary auroral electrons during a flight over a post-breakup aurora. In this case, however, auroral brightness pulsations were observed photometrically from the ground to be of roughly the same amplitude and *in phase* with the variations observed directly in the precipitating electrons. This correlation proves the temporal nature of the phenomena and vindicates the assumptions made in the earlier paper. Bryant's (1971) analysis of this second observation yielded source distances of some 10 R_E which were quite consistent with his earlier results.

While the number of direct observations of temporal variations in auroral electrons is as yet limited, a trend does seem to be appearing. That is that the very rapid fluctuations ($\tau < 1$ sec) are associated with processes close to or within the auroral ionosphere, while the slower pulsations ($\tau >$ sec) seem to originate at large distances, probably at the equatorial plane of the auroral zone line of force.

Some of this bias may be due to the fact that rapid fluctuations introduced into the electron population at large distance from the atmosphere could be 'washed out' by velocity dispersion effects during transit to the atmosphere (Omholt and Petterson, 1967) and, thus, be undetectable by the rocket borne instrument.

However, that there exists processes close to the atmosphere that can introduce rapid variations in a beam of energetic electrons is now firmly established.

In contrast with studies of the temporal behavior of auroral electrons – in which the equivalent of many of the optical and X-ray pulsations have been observed directly in the electron precipitation – there has been no reported observation of spatially thin sheets of precipitating electrons, the analog of the fine scale structure seen in the visual aurora. Evans (1969) commented that the total electron energy influx, as measured onboard a moving rocket, appeared to have far less spatial structure than the TV studies of the visual aurora would have suggested, but he presented no data in support of his point.

Thus far, the only relevant observation in the literature is that of Cloutier *et al.* (1970) who show a rocket flight through a relatively stable aurora in which the low energy auroral electrons were precipitated in a sheet roughly 1.5 km thick. The dimension is well to the high side of the occurrence plot of Maggs and Davis (1968).

In order to find more observations of this sort, an examination was made of data

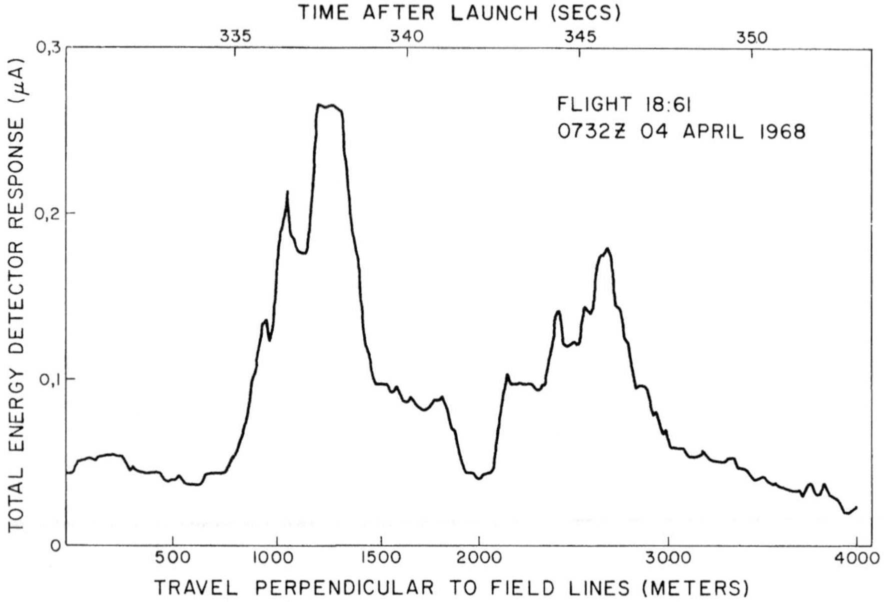

Fig. 7. The spatial response of a total electron energy flux detector as it was carried over an active break up aurora.

obtained during the entire GSFC auroral sounding rocket program. As a rule these payloads included a 'total energy detector' (a plastic scintillator-phototube detector in which the anode current was monitored as a measure of the total charged particle energy deposition in the scintillator). The output of this detector should give a reasonable notion as to the spatial extent of the energy flux into the atmosphere and thus also of the visual aurora. Figure 7 is the response of this detector during a flight over a breakup, and thus evidently a very active aurora. The ordinate in Figure 7 is the phototube anode current in micro amps, which to first approximation is proportional to the total energy flux incident on the scintillator. The abscissa is the position of the payload relative to magnetic lines of force which, because the rocket was fired southward in the magnetic meridian, will mark off the N-S dimensions of the auroral electron influx.

The apparent N-S extent of the gross peak in precipitation is about 800 m, about 3 times the median luminosity width given by Maggs and Davis (1968), although lesser changes may be seen on a smaller spatial scale. This particular example is the most structured observed during the program, and other more typical cases yielded dimensions on the order of km. It should be pointed out that the apparent dimension is obtained assuming a stationary and stable auroral form, the only motion being on the part of the rocket. The effect of motions and temporal variations on the part of the precipitation would generally (but not invariably) produce more structure than would actually be present so that the dimension could be regarded as a *lower* limit.

Hunten (1969) in reply to Evans' comments at Ås (Evans, 1969) pointed out that many of the instruments used to view the aurora (including the eye, all-sky photography, and TV systems) tend to over emphasize the contrast between a homogeneous background glow and a slightly brighter auroral structure. He believed that many of the fine scale structures studied by Maggs and Davis (1968) represented only a 15% or so enhancement in luminosity over the background. If this is indeed the case, then the fine structure in energy flux seen as the rocket passed over the gross form (Figure 7) could easily be the analog of the fine structure in luminosity revealed by the TV studies.

On the other hand, estimates in the TV field of the difference in brightness between a fine scale feature and the background range from a factor of 2 to 5 (Davis, 1969; Parsons, 1969). If this is the case, then the lack of any direct observations of very thin sheets of precipitating kilovolt energy electrons represents a true mystery. The discrepancy, incidentally, appears *not* to be caused by insufficient samples on the part of rocket flights because such launches are made most often during bright and very active aurora as evidenced by the fact that the relatively rare flicker aurora has been observed directly on at least three occasions by sounding rocket instrumentation.

The final determination as to whether there is a true problem explaining fine scale auroral structures by means of fine scale particle precipitation rests with further ground based observations, preferably with truly linear optical detectors (such as phototubes) viewing only allowed emission lines which are excited only by energetic electrons.

References

Anderson, K. A., Chase, L. M., Hudson, H. S., Lampton, M., Miton, D. W., and Parks, G. K.: 1966, *J. Geophys. Res.* **71**, 4617.
Arnoldy, R. L.: 1970, *J. Geophys. Res.* **75**, 224.
Barcus, J. R. and Christensen, A.: 1965, *J. Geophys. Res.* **70**, 5455.
Barcus, J. R., Brown, R. R., and Rosenberg, T. J.: 1966, *J. Geophys. Res.* **71**, 125.
Beach, R., Cresswell, G. R., Davis, T. N., Hallinan, T. J., and Sweet, L. R.: 1968, *Planetary Space Sci.* **16**, 1525.
Bryant, D. A., Courtier, G. M., and Bennett, G.: 1971, this volume, p. 170.
Bryant, D. A., Collin, H. L., Courtier, G. M., and Johnstone, A. D.: 1967, *Nature* **215**, 45.
Cloutier, P. A., Anderson, H. R., Park, R. J., Vondrak, R. R., Spiger, R. J., and Sandel, B. R.: 1970, *J. Geophys. Res.* **75**, 2595.
Cresswell, G. R. and Davis, T. N.: 1966, *J. Geophys. Res.* **71**, 3155.
Davis, T. N.: 1969, private communication.
Evans, D. S.: 1967, *J. Geophys. Res.* **72**, 4281.
Evans, D. S.: 1969, in *Atmospheric Emissions* (ed. by B. M. McCormac and A. Omholt), Van Nostrand Reinhold Company, New York, p. 550.
Hunten, D. M.: 1969, in *Atmospheric Emissions* (ed. by B. M. McCormac and A. Omholt), Van Nostrand Reinhold Company, New York, p. 553.
Kim, J. S. and Volkman, R. A.: 1963, *J. Geophys. Res.* **68**, 3187.
Lampton, M.: 1967, *J. Geophys. Res.* **72**, 5817.
Maggs, J. E. and Davis, T. N.: 1968, *Planetary Space Sci.* **16**, 205.
Mozer, F. S.; 1968, *J. Geophys. Res.* **73**, 999.
Omholt, A. and Petterson, H.: 1967, *Planetary Space Sci.* **15**, 347.
Parks, G. K., Milton, D. W., and Anderson, K. A.: 1967, *J. Geophys. Res.* **72**, 4587.
Parsons, N. R.: 1969, private communication.
Winiecki, T.: 1967, M.Sci Thesis, Rice University, Houston, Texas.

INSTITUTE DISCUSSION

S. Mende reported that his two channel image intensifier TV experiment was used to measure the emissions of 4278 Å N_2^+ and of 5577 Å O I. Within the resolving limit of the system the 4278 Å auroral emission had a similar ray structure to 5577 Å auroras suggesting that the incoming electrons are structured and there is no need for atmospheric structure generation.

It was noted by M. Walt that at the conference two years ago cases were reported of anomalous angular distributions (too many back scattered particles). He wondered if these cases have survived careful analysis. D. Evans indicated that not so many new cases are being reported.

PRELIMINARY REPORT ON THE OBSERVATION OF LOW ENERGY IONS IN THE AURORA

B. A. WHALEN, J. R. MILLER, and I. B. McDIARMID

Division of Physics, National Research Council, Ottawa, Canada

Abstract. A rocket borne ion spectrometer capable of separating and detecting H^+, He^+, and He^{++} with energies between 2 and 24 keV has been flown into an early evening proton aurora from Fort Churchill. The measurements made with a Black Brant IV at altitudes to 775 km enabled the positive ion composition of the primary auroral precipitation to be determined above the masking effects of atmospheric interactions. The proton differential energy spectrum was found to peak at 8 keV with an intensity of 1.5×10^5 (cm^2 sec sr keV)$^{-1}$. No statistically significant flux of He^+ was observed. The He^{++} intensity at 15 keV was measured to be $7 \pm 3.5 \times 10^3$ (cm^2 sec sr keV)$^{-1}$ and an upper limit of 5×10^3 was set at 7.5 keV. The H^+ and He^{++} measurements were used to calculate an integrated He^{++}/H^+ flux ratio of approximately 5%.

The observation of auroral He^{++} and H^+ ions of approximately the same ratio as in the solar wind, and the apparent lack of He^+ ions, is interpreted as strong evidence for a direct solar wind origin of auroral primaries.

1. Introduction

It is generally believed that the solar wind is the energy source for auroral phenomena; however, it is unclear whether particles precipitating into the atmosphere come directly from the solar wind or are accelerated particles of terrestrial origin. Attempts at resolving this problem were reported by Whalen and McDiarmid (1968) and later by Reasoner *et al.* (1968). They measured the proton to α-particle ratio of energetic ions in the aurora and found values which were consistent with a solar wind origin.

Axford (1970) observed that the measurement of the ratio of α-particle to proton intensities is not a completely convincing method of settling this question since the expected α/p ratio is not significantly different in the solar wind (~4%) and the ionosphere (~1%).

A more definitive ion measurement which is reported here involves the determination of the relative intensities of H^+, He^+, and He^{++}. Measurements in the solar wind (Bame *et al.* 1968) show that the solar wind He flux is almost completely ionized ($^4He^+/^4He^{++} < 10^{-3}$) whereas charge composition measurements of ambient terrestrial ions (Taylor *et al.*, 1965) indicate that the dominant He charge state is singly ionized. Clearly, He precipitation resulting from acceleration of solar wind ions will be predominantly He^{++} whereas precipitation of terrestrial origin is expected to be mostly He^+.

2. Experimental

An ion spectrometer was designed which had sufficient resolution to measure H^+, He^+, and He^{++} ions in the energy range 2 to 20 keV. Figure 1 is a schematic diagram of the spectrometer which consists of two sections. The first section performs an electrostatic analysis on the incoming ion beam thereby selecting particles of the same

B. M. McCormac (ed.), The Radiating Atmosphere, 279–287. *All Rights Reserved.*

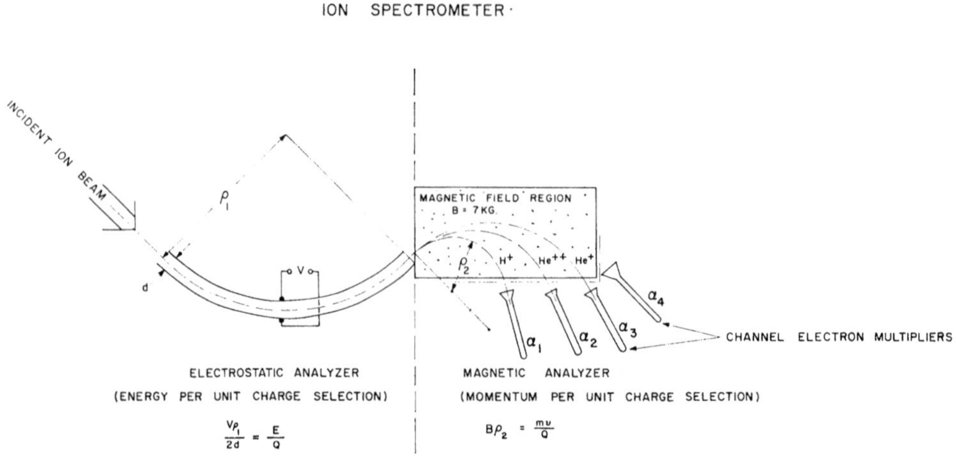

Fig. 1. Schematic diagram of the auroral ion spectrometer. Shown in this diagram are the electrostatic and magnetic analysis sections which define the velocity and mass per unit charge of the ion being detected by the electron multipliers.

energy/unit charge. This selected beam of ions then passes into a constant magnetic field region where magnetic analysis separates the beam into components with the same momentum per unit charge. This combination of electrostatic and magnetic analysis allows one to determine the velocity and mass per unit charge of the incident ions.

In detail, the electrostatic analyzer consisted of two concentric cylindrical plates across which a sawtooth wave form with a 2 sec period and 0 to 3.8 kV voltage amplitude were applied. The geometry was such that the energy of ions selected by the plates varied between 0 and 24 keV/unit charge. Magnetic analysis was performed between the pole faces of a permanent magnet with a 7 kG field strength. Four channel electron multipliers were arranged around the magnet as shown in Figure 1 and were centered on the end points of the average trajectories followed by 2.25 keV, 3.75 keV, 7.5 keV and 15 keV protons. These four electron multipliers will be referred to as α_1, α_2, α_3, and α_4, respectively. This detector geometry was selected to give reasonably large geometric factors while still retaining the resolution necessary to separate H^+, He^+, and He^{++} ions.

Trajectories of various ions inside the analyzer are indicated in Figure 1 corresponding to a deflection plate voltage which would select 2 keV protons. In this example the beam is assumed to be composed of H^+, He^+, and He^{++} and, therefore, the electrostatic analyzer would be selecting 2 keV H^+, He^+ and 4 keV He^{++}. In this configuration α_1 would be detecting 2 keV protons, α_2 4 keV He^{++} and α_3 2 keV He^+.

Figure 2 shows the measured response function or equivalent geometric factor for detector α_4. The count rate profile observed by detector α_4 during one sweep of the electrostatic analyzer would appear as a sum of the three curves (Figure 2) if the

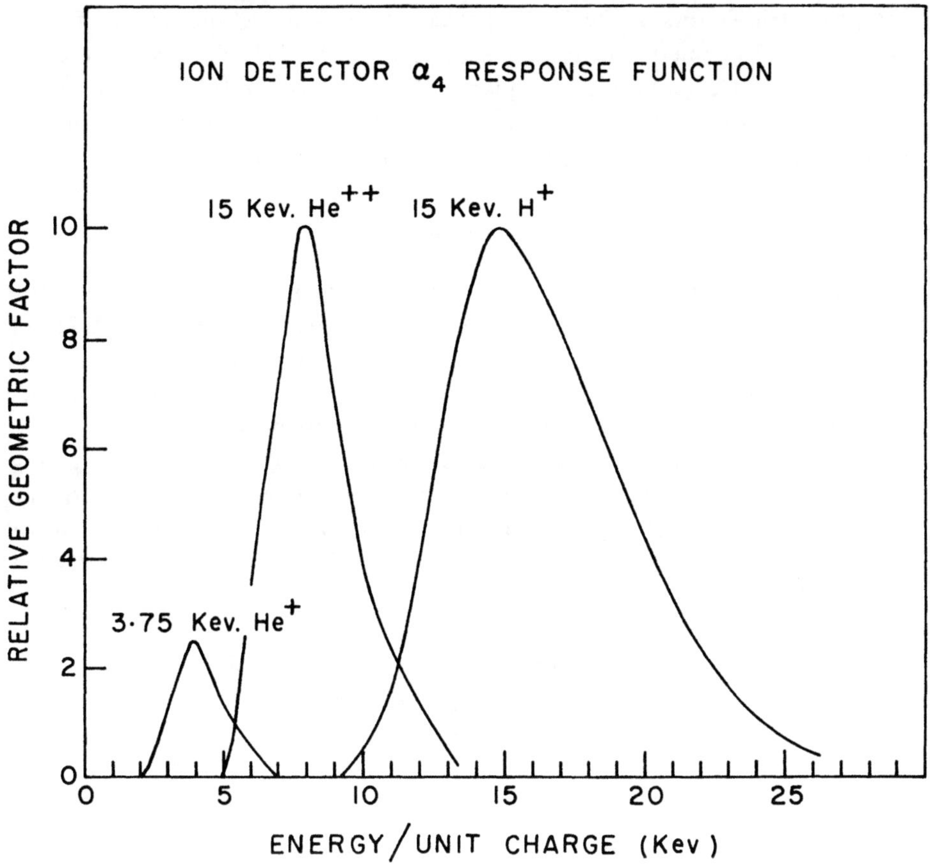

Fig. 2. Ion detector α_4 response function. The count rate profile observed in detector α_4 during one sweep of the electrostatic analyzer would appear as the sum of the above curves if the spectrometer was detecting an ion beam composed of H^+ and He^{++} with equal, flat differential energy spectra.

differential energy spectra between 0 and 25 keV were flat and equal for all three ion components. The response functions for the other three detectors are similar to that shown for α_4 (Figure 2).

The energies and geometric factors which correspond to the H^+ peaks in the profiles for α_1 to α_4 are as follows:

Detector	H^+ Energy (keV)	Geometric Factor (cm^2 sr keV)
α_1	2.25	2.46×10^{-4}
α_2	3.75	3.93×10^{-4}
α_3	7.5	5.79×10^{-4}
α_4	15.0	1.08×10^{-3}

The geometric factors at the He^{++} peaks are the same as those listed for H^{++}, whereas the He^{+} geometric factor is one quarter that for H^{+}. The peak in the profile corresponding to He^{+} occurs at one quarter the energy of the H^{+} peak and since the energy bandpass of the electrostatic analyzer is proportional to the energy of the ion being selected, the He^{+} geometric factor is one quarter smaller.

3. Results

A Black Brant IV (AAF-IV-20) sounding rocket carrying the experiment described in Section 2 was launched from Churchill Research Range at 0227:30 UT (2027:30 LT) on December 15, 1969, into an early evening H aurora. Ground based measurements of the Hβ emission (Creutzberg, 1970) showed 15 to 25 R on trajectory at liftoff. No detectable magnetic activity or radio absorption accompanied the event.

The ion spectrometer was turned on at 51 sec after liftoff; however, outgasing of the experiment and the resultant high voltage breakdown continued until about 0232:30 UT when the rocket was at an altitude of 676 km. The spectrometer operated satisfactorily from this time until the rocket reentered the atmosphere (about 0242 UT).

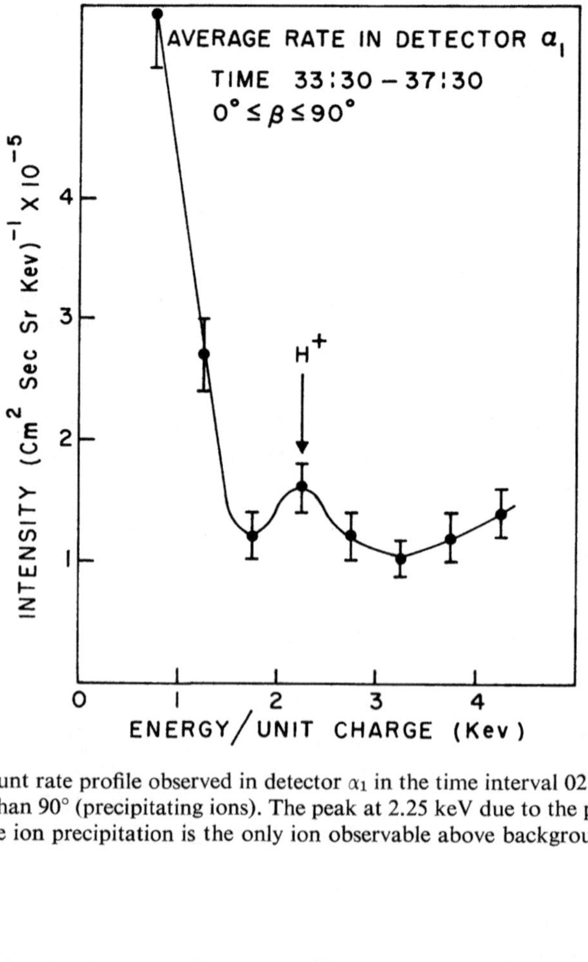

Fig. 3. Average count rate profile observed in detector α_1 in the time interval 0233:30–0237:30 UT at pitch angles less than 90° (precipitating ions). The peak at 2.25 keV due to the presence of protons in the ion precipitation is the only ion observable above background.

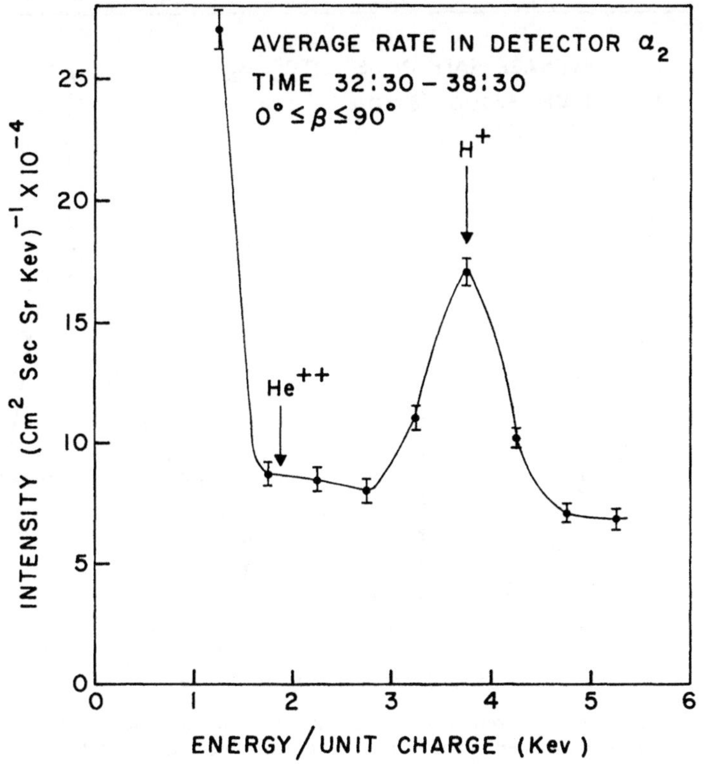

Fig. 4. Average α_2 profile observed at small pitch angles in the time interval 0232:30–0238:30 UT. The H^+ peak in the profile at 3.75 keV is clearly visible above the background count rate whereas no peak is observed where He^{++} is expected.

The rates recorded in the ion detector were found to be too low to produce statistically significant count rate profiles for each sweep of the electrostatic analyzer and therefore the rates were averaged over several sweeps. Figures 3 and 4 show the average rates recorded by detectors α_1 and α_2 for times when the detectors were sampling particles with pitch angles less than 90°. The arrows in these Figures indicate where peaks in the profiles due to H^+ and He^{++} are expected. Although these profiles show the proton peaks, a high background rate, particularly at energies less than 1.5 keV, makes the detection of a low flux of He^{++} or He^+ in these detectors impossible. These figures were used to determine the H^+ flux and set an upper limit on the intensity of 3.75 keV He^{++} ions at:

$$I(2.25\text{ keV } H^+) = 7 \pm 3 \times 10^4 (\text{cm}^2 \text{ sec sr keV})^{-1}$$
$$I(3.75\text{ keV } H^+) = 1.0 \pm 0.2 \times 10^5 (\text{cm}^2 \text{ sec sr keV})^{-1}$$
$$I(3.75\text{ keV } He^{++}) < 2 \times 10^4 (\text{cm}^2 \text{ sec sr keV})^{-1}.$$

The average profile observed in detector α_3 early in the flight (0233 to 0234 UT) is shown in Figure 5. The H^+ peak is indicated on the curve as well as the regions where

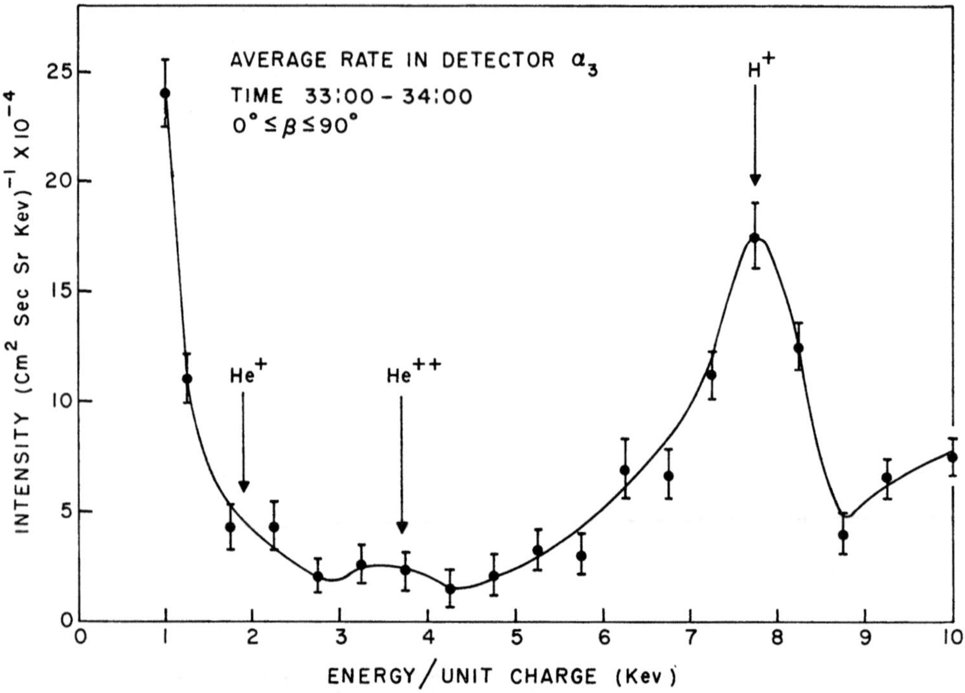

Fig. 5. Average α_3 profile measured at small pitch angles in the time interval 0233:00–0234:00 UT An H^+ peak and an enhancement in the He^{++} region is observed above the background which increases below 1.5 keV and above 8 to 9 keV.

He^{++} and He^+ peaks are expected. At this time there is a slight enhancement in the region where the He^{++} peak is expected. The rise in count rates at low energies is similar to that observed in α_1 and α_2. An increase in the background rate is also observed above 8 to 9 keV.

Similar profiles for detector α_3 at intervals later in the flight were studied to determine if there were any times when statistically significant fluxes of 8 keV He^{++} were detected. No such interval was found. The profiles were averaged over the 7 min period 0232:30 UT to 0239:30 UT, an interval when the rocket was above 500 km; the results appear in Figure 6. No significant enhancements are observed at 4 or 2 keV per unit charge where one expects to observe He^{++} and He^+ peaks. Using Figure 6 average intensities and upper limits on 7.5 keV H^+, He^{++} and 1.9 keV He^+ were determined to be:

$$I(7.5\text{ keV } H^+) = 1.4 \pm 0.1 \times 10^5 (\text{cm}^2\text{ sec sr keV})^{-1}$$
$$I(7.5\text{ keV } He^{++}) < 5 \times 10^3 (\text{cm}^2\text{ sec sr keV})^{-1}$$
$$I(1.9\text{ keV } He^+) < 2 \times 10^4 (\text{cm}^2\text{ sec sr keV})^{-1}.$$

The average profile for detector α_4 in the time interval 0233:35 to 0235:03 UT is shown in Figure 7. An H^+ peak at 15 keV is observed as expected and a second peak

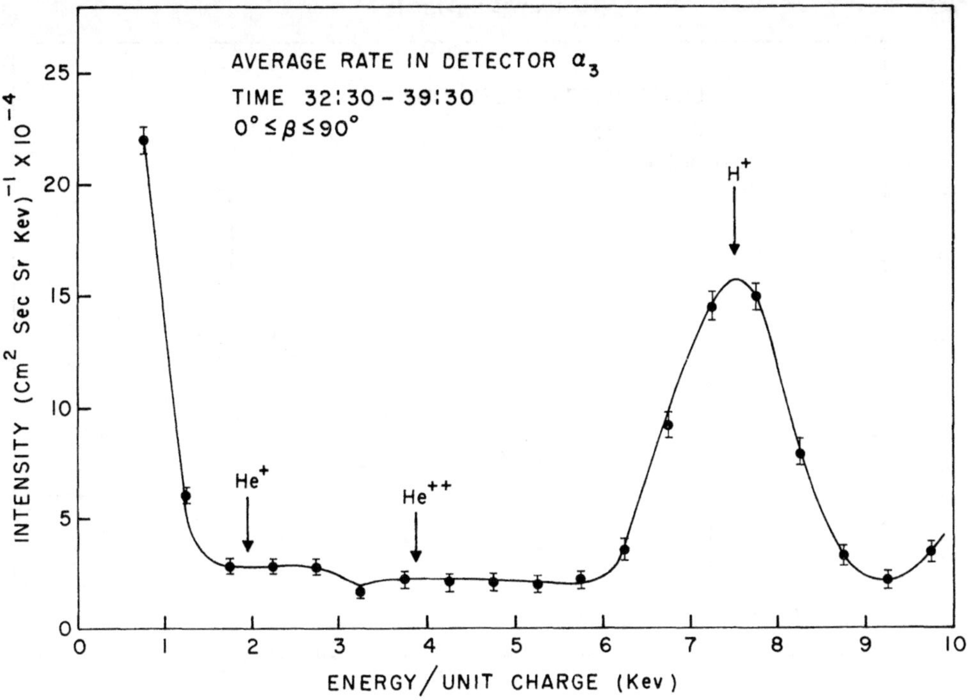

Fig. 6. Average α_3 profile of precipitating ions in the time period when the rocket was above 500 km (0232:30–0239:30 UT). No indication of the presence of He^{++} or He^{+} was observed in this detector.

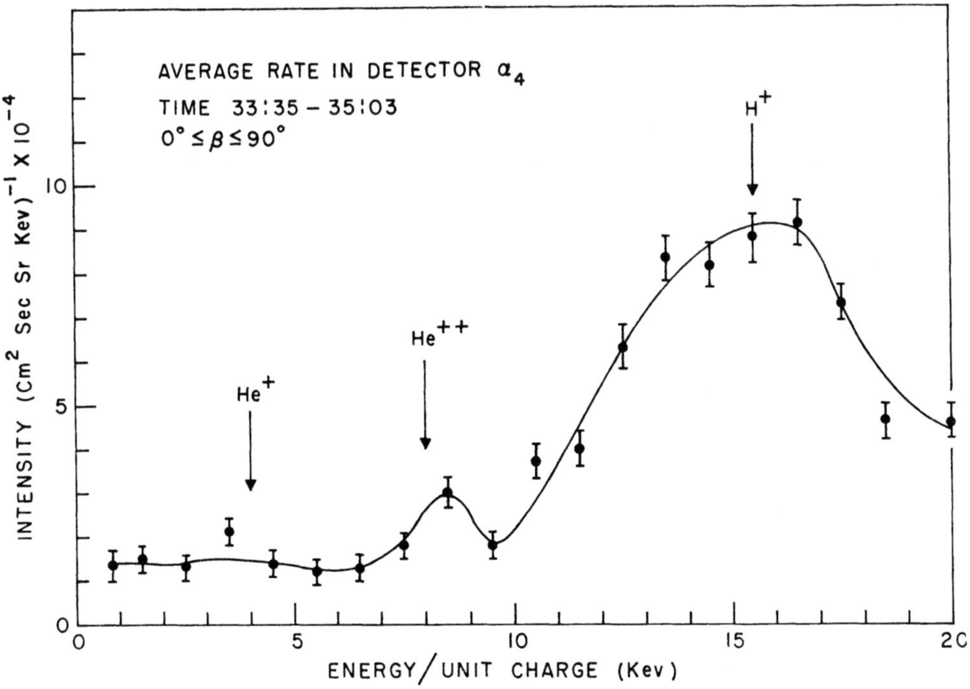

Fig. 7. Detector α_4 profile early in the flight (0233:35–0235:03 UT) at pitch angles less than 90°. The presence of H^{+} and He^{++} and possibly some He^{+} in the incident ion beam is indicated in this profile.

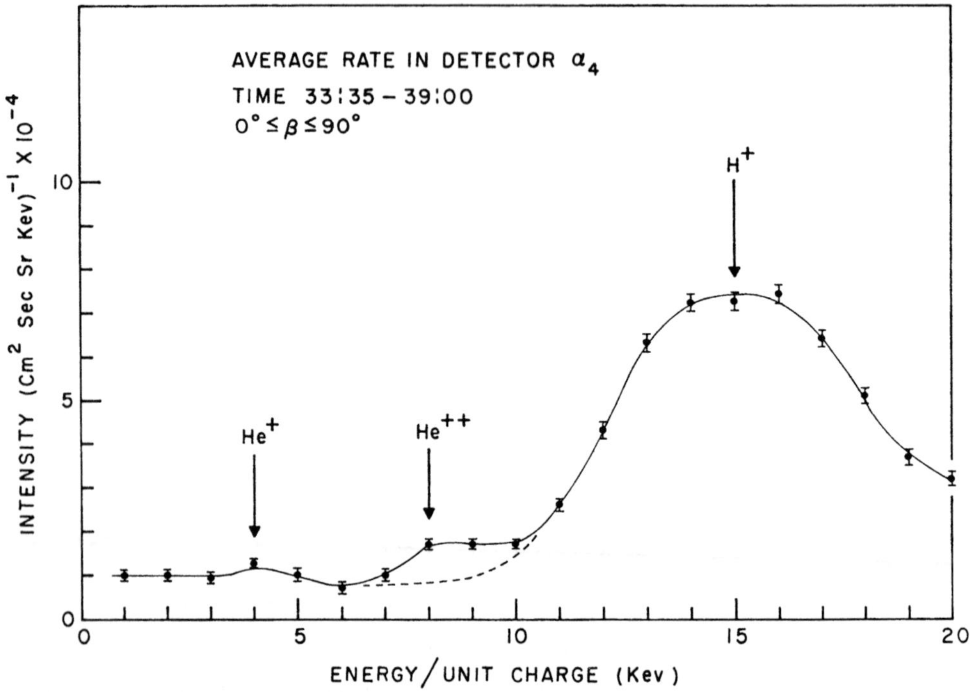

Fig. 8. α_4 profiles averaged over the time period 0233:35–0239:00 UT when the rocket was above 500 km. The dashed line indicates the profile expected if the ion precipitation was pure H^+. The remainder above this line is due to the He^{++} component in the beam. Although the slight enhancement observed at 4 keV is not statistically significant it occurred in the energy interval where one expects a He^+ peak.

appears near 8 keV which is where He^{++} is expected. A slight enhancement is also observed near the point where one expects the He^+ peak. Profiles in various time intervals later in the flight were examined; however, at no time was a statistically significant flux of He^+ detected.

The average profile for the interval 0233:35 to 0239:00 UT appears in Figure 8. The dashed line in the energy region 6 to 10 keV is the sum of the background, as determined in the 1 to 6 keV interval, and the response of the detector to a pure H^+ beam. That is, since protons do not contribute to the count rate below 9 keV the enhancement in the 7 to 9 keV region is assumed to be due to He^{++} plus a flat background. The average ion fluxes with background subtracted are:

$$I(15\text{ keV } H^+) = 7.5 \pm .5 \times 10^4 (\text{cm}^2 \text{ sec sr keV})^{-1}$$
$$I(15\text{ keV } He^{++}) = 7 \pm 3.5 \times 10^3 (\text{cm}^2 \text{ sec sr keV})^{-1}$$
$$I(3.75\text{ keV } He^+) = 1 \pm 1 \times 10^4 (\text{cm}^2 \text{ sec sr keV})^{-1}.$$

4. Discussion

In the preliminary analysis of our auroral ion measurements the only ions identified in the precipitation were H^+ and He^{++}. The He^+ intensity was always below the limit of detectability. Peak intensities of the two identifiable constituents of the precipitation were:

$$I_{He^{++}} = 7 \pm 3.5 \times 10^3 \ (\text{cm}^2 \text{ sec sr keV})^{-1} \quad \text{at} \quad 15 \text{ keV}$$

and

$$I_{H^+} = 1.4 \pm 0.1 \times 10^5 \ (\text{cm}^2 \text{ sec sr keV})^{-1} \quad \text{at} \quad 8 \text{ keV}.$$

If it is assumed that the peak intensities are proportional to the total fluxes in the auroral ionosphere one finds a H^+ to He^{++} flux ratio of about 5%. This value, of course, is very close to the solar wind value.

In summary the observation of He^{++} and H^+ ions of approximately the same ratio as in the solar wind, and the apparent lack of He^+ ions, is the most compelling evidence yet presented in favor of a direct solar wind origin of auroral primaries.

Acknowledgment

The authors are grateful for the ground based photometric support kindly provided by Dr. F. Creutzberg.

References

Axford, W. I.: 1970, in *Particles and Fields in the Magnetosphere* (ed. by B. M. McCormac), D. Reidel Publishing Company, Dordrecht-Holland, p. 46.

Bame, S. J., Hundhausen, A. J., Asbridge, J. R., and Strong, I. B.: 1968, *Phys. Rev. Letters*, **20**, 393.

Creutzberg, F., 1970, private communication.

Reasoner, D. L., Eather, R. H., and O'Brien, B. J.: 1968, *J. Geophys. Res.* **73**, 4185.

Taylor, H. A., Jr., Brinton, H. C., and Smith, C. R.: 1965, *J. Geophys. Res.* **70**, 5769.

Whalen, B. A. and McDiarmid, I. B.: 1968, *J. Geophys. Res.* **73**, 2307.

INSTITUTE DISCUSSION

L. Kavanagh suggested that particles of equal momentum should be compared rather than equal energy if the ratio is determined in solar wind.

The entry of electrons, protons, and α-particles was discussed. It was pointed out by B. McCormac that electrons and protons enter the magnetosphere at different locations and while the efficiency of electron injection may on occasion be about 100 %, the efficiency for protons, whenever they seem to enter without energy change, is quite low. There are also arguments that the location and efficiency of entry of α-particles should still be different from protons.

D. Winningham gave a preliminary report of some very interesting electron and proton precipitation data for energies of 100 eV to 12 keV. At high latitudes well above the auroral oval, low energy protons and electrons seem to have direct access. The particle data agree with the Vela 4 observations in the magnetospheric tail. The peak energy for electrons is approximately 100 eV. The particles are conducted downward by convection in the magnetosphere in both polar regions. The peak energies seem to be the same in each hemisphere, but the distribution spread indicates a winter-summer effect.

X-RAYS ASSOCIATED WITH AURORAS

HARALD TREFALL
Dept. of Physics, University of Bergen, Norway

1. Introduction

X-rays recorded with balloon borne detectors at atmospheric depths of 5 to 15 g cm^{-2} are generated by $\gtrapprox$30 keV electrons stopping at heights of $\gtrapprox$100 km. For a general review of results obtained to the end of 1965 see Brown (1966). Short reviews summarizing more recent developments have been given by Kremser (1967, 1969).

The available new results on the direct relationship between aurora and X-rays will be discussed, and then more generally the morphology and dynamics of auroral zone X-ray events, especially those relating directly to Akasofu's picture of the auroral substorm.

2. Simultaneous Measurements of Aurora and X-Rays

There have been several attempts in the past to compare X-ray measurements with photometric and all-sky camera observations of aurora. These observations show a loose correspondence on a gross time scale, with the possible exception of X-rays associated with widespread diffuse glows (Anderson and DeWitt, 1963) and with auroral forms exhibiting a type B red lower border and associated with the onset of an auroral substorm (Clark and Anger, 1967). It thus seems that normally two separate acceleration mechanisms are at work, one for low and another for high energy electrons.

The situation is quite different with regard to those time variations which seem attributable to some kind of modulation mechanism. Rosenberg *et al.* (1967) found a clear correlation between simultaneous pulsations in X-rays and aurora. An example is shown in Figure 1. Even though the X-ray flux is weak and the pulsation amplitudes small, the correlation is clearly seen, with the auroral pulsations being somewhat delayed with respect to those in the X-rays, because the auroral emission is from a metastable state with a lifetime of 0.6 sec (Paulson and Shepherd, 1965). A more detailed account of this work has been published elsewhere (Rosenberg *et al.*,1971).

All this is in marked contrast to the gross behavior of the light and X-ray fluxes, and no case has been found where X-ray pulsations occurred in the absence of pulsating aurora during the above mentioned measurements. Even though the data available are still very limited, it seems that whatever modulation mechanism is active here, it affects the whole energy range of precipitating electrons.

Similar conclusions have been reached by Scourfield *et al.* (1970).

3. Morphology and Dynamics of Auroral Zone X-Ray Events in the Midnight Sector

The dynamics of electron precipitation events associated with local magnetic bay

B. M. McCormac (ed.), The Radiating Atmosphere, 288–298.

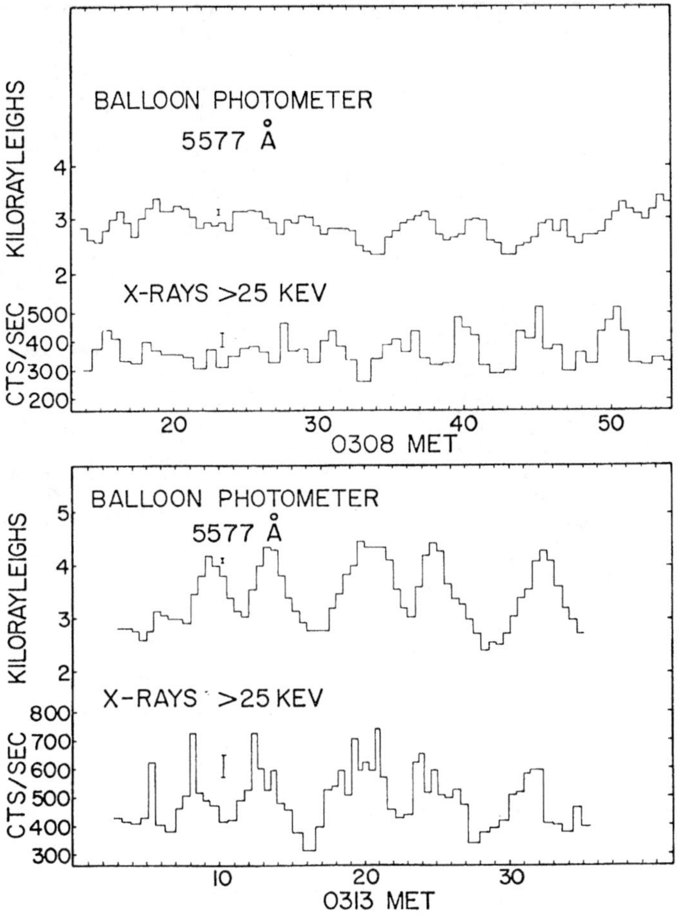

Fig. 1. Simultaneous modulation of X-rays and visual aurora.

disturbances in the geomagnetic midnight sector have been studied by Barcus and Brown (1962), Brown and Campbell (1962), Gosling (1966) and Clark and Anger (1967). With the westward electrojet there seemed to be associated a region of electron precipitation aligned E-W which took part in the N-S movements executed by the electrojet. Furthermore, it was found that bay associated precipitation events may have a large longitudinal extension (Brown *et al.*, 1965).

A comprehensive study of such events has recently been completed (Bjordal *et al.*, 1971). The launching sites used are shown in Figure 2. Some preliminary results of this work were reported by Kremser (1969).

A typical recording showing the movements of such X-ray events and their association with the auroral electrojet is given in Figure 3. Assuming that the observed magnetic variation is due to a line current flowing from east to west, the ratio $\Delta Z/\Delta H$ is a measure of the position of this line current with respect to the observatory. [If there

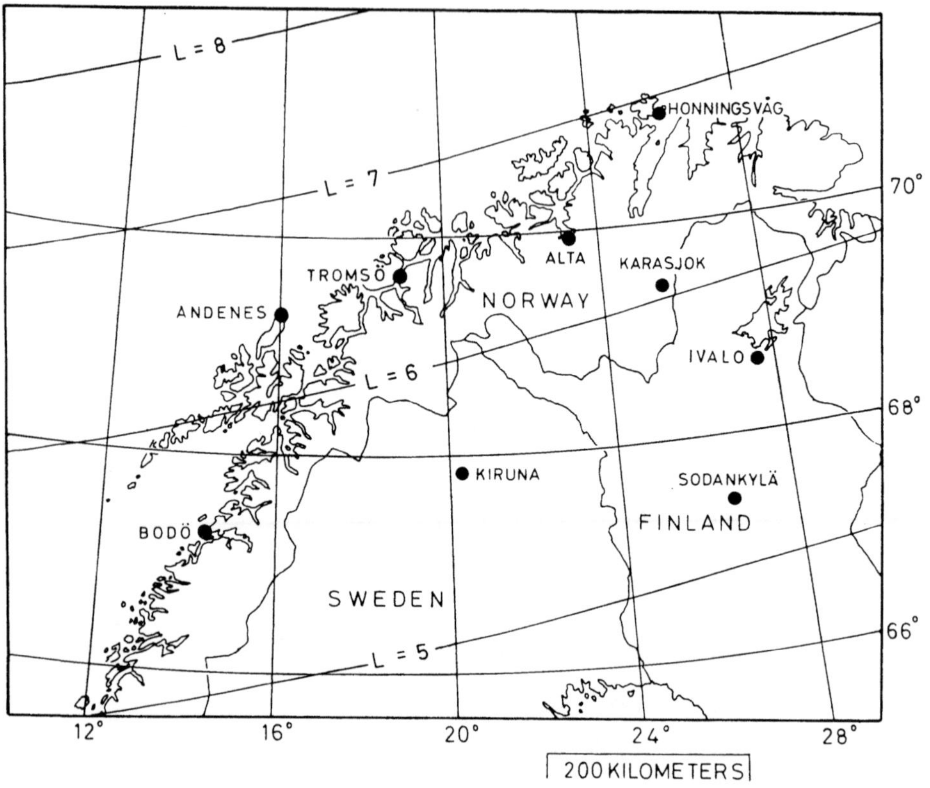

Fig. 2. Launching sites in northern Scandinavia.

were no induction effects giving rise to earth currents, and if ionospheric currents do flow at a height of 100 km, the line current would be located 100 km $(\Delta Z/\Delta H)$ north of the magnetic observatory. We see however that a displacement of the current system from Tromsø to Kiruna or vice versa corresponds on the average to a change of about 0.8 in $\Delta Z/\Delta H$; thus $\Delta Z/\Delta H = 1$ corresponds to a distance of about 250 km.]

In Figure 3 we see, after impulsive precipitation events at about 2347 UT and 0006 UT, clear latitudinal movements of the precipitation region in association with an apparently well defined line current. The disappearance of precipitation over one station and the appearance of it over another station is in all cases associated with a corresponding movement of the electrojet. Also, the lack of precipitation over both balloons at about 0050 UT agrees well with the electrojet which is then far south of even Kiruna. Finally, the good correlation between the measurements by the Andenes and Alta balloons demonstrates the large E-W extension of these radiation events.

Another case of electron precipitation in the midnight sector is shown in Figure 4, recorded by 4 balloons at different locations. Here an impulsive event occurred at 2102 UT, just at the onset of a polar magnetic substorm. Especially noteworthy is the gradual increase of precipitation observed over Kiruna before the onset of the

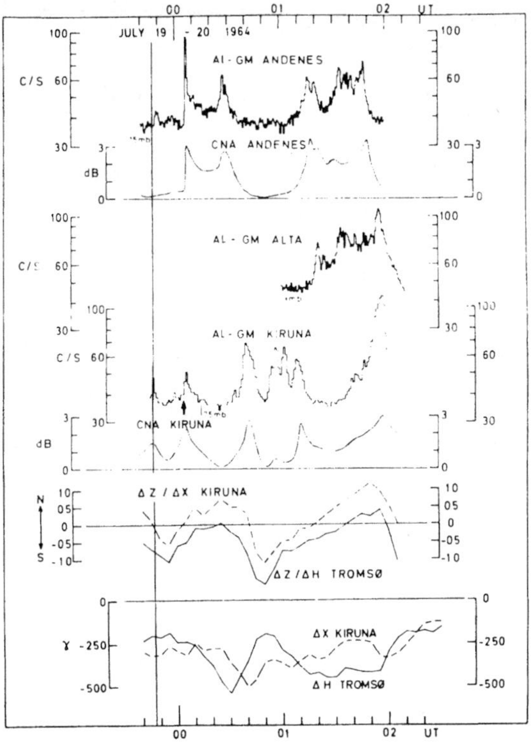

Fig. 3. Typical midnight event.

negative bay, and that the impulsive precipitation events observed over Karasjok, Honningsvåg, and Andenes were followed by a sudden decrease of electron precipitation over Kiruna. (The peak at Andenes is very small because the balloon was still rising at a depth of about 30 g cm^{-2} at the time.)

There is a clear time difference of about 50 sec between the impulsive X-ray events over Karasjok and Honningsvåg, which corresponds to a northward velocity of 3 to 4 km/sec. The corresponding peak at Andenes also seems to have occurred somewhat later than the peak at Honningsvåg. There thus appears to have been a sudden northward and westward motion of the precipitation region as the bay started, which swept most of the precipitation away from Kiruna and gave rise to the impulsive event observed farther north and west.

From detailed studies of a considerable number of X-ray events occurring around geomagnetic midnight and associated with local magnetic bay disturbances, the following general results have been obtained:

(a) In many cases an impulsive precipitation event is seen near onset of a substorm, and may move or surge quickly northwards or westwards. Over southern stations one sometimes sees a sudden decrease of previously existing precipitation at about the same time. The development during this initial phase seems explosive in character.

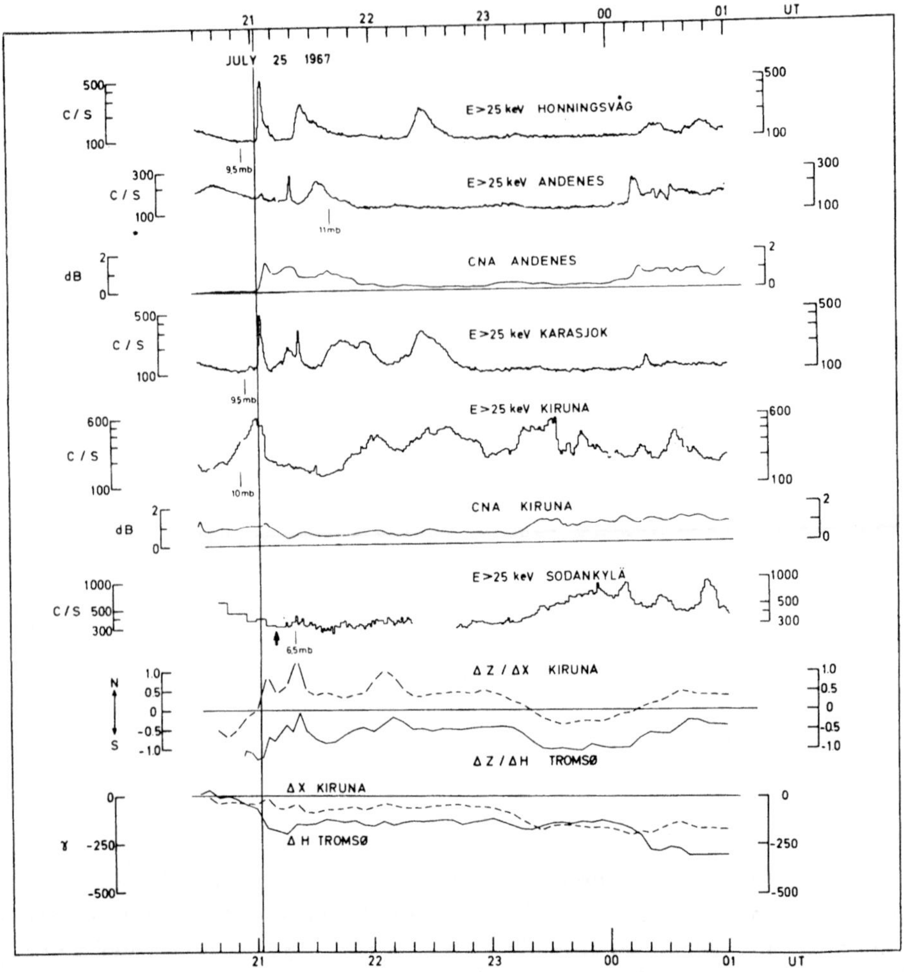

Fig. 4. Impulsive midnight event.

(b) During the substorm the precipitation region appears to have a large longitudinal extension, following the electrojet (when this is well defined) and its N-S extension is then very limited.

(c) When an apparently well defined line current exists, there is usually good correlation between the motion of the precipitation region and the motion of the electrojet.

(d) There is no preferred direction of movement of precipitation during the early part of the substorm.

(e) Towards the end of a substorm there is a general tendency for northward movement of the precipitation region.

In the framework of Akasofu's picture of the auroral substorm, these findings can be described as follows:

(a) When the balloons are in the right positions, one may observe a buildup of intensity over the southernmost balloon before the bay starts, corresponding to the brightening of the auroral arc in Akasofu's picture, but on a time scale of 15 to 20 min rather than the approximately 5 min which (according to Akasofu) is typical for the auroral case.

(b) Also in the X-ray measurements we see an explosive phase with a rapid northward or westward movement associated with the onset of the magnetic substorm. This phase seems to correspond closely to the explosive phase of the auroral substorm, when Akasofu has systematically observed similar movements.

(c) During the next phase, which comprises the main part of the magnetic bay, northward as well as southward motions of X-ray events can be seen. This indicates a general expansion of the active region.

(d) During the recovery phase there is a general tendency for the precipitation region to retire to the north, which however may merely be due to the fact that a station located in the auroral zone will tend to move out of the oval after local geomagnetic midnight.

The X-ray substorm as observed in the midnight sector thus seems to fit into the same picture as the auroral substorm, as far as morphology and dynamics are concerned.

4. X-Ray Events at Other Local Times

A. EARLY MORNING EVENTS

Early morning events are classified as those events which are observed after local geomagnetic midnight, but before local dawn. We shall only briefly mention the results of a preliminary investigation of such events (Kremser *et al.*, 1970) which are as follows:

1. In the auroral zone, the most intense precipitation events at these local times are normally not directly associated with the electrojet, but occur farther south. Such events may be referred to as 'zone precipitation'.
2. These events are apparently related to precipitation events and substorm activity in the local midnight sector, but show a delayed onset with respect to the midnight activity farther west.
3. However, if the auroral oval is sufficiently expanded, and the electrojet extends sufficiently far to the east, one can see, in the same local time sector X-ray events apparently directly associated with the electrojet, which are then referred to as 'oval' precipitation events.

'Oval' events thus represent precipitation directly associated with the auroral electrojet. 'Oval' precipitation observed in the morning hours therefore is presumably of basically the same kind as that of 'midnight' events, being the continuation of these events along the oval into the morning sector, whereas 'zone' events signify precipitation occurring south of the auroral oval. This suggests that early morning precipitation is fed by electrons drifting out of an acceleration region following the electrojet eastwards from the local midnight sector.

Still unpublished observations over large distances ($\simeq 2000$ km) in the midnight morning sector seem to confirm this relationship between midnight and early morning events.

B. LATE MORNING EVENTS

Late morning events previously have been studied by Bewersdorff *et al.* (1968). It was shown then that there is an apparent movement of the precipitation region from north to south during such events. A good example of this kind of event is seen at about 0430 UT in Figure 5 which shows simultaneous recordings from 3 different balloons. A closer inspection of the recordings shows that X-rays were indeed observed somewhat earlier by the Andenes and Alta balloons than by the Bodø balloon, indicating a general southward movement of the precipitation region in accordance with the results referred to above. There appears also however to be a time delay from

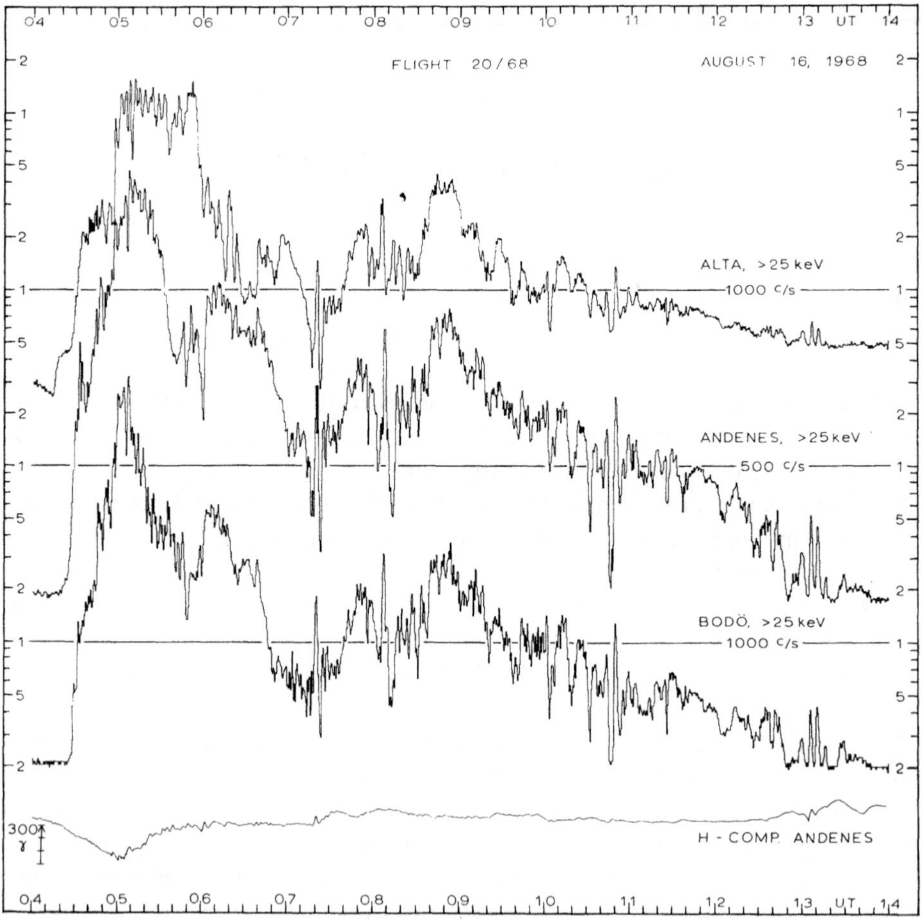

Fig. 5. Dayside X-ray events.

the onset at Andenes to the onset at Alta of about 2 min, indicating that an eastward movement of the precipitation region took place at the same time as the movement to the south.

Rather than interpret the general motion as a relatively slow one from the north to the south, it seems now in the light of these measurements and the results mentioned above, more probable that what we see is the effect of a more rapid eastward motion of a very elongated region of electron precipitation, the southern front of which is slightly inclined to the auroral zone, presumably parallel to the auroral oval. If this is so, the eastward motion may again be caused by gradient drift of electrons out of the auroral oval.

C. LARGE SCALE PULSATING EVENTS

A special pulsation train started at about 0806 UT (Figure 6). The interesting feature here is that whereas the pulsations recorded by the Bodø and Andenes balloons were nearly in phase, they showed a clear delay with respect to the Alta balloon. We conclude that we have here another case of a westward moving pulsation train of the kind reported by Brønstad and Trefall (1968). The observed delay corresponds to a speed of propagation of about 100 km/min, which agrees well with the 120 to 150 km/min found in the previous cases. The present recordings show that this event extended for a considerable distance in the N-S direction.

From around 0840 UT we begin to see large scale stationary pulsations over the whole area covered by the three balloons, apparently of the same basic type as reported by Ullaland *et al.* (1967). This event extended over at least 500 km (Alta-Bodø) and showed within a time resolution of $\frac{1}{2}$ min remarkably simultaneous intensity variations at all balloons over a period of almost 5 hr. It can now be considered therefore as well established that such events are typical for the noontime regime of electron precipitation.

5. General Relationship between X-Ray Events and Polar Magnetic Substorms

If 'zone' events are caused by electrons which have drifted out of an acceleration region near local midnight, one should on proceeding eastward towards later local times expect systematically increasing delays between the direct precipitation event in the midnight sector, and associated precipitation events observed in the morning sector, or on the dayside. Such a study has already been made (Sletten *et al.*, 1971), and the results are presented in Figure 6. There is a clear tendency toward increasing delays with increasing separation from the midnight sector to the X-ray event.

The spread of data in Figure 6 indicates however that for each local geomagnetic time most of the observed delays lie between a maximum and a minimum value characteristic for that local time, represented by the two dashed lines in the figure. In particular, the occurrence of 'prompt X-rays' with respect to the midnight bay until as late as 0600 LMT at the balloon position, suggests that acceleration and direct precipitation of electrons take place not only near geomagnetic midnight, but simultaneously along the auroral electrojet all the way from midnight to dawn. Whether we shall see 'prompt X-rays' or 'delayed X-rays' near dawn then depends on the location of the instantaneous auroral oval. If the oval is sufficiently expanded, the electrojet may well cross the statistically determined auroral zone near dawn, in which case a balloon recording at that point in the zone will see 'prompt X-rays' if a substorm

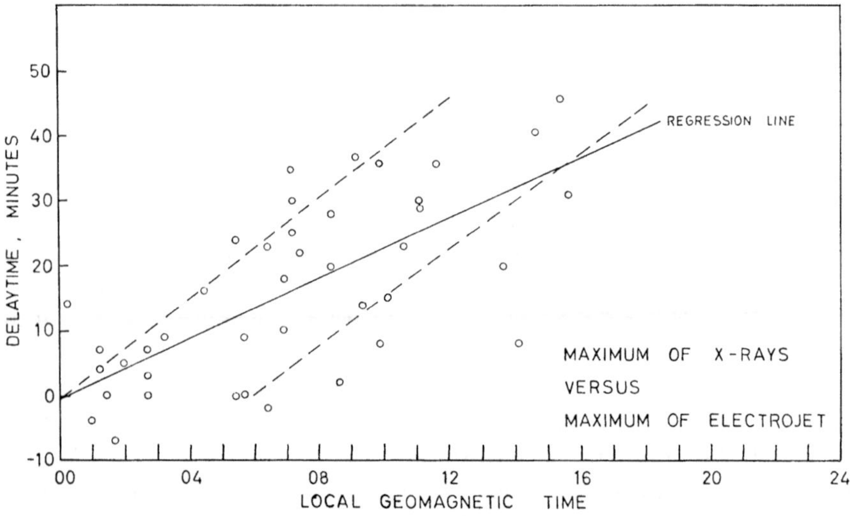

Fig. 6. Time delays of X-ray events with respect to magnetic substorm onsets.

occurs. The line of minimum delays then suggests that the region of acceleration and associated prompt precipitation does not normally extend past local dawn. Intermediate delays must represent cases when the auroral electrojet crossed the zone between midnight and dawn.

The slope of the regression line in Figure 6 corresponds to a total drift period of 57 min, which requires $E \simeq 170$ keV if the electrons mirror close to the atmosphere (Hamlin *et al.*, 1961), which seems somewhat high. However, precipitation from a drifting cloud of electrons must require some efficient mechanism for changing the pitch angles of the electrons, and electrons may then drift with large as well as small pitch angles and still become precipitated. As particles mirroring close to the equator drift faster by a factor of 1.5, this lowers the electron energy correspondingly. However, the lines representing the approximate maximum and minimum delay times as a function of local geomagnetic time have a slope corresponding to a total drift period of 90 min, which corresponds to 105 keV for electrons mirroring near the top of the atmosphere, and only 70 keV for electrons mirroring near the equatorial plane. The energies then obtained do not seem unreasonable.

Our picture is basically in agreement with Wilhelm's (1968) observations of longitude dependent time delays of Pc-1 pulsations with respect to electron precipitation, which were interpreted as showing drift effects of both electrons and protons. Of particular interest in this connection, however, are the results of Arnoldy and Chan (1969), who investigated the temporal relationship between energetic particle events at the geosynchronous orbit and magnetic substorms as observed by ground based instruments in the midnight sector. These authors also found time delays which increased systematically with increasing distance between the satellite and the magnetic midnight meridian in a manner corresponding to the drift of 100 keV electrons mirroring in the equatorial plane.

6. Summary and Discussion

The following phenomenological description enables us to summarize the majority of available observations: There is a region of electron acceleration following the electrojet along the auroral oval eastwards from the midnight sector towards dawn, over which the acceleration process spreads very rapidly. Ordinary midnight events would then be associated with this acceleration region and be caused by freshly accelerated electrons. Other events would be caused by particles drifting eastwards from the acceleration region along the auroral zone, where they are gradually precipitated by such mechanisms as pitch angle scattering, which is again modulated by magnetic pulsations and similar disturbances, thus giving rise to the many different types of intensity variations found in X-ray recordings from the morning sector onwards.

This description is consistent with the conclusions reached by Pfitzer and Winckler (1969) on the basis of satellite measurements at about 6.6 R_E, even though the acceleration region seems to be more extended in longitude and more limited in latitude then suggested by those authors.

Acknowledgments

Mr. J. Bjordal and Mr. K. Brønstad have rendered very valuable assistance during the preparation of the manuscript and the figures.

References

Anderson, K. A. and DeWitt, R.: 1963, *J. Geophys. Res.* **68**, 2669.
Arnoldy, R. L. and Chan, K. W.: 1969, *J. Geophys. Res.* **74**, 5019.
Barcus, J. R. and Brown, R. R.: 1962, *J. Geophys. Res.* **67**, 2673.
Bewersdorff, A., Kremser, G., Stadsnes, J., Trefall, H., and Ullaland, S.: 1968, *J. Atmospheric Terrest. Phys.* **30**, 591.
Bjordal, J., Trefall, H., Ullaland, S., Bewersdorff, A., Kangas, J., Tanskanen, P., Kremser, G., Saeger, K. H., and Specht, H.: 1971, *J. Atmospheric Terrest. Phys.*, to be published.
Brown, R. R.: 1966, *Space Sci. Rev.* **5**, 311.
Brown, R. R. and Campbell, W. H.: 1962, *J. Geophys. Res.* **67**, 1357.
Brown, R. R., Barcus, J. R., and Parsons, N. R.: 1965, *J. Geophys. Res.* **70**, 2579.
Brønstad, K. and Trefall, H.: 1968, *J. Atmospheric Terrest. Phys.* **30**, 205.
Clark, T. A. and Anger, C. D.: 1967, *Planetary Space Sci.* **15**, 1287.
Gosling, J. T.: 1966, *J. Geophys. Res.* **71**, 835.
Hamlin, D. A., Karplus, R., Vik, R. C., and Watson, K. M.: 1961, *J. Geophys. Res.* **66**, 1.
Kremser, G.: 1967, in *Aurora and Airglow* (ed. by B. M. McCormac), Reinhold Publishing Corporation, New York, p. 477.

Kremser, G.: 1969, in *Atmospheric Emissions* (ed. by B. M. McCormac and A. Omholt), Van Nostrand Reinhold Company, New York, p. 181.
Kremser, G., Wilhelm, K., Riedler, W., Brønstad, K., Trefall, H., Ullaland, S. L., Bewersdorff, A., Legrand, J. P., and Kangas, J.: 1970, Mitteilungen, Max-Planck-Institut für Aeronomie **40** (S).
Paulson, K. V. and Shepherd, G. G.: 1965, *J. Atmospheric Terrest. Phys.* **27**, 831.
Pfitzer, K. A. and Winckler, J. R.: 1969, *J. Geophys. Res.* **74**, 5005.
Rosenberg, T. J., Bjordal, J., and Kvifte, G. J.: 1967, *J. Geophys. Res.* **72**, 3504.
Rosenberg, T. J., Bjordal, J. Trefall, H., Kvifte, G. J., Omholt, A., and Egeland, A.: 1971, *J. Geophys. Res.* **76**, 122.
Scourfield, M. W. J., Cresswell, G. R., Pilkington, G. R., and Parsons, N. R.: 1970, *Planetary Space Sci.* **18**, 495.
Sletten, A., Stadsnes, J., and Trefall, H.: 1971, *J. Atmospheric Terrest. Phys.*, to be published.
Ullaland, S. L., Trefall, H., Kremser, G., and Bewersdorff, A.: 1967, *J. Atmospheric Terrest. Phys.* **29**, 395.
Wilhelm, K.: 1968, *J. Geophys. Res.* **73**, 7491.

INSTITUTE DISCUSSION

It was indicated by T. Jørgensen that H. Trefall's model seems different from the Akasofu model. They are different; however, Akasofu describes optical phenomena produced by low energy particles while Trefall describes phenomena associated with higher energy electrons.

Discussions of the average electron energy as a function of time show that some observables reflect the duration of the event. The most intense X-ray events occur about midnight, but the longest duration events occur in the morning. The CNA observations in the morning seem to reflect the longer duration.

PART VI

RADIO OBSERVATIONS

SURVEY OF RADIO REFLECTIONS FROM AURORA

A. G. McNAMARA
National Research Council, Canada

1. Introduction

This paper is a survey of VHF and UHF radio reflections and the mechanisms by which these observations may be explained in terms of the auroral plasma structure.

2. Observational Characteristics of Radio Aurora

A. MAGNETIC ASPECT

Almost all the theories of the reflection process demand nearly exact perpendicularity of the line-of-site to the earth's field lines at the reflection point. This appears to hold true for lower latitude observations but is not nearly so strictly obeyed in the intense activity of the auroral zone. Figure 1 shows a network of 48 MHz radars which was operating in Canada during the IGY. The network was specifically designed to study the aspect effect with overlapping coverage from different directions. For example, the Baker Lake station north of the auroral zone was nowhere less than 10° from perpendicularity. The best aspect angle (10° off) was obtained to the north at a range of 700 km. In the auroral zone to the south the aspect angles were 20° off perpendicularity. However, reference to the azimuth occurrence distributions of Figure 2 shows that the high occurrence of echoes to the south outweighed the much more favorable aspect angles to the north. A detailed analysis of all the simultaneous echoes obtained from Saskatoon and Baker Lake in the 2 yr period was made by McDiarmid and McNamara (1967) to obtain a mean value of the aspect attenuation. In the interval 5 to 20° off-perpendicularity, the aspect attenuation was found to be about 1.2 dB/°. Other data on the aspect dependence of echoes obtained by Leadabrand *et al.* (1965) who give values of 2 dB/° at 30 MHz, and 10 dB/° at 400 and 800 MHz, for off-perpendicular angles in the ranges 0 to 7° and 0 to 1° respectively.

B. DOPPLER SPECTRUM

A distinctive characteristic of the auroral echo is that the returned signal may be broadened and/or Doppler shifted, positively or negatively, by frequencies corresponding to target velocities of up to several hundred m sec^{-1} and even as high as 2000 m sec^{-1} (Bowles, 1954; McNamara, 1955; Nichols, 1957; Leadabrand *et al.*, 1959).

C. POLARIZATION

The plane of polarization of the returned signal is usually the same as the transmitted signal, but sometimes it may change rapidly to partial or complete depolarization

B. M. McCormac (ed.), The Radiating Atmosphere, 301–313.

Fig. 1. Network of 48 MHz auroral radars operated in Canada during the IGY.

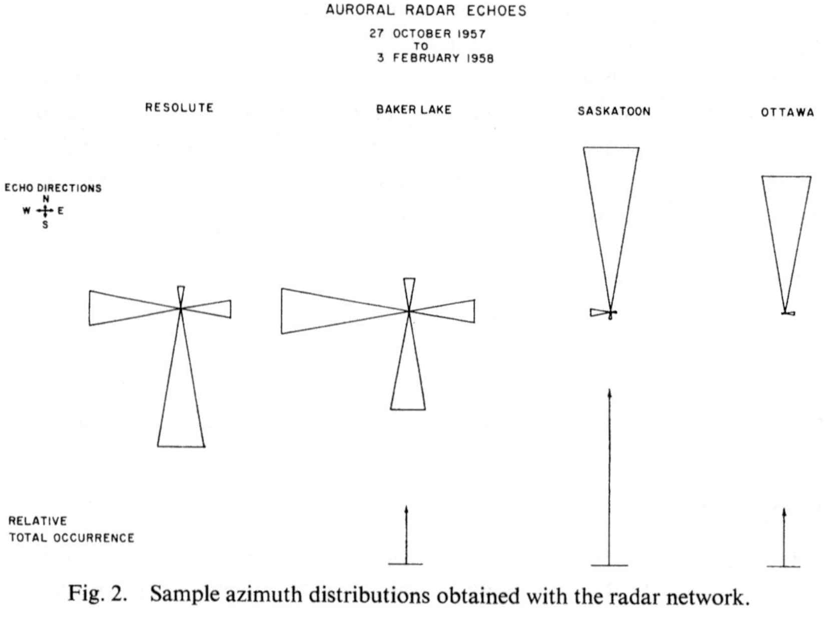

Fig. 2. Sample azimuth distributions obtained with the radar network.

(McNamara and Currie, 1954; Kavadas and Glass, 1959; Harang and Troim, 1970). The occurrence of depolarization increases with decreasing frequency, and may possibly be due to Faraday rotation or it could be a property of the reflection mechanism.

D. THRESHOLD EFFECT

With sufficiently high sensitivity radars it is observed that when echoes are seen they are never below a particular amplitude (Bowles *et al.*, 1963; Leadabrand *et al.*, 1965). These observations are strong evidence in favor of an instability mechanism for the generation of the echoes.

E. ASSOCIATION WITH VISUAL AURORA

Radio echoes have been shown to occur in association with visual auroras and ionospheric current systems. It is beyond the scope of this paper to discus the detailed observations. Lange-Hesse (1967) and Czechowsky and Lange-Hesse (1971) have presented observations of these relationships.

F. FREQUENCY DEPENDENCE

All of the above characteristics of radio aurora are a strong function of the radio frequency used in the observations. Reflections have been observed up to 3000 MHz. Care must be used in comparing results from backscatter and forward scatter transmission paths since the scattering mechanisms may be very sensitive to the scattering angle. Most of the comments in this paper are with reference to the backscatter case.

3. The development of Radio Aurora Models

A. UNDER DENSE SCATTERING FROM NONISOTROPIC IRREGULARITIES

Booker (1956) developed the first auroral model which satisfactorily related a spatial structure of a turbulent underdense scattering medium to the observed radio wave angular scattering spectrum, and to its variation with frequency. A threedimensional autocorrelation function of the medium was assumed, and then by a Fourier transform the scattering coefficient was obtained. The autocorrelation function chosen did not correspond to any physical mechanism, but was chosen for mathematical convenience to permit the transform to be readily solved. Booker (1956) used an exponential autocorrelation function of the form

$$\varrho(r, \theta, z) = \exp\{-\tfrac{1}{2}(r^2/T^2 + z^2/L^2)\} \tag{1}$$

which is axially symmetric about the z axis (taken to be coincident with earth's magnetic field lines) and with transverse and longitudinal correlation distances of T and L, respectively.

Thus, they obtained the backscattering coefficient for elongated, axially symmetric, random irregularities

$$\sigma = (2\pi)^{3/2}\pi^2 \frac{1}{\lambda_N^4}\overline{\left(\frac{\Delta N}{N}\right)^2}\, T^2 L \exp\left(-\frac{8\Pi^2 T^2}{\lambda^2}\right)\exp\left(-\frac{8\Pi^2 L^2}{\lambda^2}\Psi^2\right) \mathrm{m}^2/\mathrm{m}^3 \tag{2}$$

for small values of the off-perpendicular magnetic aspect angle Ψ where

λ = the radar wavelength,

λ_N = the free-space wavelength of the plasma frequency in the auroral ionization, and

$\overline{\left(\frac{\Delta N}{N}\right)^2}$ = rms fractional variation in electron density.

Nichols' (1957) observations of aurora showed that $\Delta N/N \sim 2.5\%$ were required to reproduce auroral signal strengths, and that radar scattering to at least 400 MHz was possible. The turbulence model was obviously incapable of explaining the large Doppler frequencies observed, and it was later shown that it was not possible for the neutral medium to generate elongated irregularities of the structure, scale size, and intensity required.

B. RANDOM ASSEMBLIES OF UNDER DENSE AND OVER DENSE GAUSSOIDS

A generalized theory of scattering from an elongated axially symmetric prolate spheroids with a gaussian distribution of ionization was developed by Moorcroft (1961a, b). It permits synthesis of any arbitrary random ensemble of a distribution of under dense and over dense scatters with any size distribution. While this more general theory was successful in fitting more of the auroral observations, Moorcroft (1966) in a final critical review of the model in terms of the signal strength vs frequency data of Lyon and Forsyth (1962) and Lyon (1962), concluded that no possible combination of weak and strong random scatterers, scale size, absorption, etc., could explain the scaled multifrequency data. The only ionization structure that was capable of producing such data was a medium in which some periodic structure was present.

C. PLASMA INSTABILITIES

Observations of VHF backscatter from the equatorial electrojet led Bowles *et al.* (1960, 1963), and Cohen and Bowles, (1963) to postulate the presence of acoustic plasma waves and to point out the resemblance to auroral echoes. Farley (1963a, b) and Buneman (1963) developed theories of the two-stream plasma instability as a mechanism for the creation of the acoustic plasma waves.

1. *Generation of Plasma Waves*

One mechanism which was known to generate plasma waves was the streaming of charged particles through a counter-streaming or stationary background of the opposite charge. Observations of the equatorial electrojet indicated that when the magnetic disturbance passed some threshold value, plane waves were generated which propagated with wavefronts parallel to the magnetic field lines and with the propagation vector **k** approximately in the electron streaming direction. The phase velocity of this longitudinal density wave was nearly that of an acoustic wave (of the order of 400 m sec^{-1}), and this imposed a corresponding Doppler shift on electromagnetic waves scattered from it.

In the auroral case, the geometry for the two-stream instability and its acoustic waves is shown in Figure 3. Landau damping will be extemely effective in damping any waves propagating more than a degree or so from magnetic perpendicularity, and enormous streaming velocities would be required to sustain them. Hence, a very strong magnetic aspect control α is built into the linear theory. In addition, a weaker aspect control is produced by β, the angle between the streaming direction and the propagation vector.

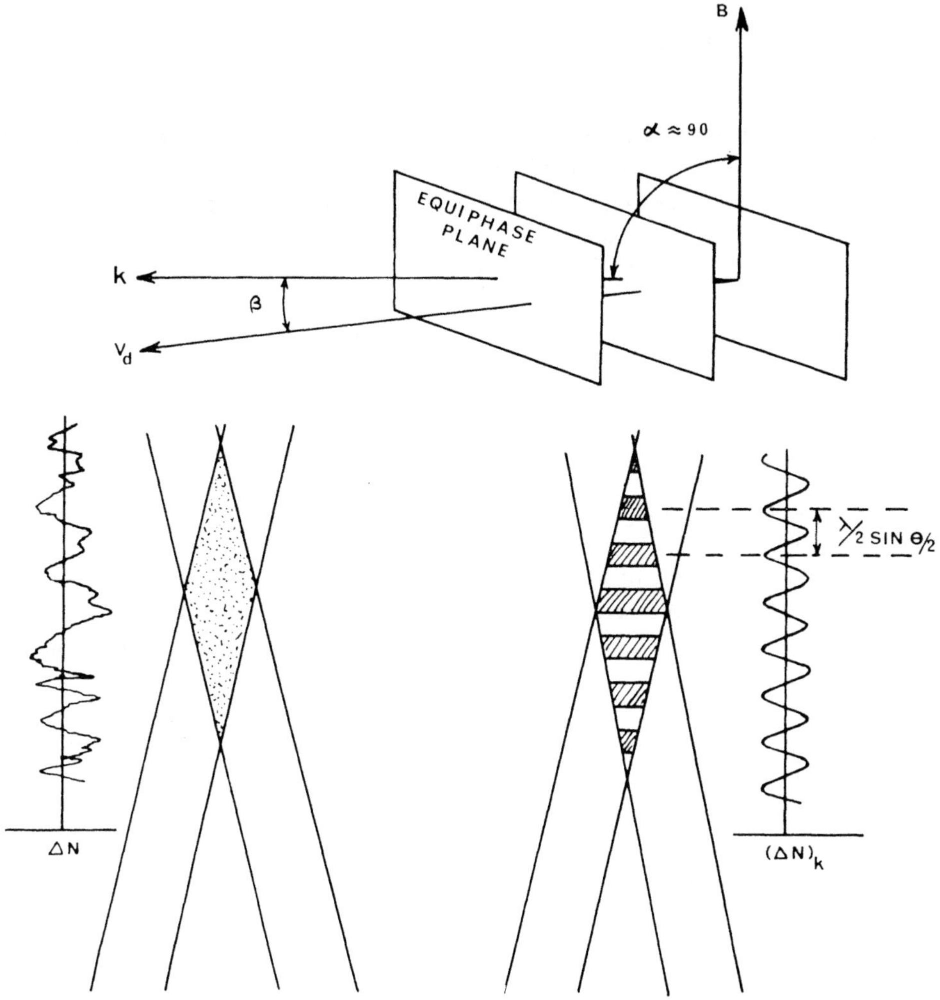

Fig. 3. Geometry of radar reflection from ion acoustic waves generated by the twostream instability. The angle θ is the scattering angle; 0° for forward scatter and 180° for backscatter.

On physical arguments it would appear that plasma instabilities could generate waves down to very short wavelengths. When the wavelength becomes comparable to the electron gyroradius, strong Landau damping will occur. When the wavelength approaches the Debye length, collective interaction as a plasma will cease. In the auroral ionosphere the electron gyroradius and Debye length are of the order of 2 cm and 0.3 cm, respectively.

2. *Plasma Waves as Radar Scatterers*

The analyses of Booker (1956) and Flood (1967) showed that the autocorrelation function of any scattering medium is related to the spectrum of its spatial fluctuations by the Fourier transform

$$\varrho(\mathbf{r}) = \int_{-\infty}^{+\infty} \exp(i\mathbf{k}\cdot\mathbf{r})\, S(\mathbf{k})\, \mathrm{d}\mathbf{k} \tag{3}$$

where, $\varrho(\mathbf{r})$ = three-dimensional autocorrelation function, $S(\mathbf{k})$ = three-dimensional spectral density, and $\mathbf{k}$ = vector spatial wave number.

In the Born approximation, the scattering cross section per unit volume is

$$\sigma_v = 4\pi\sigma_e N^2 \overline{\left(\frac{\Delta N}{N}\right)^2}\, S(\mathbf{k}_r)\ \mathrm{m}^2/\mathrm{m}^3 \tag{4}$$

where,

k_r = radar wave number = $2\pi/\lambda$,

$S(\mathbf{k}_r)$ = three-dimensional spectral density function,

$\frac{\Delta N}{N}$ = fractional variation in electron density, and

σ_e = Thompson radar scattering cross section for a single electron ($0.985\times \times 10^{-28}$ m).

A radar does not respond to the entire spectrum of density fluctuations. It is an instrument which directly selects and measures the particular spatial Fourier component, along the line-of-site, which corresponds to the radar wavelength. For backscatter (scattering angle $\theta = 180°$) the Fourier wavelength is equal to one-half of the radar wavelength. As indicated in Figure 3, the radar resolves only a particular spacing of plane wave fronts imbedded in an apparently random medium. The two-stream instability is a particular mechanism for generating a more or less limited spectrum of such waves. Farley's (1963a, b) theory is linear and hence predicts only the conditions for onset of the instability. It cannot predict, therefore, the non-linear limiting effects which must occur, and hence cannot determine the magnitude $\Delta N/N$ of the fluctuations nor its spectral distribution, $S(\mathbf{k})$.

3. *Radio Observations of Plasma Instabilities*

A number of recent radio studies have been directed toward a search for evidence of the two stream instability and of plasma waves in the aurora. Leadabrand (1965),

Leadabrand *et al.* (1965), and Chesnut *et al.* (1968) have obtained aurora on frequencies from 50 to 3000 MHz in Scotland and in Alaska. The echoing volume had a thickness of 10 to 20 km and the maximum signal occurred at 100 to 120 km altitude. An aspect angle dependence of 10 dB per ° off-perpendicular angle was found at 800 MHz, and 10 dB per 5° off-perpendicular at 30 MHz. The latter value is somewhat different from our own 48 MHz determination of 6 dB per 5° in the interval 5 to 20° away from perpendicularity. The combined evidence indicates that the rate of aspect attenuation is decreasing as one moves away from perpendicularity. Their evidence for ion acoustic waves is described as suggestive but not conclusive.

McDiarmid and McNamara (1969) have analyzed a particular event on the 48MHz network in terms of the two stream model with an external applied electric field. The echo occurrences seen by two separated radars are consistent with currents flowing under the action of a westward directed electric field which should give the required streaming.

Newell and Abel (1968) have analyzed the Doppler shifts of afternoon echoes seen on a 1295 MHz radar at Millstone. They found that echoes occur in the height region 90 to 130 km with Doppler shifts varying with height and azimuth in a manner which is consistent with an ionospheric current system composed of Pedersen and Hall currents driven by a north directed electric field.

Hofstee and Forsyth (1969) used a 40 MHz bistatic system to observe the displaced line spectrum which is characteristic of the ion-acoustic wave and found that it does indeed occur in radio aurora. However, it is important to note that they also find that most of the recordings made in their brief study show only a broad spectral spreading with little or no Doppler shift (see also McNamara, 1955). Even at times when the line spectrum was present, a large fraction of the signal power was contained in the spread part of the spectrum. The Doppler shift of the line corresponds to a velocity of 320 to 480 m sec^{-1}.

The data presented on their event have been analyzed further by McDiarmid (1970) with the addition of data obtained simultaneously at Ottawa on backscatter radar and incorporated into the model of McDiarmid and McNamara (1969). The radar backscatter echoes recorded on film from a rangetime cathode ray tube display were identical in appearance during the intervals in which Hofstee and Forsyth (1969) observed the spread and the line spectra. The two-stream instability fits the echo data for an electric field of 30 mV m^{-1} oriented W10°S geomagnetic, and is consistent with a strong Hall current observed by ground based magnetometers at Great Whale and Ottawa.

Recent Doppler studies at 144 MHz by Camnitz *et al.* (1969) at Buffalo, New York have also confirmed the presence of spectral peaks centered on Doppler velocities of 325 to 800 m sec^{-1}, with occasional values as great as 1800 m sec^{-1}. However, they too observe a wide variety of line and spread spectral forms, some with velocities both less than and greater than the expected Doppler shift. The presence of secondary irregularities is also inferred.

Both equatorial and auroral echoes show ample evidence of signal components which are not characteristic of the primary two stream instability. These signals can have zero or variable Doppler shifts and can be observed at large angular deviations from the usual aspect conditions. These irregularities may be the result of nonlinear effects in the two stream process or they may be generated by some completely different plasma instability (Balsley, 1965, 1969; Cohen and Bowles, 1967; Dougherty and Farley, 1967; Skadron and Weinstock, 1969; Unwin and Knox, 1970).

4. Direct Observation of Auroral Ionization Microstructure with Rockets

A rocket program has been in progress for some time in Canada for the direct measurement of the auroral plasma (McNamara, 1969) and its correlation with ground based radio, optical, and magnetic measurements. Figure 4 shows the large scale structure and electron density determined by the passage of a probe through a strong (70 kR) visual band. The maximum electron densities occur within the visual band and the greatest densities seen on any of the flights are approximately 10^6

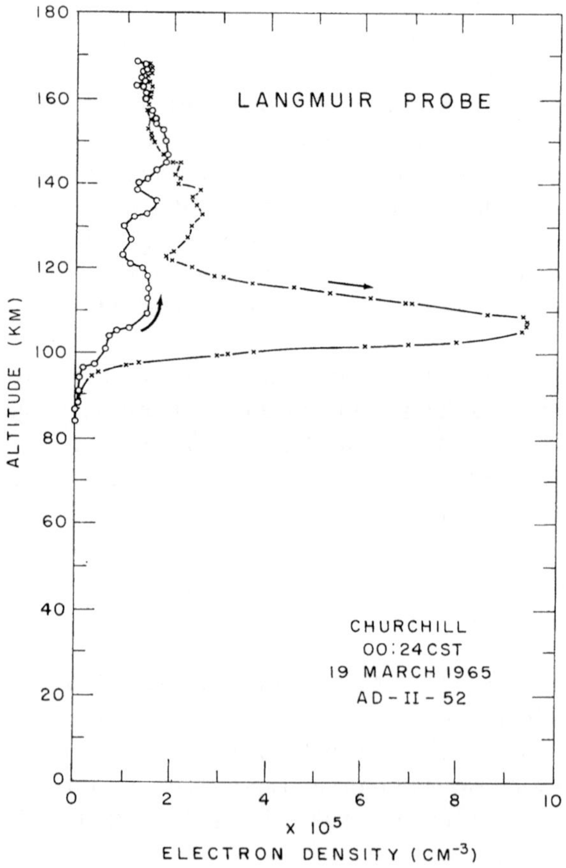

Fig. 4. Height profile of electron density in a visual auroral band by direct measurement.

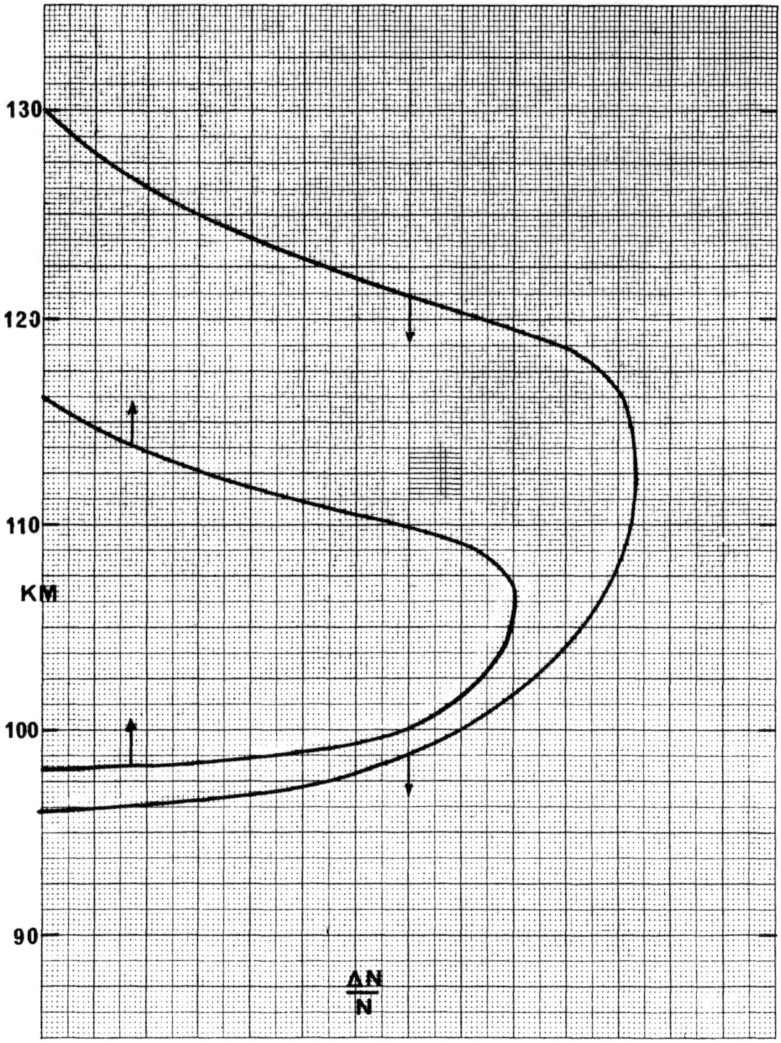

Fig. 5. Height profile of microstructure irregularities causing radio aurora measured on ascent and descent by a rocket borne probe. The maximum value of $\Delta N/N$ is approximately 5%.

electrons cm^{-3}. These flights indicate that it is very unlikely that over dense regions ever occur with critical frequencies much greater than about 10 MHz.

The rocket experiments were designed to measure directly the microstructure irregularities of the ionization, i.e., the density fluctuations ΔN, and the spectral distribution $S(\mathbf{k})$. The inherent limiting spatial resolution of the experiment is approximately the Debye length.

Typical microstructure profiles on ascent and descent from a rocket in 1964 are shown in Figure 5. Values of $\Delta N/N$ up to about 5% are typically observed. The microstructure is seen only within the height interval 96 to 130 km; the same interval in

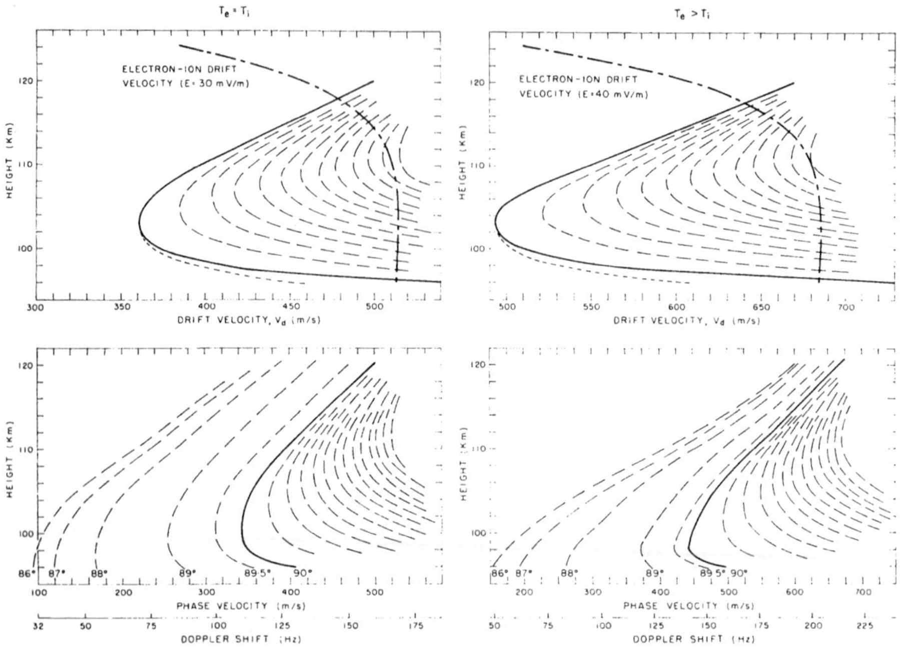

Fig. 6. Drift velocity and Doppler shift profiles for 48 MHz auroral radar backscatter at Fort Churchill. The solid line is the threshold growth curve (From McDiarmid, 1970).

which radio auroral echoes occur. Temporal and spatial correlation with radar echoes has been very good on those occasions when data were available for correlation. The ΔN height profiles of Figure 5 compare very well with the two stream instability region of Figure 6 which was calculated by McDiarmid (1970).

Figure 7 shows examples of the spectra of the irregularities obtained by computer analysis of the digitalized density fluctuations which were observed by the rocket probe. Experiment band widths and digitizing sampling rates were sufficient to allow detection of frequency components up to at least 3 kHz. From these data the shapes of the spatial spectra can be seen and estimates made of the magnitudes of the spectral components $S(k)$. At the low end of the scale the spectrum must turn over and decrease but the data processing was not able to resolve this region. Over most of the spectrum the data can be fitted very well by a function of the form

$$S(k) : e^{-f/f_0} : e^{-\lambda_0/\lambda} : e^{-k/k_0}. \tag{5}$$

To convert the density-frequency spectra to spatial spectra some assumptions must be made about the type of instability being measured in order to establish a spatial scale for the abscissa.

(a) If the irregularities are assumed to be isotropic and stationary (compared to the rocket velocity), then the abscissa scale (f) need only be transformed to a wave number or scale size by the relation

$$k = 2\pi f / v_{\mathrm{T}} \quad \text{or} \quad l = v_{\mathrm{T}}/f \tag{6}$$

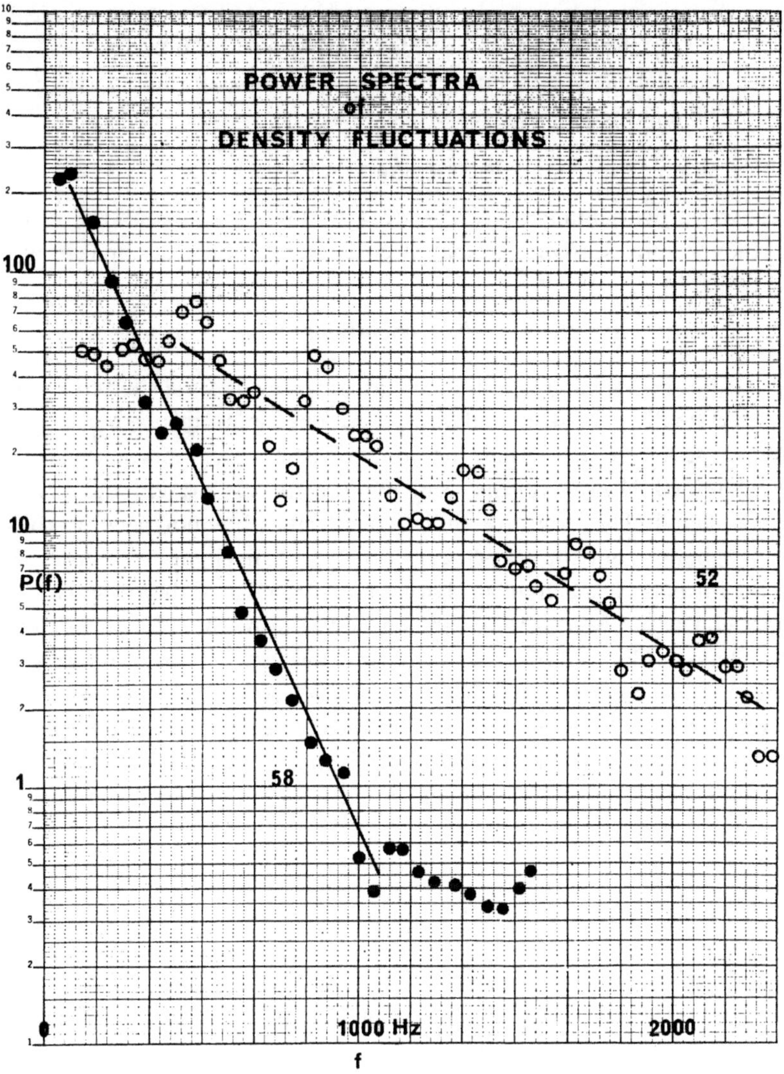

Fig. 7. Spatial power spectra of the microstructure irregularities observed on two rocket flights.

where v_T is the total rocket velocity. Typically, the minimum measurable l is determined by the rocket velocity (≈ 1000 m sec^{-1}) and the bandwidth of the system (≈ 3000 Hz). In this case, $l_{min} = 33$ cm which corresponds to a structure capable of reflecting 450 MHz radar waves.

(b) If the irregularities are assumed to be a spectrum of stationary plane surfaces or cylinders with surfaces nearly vertical along the magnetic field and oriented transversely to the trajectory plane, then

$$k = 2\pi f / v_H \quad \text{or} \quad l = v_H / f \tag{7}$$

where v_H is the horizontal component of the rocket velocity. For a typical value of $v_H \approx 200$ m sec^{-1}, $l_{min} = 6.7$ cm and the radar frequency is 2250 MHz.

(c) If the irregularities are assumed to be a spectrum of plane waves propagating normal to the field at the acoustic velocity $\mathbf{v}_a$ then

$$k = 2\pi f/(\mathbf{v}_H - \mathbf{v}_a) \quad \text{or} \quad \lambda = (\mathbf{v}_H - \mathbf{v}_a)/f\,. \tag{8}$$

On a near vertical launch, $\lambda \approx v_a/f$. If $v_a \approx 500$ m sec^{-1} and the bandwidth is 3000 Hz, then λ min ≈ 16 cm, a spatial component which is capable of radar reflections at 1000 MHz.

5. Conclusions

The irregularities responsible for radio aurora are clearly established to be fluctuations reaching very small scale sizes and caused by one or more plasma instability processes. Wave structure matching the ion acoustic wave characteristics has been positively identified. However, other echo components are present and often dominate the returned signal. The origin of these irregularities has not been established, and they may originate from the non-linear two stream mechanism or from some completely different plasma mechanism. Direct measurement of the irregularity spectrum and its location with rocket probes adds a powerful method for measuring some of the parameters in the radio wave scattering process. Theory suggests that there should be a close association with the local electric and magnetic fields in the ionosphere. Direct measurements with chemical releases and electric and magnetic field probes should be very helpful in establishing the excitation conditions for the plasma instabilities. These measurements are also helpful in establishing the relationship with optical aurora.

References

Balsley, B. B.: 1965, *J. Geophys. Res.* **70**, 3175.
Balsley, B. B.: 1969, *J. Geophys. Res.* **74**, 2333.
Booker, H. G.: 1956, *J. Atmospheric Terrest. Phys.* **8**, 204.
Bowles, K. L.: 1954, *J. Geophys. Res.* **59**, 553.
Bowles, K. L., Cohen, R., Ochs, G. R., and Balsley, B. B.: 1960, *J. Geophys. Res.* **65**, 1853.
Bowles, K. L., Balsley, B. B., and Cohen, R.: 1963, *J. Geophys. Res.* **68**, 2485.
Buneman, O.: 1963, *Phys. Rev. Letters* **10**, 285.
Camnitz, H. G., Tsunoda, R. T., and Barczys, D. A.: 1969, Cornell Aeronautical Laboratory Report CAL No. RG-2460-P-1.
Chesnut, W. G., Hodges, J. C., and Leadabrand, R. L.: 1968, Final report of SRI Project 5535, Stanford Research Institute, Menlo Park, California.
Cohen, R. and Bowles, K. L.: 1963, *J. Geophys. Res.* **68**, 2503.
Cohen, R. and Bowles, K.: 1967, *J. Geophys. Res.* **72**, 885.
Czechowsky, P. and Lange-Hesse, G.: 1971, this volume, p. 314.
Dougherty, J. P. and Farley, D. T.: 1967, *J. Geophys. Res.* **72**, 895.
Farley, D. T.: 1963a, *Phys. Rev. Letters* **10**, 279.
Farley, D. T.: 1963b, *J. Geophys. Res.* **68**, 6083.
Flood, W. A.: 1967, in *Aurora and Airglow* (ed. by B. M. McCormac), Reinhold Publishing Corporation, New York, p. 563.
Harang, L. and Troim, J.: 1970, *Planetary Space Sci.* **18**, 655.
Hofstee, J. and Forsyth, P. A.: 1969, *Can. J. Phys.* **47**, 2797.
Kavadas, A. and Glass, D. C.: 1959, *Can. J. Phys.* **37**, 690.

Lange-Hesse, G.: 1966, in *Aurora and Airglow* (ed. by B. M. McCormac), Reinhold Publishing Corporation, New York, p. 519.
Leadabrand, R. L.: 1965, *Radio Sci.* **69D**, 959.
Leadabrand, R. L., Schlobohm, J. C., and Baron, M. J.: 1965, *J. Geophys. Res.* **70**, 4235.
Leadabrand, R. L., Presnell, R. I., Berg, M. R., and Dyce, R. B.: 1959, *J. Geophys. Res.* **64**, 1197.
Lyon, G. F.: 1962, *Can. J. Phys.* **40**, 739.
Lyon, G. F. and Forsyth, P. A.: 1962, *Can. J. Phys.* **40**, 749.
McDiarmid, D. R.: 1970, *Can. J. Phys.* **48**, 1863.
McDiarmid, D. R., and McNamara, A. G.: 1967, *Can. J. Phys.* **45**, 3009.
McDiarmid, D. R. and McNamara, A. G.: 1969, *Can. J. Phys.* **47**, 1271.
McNamara, A. G.: 1955, *J. Geophys. Res.* **60**, 257.
McNamara, A. G.: 1969, *Can. J. Phys.* **47**, 1913.
McNamara, A. G. and Currie, B. W.: 1954, *Nature* **174**, 1153.
Moorcroft, D. R.: 1961a, *Can. J. Phys.* **39**, 677.
Moorcroft, D. R.: 1961b, *Can. J. Phys.* **39**, 695.
Moorcroft, D. R.: 1966, *Planetary Space Sci.* **14**, 269.
Newell, R. E. and Abel, W. G.: 1968, *Nature* **218**, 454.
Nichols, B.: 1957, Scientific Report No. 1, University of Alaska.
Skadron, G. and Weinstock, J.: 1969, *J. Geophys. Res.* **74**, 5113.
Unwin, R. S. and Knox, F. B.: 1970, presented at the Symposium on Upper Atmospheric Currents and Electric Fields, Boulder, Colorado.

INSTITUTE DISCUSSION

The merits of backscatter vs. bistatic radar observations were discussed. A. McNamara prefers backscatter because forward scatter is more sensitive to the low wave number part of spectrum while backscatter is more sensitive to high wave number part of spectrum.

OPTICAL AND RADIO OBSERVATIONS OF AURORA

P. CZECHOWSKY and G. LANGE-HESSE
Max-Planck-Institut für Aeronomie, Lindau/Harz, West Germany

1. Introduction

The Max-Planck-Institute for Aeronomy at Lindau/Harz, West Germany has operated a network of VHF beacon transmitters and receiving stations in Scandinavia and northern Germany since autumn 1967 to study VHF continuous wave bistatic auroral backscatter communications. This is done in cooperation with the Royal Swedish Telecommunication Board, the Swedish Radio Amateur Organization (SSA), the Norwegian Defence Research Establishment, the Research Institute of Swedish National Defence, the University of Umeå in northern Sweden and the German Federal Post. Since September 1969 simultaneous optical observations of the aurora have been carried out by all-sky cameras and meridian scanning photometers in northern Scandinavia in cooperation with the Auroral Observatory Tromsø and the Geophysical Observatory Sodankylä in northern Finland to study the correlation between optical and radio observations of the aurora. The whole network is shown in Figure 1.

According to theory the backscattering of VHF radio waves by aurora is controlled not only by the visual aurora but obviously on a larger scale by the polar electrojet (PEJ). Some years ago Buneman (1963) and Farley (1963) independently pointed out that in an ionospheric current system, e.g., the equatorial electrojet (EEJ) plasma instabilities of a 'two-stream' type can occur so that acoustic plasma waves are generated. The plasma instability appears when the relative drift velocity between the ions and electrons in the electrojet (EJ) exceeds a certain critical speed which is close to thermal velocity of the ions. It is said that the electrojet at this phase has exceeded the threshold. Acoustic plasma waves then are generated in the shape of longitudinal density waves, which propagate along the electrojet transverse to the geomagnetic lines of force. The periodical density oscillations caused by the plasma waves (which represent a special fine structure of the EJ) are the field aligned centers (field aligned to the geomagnetic lines of force) which give rise to the backscattering of VHF radio waves. Bowles *et al.* (1963) and Cohen and Bowles (1963) have shown that the VHF waves backscattering centers in the EEJ are caused by acoustic plasma waves. In the same publication the authors pointed out that the VHF waves backscattering centers in the PEJ obviously are caused by the same mechanism. The purpose of the observation network shown in Figure 1 is among others to check whether this assumption is correct and also to study the correlation between optical and radio aurora.

The maximum of the current density of the PEJ fits together in a first approximation with the visual aurora. Previously, therefore, it was assumed that the normal

B. M. McCormac (ed.), The Radiating Atmosphere, 314–326.

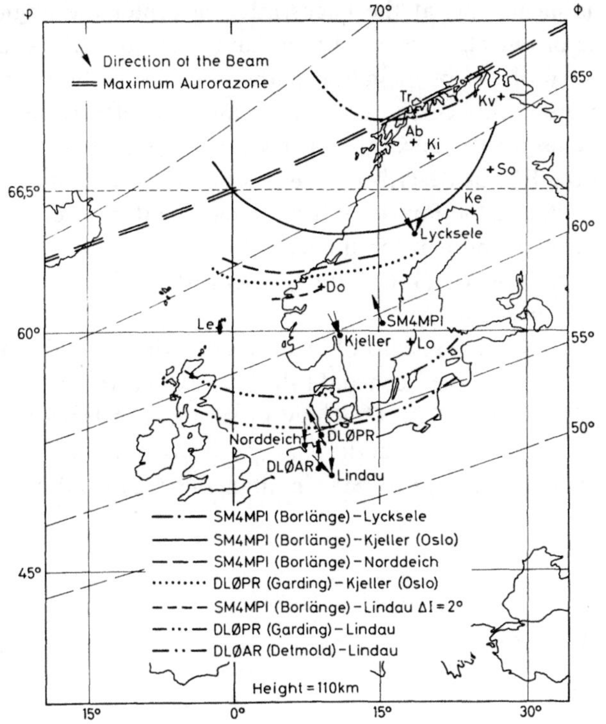

Fig. 1. Position of the VHF beacon transmitters (SK 4 MPI near Borlänge, 145.960 MHz; DL Ø PR, Garding, 145.971 MHz; DL Ø AR, Detmold, 29.0 MHz) and of the receiving stations (Lycksele, Kjeller, Norddeich, and Lindau) with the appertaining backscatter curves for the pairs of stations mentioned in the figure. The arrow at the stations indicates the direction of the antenna beam (half-power width $\pm 30°$). The backscatter curves represent the location at the 110 km height level (frequency maximum of the auroras) where the ideal backscatter conditions are fulfilled (Lange-Hesse and Czechowsky, 1966; Lange-Hesse, 1967) meaning where the backscattering centers (which obviously are spatially and temporally correlated with the PEJ beyond the threshold) must be located in case of the occurrence of auroral backscatter communications between the pairs of stations specified in the Figure. Tr = Auroral Observatory Tromsø, Ki = Geophysical Observatory, Kiruna; So = Geophysical Observatory Sodankylä; Ab, Lo, Do, Le = Geomagnetic Observatories Abisko, Lovö, Dombås and Lerwick, respectively. Φ = geomagnetic latitude.

ionization in the region of the visual aurora directly causes the backscattering of VHF radio waves. On the assumption that the theory of Buneman and Farley is also valid for auroral regions the fine structure of the PEJ, however, would be the controlling element of the VHF auroral backscatter phenomenon. If the PEJ (in the phase beyond the threshold) intersects a backscatter curve in Figure 1 then VHF auroral backscatter communication ought to occur between the pairs of stations belonging to the backscatter curve. For a pair of stations the appertaining backscatter curve represents the center of a region where the most favorable geometrical conditions (also called 'ideal backscatter conditions') are fulfilled for VHF auroral backscatter communications. These curves are calculated in Figure 1 for a height of 110 km. Ideal backscatter conditions mean that at the backscattering center the angle of incidence is equal to the angle of reflection referred to a plane vertical to the geomagnetic line of force. The geometry of the propagation paths referred to the geomagnetic lines of force is much more complicated for bistatic auroral backscatter communications between stations close to the auroral zone (Figure 1) compared to radar backscatter measurements at the EEJ as described by Bowles *et al.* (1963) and Cohen and Bowles (1963). This is why the reduction of all bistatic backscatter observations with respect to other geophysical phenomena is difficult. For more details about VHF auroral backscatter see Czechowsky (1966, 1969), Lange-Hesse and Czechowsky (1966), Lange-Hesse (1967, 1968, 1969).

First reduction results from radio observations carried out with a part of the network shown in Figure 1 have been published by Lange-Hesse (1969), Czechowsky (1969), Czechowsky *et al.* (1970), Czechowsky and Lange-Hesse (1970). The events discussed below are results from additional radio observations of the network and results from correlation studies between simultaneous radio and optical observations.

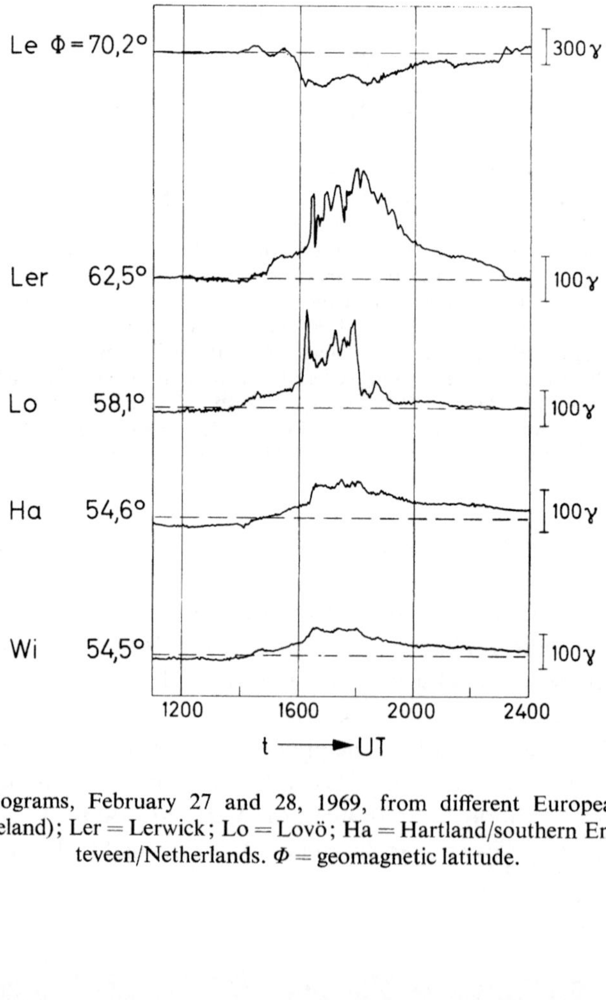

Fig. 2. *Z* magnetograms, February 27 and 28, 1969, from different European Observatories: Le = Leirvogur (Iceland); Ler = Lerwick; Lo = Lovö; Ha = Hartland/southern England; Wi = Witteveen/Netherlands. Φ = geomagnetic latitude.

2. Examples for the Control of Radio Aurora by the Polar Electrojet

At about 1400 UT on February 27, 1969, a geomagnetic storm started (Figure 2) and reached its peak disturbance of $K_p = 6+$ from about 1600 to 2000 UT. During this time VHF auroral backscatter echoes were recorded with the network in northern Germany and Scandinavia (Figure 1). These observations are now compared with simultaneous geomagnetic records i.e., with the location, direction and intensity variation of the PEJ. The results of this comparison are discussed on the basis of the ion acoustic wave theory according to Farley (1963). In the magnetograms of Figure 2 bays occurred in the afternoon and early evening of February 27. From the sign of the bays in *Z* and from the positive sign in *H* (not shown here) it can be derived that from about 1400 to 1900 UT the current is located north of Lerwick and south of Leirvogur and flows from west to east. If one takes the PEJ model after Heppner (1967) the observed current must be the eastward flowing current of the PEJ system. This part of the PEJ between Lerwick and Leirvogur intersects the region of the Borlänge-Kjeller backscatter curve (solid curve in Figure 1). The current is beyond the threshold at about 1420 UT when the first backscatter signals are recorded on the

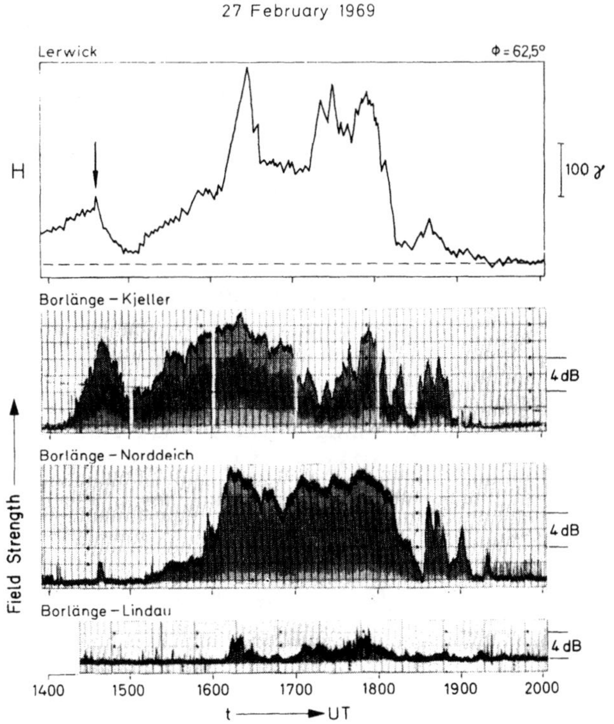

Fig. 3. VHF auroral backscatter recordings from the beacon station Borlänge, SK 4 MPI, 145.960 MHz at three different stations compared with the simultaneous *H* magnetogram from Lerwick (see Figure 1). The northern most receiving station (Lycksele) was not functioning at this time.

Borlänge-Kjeller line. At this moment the amplitude of the bay in the H magnetogram from Lerwick (Figure 3) is of the order of 100 γ, which indicates the threshold. The peak of the first smaller bay in the H magnetogram in Figure 3 (Lerwick) occurs at about 1440 UT simultaneously with the first peak in the Borlänge-Kjeller backscatter signal. Exactly during this peak a short backscatter signal with small amplitude appears on the southernmost Borlänge-Norddeich line indicating that the PEJ extends to the south so that it could just intersect the region of the Borlänge-Norddreich backscatter curve. No backscatter signal, however, occurs during this time on the southernmost line in Figure 3 (Borlänge-Lindau).

The strongest bay with about 400 γ amplitude appears in the magnetogram in Figure 3 at about 1630 UT. Simultaneous peaks in the backscatter signals now occur on all three lines of Figure 3. The amplitude of the peaks on the two northern lines now are of the same order, however, the amplitude on the southernmost Borlänge-Lindau line is much smaller. This indicates that compared to 1445 UT the current has extended further to the south which now enables it to intersect the region of all three appertaining backscatter curves.

A second stronger bay with an amplitude of about the same order as the first appears from about 1720 to 1815 UT. This bay is bifurcated and shows two peaks. Simultaneously with the magnetic peaks, peaks in the backscatter signals occur on the lines from Borlänge to Norddeich and Lindau. The magnetic bifurcation is best reproduced in the Borlänge-Lindau backscatter signal trace. Strong backscatter signals are also recorded during the bifurcated bay on the northernmost line (Borlänge-Kjeller). If the VHF waves backscattering centers are produced in the PEJ by acoustic plasma waves, according to the theory of Buneman (1963) and Farley (1963), a broad eastward flowing current beyond the threshold must have intersected the region of the three appertaining backscatter curves simultaneously from about 1720 to 1815 UT.

The last bay which is smaller in amplitude occurs in the Z magnetogram at about 1840 UT. Simultaneously, a peak in the backscatter signal appears on the Borlänge-Kjeller and Norddeich lines but barely a signal appears on the southernmost Borlänge-Lindau line. This indicates that the PEJ and the backscattering centers show a contraction to the north with declining substorm. The results shown in Figure 3 exhibit that the centers backscattering the VHF waves show a similar behavior during a substorm as the visual aurora: extension to the south just after the start of the substorm and a contraction to the north towards the end. In a later figure it will be shown that obviously there is also an extension to the north of the backscattering centers like the visual aurora just after the start of a substorm. A more detailed study of the magnetograms from a chain of stations in the N-S direction indicates that during a substorm the PEJ also extends to the north and south and reaches its peak extension during the peak of the substorm (Boström, 1971).

In the following section the close relation between VHF auroral backscatter echoes and the geomagnetic disturbance is investigated in detail. In this special case the H magnetogram from Lerwick and the VHF backscatter recordings from Figure

3 are used. The variation of the backscattered echo signals (in scale units) as a function of the horizontal perturbation vector H_d (in γ) from Lerwick is shown in Figure 4. The H_d vector is computed from the values ΔH, ΔD and the base line values H_0. It can be seen from the upper diagram in Figure 4 (Borlänge-Kjeller) that a threshold value of the order of 100 γ must be exceeded before the VHF auroral backscatter process starts at about 1420 UT in Figure 3. In the two lower diagrams of Figure 4 no threshold value of this order can be recognized. This means that in these lower latitudes no new irregularities in the form of plasma acoustic waves are generated. Waves of this kind excited in the region of the backscatter Borlänge-Kjeller curve must therefore have extended to the south during the development of the geomagnetic storm. A hysteresis like behavior can also be seen in Figure 4, especially in the upper diagram, which means that a stronger PEJ current (higher H_d value) is necessary for the generation of the plasma waves than for the maintenance of the waves which gives rise to the backscattering process. A hysteresis like behavior was already observed at the EEJ by Bowles *et al.* (1963). Hysteresis like behavior of the PEJ also was observed (e.g., Czechowsky, 1969; Lange-Hesse, 1968).

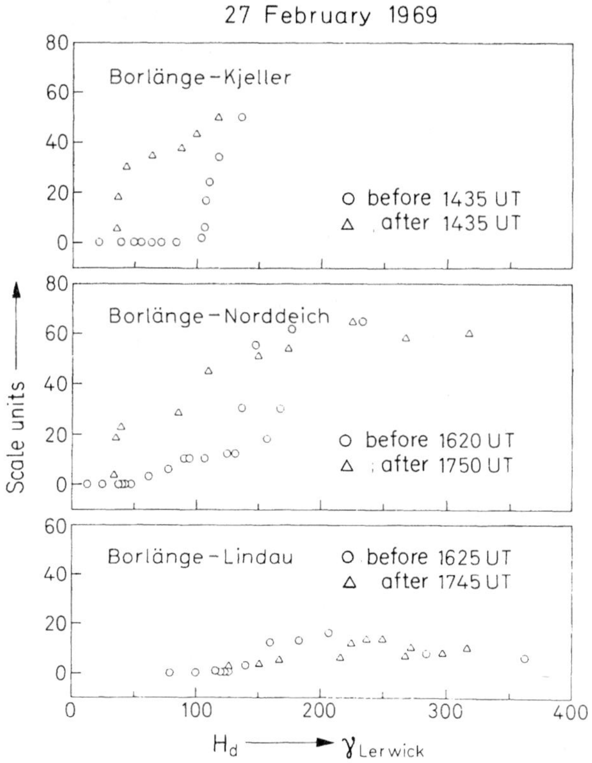

Fig. 4. Relation between the amplitude of auroral backscatter echoes and the variation of the horizontal disturbance vector H_d.

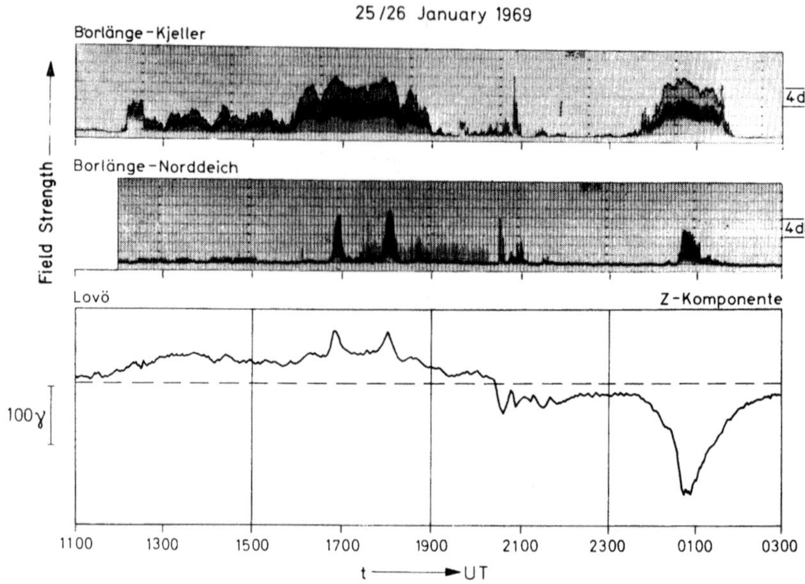

Fig. 5. *Z* magnetogram from the Observatory Lovö and the simultaneous VHF auroral backscatter recordings from the Borlänge beacon station, SK 4 MPI, 145.960 MHz at Kjeller and Norddeich (see Figure 1). The northern most receiving station (Lycksele) was not functioning at this time.

The *Z* magnetogram from Lovö in Figure 5 and other magnetograms (not shown here) indicates an eastward flowing current on January 25, 1969, north of Lovö and Lerwick in the afternoon and early evening. According to the PEJ model of Heppner (1967) this current is (as in Figure 3) the eastward flowing current of the PEJ system. The current starts to flow just after noon. Just after 1200 UT the threshold is reached revealed by the onset of the backscatter signal on the northernmost line (Borlänge-Kjeller). The appertaining backscatter curve of this line is intersected only by the current since no backscatter signal occurs on the southernmost line (Borlänge-Norddeich) at this time. Two positive bays occur in the magnetogram in Figure 5 at about 1700 and 1800 UT. During the bays the current extends to the south so that it could intersect the backscatter Borlänge-Norddeich curve indicated by the occurrence of backscatter signals on this line. The peaks of the signals coincide exactly with the peaks of two magnetic bays. The backscatter signal on the Borlänge-Kjeller line shows a strong increase during the time around the two bays and the signal peaks also coincide exactly with the bay peaks. The eastward flowing back current decreases around 1900 UT (indicated by the magnetogram and the disappearance of the backscatter signal on both lines in Figure 5).

The current reverses the direction at about 2030 UT. According to the PEJ model after Heppner (1967) this is caused by the westward flowing PEJ. Two smaller bays occur from about 2030 to 2100 UT. During this time the backscatter signals reappear on both lines and the peaks of the magnetic bays coincide with the peaks of the backscatter signals on the Borlänge-Norddeich line. The signals disappear after 2100 UT and reappear again about 1 hr after midnight during the strong negative bay of the order of 200 γ. The signal on the Borlänge-Kjeller line starts just before 2400 UT simultaneously with the onset of the bay and disappears at about 0200 UT at the end of the bay. The signal on the Borlänge-Norddeich line occurs only during the peak of the bay from about 0030 UT to a few minutes after 0100 UT. During the peak of the bay the westward flowing PEJ has extended again to the south so that it could intersect the Borlänge-Norddeich backscatter curve.

3. Correlation between Radio Aurora and Optical Aurora

A comparison between the occurrence of radio aurora and optical aurora is shown in Figure 6. At the top of the upper diagram the brightness of the 5577 Å emission is shown as observed at Tromsø by a scanning photometer. A brightness of 100%

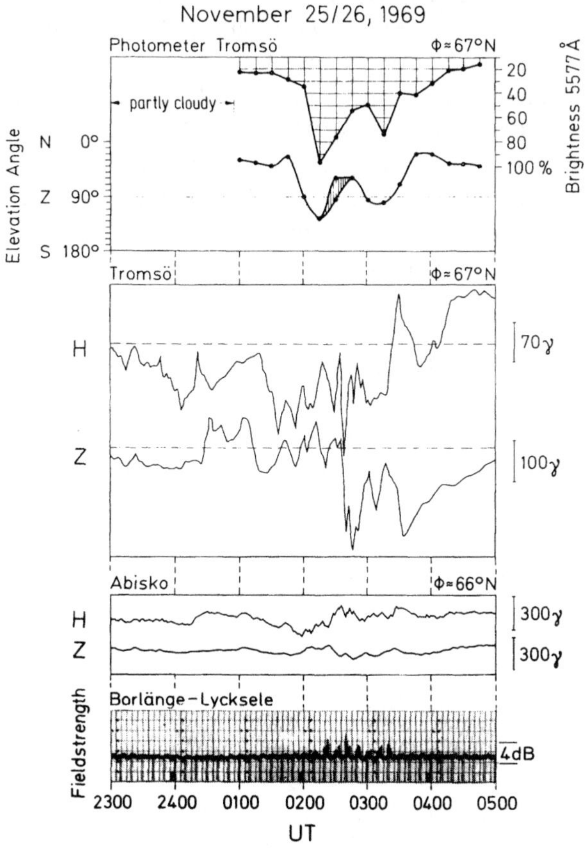

Fig. 6. Comparison of optical auroral observations at Tromsø (upper diagram) with VHF auroral backscatter (radio aurora) recordings on the Borlänge-Lycksele line (145.960 MHz), and with geomagnetic recordings from Tromsø and Abisko (see Figure 1).

means that the recorder has an amplitude over the whole scale. At the bottom of the upper diagram the elevation angle in given in the N-S meridian of the 5577 Å maximum emission. At 0230 UT the maximum is spread over an angle interval of 40° (hatched region), meaning that at this time the breakup phase of the visual aurora occurs. From 0100 to 0145 UT the aurora is located north from Tromsø and backscatter signals do not occur during this time. The aurora then moves to the south and passes the zenith to the south at 0200 UT. Just after that time the first backscatter signal occurs on the Borlänge-Lycksele line which is the northernmost one and the appertaining backscatter curve is located close to Tromsø to the south (Figure 1). At about 0215 UT the breakup phase of the aurora starts and reaches its peak at about 0230 UT at the zenith. From the diagrams in Figure 6 it can be seen that there is a close correlation in time between the peak of the visual breakup phase and the peak of the geomagnetic disturbance at Tromsø and the peak of the backscatter signal on the Borlänge-Lycksele line. At 0315 UT the aurora starts to move to the north and from about 0330 UT, when it is well north of the zenith, the backscatter signal disappears for the rest of the observation time. The visual aurora over Tromsø observed by an all-sky camera at 0239 UT is shown in Figure 9a. It shows the breakup phase at the zenith as also indicated by the photometer (Figure 6).

The event shown in Figure 6 is a relatively weak one. The planetary geomagnetic activity degree for this time (0000 to 0300 UT) is $K_p=30$. The geomagnetic disturbance and the simultaneous VHF backscatter is restricted to places close to the maximum auroral zone, meaning that geomagnetic bays occur only at Kiruna ($\Phi \approx 65°$), Abisko ($\Phi \approx 66°$), and Tromsø ($\Phi \approx 67°$) but not e.g., at Lovö ($\Phi \approx 58°$). Backscatter signals occur only on the northernmost line (Borlänge-Lycksele) and in exceptional cases on the next line to the south (Borlänge-Kjeller).

The events shown in Figures 3 and 5 are of stronger degree than those in Figure 6. The maximum K_p index for the time given in Figure 3 is $K_p=6+$ and in Figure 5 $K_p=6-$. In these cases strong bays occur in the magnetograms from Lovö ($\Phi \approx 58°$) of the order up to 200 γ and backscatter signals can be observed down to the Borlänge-Norddeich line and also during the peaks down to Borlänge-Lindau (Figure 3).

During very strong events with maximum K_p index of the order of 8 to 9 VHF backscatter signals can be observed down to the southernmost line (Bielstein (DL Ø AR)-Lindau) (Figure 1) and strong bays occur in the magnetograms from observatories in the latitudes of Denmark (e.g., Rude Skov near Copenhagen) and northern Germany. An example of such a strong event is shown by Czechowsky *et al.* (1970).

Another comparison between optical and radio aurora is illustrated in Figure 7. During the time of optical observations from 1900 to 2300 UT two substorms occur as indicated by the bays in the magnetograms from Abisko and the photometer observations at Tromsø (hatched region in the photometer elevation angle diagram). The maximum K_p index at this time is 4+. The peaks of the amplitude of the backscattered signal (Borlänge-Lycksele and Borlänge-Kjeller) again coincide with the peaks of the visual breakup phase and the peaks of the geomagnetic bays. During the

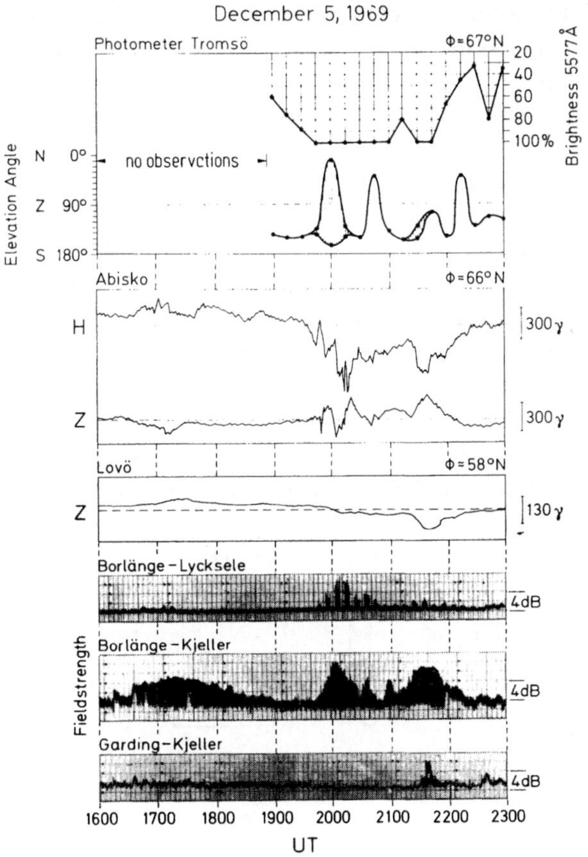

Fig. 7. Comparison of optical auroral observations at Tromsø (upper diagram) with VHF auroral backscatter (radio aurora) recordings on the lines from Borlänge (SK 4 MPI, 145.960 MHz) to Lycksele and to Kjeller, and from Garding (DL Ø PR, 145.971 MHz) to Kjeller, and with geomagnetic recordings from Abisko and Lovö (see Figure 1).

substorm at about 2130 UT the PEJ is extended more to the south than at about 2000 UT as indicated by the bay in the *Z* magnetogram from Lovö ($\Phi \approx 58°$). Therefore, the PEJ had been able to intersect the backscatter curve of the southernmost line in Figure 7 (Garding-Kjeller) indicated by the occurrence of a backscatter signal during the bay in the *Z* magnetogram from Lovö. More information about the optical situation in Tromsø during the two substorms illustrated in Figure 7 presented by all-sky cameras is given in Figure 9b and c. Figure 9b shows the breakup phase in full development at 2009 UT. The postbreakup or contraction phase with only one arc in the south and drifting patches at 2030 UT is illustrated in Figure 9c.

The signal on the Borlänge-Kjeller line is most frequent and the amplitude is the strongest from all three lines presented in Figure 7. During the peaks of the substorms signals also occur on the northern (Borlänge-Lycksele) and southern (Garding-Kjeller) lines from the Borlänge-Kjeller line. It is known that the visual aurora and the

PEJ extends in latitude to the north and south during the breakup phase of a substorm. The results shown in the backscatter recordings of Figure 7 indicate that obviously the backscattering centers also show a simultaneous extension to the north and south. However, the authors feel this assumption must be proved by additional observations.

Finally a last comparison between optical and radio aurora is given in Figure 8. The maximum K_p index during the time interval given in the figure is 4+. The first substorm occurs around 2400 UT and the center of the PEJ is located south of Abisko (and also south of Tromsø) as indicated by the reversed sign of the geomagnetic bays at that time. The breakup phase of the visual aurora occurs therefore south of Tromsø at a low elevation angle. During the second substorm, between about 0200 and 0300 UT, the center of the PEJ is located at a first approximation over Abisko and the breakup phase of the visual aurora is well developed at the zenith and to the south of Tromsø. On the northernmost line (Borlänge-Lycksele) the backscatter signals occur only during the peaks of the substorm. On the southernmost (Borlänge-Kjeller)

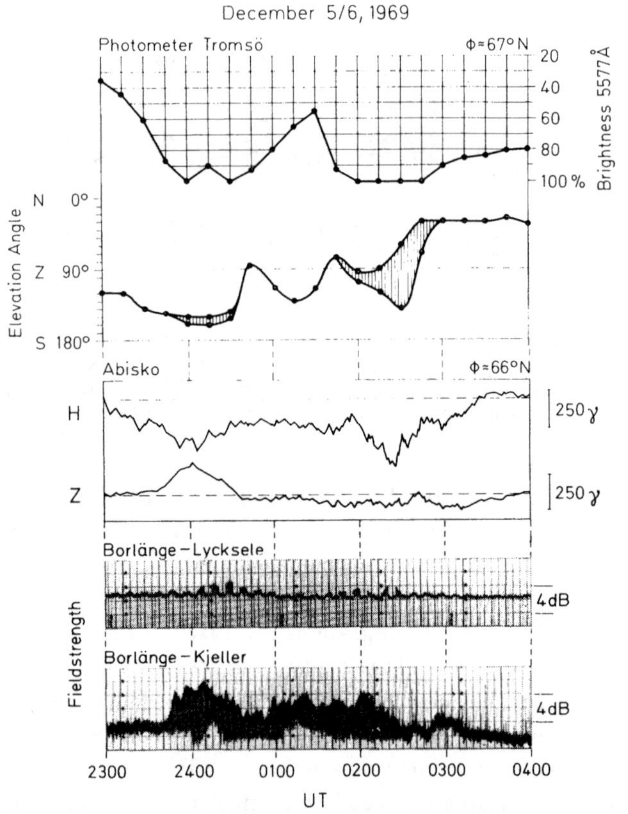

Fig. 8. Comparison of optical auroral observations at Tromsø (upper diagram) with VHF auroral backscatter (radio aurora) recordings on the lines from Borlänge (SK 4 MPI, 145.960 MHz) to Lycksele and to Kjeller, and with geomagnetic recordings from Abisko (see Figure 1).

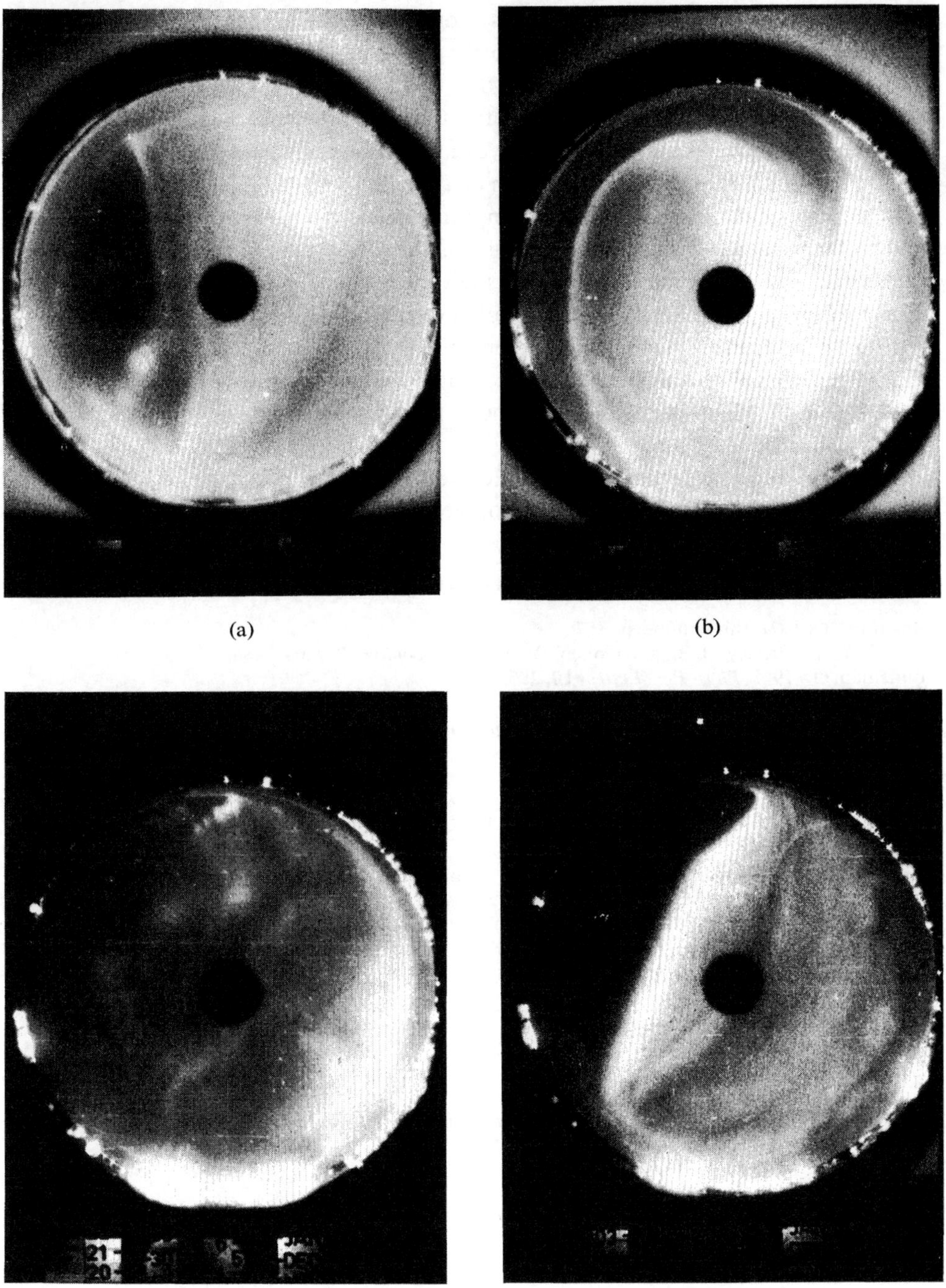

(a) (b)

(c) (d)

Fig. 9a. All-sky camera pictures from Tromsø; (a) Nov. 26, 1969, 0239 UT; (b) Dec. 5, 1969, 2009 UT; (c) Dec. 5, 1969, 2030 UT; and (d) Dec. 6, 1969, 0215 UT. The bright spot in the upper right part of Figure 9a is the moon. An all-sky camera picture is north to the left.

line, however, the backscatter signal is more frequent. Amplitude peaks are also observed during the peaks of the substorms but in addition to this a third amplitude peak occurs at about 0115 UT when the visual aurora is south of Tromsø. The amplitude shows a decrease before and after that time when the aurora is located more to the north. Figure 9d shows an all-sky camera picture from the peak of the breakup phase at 0215 UT at the zenith and south of Tromsø.

In summary it can be said that the amplitude peak of the radio aurora coincides in time with the peak of the breakup phase of the visual aurora and the peaks of the geomagnetic bays. Furthermore, there is fairly good evidence that the backscattering centers which give rise to the backscattering of radio waves show an extension to the north and south during the development of the substorm simultaneous with a N-S shift of the visual aurora and the PEJ. Finally, all VHF auroral backscatter events presented in the Figures are in accordance with a part of the results (threshold) of the plasma acoustic wave theory of Farley (1963) and Buneman (1963). More detailed information about accordance with other points of the acoustic wave theory will be presented in the near future when the Doppler shift recordings at the receiving stations shown in Figure 1 are in action.

References

Boström, R.: 1971, this volume, p. 357.

Bowles, K. L., Balsley, B. B., and Cohen, R.: 1963, *J. Geophys. Res.* **68**, 2485.

Buneman, O.: 1963, *Phys. Rev. Letters* **10**, 285.

Cohen, R. and Bowles, K. L.: 1963, *J. Geophys. Res.* **86**, 2503.

Czechowsky, P.: 1966, Master's Thesis, University of Göttingen, Germany.

Czechowsky, P.: 1969, *Kleinheubacher Berichte* Vol. 13, published by Fernmeldetechnisches Zentralamt, D 33, Darmstadt, Germany, p. 97.

Czechowsky, P. and Lange-Hesse, G.: 1970 in *Intercorrelated Satellite Observations Related to Solar Events* (ed. by V. Manno and D. E. Page), D. Reidel Publishing Company, Dordrecht-Holland, p. 405.

Czechowsky, P., Kochan, H., Lange-Hesse, G., Lauche, H., and Möller, H. G.: 1970, *Z. Geophys.* **36**, 77, and in *Report UAG 8, Part 1, Data on Solar-Geophysical Activity, Oct. 24-Nov. 6, 1968* (compiled by J. V. Lincoln), World Data Center A, Upper Atmosphere Geophysics, Boulder, Colo., p. 296.

Farley, D. T., Jr.: 1963, *J. Geophys. Res.* **68**, 6083.

Heppner, J. P.: 1967, in *Aurora and Airglow* (ed. by B. M. McCormac), Reinhold Publishing Corporation, New York, p. 75.

Lange-Hesse, G.: 1967, in *Aurora and Airglow* (ed. by B. M. McCormac), Reinhold Publishing Corporation, New York, p. 519.

Lange-Hesse, G.: 1968, in *Ionospheric Radio Communication* (ed. by K. Folkestad), Plenum Press, New York, p. 174.

Lange-Hesse, G.: 1969, in *Atmospheric Emissions* (ed. by B. M. McCormac and A. Omholt), Van Nostrand Reinhold Company, New York, p. 201.

Lange-Hesse, G. and Czechowsky, P.: 1966, *Archiv Elektr. Übertragung (AEÜ)* **20**, 365.

PROGRESS IN ELF AND VLF EMISSION STUDIES AT HIGH LATITUDES

T. STOCKFLET JØRGENSEN

The Danish Meteorological Institute, Charlottenlund, Denmark

1. Introduction

Electromagnetic phenomena associated with aurora have been observed in the very large frequency range from about 10^{-3} Hz to more than 10^9 Hz. Recently Ellyett (1969) reviewed the observations at frequencies above 10^5 Hz.

In this paper only the frequency range from about 10^2 to 10^5 Hz (ELF and VLF) will be considered, and furthermore only the most commonly occurring emissions known as chorus, ELF hiss, and auroral hiss will be considered. Mainly the research made during the last couple of years will be reviewed, and this paper therefore may be regarded as a kind of progress report. For more information on earlier work in the field of high latitude ELF and VLF emissions the reader is referred to the ELF and VLF papers in the proceedings from the previous Aurora and Airglow institutes (McCormac, 1967, McCormac and Omholt, 1969).

Chorus is observed at middle and high latitudes at frequencies from a few hundred Hertz to a few kilo Hertz. The distinguishing characteristics of this class of emissions is the presence of well defined short duration rising tones. These discrete elements often appear in a background of steady ELF hiss or at the upper edge of an ELF hiss band.

Auroral hiss is observed in the polar regions in close association with visible aurora. Its spectrum resembles band-limited white noise, which may or may not show rapid fluctuations with time and space. The spectrum tends to peak in the vicinity of 10 kHz with a low frequency cutoff at about 2 kHz and an upper frequency cutoff as high as several hundred kilo Hertz (Jørgensen, 1968). The Alouette satellites have observed noise at high latitudes which may be the high frequency extension of auroral hiss. The high frequency limit approaches, but is always less than, the electron gyrofrequency at the position of the observer (Hartz, 1969). It is possible that auroral hiss may occur at still higher frequencies, e.g., Harang (1969a) has observed several cases of emissions at 8 kHz and 28 MHz simultaneously, but it is not known if there is one wide band of emissions including both frequencies.

Also Gregory (1969) has reported on noise at HF frequencies which may be part of the auroral hiss. Using the Ariel 3 radio astronomy experiment, he observed noise between 1.94 and 4.24 MHz in a narrow range of IN Lat which coincides very closely with the auroral oval.

2. Chorus and ELF Hiss

A. OCCURRENCE

Magnetospheric observations of whistler mode emissions were made by the OGO 1

B. M. McCormac (ed.), The Radiating Atmosphere, 327–335.

satellite over the frequency range 0.3 to 100 kHz (Dunckel and Helliwell, 1969). Emissions are found within $L \sim 10$ at all local times except for $L > 5$ in the midnight-to-dawn sector, with intensity peaking at $L \sim 4$ and $L \sim 9$ near 10 hr Local Time. Their spectra resemble those of chorus and mid-latitude hiss observed on the ground.

An important new result is the observation that the upper frequency limit of most of these emissions is proportional to the minimum electron gyro frequency along the magnetic fieldline passing through the satellite. This is interpreted to mean that the source of the emissions lies close to the equatorial plane.

Emissions observed on the dayside beyond $L = 8$ are burst-like, and their occurrence tends to decrease with increasing radial distance. No emissions were observed beyond the estimated position of the shock boundary.

The occurrence of chorus and ELF hiss has been recorded at low altitudes by Injun 3 (Taylor and Gurnett, 1968). Many of the features observed by Dunckel and Helliwell (1969) appear in their map of IN Lat vs LT, including the sudden onset near 06 hr, the active day region from 06 to 14 hr, and the quiet night-side. However, little activity is found at IN Lat greater than 75° near the noon meridian. One explanation is the diminished sensitivity of the Injun 3 receiver at the low frequencies where the emissions at high latitudes are most likely to occur. Another is that the burst-like nature of the emissions at high latitudes will tend to reduce the occurrence index even though the intensity may be high.

In an investigation of the occurrence of ELF emissions in the magnetosphere on OGO 3 (Russell *et al.*, 1969), emissions are categorized as either bursts or steady noise. Bursts are defined as signals that rise and fall in less than 2 min and reach a level above 2 V at either 30, 100, 300, or 800 Hz. Steady noise is defined as noise that persists (or changes intensity slowly) over 2 min at either 100, 300, or 800 Hz. The long decay time of the detector (30 sec) is likely to cause chorus with closely packed elements or with a hiss background to be identified as steady noise but no high resolution spectra are available to check this possibility.

For burst noise the same conclusions are drawn regarding equatorial control as from the observations of Dunckel and Helliwell (1969). However, from the increased occurrence of steady noise on the dayside in the region $6 \leqslant L \leqslant 10$ at magnetic latitudes greater than 30°, Russell *et al.*, (1969) conclude that steady noise is generated at a latitude of about 45°. This conclusion differs from that of Dunckel and Helliwell (1969) based on their upper cutoff data.

A peak in 'steady' noise intensity at 45° latitude might also be explained in other ways than in terms of local generation. If we assume that generation occurs at the equator, as suggested by Dunkel and Helliwell (1969), the following interpretations of the 45° peak seem possible: (1) Emissions generated at the equator are amplified further down the field line; (2) an increase in the relative amount of nonducted propagation from higher magnetic latitudes could concentrate the energy at 45°; and (3) the scattering of ducted emissions near the base of the field line could cause a maximum in occurrence at high latitudes.

B. PROPAGATION

Satellite observations of ELF and VLF radio noise have revealed that noise bands in the frequency range from a few hundred Hertz to several kilo Hertz often have a very sharp lower cutoff frequency near the proton gyrofrequency.

Gurnett and Burns (1968) propose an explanation for the cutoff based on the reflection of downgoing, extraordinary mode waves near the two-ion cutoff between the proton and He (or O) gyrofrequencies. This explanation indicates that chorus and ELF hiss are generated at high altitudes (above 3000 km) and not near the base of the ionosphere. Measurements of the Poynting flux of these emissions with the Injun 5 satellite (Mosier and Gurnett, 1969) provide direct evidence for this.

Ground based observations of 700 Hz noise bands near the auroral zone, previously attributable to proton cyclotron radiation at low altitudes in the ionosphere, can now be interpreted as being due to a combination of the low frequency cutoff in the transmission of the noise to the ground and the frequency spectrum of the emitted radiation.

C. RELATION TO OTHER PHENOMENA

Observations made with the Injun 3 satellite of bursts of precipitating $E \geqslant 40$ keV electrons and of VLF chorus emissions have revealed their simultaneous occurrence (Oliven and Gurnett, 1968). Observed microbursts are always accompanied by a group of VLF chorus emissions; chorus is not necessarily accompanied by microbursts. The maximum region of microbursts occurrence from $04 \leqslant$ IN LT $\leqslant 13$ and $65° \leqslant$ IN Lat $\leqslant 70°$ lies well within the maximum region of chorus emissions from $03 \leqslant$ IN LT $\leqslant 15$ and $55° \leqslant$ IN Lat $\leqslant 75°$. Generally it is not possible to find a one-to-one (burst to burst) correspondence between individual microbursts and VLF chorus bursts.

Etcheto *et al.* (1969) have also reported on a relation between VLF emissions and electrons. They launched rockets from Kerguelen (IN Lat $= 58°$) and detected a fluctuation ($T = 80$ sec) of the trapped electron ($E > 55$ keV) intensities in coincidence with a periodic VLF emission.

Troitskaya *et al.* (1968) and Vigneron *et al.* (1969) have studied the relationship at substorm conditions between irregular geomagnetic field pulsations called AA (auroral agitation) with periods from 5 to 10 sec and amplitudes of several tens of milligammas, pulsations of auroral light intensity in the range 3800 to 4200 Å and VLF chorus at Sogra (IN Lat $= 57°$). Auroral absorption was only recorded when the local K index was high ($\geqslant 3$). All cases of auroral pulsations were accompanied by AA (Figure 1), but not all AA events were accompanied by auroral pulsations.

After dawn, when the aurora is no longer visible and the intensity of AA decreases, chorus was often recorded. There were 20 days with chorus at Sogra in February to March 1968. They were all preceded at night by AA series, and in 11 cases chorus and AA were observed simultaneously. It is important to emphasize that all cases of pulsating auroras at Sogra were accompanied in the morning hours by VLF emissions of the chorus type. Such a coincidence was reported by Kleymenova *et al.* (1969) for Sogra and Loparskaya (IN Lat $= 64°$).

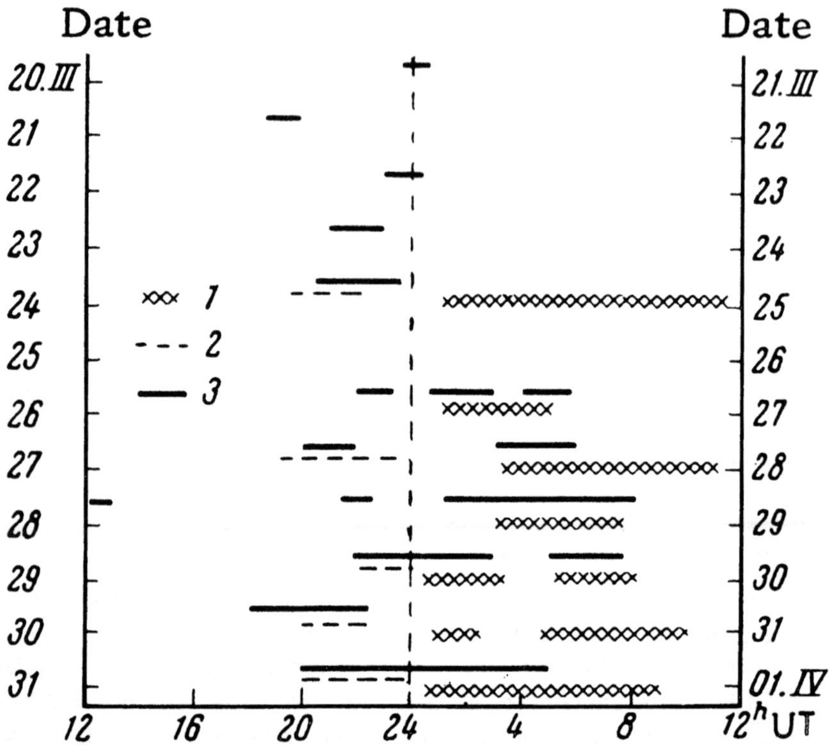

Fig. 1. Occurrence of VLF chorus (1), pulsations in aurora (2), and auroral agitation (3) during March 20 to 31, 1968, at Sogra. Due to dawn around 24 UT the photometric observations (2) are stopped at this time (Troitskaya *et al.*, 1968).

A general picture of the development of geophysical phenomena recorded at Sogra is shown in Figure 1. We can see clearly the good correlation between the excitation of AA, pulsations in the intensity of auroral emissions, and chorus.

Hayashi *et al.* (1968) made comparisons between VLF chorus observed at Syowa Base, Antarctica (IN Lat = 66°) and magnetograms from low latitude stations. They found that variations in the chorus intensity are closely related to sudden impulses in the geomagnetic field. A positive impulse is accompanied by a sudden enhancement or commencement of chorus, and a negative impulse is associated with a sudden decrease or fade-out of chorus emissions.

D. MECHANISM OF GENERATION

Troitskaya *et al.* (1968) and Vigneron *et al.* (1969) suggest that plasma blobs drift from the tail of the magnetosphere toward the earth. Due to gradient drift the electrons will move from the night part of the magnetosphere into its morning side where they cause the observed phenomena.

It is believed that chorus emissions are generated by plasma instabilities within the magnetosphere. A plasma instability acts to coherently bunch or organize the charged particles, so that large intensities can be obtained. On general grounds Brice (1964) and Kennel and Petschek (1966) show that the generation of whistler mode wave energy by cyclotron interaction with electrons gives rise to a decrease in the pitch angle of the resonant electrons and to precipitation if the mirror altitude is sufficiently decreased. Thus, the plasma instability that produces VLF chorus may also cause precipitation of the $E \geqslant 40$ keV electrons that are observed as microbursts (Oliven and Gurnett, 1968).

The result of sudden impulse effects on polar chorus seems to support the possibility of electron cyclotron generation. Since a sudden impulse is interpreted as a manifestation of magnetospheric compression or expansion resulting from changes in the state of the solar wind, a sudden compression at the time of a positive sudden impulse would tend to create an enhanced anisotropy in pitch angle distribution of trapped electrons in a direction perpendicular to the magnetic field through the betatron acceleration mechanism. Then, the enhanced anisotropy produces an increase in wave energy through the cyclotron instability process. The fact that polar chorus enhancements correspond to positive impulses may be explained by the above process. Decreases in chorus intensity at times of negative sudden impulses are also explained by a reduced pitch angle anisotropy through deceleration associated with a decrease in the magnetic field.

Estimates of the average intensity of the emissions observed by Dunckel and Helliwell (1969) agree with the intensity required to explain the precipitation of electrons through pitch angle scattering by whistler mode waves.

Bullough *et al.* (1969a) arrived at similar conclusions regarding mechanism of generation from observations by the Ariel 3 satellite.

Cauffman and Gurnett (1969), on the other hand found from Injun 3 observations of VLF noise and energetic electrons that the noise levels predicted in the equitorial plane by Kennel and Petschek (1966) are not observed at the altitude of the satellite (less than 2785 km).

Swift (1968) has proposed a theory in which chorus is generated at low altitudes by an electrostatic loss cone instability associated with ring current protons, but the observation by Mosier and Gurnett (1969) that the Poynting flux direction of chorus is downgoing at an altitude of approximately 2500 km seems to reject this idea.

3. Auroral Hiss

A. OCCURRENCE

The hiss zone is apparently the same as the zone of auroral precipitation. It is approximately circular shaped with its center situated close to the magnetic midnight meridian a few degrees from the magnetic pole. Hiss occurs at all IN LT but is most frequently observed between noon and midnight. Observations from the Ariel 3 satellite (Bullough *et al.*, 1969b) show that hiss is a very common phenomenon, as it is observed on the majority of the satellite revolutions.

Hughes *et al.* (1970) have found an asymmetry in the occurence of hiss in the two polar regions from the Ariel 3 observations at 3.2 and 9.6 kHz. At midnight the maximum frequency of occurrence is approximately the same in both hemispheres, but hiss is observed 40% more frequently in the afternoon in the north (summer) than in the south (winter). This could indicate that the emissions on the sunward side of the earth are due to particles entering the magnetosphere through the neutral points and the nighttime emissions due to particles accelerated in the tail of the magnetosphere.

B. PROPAGATION

Mosier and Gurnett (1969) have determined the direction of the Poynting flux of auroral hiss from observations with the Injun 5 satellite. Auroral hiss is observed to be progagating down the geomagnetic field lines. However, in all but 2 of the 16 cases observed, there are also some brief impulsive bursts propagating up the geomagnetic field lines from below the satellite. The upgoing signals are much less intense than the downgoing signals, and the impulsive bursts that characterize many of the hiss events are much less frequent for the upgoing waves than for the downgoing waves.

These observations suggest that generally, auroral hiss is generated above the altitude range (677 to 2528 km) of the Injun 5 satellite.

C. RELATION TO OTHER PHENOMENA

The occurrence of auroral hiss observed on the ground in the polar regions has been found to be independent of the magnetic activity expressed by the auroral electrojet

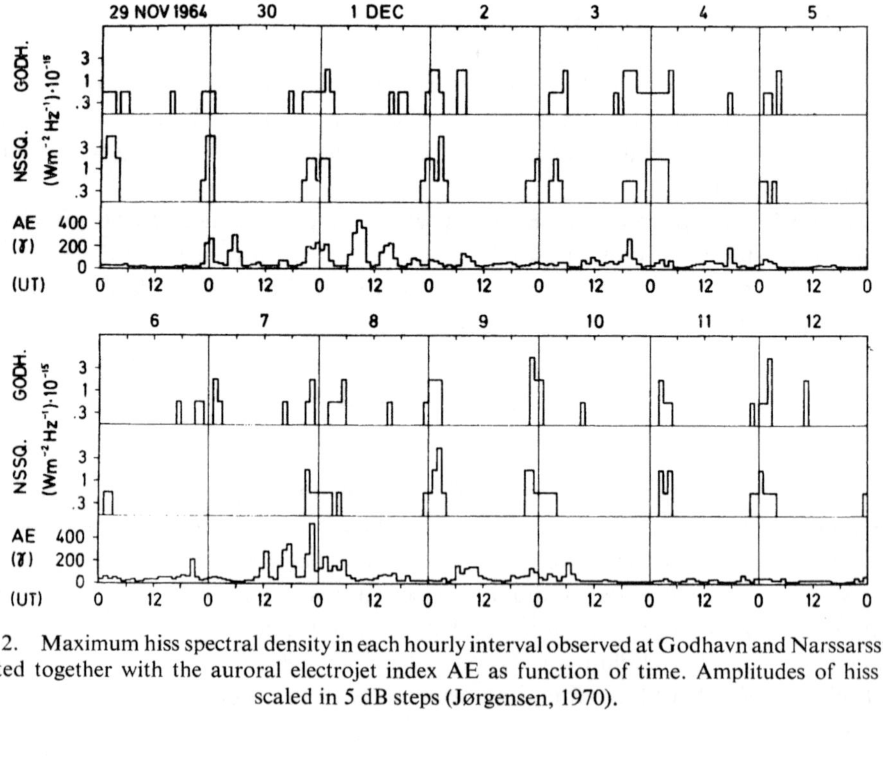

Fig. 2. Maximum hiss spectral density in each hourly interval observed at Godhavn and Narssarssuaq plotted together with the auroral electrojet index AE as function of time. Amplitudes of hiss are scaled in 5 dB steps (Jørgensen, 1970).

index AE (Jørgensen, 1970). Hiss is usually observed in connection with substorms and normally at very quiet magnetic conditions (Figure 2). Also Bullough *et al.* (1969b) found from Ariel 3 observations that the occurrence and intensity of high latitude hiss does not appear to correlate strongly with magnetic disturbances.

This indicates that when looking for particles which generate hiss rather than particles newly accelerated by a substorm we should find a more or less continuous particle flux, the existence of which is relatively independent of substorms. The plasma sheet electrons seem to make up such a particle population.

In order to study the behavior of hiss emissions during substorms, hiss records from Godhavn (IN Lat=77°), Narssarssuaq (IN Lat=68°) and Tromsø (IN Lat=66°) have been closely examined in eight substorm periods (Jørgensen, 1970). The result is that for hiss no clear effect of substorms is found. There is no indication of drift of hiss sources in any direction within the auroral region caused by a substorm. This is contradictory to the result obtained by Akasofu (1968), who found that the hiss sources early in a substorm appeared in the midnight section of the auroral oval and then propagated west along the oval together with the so-called westward traveling surge.

The studies mentioned above may give the impression that the occurrence of auroral hiss is completely independent of magnetic disturbances, but this is not the case. Harang and Larsen (1965) and Harang (1969b) have demonstrated that the moderate positive disturbances in the horizontal component of the geomagnetic field appearing early in the night at the auroral zone station very often are accompanied by hiss bursts, and that sudden and strong magnetic disturbances which appear as so-called negative bays in the horizontal component near midnight usually are accompanied by hiss emissions.

The interpretation of these seemingly conflicting results discussed here may be that auroral hiss can be generated during quiet as well as disturbed periods, and that the ionospheric disturbances are modulating the production of hiss and/or the propagation conditions between the source and the observer.

A correlation between auroral hiss and X-rays has been observed by Rosenberg (1968). The observation indicates that on occasion electrons with energies reaching at least 150 keV are precipitated when hiss is recorded simultaneously, but it was estimated that less than 20% (and probably no more than 5%) of the energy deposited in the ionosphere was brought there by electrons with energies greater than 25 keV. This observation, thus, also seems to show that hiss is associated with intense fluxes of low (few kilo electron Volt) energy electrons (Gurnett, 1966).

D. MECHANISM OF GENERATION

Cerenkov radiation from electrons with energies near 1 keV in the high latitude magnetosphere has been studied by Jørgensen (1968), and it was found that a calculated noise spectrum between 1 and 100 kHz as produced by an incoherent Cerenkov process is comparable in magnitude and shape to observed noise spectrums.

Singh and Singh (1969) also calculated the power radiated from electrons in the magnetosphere by the Cerenkov mechanism. They found that the highest intensities should occur in the auroral zone and concluded that the Cerenkov mechanism is a potential source of VLF wave generation.

The observations of the Poynting flux direction of auroral hiss (Mosier and Gurnett, 1969) are consistent with the theory of Cerenkov radiation as this implies that the most important source regions should be at altitudes of several thousand kilometers.

Cerenkov radiation may occur when the refraction index is higher than one. Besides for whistler mode waves this condition also is fulfilled for waves above the plasma cutoff and below the upper hybrid resonance. The observations from Ariel 3 between 1.94 and 4.24 MHz were in this last domain and Gregory (1969) concluded that the noise could be due to Cerenkov radiation.

4. Summary and Conclusions

The regions of generation of ELF and VLF noise observed at high latitudes now seem to be approximately known. All experiments indicate that chorus and ELF hiss are generated at several thousand kilometers altitude. There is much evidence that the source regions for these noises are in and near the geomagnetic equatorial plane, but there also exist observations which have been interpreted to show that the source region is around 45° latitude. Auroral hiss is generated in auroral regions of the ionosphere and magnetosphere. The source region probably extends from the bottom of the ionosphere and up to an altitude of a few R_E.

The noise producing mechanism has not been completely determined, but it appears likely that an electron cyclotron instability is responsible for chorus and ELF hiss, and that auroral hiss is due to Cerenkov radiation.

A few studies concerning the behavior of VLF emissions during substorms have been carried out. While a positive correlation between substorms and chorus is observed, the relation between substorms and auroral hiss is not very clear, as rather conflicting results have been obtained. It is strongly recommended to investigate the substorm-VLF emission relationship further, as such a study may not only throw light on the conditions required in the magnetosphere for the VLF emission mechanisms to work, but also give information on drift of the source regions and thereby eventually on electric fields in the magnetosphere. A study of this kind could probably best be made from a chain of ground based VLF stations dispersed in longitude around the auroral zone.

References

Akasofu, S.-I.: 1968, *Polar and Magnetospheric Substorms*, D. Reidel Publishing Company, Dordrecht, Holland.
Brice, N. M.: 1964, *J. Geophys. Res.* **69**, 4515.
Bullough, K., Hughes, A. R. W., and Kaiser, T. R.: 1969a, *Planetary Space Sci.* **17**, 363.
Bullough, K., Hughes, A. R. W., and Kaiser, T. R.: 1969b, *Proc. Roy. Soc.* **A311**, 563.
Cauffman, D. P. and Gurnett, D. A.: 1969, *J. Geophys. Res.* **74**, 1144.
Dunckel, N. and Helliwell, R. A.: 1969, *J. Geophys. Res.* **74**, 6371.
Ellyett, C. D.: 1969, *J. Atmospheric Terrest. Phys.* **31**, 671.

Etcheto, E., Gendrin, R., and de la Porte des Vaux, B.: 1969, Paper presented at IAGA meeting, Madrid.
Gregory, P. C.: 1969, *Nature* **221**, 350.
Gurnett, D. A.: 1966, *J. Geophys. Res.* **71**, 5599.
Gurnett, D. A. and Burns, T. B.: 1968, *J. Geophys. Res.* **73**, 7437.
Harang, L.: 1969a, *Planetary Space Sci.* **17**, 869.
Harang, L.: 1969b, *Planetary Space Sci.* **17**, 1565.
Harang, L. and Larsen, R.: 1965, *J. Atmospheric Terrest. Phys.* **27**, 481.
Hartz, T. R.: 1969, *Proc. IEEE* **57**, 1042.
Hayashi, K., Kokuban, S., and Oguti, T.: 1968, *Rep. Ionosphere Space Res. Japan* **22**, 149.
Hughes, A. R. W., Kaiser, T. R., and Bullough, K.: 1970, Paper presented at COSPAR, Leningrad.
Jørgensen, T. S.: 1968, *J. Geophys. Res.* **73**, 1055.
Jørgensen, T. S.: 1970, Paper presented at COSPAR, Leningrad.
Kennel, C. F. and Petschek, H. E.: 1966, *J. Geophys. Res.* **71**, 1.
Kleymenova, N. G., Roldugin, V. K., and Vigneron, J.: 1969, *Geomag. Aeron.* **9**, 148.
McCormac, B. M.: 1967, *Aurora and Airglow*, Reinhold Publishing Corporation, New York.
McCormac, B. M. and Omholt, A.: 1969, *Atmospheric Emissions*, Van Nostrand Reinhold Company, New York.
Mosier, S. R. and Gurnett, D. A.: 1969, *J. Geophys. Res.* **74**, 5675.
Oliven, M. N. and Gurnett, D. A.: 1968, *J. Geophys. Res.* **73**, 2355.
Rosenberg, T. J.: 1968, *Planetary Space Sci.* **16**, 1419.
Russell, C. T., Holzer, R. E., and Smith, E. J.: 1969, *J. Geophys. Res.* **74**, 755.
Singh, R. N. and Singh, R. P.: 1969, *Ann. Geophys.* **25**, 629.
Swift, D. W.: 1968, *J. Geophys. Res.* **73**, 7447.
Taylor, W. W. L. and Gurnett, D. A.: 1968, *J. Geophys. Res.* **73**, 5615.
Troitskaya, V. A., Kleymenova, N. G., Raspopov, O. M., Geller, L. A., Roldugin, V. K., Vigneron, J., Karchevskiy, Zh., and Loran, J.: 1968, *Geomag. Aeron.* **8**, 872.
Vigneron, J., Karczewski, J. F., Laurent, G., Lacourly, S., Kleimenova, N., Raspopov, O., Roldugin, V., and Troitskaya, V.: 1969, *Ann. Geophys.* **25**, 385.

INSTITUTE DISCUSSION

Auroral hiss may be absorbed during a substorm. In particular, ground observations taken during an active time may be misleading and may show no hiss due to ionospheric absorption. L. Kavanagh reported that Injun 5 observations show VLF noise everytime particle precipitation occurs, but only a few cases have been published.

VLF RADIO EMISSIONS AND THE AURORA

A. R. W. HUGHES and T. R. KAISER
Dept. of Physics, University of Sheffield, U.K.

1. Introduction

Very low frequency radio emissions have been observed at high latitudes by all satellite borne VLF experiments. From observations made on Injun 3 Gurnett (1966) showed that hiss was observed mainly between noon and midnight in a band about 7° wide centered on 77° Λ (IN Lat) at 1200 hr and decreasing to 72° Λ at 2300 hr MLT. More recently comprehensive statistics of VLF emissions have been compiled from observations made on Ariel 3. On this satellite the high receiver sensitivity revealed emissions at all MLT and the onboard tape recording of data made it possible to make observations for approximately 17 hr per day for a 9 mo period. The data presented here refer from May 5 to July 24, 1967, during which complete coverage in LT is obtained at all IN Lat.

There is now considerable observational and theoretical evidence (see Gurnett, 1966; Jørgensen, 1968) that these high latitude VLF emissions are produced by Cerenkov radiation from low energy electrons, probably with energies of 1 keV or less. Preliminary observations on Injun 5 of the direction of the Poynting flux are also consistent with this view (Moiser and Gurnett, 1969). It is clear therefore that a study of the pattern of occurrence of VLF emissions will yield information about low energy electron fluxes. Of particular interest here is the relation of the VLF emissions and low energy electrons to other auroral phenomena and especially to the visible aurora.

2. The Ariel 3 VLF Experiment

The Ariel 3 satellite was launched on May 5, 1967, into a low altitude (500 to 600 km) orbit of high inclination (80°) and 96 m. By means of a VLF receiver with a loop aerial, measurements of the peak, mean and minimum VLF magnetic field intensities are made at three frequencies (3.2, 9.6 and 16 kHz) in 28 sec periods ($\simeq 2°$) around the orbit. At each frequency there is a wide-band ($\Delta f = 1$ kHz) channel and at 16 KHz there is also a narrow band ($\Delta f = 100$ Hz) channel to make it possible to identify ground transmitters at this frequency. Measurements were stored on the satellite tape recorder which was replayed once per revolution. The receiver sensitivity ranges from approximately $10^{-13}\ \gamma^2\ \mathrm{Hz}^{-1}$ at 16 kHz to $10^{-11}\ \gamma^2\ \mathrm{Hz}^{-1}$ at 3.2 kHz. A detailed description of the experiment is given by Bullough *et al.* (1968).

A typical 9.6 kHz record from a single satellite revolution is shown in Figure 1 in which high latitude ($\Lambda > 60°$) emissions may be seen centered at times 9, 17, 57 and 64 min. The emission is almost always sharply defined with an average width (to the 10 dB points) in latitude of about 7°. It is present at all three frequencies on most

B. M. McCormac (ed.), The Radiating Atmosphere 336–344.

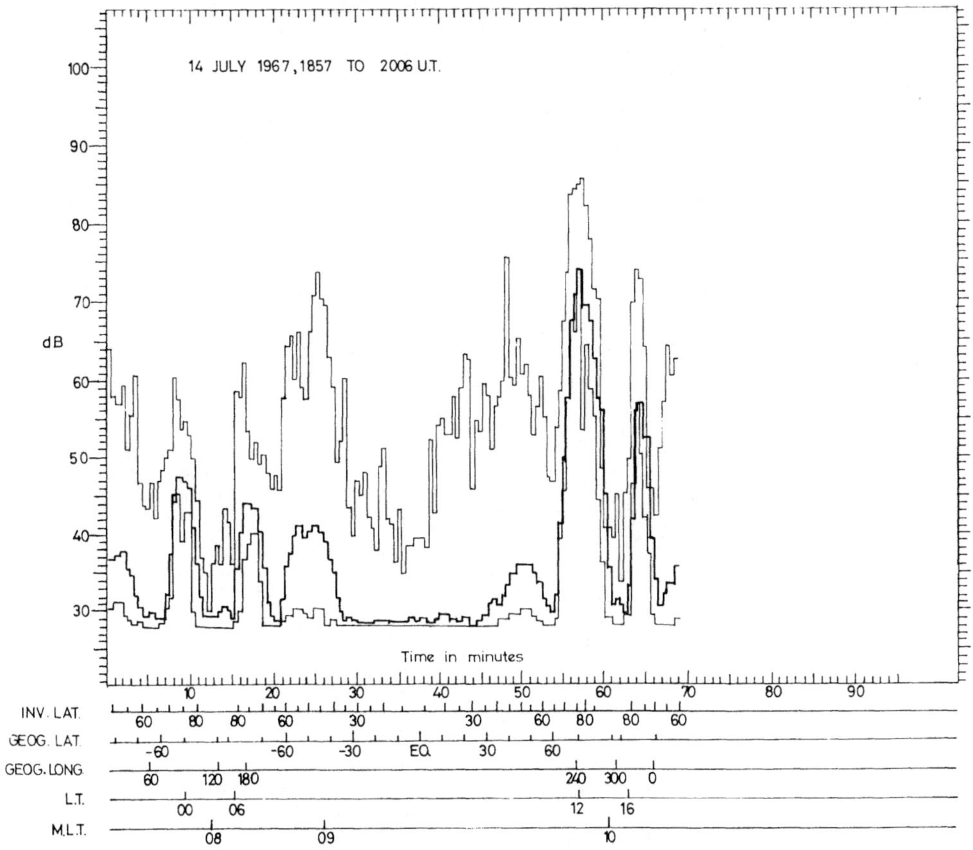

Fig. 1. A typical 9.6 kHz record of peak, mean and minimum signals for a single satellite revolution. The ordinate gives the measured intensity in dB above $10^{-15}\ \gamma^2\ \mathrm{Hz}^{-1}$.

Ariel 3 records and its latitude of occurrence is a function of MLT and magnetic disturbance. It is characteristic of the high latitude emissions that, at the latitude of maximum intensity, the ratio of the mean to minimum signal recorded over one sampling interval is less than 3 dB and the peak signal is some 10 to 20 dB larger.

3. Frequency of Occurrence of Emissions

For IN Lat greater than 60° the data were subdivided into intervals of 1° latitude and 1 hr MLT (centered dipole). The frequency of occurrence was then defined as the ratio (expressed as a percentage) of the number of times the satellite observed signals above a given level to the total number of times the satellite intersected that interval. The frequency of occurrence was determined above different signal intensities ranging in 10 dB steps from 20 to 60 dB above $10^{-15}\ \gamma^2\ \mathrm{Hz}^{-1}$ (free space equivalent for given power flux). The correction for the effect of the refractive index on the magnetic component of the wave was made using the quasi-longitudinal approximation; its

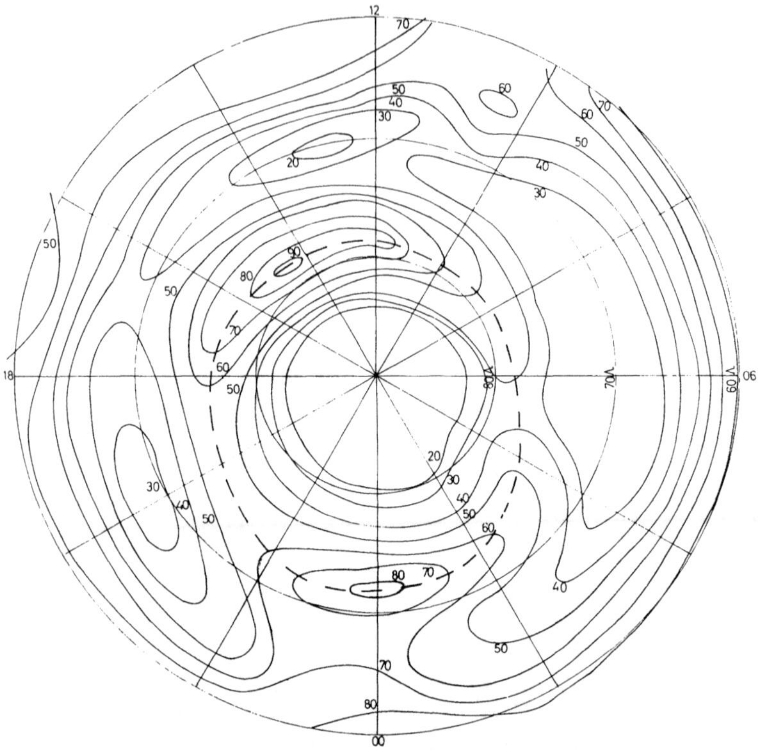

Fig. 2. Frequency of occurrence of VLF emissions at 3.2 kHz (mean intensity $> 10^{-12}\ \gamma^2\ \mathrm{Hz}^{-1}$, free space equivalent) in the Northern Hemisphere from May 5 to July 24, 1967. The coordinates are IN Lat and MLT (centered dipole).

value was determined by *in situ* measurements of electron density (Sayers *et al.*, 1969) combined with a magnetic field model (Cain *et al.*, 1964). The lowest signal intensity threshold (20 dB) therefore represents a power flux of $4.8 \times 10^{-17}\ \mathrm{Wm}^{-2}\ \mathrm{Hz}^{-1}$.

Contours of constant frequency of occurrence of the 3.2 kHz mean signals greater than $10^{-12}\ \gamma^2\ \mathrm{Hz}^{-1}$ are given in Figure 2 from May 5 to July 24, 1967, in the Northern Hemisphere. Because of the inclination (80°) of the satellite orbit the coverage is a function of latitude and this is illustrated in Figure 3. These emissions have maxima in occurrence (greater than 80°) at magnetic midnight and between 12 and 16 hr MLT at IN Lat of 72 and 78°, respectively. There are two well defined sectors of minimum occurrence between 0500 to 0600 hr and 1900 to 2100 MLT, separating the active regions of the afternoon and late evening. There is also a band of minimum occurrence, varying in position from about 68° at midnight to 70° at noon, separating the high latitude emissions from those at medium latitudes. It is possible to define a locus of maximum occurrence (the broken curve in Figure 2) which varies in position from 79° Λ at noon, 76° Λ at 1800 hr, 78° Λ at 0600 hr to 72° Λ at midnight, giving the familiar oval shape which is characteristic of many auroral phenomena.

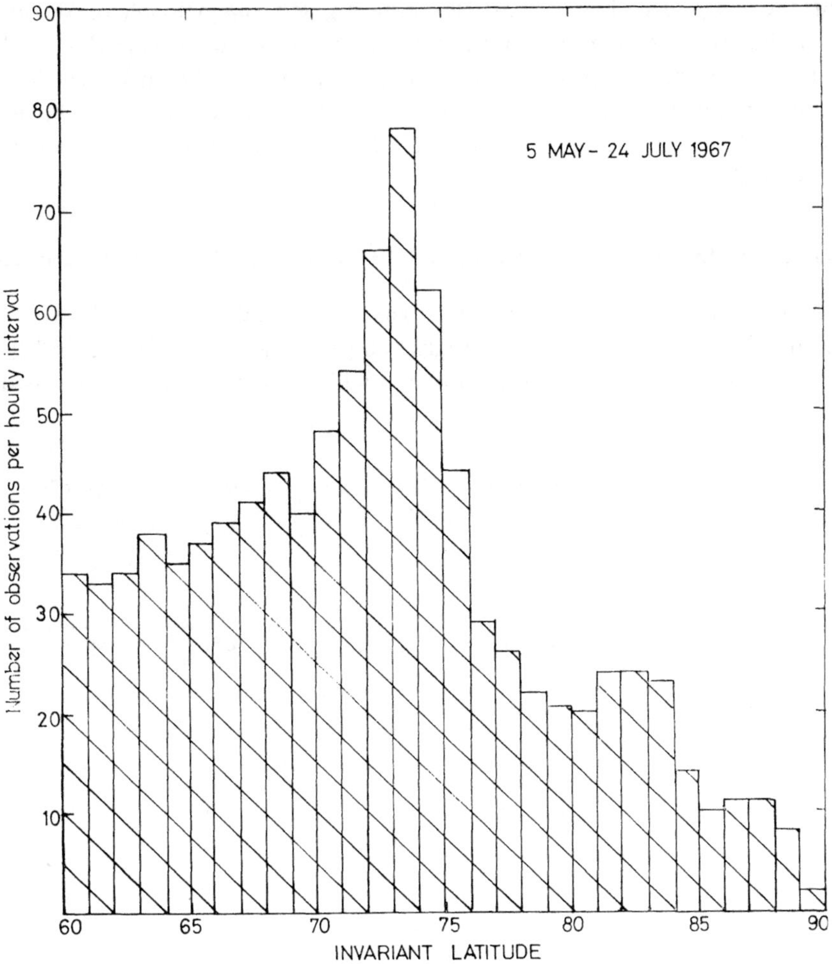

Fig. 3. The distribution of the observations with IN Lat in hourly intervals of MLT.

Similar contours for the same period have been derived for the southern hemisphere (Hughes *et al.*, 1970a). In these the locus of maximum occurrence is more circular than in the north being at 75° Λ at 1200 and 2400 hr, and 74 Λ and 72° Λ at 0600 and 1800 hr, respectively. The different shape of the two loci (north and south) may be due in part to the orientation of the earth's dipole axis relative to the earth–sun line, during the period of the observations (Hughes *et al.*, 1970b). The midnight and afternoon maxima are also present in the Southern Hemisphere but, while the frequency of occurrence at midnight in the north and south is approximately the same, in the afternoon it is about 40% less in the south (winter) than in the north (summer). It is notable that these differences occur on the sunward side of the earth, whereas on the midnight side there is greater similarity between the two regions of maximum occurrence. This would be consistent with the idea that emissions on the sunward side of the earth are due to particles entering the magnetosphere through the neutral points (or cusp regions) and the nighttime emissions to particles accelerated in the neutral sheet. The position of the neutral points is known to be affected by dipole orientation (Spreiter and Briggs, 1962). It is also known that particle flux densities leaking from the cusp regions are influenced by dipole orientation (Spreiter and Summers, 1967).

4. Variation in Intensity and Location of Emissions with Magnetic Disturbance (K_p)

For this study periods were selected which gave a reasonable range of K_p values. The data were divided into four MLT ranges: 00±0300 hr, 12±0300 hr, 06±0300 hr, and 18±0300 hr. The location of an emission within these time intervals was defined

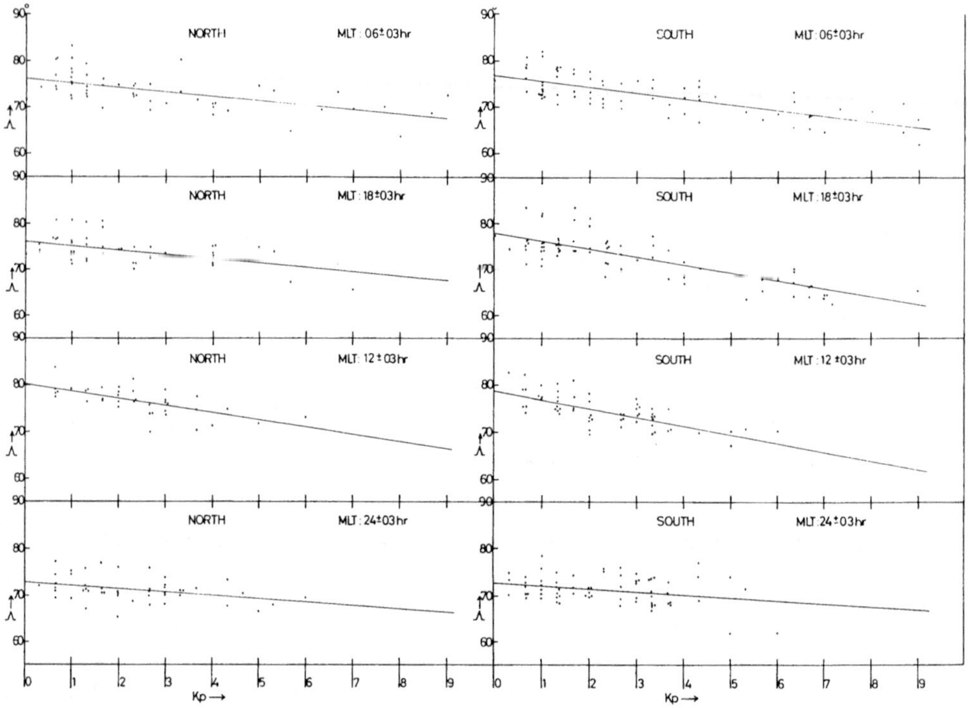

Fig. 4. The location of the high latitude emissions as a function of Λ and magnetic disturbance (K_p) for various MLT.

as the IN Lat at maximum intensity. Figure 4 shows its variation with K_p for the Northern and Southern Hemispheres with, in each case, the solid line giving the linear least squares fit. It is clear that there is a significant equatorwards displacement of the location of the emissions with increasing level of magnetic disturbance. The displacement ($\Delta\Lambda$) per unit increase in K_p may be summarized as shown in Table I.

TABLE I

MLT ($\pm$0300 hr)	$\Delta\Lambda$ (north)	$\Delta\Lambda$ (south)
0000 hr	$-0.7°$	$-0.7°$
1200 hr	$-1.6°$	$-1.9°$
0600 hr	$-1.0°$	$-1.3°$
1800 hr	$-1.0°$	$-1.7°$

It may be noted that the displacement of the emissions towards the equator at local noon appears to be significantly greater than at midnight. This behavior is similar to that reported by Feldstein and Starkov (1967) for the mean position of the auroral belt.

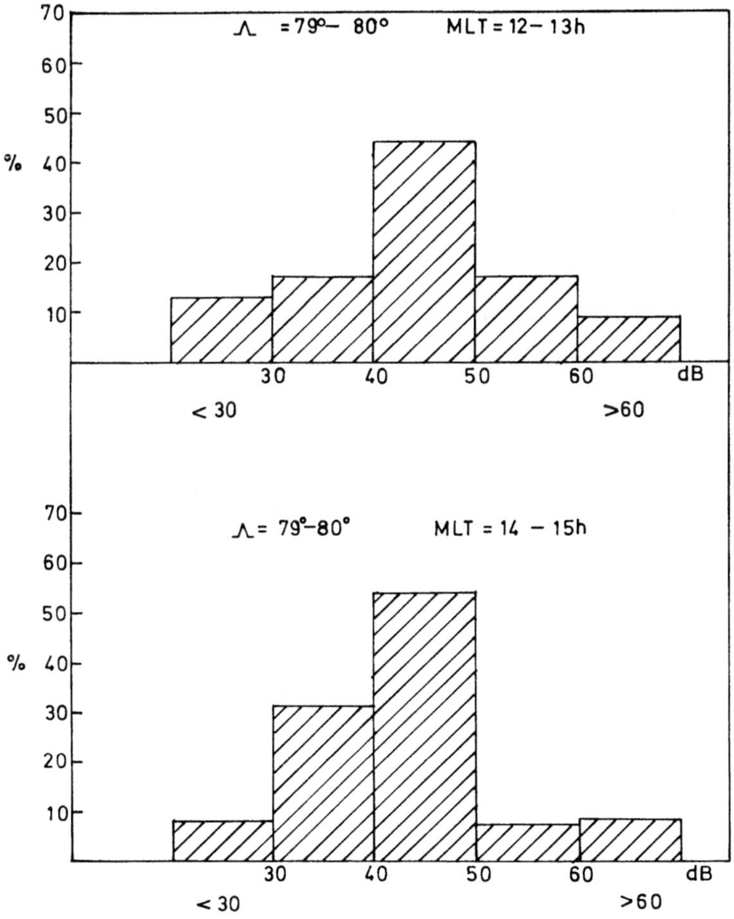

Fig. 5. Distribution of signal intensities for specific intervals of IN Lat and MLT. The abscissa gives the intensity in dB above $10^{-15}\ \gamma^2$ Hz (free space equivalent) and the ordinate the percentage occurrence in the given intensity intervals.

No general pattern for the change in intensity of high latitude emissions with K_p has been established though there appears to be a tendency for the more intense emissions ($\geqslant 60$ dB above $10^{-15}\ \gamma^2\ Hz^{-1}$) to be relatively less frequent when K_p exceeds 4. This behavior is consistent with other high latitude disturbance phenomena such as the radio aurora where, in the inner zone, a close correlation with the local K index and an anticorrelation with the K_p index has been found (Bullough, 1961). It is interesting here to record that Jørgensen (1970) found no clear association between the occurrence of auroral hiss and substorms.

The intensity of VLF emissions may vary by more than 5 orders of magnitude (10^{-12} to $10^{-7}\ \gamma^2\ Hz^{-1}$ at 3.2 kHz) but, if the distribution in intensity is considered for a given interval of 1° IN Lat and 1 hr MLT the variation is considerably less. Two representative examples are shown in Figure 5; in the upper example 61% of the

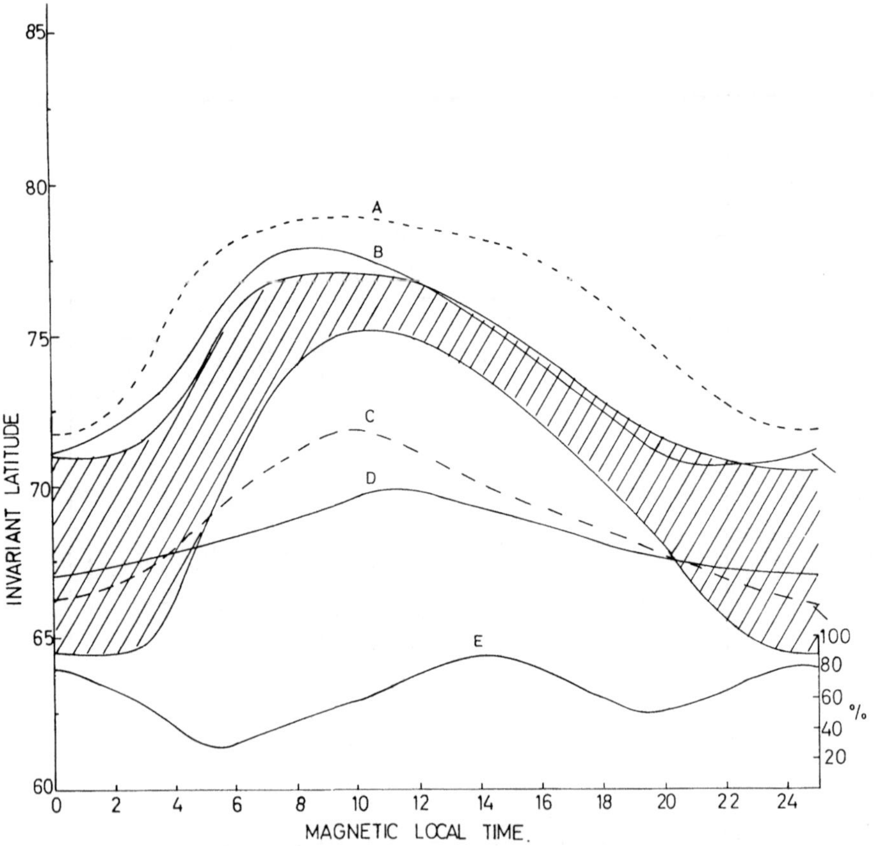

Fig. 6. Location and occurrence of 3.2 kHz emissions in relation to other high latitude phenomena. (a) Locus of maximum occurrence of emissions obtained from Figure 2. (b) Limit of closed field lines, 35 keV electron 'background' boundary, Alouette 2. (c) Locus of minimum occurrence of emissions (dividing the high and medium latitude activity). (d) Stable trapping region, 280 keV electron boundary (Williams and Mead, 1965). (e) Percentage frequency of occurrence of VLF emissions ($> 10^{-12}\ \gamma^2\ Hz^{-1}$ free space equivalent) at the locus of maximum occurrence. Shaded region is the auroral oval (Feldstein, 1966).

signals is between 10^{-12} and 10^{-10} γ^2 Hz^{-1} and in the lower example 85% lies in the same range. While this variation is comparable with changes observed in the solar wind flux density, the emission intensity does not appear to correlate positively with the K_p index (an increase in K_p is normally indicative of an increase in solar wind flux).

5. Discussion

The relationship between the occurrence of VLF emissions at high latitudes and the auroral oval is shown in Figure 6. It may be seen that the locus of maximum occurrence lies a few degrees poleward of Feldstein's auroral oval (Feldstein, 1966) and of the boundary of the closed magnetic field lines given by McDiarmid and Burrows (1968). This implies that low energy electrons (1 keV or less) are primarily found in a band of latitudes poleward of the auroral oval. This would agree with O'Brien's (1962) observation that the spectra of auroral electrons are softer towards higher latitudes and with similar conclusions by Sharp and Johnson (1968), Burch (1968), and Hoffman (1969). Hoffman (1969) also shows that there appears to be a hole in the electron precipitation pattern, above 85° geomagnetic at noon and 78° at night, where fluxes are less by one or two orders of magnitude. This feature may also be seen in Figure 2 and agrees with the fact that the VLF emissions (above 10^{-12} γ^2 Hz^{-1}) have a frequency of occurrence of only 10% (Hughes *et al.*, 1970a) in the region of the magnetic pole ($\Lambda > 85°$).

The location of the maxima in occurrence of the VLF emissions requires some further discussion. The midnight maximum seems likely to be associated with particles which originate in the plasma sheet (Hughes *et al.*, 1970b; Jørgensen, 1970). On the sunward side the emissions would appear to be generated by particles from the neutral points but the fact that the afternoon maximum occurs at about 1400 hr MLT is difficult to reconcile with this view. A study by Walters (1964) of the shape of the magnetosphere, including the effect of the interplanetary magnetic field, indicates that the apparent direction of the solar wind is shifted by 10 to 25° to the west of the sun–earth line, leading one to expect the neutral points at about 1000 to 1100 hr MLT. It may be seen, however, that the locus of maximum occurrence (curve A in Figure 6) is closest to the pole at about 1000 hr MLT and this most probably gives the position for the neutral points in our coordinate system. From curve E in Figure 6 we see that the afternoon VLF maximum is shifted from this position by about 4 hr.

References

Bullough, K.: 1961, *Ann. Geophys.* **17**, 195.

Bullough, K., Hughes, A. R. W., Hudson, T., Hickinson, D., Broomhead, P., and Tomlinson, P.: 1968, *J. Sci. Instr. (J. Phys. E.), Series 2*, **1**, 77.

Burch, J. L.: 1968, *J. Geophys. Res.* **73**, 3585.

Cain, J. C., Henricks, S. J., Daniels, W. E., and Jensen, D. C.: 1964, NASA, *Rept. X-611-64-316.*

Feldstein, Y. I.: 1966, *Planetary Space Sci.* **14**, 121.

Feldstein, Y. I. and Starkov, G. V.: 1967, *Planetary Space Sci.* **15**, 209.

Gurnett, D. A.: 1966, *J. Geophys. Res.* **71**, 5599.

Hoffman, R. A.: 1969, *J. Geophys. Res.* **74**, 2425.
Hughes, A. R. W., Kaiser, T. R., and Bullough, K.: 1970a, Paper presented at COSPAR Symposium, Leningrad in press.
Hughes, A. R. W., Bullough, K., and Kaiser, T. R.: 1970b, in 'Ionospheric Forecasting', AGARD Conference Proceedings No. 49, p. 13–1.
Jørgensen, T. S.: 1968, *J. Geophys. Res.* **73**, 1055.
Jørgensen, T. S.: 1970, Paper presented at the International Symposium on Solar Terrestrial Physics, Leningrad, in press.
McDiarmid, I. B. and Burrows, J. R.: 1968, *Can. J. Phys.* **46**, 49.
Moiser, S. R. and Gurnett, D. A.: 1969, *J. Geophys. Res.* **74**, 5675.
O'Brien, B. J.: 1962, *J. Geophys. Res.* **67**, 3687.
Sayers, J., Wilson, J. W. G., and Loftus, Brigid: 1969, *Proc. Roy. Soc.* **A311**, 501.
Sharp, R. D. and Johnson, R. G.: 1968, in *Earth's Particles and Fields* (ed. by B. McCormac), Reinhold Publishing Corporation, New York, p. 113.
Spreiter, J. R. and Briggs, B. R.: 1962, *J. Geophys. Res.* **67**, 37.
Spreiter, J. R. and Summers, A. L.: 1967, *Planetary Space Sci* **15**, 787.
Walters, G. K.: 1964, *J. Geophys. Res.* **69**, 1769.
Williams, D. J. and Mead, G. D.: 1965, *J. Geophys. Res.* **70**, 3017.

ROCKET, GROUND AND SATELLITE MEASUREMENTS OF ELF EMISSIONS DURING A PCA

J. HOLTET and A. EGELAND
Norwegian Institute of Cosmic Physics, Oslo, Norway
and
N. C. MAYNARD
NASA – Goddard Space Flight Center, Greenbelt, Md., U.S.A.

1. Introduction

A Nike-Tomahawk payload, including ac and dc electric field experiments, several particle detectors, three auroral photometers, and two VLF-Doppler experiments was launched from Andøya, Northern-Norway ($\Lambda = 65°$) on March 8, 1970, at 0017.37 UT.

The main objectives of the ac electric field experiment were to study the naturally occurring ELF and VLF emissions (10 to 10000 Hz) during disturbed conditions and to correlate these data with particle and photometer recordings, simultaneous ground observations and as closely as possible with satellite data from Alouette 2.

The rocket electric field experiment was based on the 'Long-Cylindrical-Probe-Technique' developed at GSFC by Aggson (see e.g. Aggson, 1969; Maynard and Heppner, 1970). A block diagram of the E field instrumentation is shown in Figure 1. The antennas were oriented in the rocket x-y plane. Their tip to tip length was 12.4 m, and the distance between the antenna mid points was 9.3 m as the inner half of the elements were electrically insulated from the plasma. Each probe has a separate high input impedance unity gain preamplifier which was matched for high common mode rejection. The preamplifier signals were then differentially subtracted and amplified. The potential differences between the probes and rocket ground were also recorded, checking the probe symmetry.

During the launching period ground based ELF and VLF observations were made at the field station at Lavangsdalen (140 km east of the rocket range) (Holtet and Egeland, 1969a). The VLF experiment onboard Alouette 2 (Barrington, 1969) was read out at the telemetry station at Tromsø which is located 170 km north east of Andøya. Satellite passes took place approximately 2 hr before and after the rocket launch.

2. Launch Conditions and Payload Operation

The launch took place during a PCA event. The ATS-1 records indicated a maximum proton precipitation sometime before 0020 UT March 8. The CNA (30 MHz) at Tromsø was about 1 dB and constant during the flight. The magnetometer record shows a recovery of a small substorm. At $t = 0$ sec the H component was about

B. M. McCormac (ed.), The Radiating Atmosphere, 345–354.

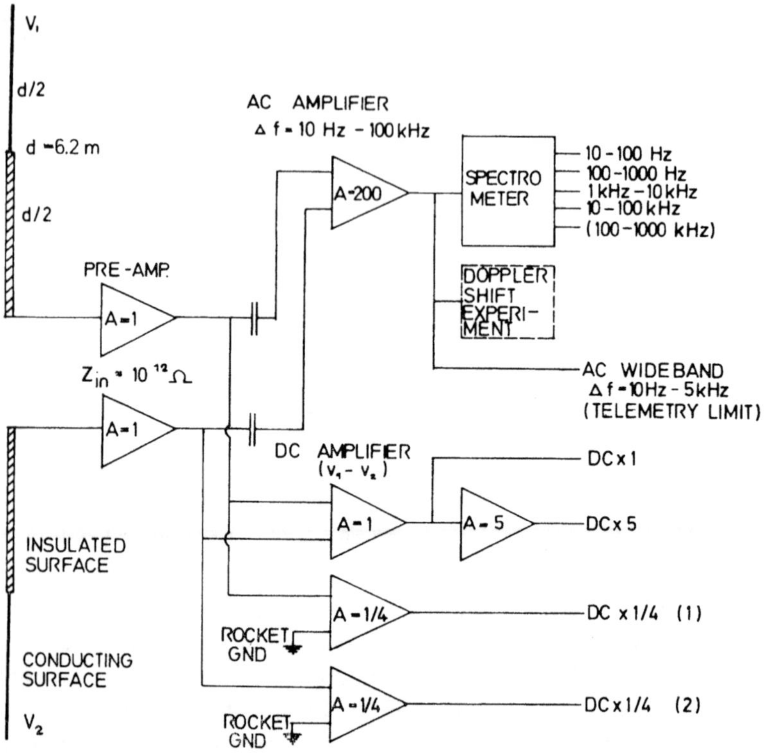

Fig. 1. Block diagram of electric field experiment.

$-300\,\gamma$ and decreased to about $-200\,\gamma$ during the flight. A homogeneous auroral glow was present. The zenith brightness at 3914 Å, as measured from the ground, decreased from 2.3 to 1.3 kR during the flight. The VLF receiving stations at Tromsø and Spitzbergen show 'day-time' propagation conditions, indicating precipitation of MeV protons.

The payload reached a peak altitude of 216 km after 228 sec. The antennas were completely deployed at 115 sec (h = 157 km) and good data were obtained from this moment until the probes started to break off at 405 sec (h = 71 km). The orientation of the 20° half-angle precession cone was such that the *E* field antenna became nearly perpendicular to the geomagnetic field once during the flight (cf. Figure 3). The spin period was approximately 1.8 sec.

3. Summary of Experimental Results

A. NATURALLY OCCURRING ELF AND VLF EMISSIONS PRIOR TO LAUNCH

The ionospheric conditions were complicated because the activity which started with a SC on March 5 had not ended before the PCA event began. As the low energy plasma cloud associated with the next SC did not arrive until 1420 UT on March 8,

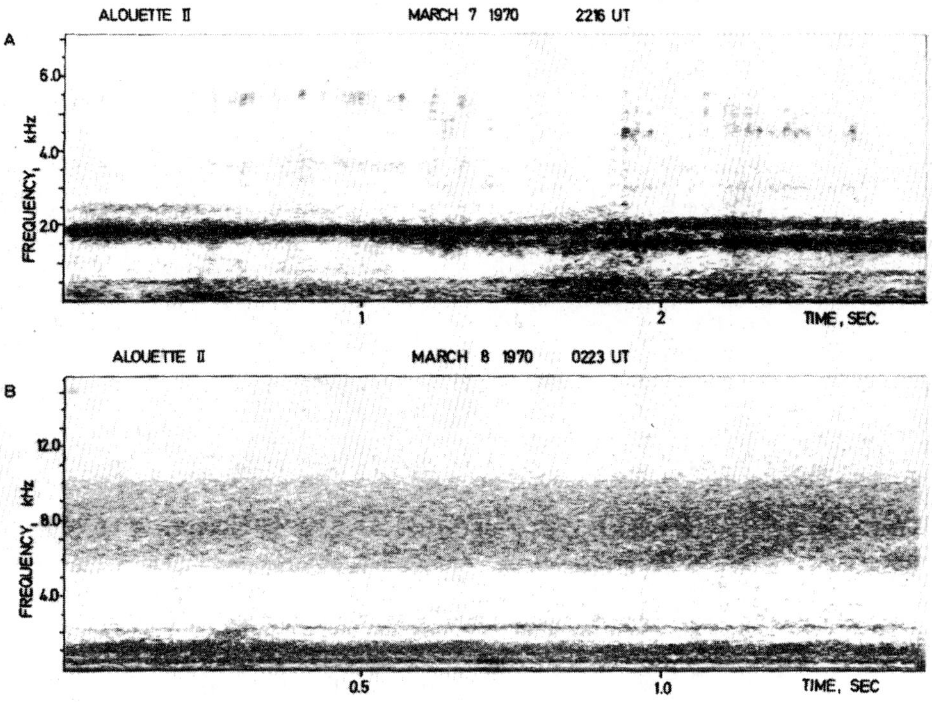

Fig. 2. Spectrograms of recordings from Alouette 2 made about 2 hr before (A) and after (B) the rocket flight. ($T_0 = 0017.37$ UT March 8th, 1970.)

it is reasonable to assume that high energy protons were more deeply involved than the auroral particles (cf. launch conditions).

The very low noise level seen by the ground based ELF-VLF equipment is in accordance with these observations. The waveguide cut off frequency was close to 3.5 kHz as compared to 1.6 kHz for normal night time conditions.

During the same period of time the Alouette data indicate several unusual characteristics. The hiss band above 5 kHz was weak. However, several strong bands between DC and 2.5 kHz were recorded with marked variations both in time and amplitude. Sometimes, a fairly uniform band may suddenly split up into two or more narrow emission bands and then be rejoined. Figure 2A shows a spectrogram from Alouette obtained during this period.

B. EXPERIMENTAL FINDINGS FROM THE ROCKET DATA

The main characteristic features of the emissions obtained from the wideband channel are listed in Table I and shown in Figure 3. Some of these features will be described in the following.

(1) A noise band between 0.8 to 2 kHz was dominant during the first part of the flight. This band has a great deal of fine structure and is partly spin modulated (cf.

TABLE I

Characteristic features of the emissions observed by the wideband channel

Type of emission			Sporadic noise	Noise band	Sporadic noise	Narrow noise band	Band ltd. noise (Partly spin mod.)	Discrete emissions (sporadic)	Hiss (spin mod.)	Spin mod. noise
Frequency range (kHz)	Center		4.5	3.8	3.0	2.5	1.5			
	Band width		0.25	0.3	0.1	0.2	1.2	$0.5 < f < 2.5$	$f > 2.5$	$f < 2.5$
Duration	From	t(s)	115	115 (Start of measurements)	115	115	115		190 280	372
		h(km)	157					Whole	209↑ 203	120
	To	t(s)	350	370	360	370	375 (400)	flight	270 375 (400)	389
		h(km)	147	123	135	123	116		208↓ 116	96
Max. intensity		t(s)	155–300	230 216	apogee	apogee	apogee	180	252 350	377 383
		h(km)	190	(apogee)					213 147	113 105

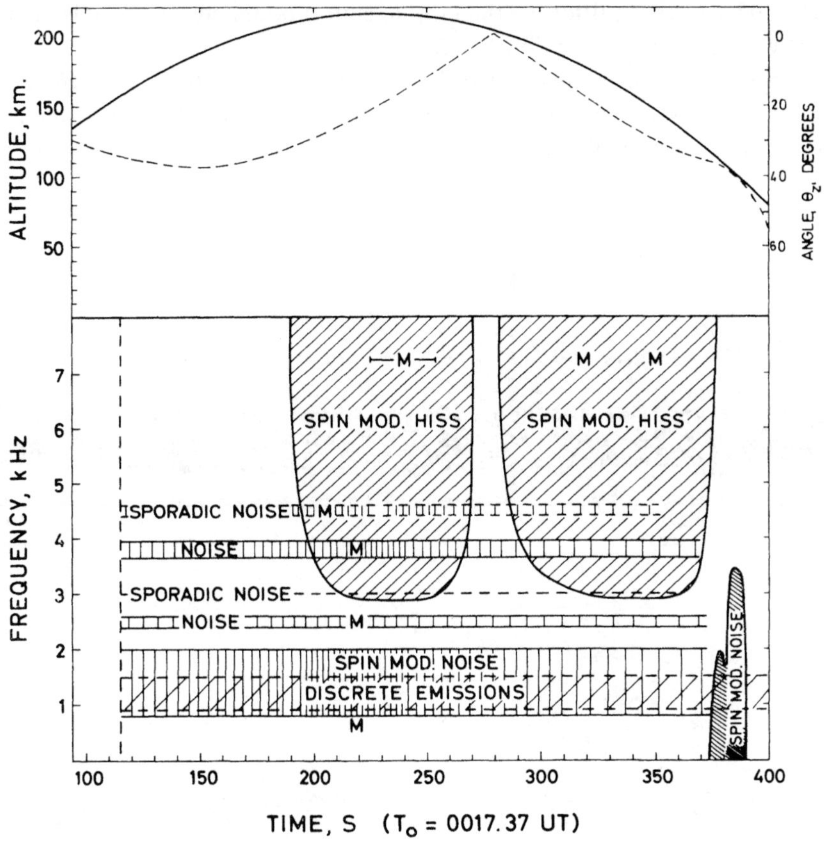

Fig. 3. Survey of the different emissions recorded by the wideband channel during the flight. M's denote points of maximum signal intensity. Altitude and angle between the rocket axis and the earth's magnetic field are also given.

Figure 4). The maximum intensity was observed around apogee, and the intensity decreased with decreasing altitude. The noise was only sporadically observed below 120 km.

(2) A few weak discrete emissions were observed below 2.5 kHz during the flight period (cf. Figure 4).

(3) Spin modulated hiss (out of phase with the noise mentioned in Point 1, cf. Figure 4) occurred above 2.5 kHz during two intervals. The first lasted from 209 km through apogee and back to 208 km on the downleg with maximum intensity occurring above 213 km. The duration of the second interval was from 203 km down to 116 km. The second burst has two maxima, one at 177 km and the other at 147 km.

(4) An interesting phenomenon quite different from the noise observed during the rest of the flight, are the emissions demonstrated in Figure 5. It consists of double-spin modulated noise bursts in the frequency range 0.1 to 4 kHz. This effect lasted from 120 km down to 96 km. It shows a marked intensity variation, with maxima at

113 and 105 km. In addition to the double-spin modulated part a noise band below 0.5 kHz with no spin modulation is present.

Examples of the spectrometer record are shown in Figure 6. Above 116 km the main part of the energy is observed in the two bands 0.1 to 1 kHz and 1 to 10 kHz and the maxima coincide with those heights marked in Figure 3. However, during some

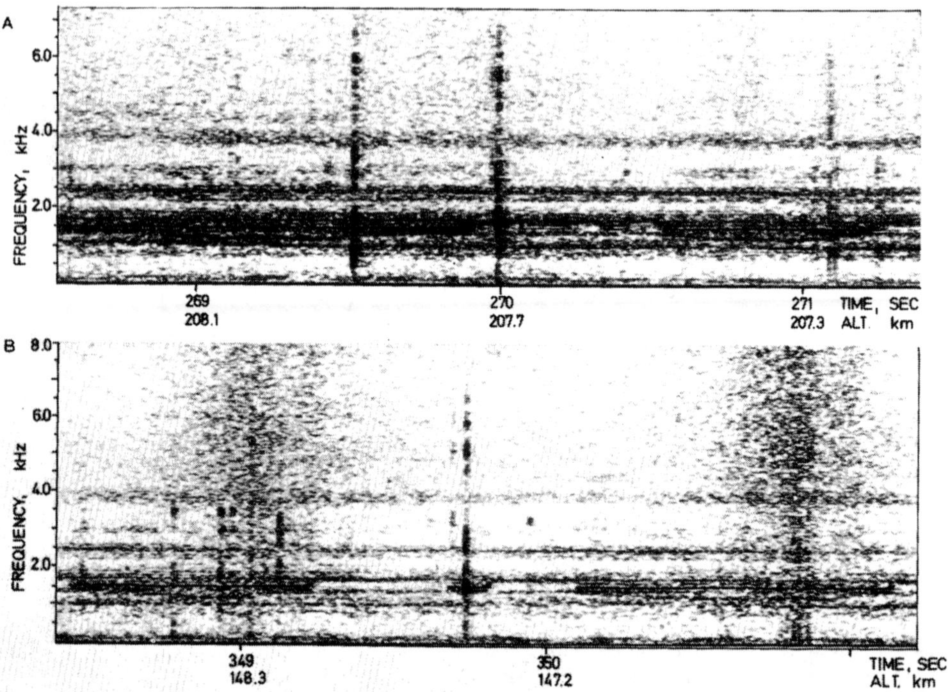

Fig. 4. Spectrograms from two different portions of the flight which include most of the features listed in Table I and Figure 3. (Note the displacement between the maxima of the spin modulated emissions in B.)

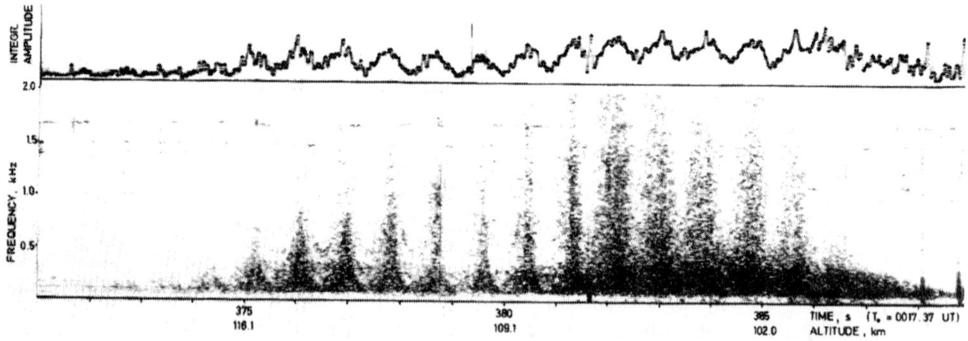

Fig. 5. Spectrogram and amplitude vs time of low frequency noise observed during the last part of the flight.

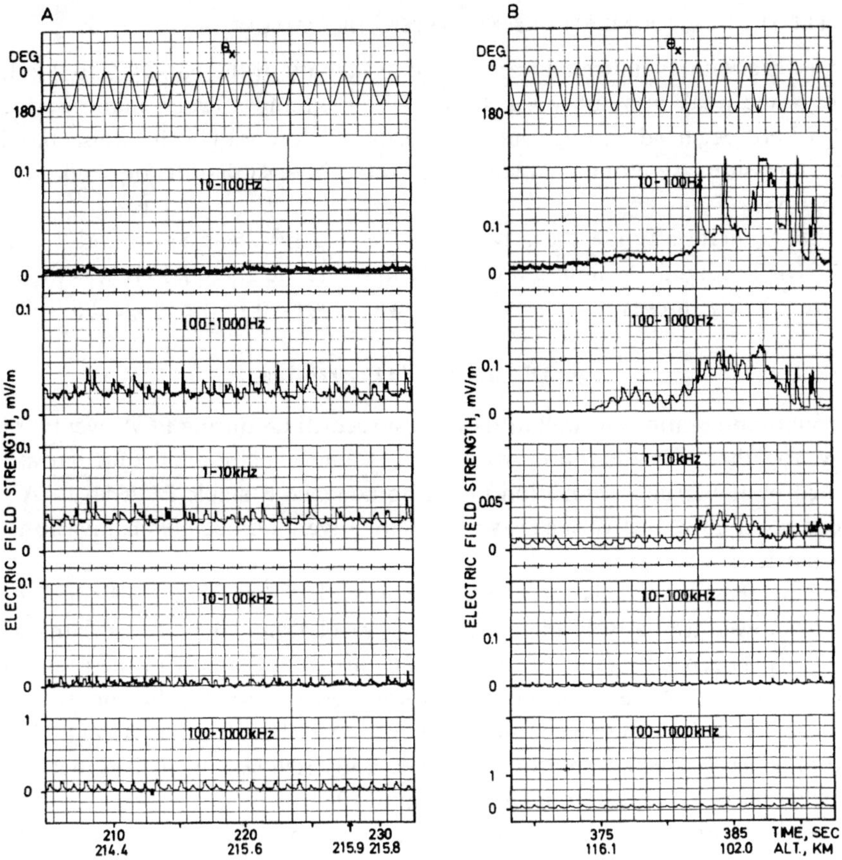

Fig. 6. Spectrometer recordings made around apogee (A) and during the time covered by the spectrogram in Figure 5 (B). The upper traces show the angle between the antennas and the magnetic meridian plane.

short burst periods the noise covers the whole frequency range from 10 Hz to 100 kHz (cf. Figure 6). Minimum intensity for all five channels is detected between 120 and 150 km. The spectrometer data obtained between 96 and 116 km are in accordance with the observations from the wideband channel. Noise bursts at twice the spin period are seen on the two channels from 0.1 to 10 kHz, while spin modulation is not found in the band between 10 to 100 Hz.

The emission activity from ground observations was still low. However, during the flight a significant increase in the noise level between 500 and 2000 Hz occurred. This noise consisted of hiss, mixed with discrete emissions. The VLF-Doppler experiments showed high electron concentration below 100 km. The photometer records indicate that a PCA glow centered at 65 km was present. The height luminosity profile shows that the auroral glow extended from 90 to 120 km, and no significant variation occurred between the ascent and descent part of the flight (Måseide, 1970) The intensity fits well with that observed from the ground.

C. SATELLITE AND GROUND RECORDINGS AFTER THE LAUNCH

Ground observations indicated an increase in activity in the band 0.5 to 2 kHz after launch. Some weak hiss above 7 kHz was observed and a broad minimum between 4 and 7 kHz occurred throughout this period. Below 4 kHz the noise bands and discrete emissions were obtained somewhat more intense than before and during the flight period.

At the Alouette altitude there was fairly strong hiss between 5 and 12 kHz (cf. Figure 2B). The band of minimum intensity in the range 2.5 to 5 kHz was the same as 2 hr before firing; however, more structure was observed in the range 0.5 to 2.5 kHz.

4. Discussion of Observations

The ground observations fit well into our past recordings during PCA events. The low activity, together with the more or less lack of fine structure and time variations, is explained by the strong absorption in the homogeneous stable PCA layer. Also, the visible auroral glow clearly suggests a weak relatively homogeneous invasion of low energy particles.

Based on these facts it is not too remarkable that the rocket and satellite borne receivers showed a relatively high degree of correlation. This also explains why the agreement between simultaneous observations on ground and with the rocket was so low. The main part of the noise above the ionosphere will be absorbed or reflected.

However, for the discrete emissions (1.5 to 0.5 kHz) there is a high degree of correspondence between both ground, rocket, and satellite data. This is in accordance with earlier measurements (Holtet and Egeland, 1969b).

Gurnett and Burns (1968) have discussed ion cutoff of ELF emissions. Their explanation is that a reflection of electromagnetic waves may occur near the point where the wave frequency approaches the two-ion cutoff below the proton gyrofrequency. If collisions are included, the angle between the wave normal and the magnetic field must exceed a critical value in order to suffer a polarization reversal (Jones, 1968). One explanation for the simultaneous occurrence of discrete emissions both above and below the ionosphere may be that these waves are strongly connected to the magnetic field lines. This indicates that the discrete emissions in the lower ELF band and the hiss often observed in connection with these (Holtet and Egeland, 1969b) are generated in different regions. While the discrete emissions probably are ducted waves generated in the equatorial region the ELF hiss has a local source. An example showing ELF hiss with cutoff at 400 Hz and discrete tones well below this frequency is presented in Figure 7.

The high frequency noise (Point 3) is of a broadband hiss character, different from the fingerprint type electrostatic VLF noise observed by Gurnett and Mosier (1969). The fact that it disappears when the antennas become perpendicular to B suggests that the noise may be electrostatic; however, that probably is coincidental. The Alouette observations would tend to support this.

The low frequency (10 to 4000 Hz) noise bursts registered on the down leg between 120 and 96 km (cf. Point 4 and Figure 5) will now be discussed. Nothing happened

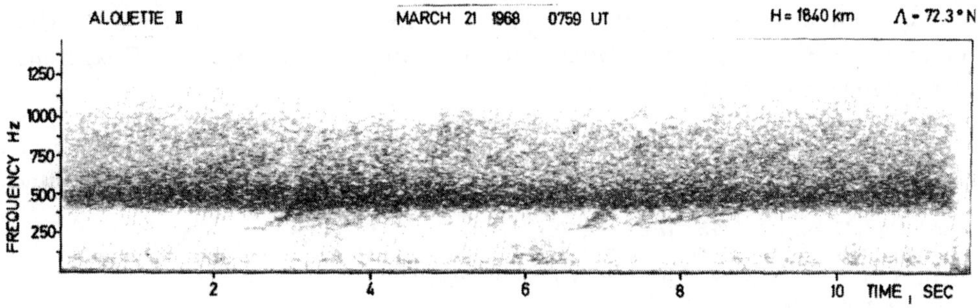

Fig. 7. Spectrogram of ELF noise recorded by Alouette 2. ELF hiss with a sharp lower cutoff frequency and discrete emissions well below this frequency can be seen.

with the payload during this period which would cause instabilities. The angle between the payload (z-axis) and the magnetic field was about 40°. Almost the same orientation occurred at 180 km on the ascent part of the flight, but no indication of similar noise was found.

One hypothesis is that this noise is propagating within the auroral layer (the E-layer) where a waveguide effect can be established by gradients in the electron density. (The rocket photometers showed that the main part of the optical emissions was located within the same height interval.)

Somewhat similar low frequency noise bursts (up to 1.5 kHz) have been observed by *E* field measurements onboard rockets in a few other cases (cf., Gurnett and Mosier, 1969). Their interpretation is that low frequency electrostatic noise may be generated by an interaction between the rocket body and the surrounding plasma. They suggest a wake effect or a plasma instability caused by the currents produced from the $\mathbf{V} \times \mathbf{B}$ potential gradient of the conducting rocket skin.

Some important differences between our results and those just mentioned should be stressed. First, our antennas are much longer, and the inner half is electrically insulated from the plasma increasing the distance of the active elements from the proximity of the vehicle. Gurnett and Mosier's (1969) data were recorded at marked higher altitudes than in this case. Our data show a double spin-modulation while the earlier results followed the precession periods. The main difference is probably the continuous noise band below 0.5 kHz (cf. Figure 5). Also differences exist in the fine structure and time variations.

One noted some similarities between the continuous part and the irregularity signals observed on OGO 6 and OV1 10 which were interpreted as probably spatial in origin (Maynard and Heppner, 1970). If these are spatial in nature the structure is significantly finer as the velocity of the rocket is about a factor of 5 less.

Acknowledgments

This project has been sponsored in part by the Royal Norwegian Council for Scientific and Industrial Research.

We would also like to express our graditude to Dr R. E. Barrington at the Communications Research Centre, Ottawa, for permission to read off the VLF experiment in the Alouette 2 satellite.

References

Aggson, T.: 1969, in *Atmospheric Emissions* (ed. by B. M. McCormac and A. Omholt), Van Nostrand Reinhold Company, New York, p. 305.

Barrington, R. E.: 1969, *Proc. IEEE* **57**, 1036.

Gurnett, D. A. and Burns, T. B.: 1968, *J. Geophys. Res.* **73**, 7437.

Gurnett, D. A. and Mosier, S. R.: 1969, *J. Geophys. Res.* **74**, 3979.

Holtet, J. and Egeland, A.: 1969a, *Phys. Norv.* **3**, 223.

Holtet, J. and Egeland, A.: 1969b, in *Atmospheric Emissions* (ed. by B. M. McCormac and A. Omholt), Van Nostrand Reinhold Company, New York, p. 175.

Jones, D.: 1968, in *Plasma Waves in Space and Laboratory*, Vol. 2 (ed. by J. O. Thomas and B. J. Landmark), Edinburgh University Press.

Maynard, N. C. and Heppner, J. P.: 1970, in *Particles and Fields in the Magnetosphere* (ed. by B. M. McCormac), D. Reidel Publishing Company, Dordrecht, Holland, p. 247.

Måseide, K.: 1970, private communication.

PART VII

AURORAL MORPHOLOGY

POLAR MAGNETIC SUBSTORMS

ROLF BOSTRÖM
Dept. of Plasma Physics, Royal Institute of Technology, Stockholm, Sweden

1. Substorm Current Systems

The polar magnetic substorm is one of several manifestations of the magnetospheric substorm. It is a magnetic perturbation with a peak magnitude of a few hundred gammas and a duration of about 1 hr. At times of high magnetic activity the substorms become more intense, occur frequently, and a sequence of substorms may build up a magnetic storm.

The global distribution of the magnetic perturbation field during a substorm may be described by means of an equivalent current system to be discussed later. The geometry of the real *physical* current system is as yet unknown, but it certainly must include the auroral electrojets in the ionosphere and Birkeland currents flowing along the geomagnetic field lines connecting the electrojets to the source regions in the outer magnetosphere. Figure 1 shows one simple model. The field computed for models of this kind shows good agreement with the field observed during isolated substorms (Birkeland, 1908, 1913; Boström, 1968, 1969; Akasofu and Meng, 1969; Bonnevier *et al.*, 1970). In fact the magnetic perturbation field from a model of this kind is equivalent to that of the classical purely horizontal ionospheric system (Birkeland, 1908, 1913; Fukushima, 1969).

In a model where the ionosphere is homogeneously conducting the field aligned Birkeland currents will be coupled only to the Pedersen currents in the ionosphere. Together these currents produce a magnetic disturbance field mainly above the ionosphere. In such a model the ground level magnetic perturbation field is produced almost solely by horizontal Hall currents circulating in the ionosphere. However, in a model similar to the one shown in Figure 1 where the ionospheric conductivity is assumed to be enhanced along the auroral oval, the Birkeland and Pedersen currents give a significant contribution to the ground level geomagnetic effect.

The current system shown in Figure 1 represents the *change* in the magnetospheric current system at the substorm onset. Before the substorm onset a *westward* current flows in the outer magnetosphere on the nightside. This westward current may be a segment of a symmetric ring current (Akasofu and Meng, 1969) or of the current in the neutral sheet. At the substorm onset a disturbance occurs in this part of the magnetosphere and a discharge to the ionosphere takes place. The magnetospheric westward current is then redirected and flows through the ionosphere. To represent this change of the state of the magnetosphere we must add the current system of Figure 1 to any already existing currents. Thus, in Figure 1, the eastward current segment in the outer magnetosphere does not represent the real currents flowing there. Initially the substorm current system has a short eastwest extent and thus the circuit has a small inductance and short time constant. During the expansive phase the upward and

B. M. McCormac (ed.), The Radiating Atmosphere, 357–365. *All Rights Reserved.*

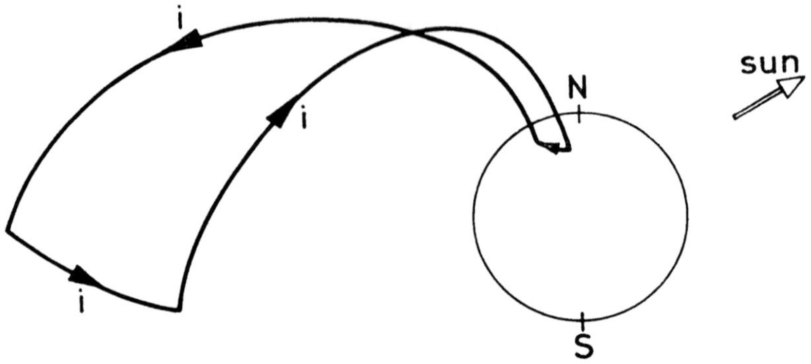

Fig. 1. Three dimensional substorm current system.

downward flowing Birkeland currents separate and the westward electrojet in the ionosphere extends in longitude, as seen from ground based observations.

Several observations indicate that changes occur in the state of the magnetosphere during some hours before the substorm onset. Pudovkin *et al.* (1968) report low latitude magnetic perturbations preceding the substorm onset. Coleman and McPherron (1970) find a general depression of the H field component on the nightside at the ATS orbit before the substorm followed by a recovery at the time of the substorm. Hones (1970) reports a thinning of the tail plasma sheet during the development of a substorm and Carpenter and Stone (1967) find an inward drift of plasma around the plasmasphere before the substorm.

A magnetospheric current system is built up before the substorm onset and at the onset a discharge to the ionosphere occurs which changes the geometry of the current system.

Field aligned Birkeland currents play an important role in the substorm phenomenon. Current densities of about 5×10^{-5} amp m^{-2} have been inferred from rocket and satellite magnetometer measurements (Armstrong and Zmuda, 1970; Cloutier *et al.*, 1970). There is evidence that the upward flowing current may be carried by the precipitating keV auroral electrons.

Indirect evidence for Birkeland currents may be found in Ba cloud electric field measurements over the polar cap (Haerendel and Lüst, 1970; Wescott *et al.*, 1970). The current that would be produced by the local electric field could not explain the magnetic perturbations seen in the polar cap. The perturbations are produced by currents flowing somewhere else, for example, along the field lines. In the magnetosphere the ATS magnetometer has recorded *D* component spikes that may be caused by field aligned currents during the expansive phase of substorms (Coleman and McPherron, 1970).

2. Space Charge Regions Above the Aurora

If the current density of the Birkeland currents exceeds a critical limit instabilities may occur which drastically decrease the effective conductivity along the field lines. The two-stream instability has been suggested (Swift, 1965) but another possibility which often occurs in laboratory plasmas is the formation of a space charge region. This has a high potential drop over a very short distance. The electric fields associated with the space charge region will accelerate auroral electrons and give rise to horizontal motions of the visual aurora (Carlqvist and Boström, 1970).

Space charge regions are expected to form when the drift velocity of the electrons exceed the thermal velocity. The critical current density estimated for the upper ionosphere, 3×10^{-5} amp m^{-2}, is of the same order as current densities inferred from rocket and satellite experiments.

It has been suggested (Alfvén and Carlqvist, 1967; Carlqvist, 1969) that the space charge regions may account for solar flares, and Akasofu (1969) suggested that the same mechanism might occur on the field lines to the auroral oval. Magnetic energy stored in the magnetospheric current circuit would then be released in the space charge region. Due to the high inductance of the pre-substorm circuit this acts as a constant current source and where the circuit is broken the energy is released. Properties of space charge regions have also been discussed by Block (1969).

Figure 2 shows a schematic cross section of the proposed space charge region. The relation between the voltage drop across the space charge region and the current density is for a one dimensional model given by a relation similar to that for a vacuum

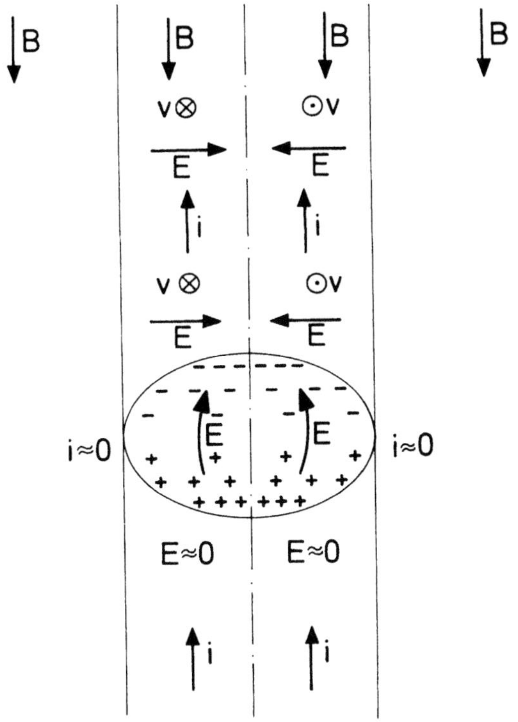

Fig. 2. Schematic cross section of the space charge region.

diode. In Figure 2 the vertical voltage drop is maximum in the center of the space charge region and zero outside of it; thus, the potential varies along the upper (cathode) boundary. Since the magnetic field lines may be considered to be equipotential lines outside of the space charge region horizontal electric fields must exist in a column above the space charge region. No intense fields are formed below the space charge region since the lower ionosphere with high Pedersen conductivity would short circuit such fields. Provided the vertical field in the space charge region is directed so it accelerates auroral electrons downward, the horizontal field will be directed from the boundaries toward the center of the column above the space charge region. These fields will set the plasma into a motion that may be seen as drifts of irregularities in visual auroral forms. For a filamentary field aligned current structure we get a rotational motion in a clockwise sense viewed in the direction of the magnetic field, while for a sheet structure we get an eastward streaming on the poleward side of the auroral form and westward on the equatorward side. Such motions with the predicted direction have been observed in TV auroral pictures (Carlqvist and Boström, 1970; Hallinan and Davis, 1970; Davis *et al.* 1971). The velocities were from 4 to 20 km sec^{-1}, corresponding to electric fields of 0.2 to 1 v m^{-1}, over regions about 10 km wide. Thus, we have a voltage drop of about 1 to 5 kV between the center and

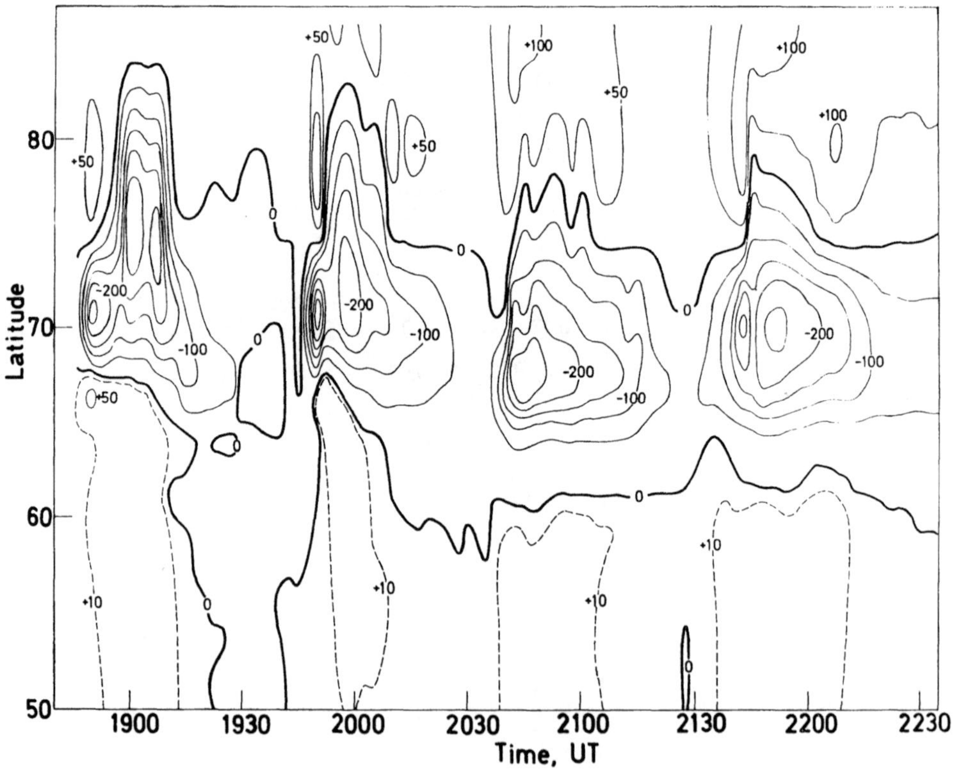

Fig. 3a.

outer boundary, and a similar voltage drop along the field lines in the space charge region.

Provided we interpret the auroral motions as electric field drifts it is necessary to introduce a mechanism by which the intense fields are prevented from penetrating into the *E* layer. This can only be done by introducing a voltage drop along the magnetic field lines, possibly in the form of a space charge region. The magnitude of the voltage drop estimated, 1 to 5 kV, falls into the range of typical energies for the monoenergetic peaks which often are seen in auroral electron fluxes. Thus the space charge region might, in addition to giving rise to the motions, also contribute to the acceleration of auroral particles.

3. Temporal Development of Substorms

For a complete description of the magnetic perturbation field observed on the ground during a substorm the three field components *H*, *D*, and *Z* should be given as function of time, latitude, and longitude. It is not possible to show the dependence on all three variables in one figure, but various plots where one or two of the variables are kept constant may be used (Zaitzev and Boström, 1971).

Figure 3 shows isointensity contours for the *H*, *D*, and *Z* components in a latitude-

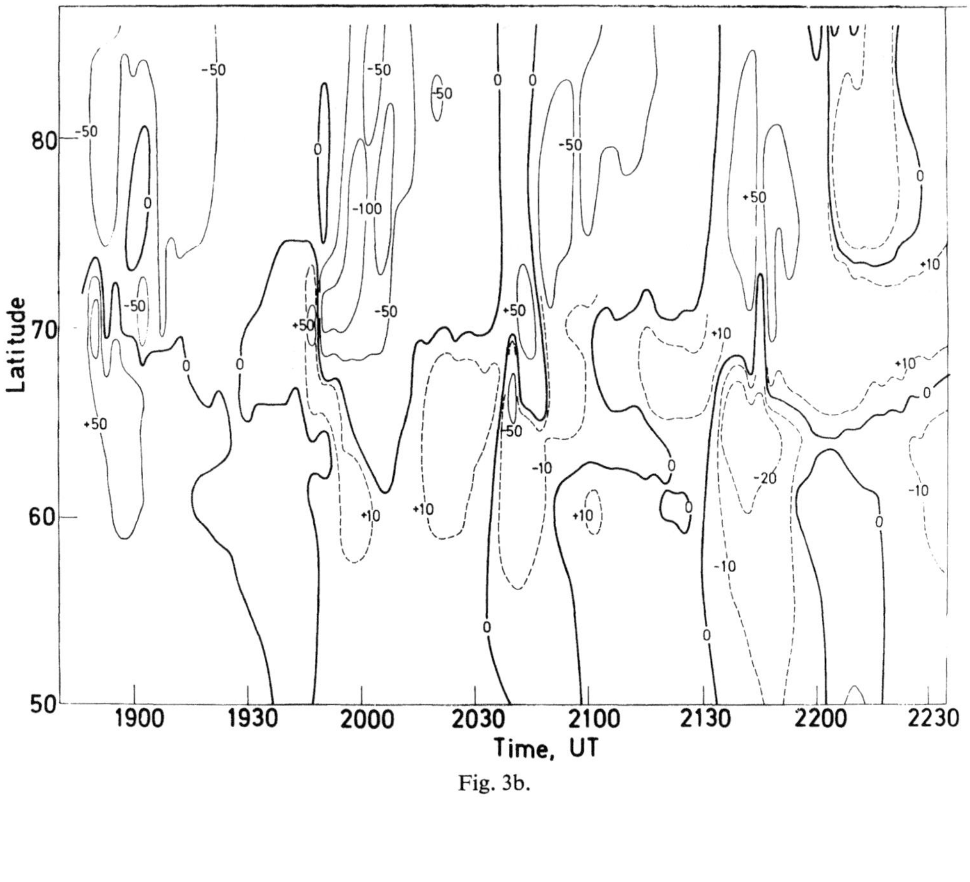

Fig. 3b.

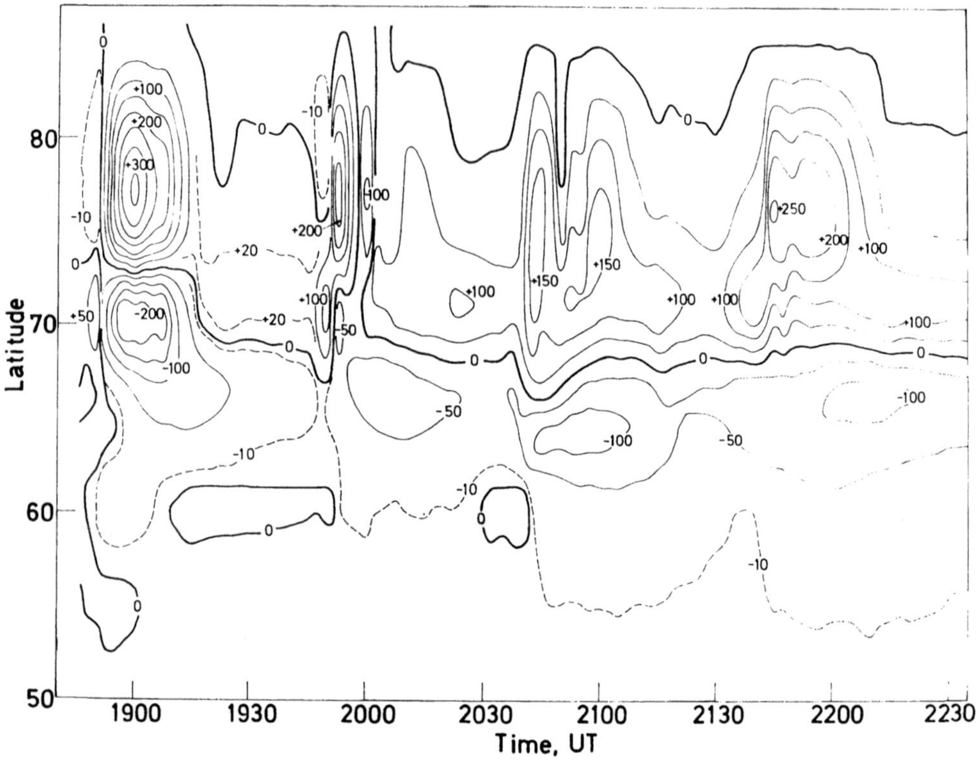

Fig. 3a–c. Isointensity contours for the three field components for a sequence of 4 substorms on September 21, 1965, observed along a N–S chain of stations. (a) Northward *H* component. (b) Eastward *D* component. (c) Downward *Z* component.

time diagram. The longitude is constant; recordings along a N-S chain of stations through the Scandinavian and Soviet countries have been used to construct the plot. The plot shows the development and decay of four substorms. One of the most noticeable features in the plot of the *H* component is the rapid northward expansions of the region of negative *H* component, indicating a broadening of the electrojet region. The real electrojet is not as broad as the region of negative *H* perturbation, but when this region expands the electrojet expands northward. Positive *H* component peaks north and south of the electrojet are most pronounced in the early phase of the substorm. These peaks are caused by the field aligned Birkeland currents at the eastern and western ends of the westward electrojet. In the later phase of the substorm, when the electrojet is extended in longitude, the Birkeland currents are further away from the N-S chain of stations and give a smaller contribution. The latitude of the peak of the negative *H* component varies with time and is lowest around 2100 UT, which is close to local magnetic midnight. This shows that the electrojet flows along the auroral oval and not at constant latitude. The vertical *Z* component is positive north of the center of the electrojet and negative to the south. The N-S motions of the center of the electrojet are not as pronounced as the northward expansion of the northern boundary seen in the Z component as well as in the H component. The D component was generally small indicating that the chain of stations was close to the center line of the current system.

Figure 4 shows in a polar plot isointensity contours for the scalar magnetic potential for various phases of a substorm. The contours are perpendicular everywhere to the horizontal component of the magnetic perturbation vector at ground level and the separation between the contours is inversely proportional to the magnitude of the horizontal perturbation. These curves are generally referred to as equivalent overhead current systems, because a purely horizontal ionospheric current system of this form would give a perturbation field similar to the one observed. Plots of this kind may be

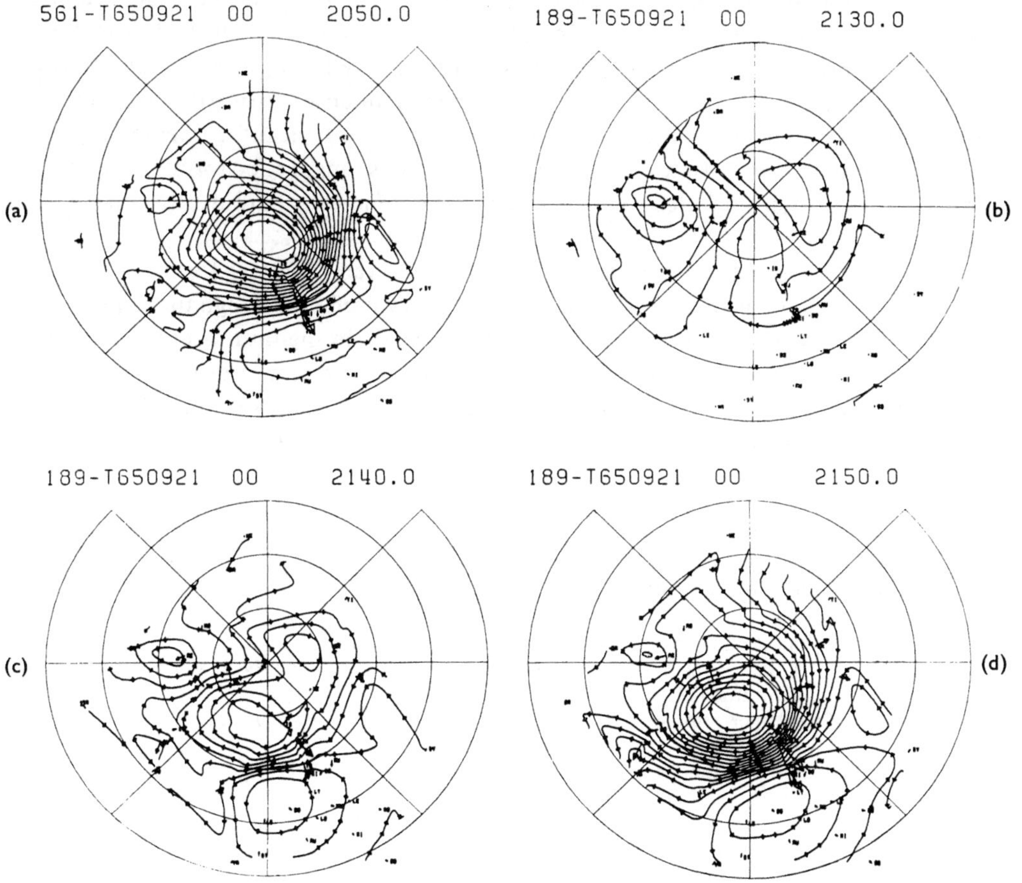

Fig. 4. Polar plots of isointensity contours for the scalar magnetic potential ('equivalent current system'). The frame of reference is geographic with 10° between latitude circles and the Greenwich meridian downwards. The UT is given at the top. The distance between isointensity contours is equal to the length of one latitude degree when the horizontal perturbation is 100 γ (about 18 kamp per equivalent current line). (a) Peak phase. (b) Weak current system between two substorms. (c) Electrojet starts to develop again. (d) Peak phase of new substorm.

used to represent both the H and D component at the same time as a function of latitude and longitude for constant time. The plots contain no observational information about the vertical Z component and should in general be supplemented by plots showing this component.

A computer program has been developed for automatic evaluation of the plots and has been used in the preparation of a movie showing the temporal development of the substorm. Figure 4 shows some frames of the movie. No smoothing of the data is performed in the program, so the contours are forced to be exactly perpendicular to the observed field at the 36 observatories used. For this reason, and due to the uneven distribution of the observatories, the contours are somewhat irregular.

In the peak phase of the substorm the current system has one cell and a westward equivalent current flows around the auroral oval. The current is most intense (electrojet) in the midnight sector. In the early phase of the substorm the westward electrojet has only a short E-W extent. In between the substorms, or in the last phase of the substorms, weak current systems with two or more cells may exist. In this study of isolated substorms no indication of any confined intense eastward electrojet was found. However, the eastward jet may possibly be more typical for intense magnetic perturbations (storms). In a recent study of an intense magnetic storm Grafe (1969) found that the eastward electrojet was dominant sometimes and at other times was absent.

Obviously the substorm current system may have very different forms. A concept like the 'typical' or 'average' system is not really very useful, although the appropriate form of this has been discussed over the last few years. In the future emphasis should be laid more on the study of individual events and the temporal development of these, including studies of possible magnetic effects of precursors to the substorm and the spatial and temporal development and decay of the electrojets.

References

Akasofu, S.-I.: 1969, *Nature* **221**, 1020.

Akasofu. S.-I. and Meng, C.-I.: 1969, *J. Geophys. Res.* **74**, 293.

Alfvén, H. and Carlqvist, P.: 1967, *Solar Phys.* **1**, 221.

Armstrong, J. C. and Zmuda, A. J.: 1970, *J. Geophys. Res.* **75**, 7122.

Birkeland, K.: 1908, 1913, *The Norwegian Aurora Polaris Expedition 1962–1963*, Vol. I, First Section (1908) and Second Section (1913) Christiania.

Block, L. P.: 1969, Paper presented at *Ninth Int. Conf. on Phenomena in Ionized Gases*, Bucharest, Rumania.

Bonnevier, B., Boström, R., and Rostoker, G.: 1970, *J. Geophys. Res.* **75**, 107.

Boström, R.: 1968, *Ann. Geophys.* **24**, 681.

Boström, R.: 1969, in *Atmospheric Emissions* (ed. by B. M. McCormac and A. Omholt), Van Nostrand Reinhold Company, New York, p. 277.

Carlqvist, P.: 1969, *Solar Phys.* **7**, 377.

Carlqvist, P. and Boström, R.: 1970, *J. Geophys. Res.* **75**, 7140.

Carpenter, D. L. and Stone, K.: 1967, *Planet Space Sci.* **15**, 395.

Cloutier, P. A., Anderson, H. R., Park, R. J., Vondrak, R. R., Spiger, R. J., and Sandel, B. R.: 1970, *J. Geophys. Res.* **75**, 2595.

Coleman, P. J. Jr. and McPherron, R. L.: 1970, in *Particles and Fields in the Magnetosphere* (ed. by B. M. McCormac), D. Reidel Publishing Company, Dordrecht, Holland, p. 171.

Davis, T. N., Hallinan, T. J., and Stenbaek-Nielsen, H. C.: 1971, this volume, p. 160.
Fukushima, N.: 1969, *Rep. Ionos. Space Res. Japan* **23**, 219.
Grafe, A.: 1969, *Planetary Space Sci.* **17**, 1395.
Haerendel, G. and Lüst, R.: 1970, in *Particles and Fields in the Magnetosphere* (ed. by B. M. McCormac), D. Reidel Publishing Company, Dordrecht, Holland, p. 213.
Hallinan, T. J. and Davis, T. N.: 1970, *Planetary Space Sci.* **18**, 1735.
Hones, E. W., Jr.: 1970. in *Particles and Fields in the Magnetosphere* (ed. by B. M. McCormac), D. Reidel Publishing Company, Dordrecht, Holland, p. 24.
Pudovkin, M. I., Shumilov, O. I., and Zaitzeva, S. A.: 1968, *Planetary Space Sci.* **16**, 881.
Swift, D. W.: 1965, *J. Geophys. Res.* **70**, 3061.
Wescott, E. M., Stolarik, J. D., and Heppner, J. P.: 1970, in *Particles and Fields in the Magnetosphere* (ed. by B. M. McCormac), D. Reidel Publishing Company, Dordrecht, Holland, p. 229.
Zaitzev, A. N. and Boström, R.: 1971, *Planetary Space Sci.*, to be published.

INSTITUTE DISCUSSION

The location of the space charge region with respect to the auroral form is uncertain; it could be 1000 km or higher, even on the equatorial plane. A. Hasegawa stated that the two stream instability may give rise to a space charge region. Proton precipitation is broader and more diffuse indicating that there is less space charge.

G. Lange-Hesse pointed out that enhanced magnetic activity and radio backscatter show an afternoon auroral electrojet. R. Boström said that it was much weaker than the westward electrojet and that he had not investigated the afternoon jet.

The Rice University vector magnetometer experiment was summarized by S. Mende. The magnetic effect showed field aligned currents; however, the vector directions indicated that the electrojet current had to be a Pedersen current to satisfy the Rice model. Most electrojets are believed to be carried by Hall currents.

B. McCormac asked if there were optical emissions that might appear different at the ends of the ionospheric currents, but there were no suggestions.

THE PROTON AURORA AND AURORAL SUBSTORM

L. E. MONTBRIAND

National Radio Propagation Laboratory, Communications Research Centre, Ottawa, Canada

Abstract. The substorm model proposed by Akasofu has been examined in detail. The basic features of the model were verified but certain refinements are necessary, of which the more important include classifying substorms according to intensity or magnitude and incorporation of features such as N-S segments, subsidiary substorms, and isolated segments of H aurora. Most behaviors of auroral H emissions can be explained in terms of these different magnitudes of the substorm.

1. Introduction

The auroral substorm model proposed by Akasofu in 1964 has been examined in detail using spectrograms, all-sky camera photographs, and magnetograms obtained from a chain of four stations straddling the auroral zone in West-Central Canada. Though all of the basic features of the model were verified, certain additional features, not previously reported in the open literature, were found. The proton aurora substorm can be described by this model when these other features are added to it.

Both the proton auroral substorm and the auroral substorm are described, and are depicted on diagrams similar to that used by Akasofu (1964) to describe his auroral substorm. Diagrams presented here cover five different magnitudes or intensities of the substorm. They have been prepared from data recorded mainly during sunspot minimum (1963–5) although some diagrams have also been prepared from data recorded during the sunspot maximum (1957–9). Use was made of over 75000 spectrograms and approximately 2 million all-sky camera photographs.

2. Presentation of Results

A. GENERAL COMMENTS

The identification of the onset of a substorm is easiest near midnight where the onsets are characterized by a brightening of one of the arcs lying along the auroral oval, the onset of negative magnetic bays at stations located directly below this arc, and the formation of the polar bulge. With the aid of all-sky camera photographs, which can cover over 90 min in time, one can identify substorm onsets at times well removed from midnight. By extensive use of such photographs it was possible to study the morphology of substorms throughout most of the night.

B. CONSTRUCTION OF SUBSTORM DIAGRAMS

The polar coordinates used for the substorm diagrams in Figures 1, 2, 4 and 5 differ somewhat from those presented by Akasofu (1964, 1965) in that corrected geomagnetic latitudes (φ_c) of Hakura (1965) and corrected geomagnetic time (CGT) of Montbriand (1970) are used instead of geomagnetic latitude and LT.

B. M. McCormac (ed.), The Radiating Atmosphere, 366–373.

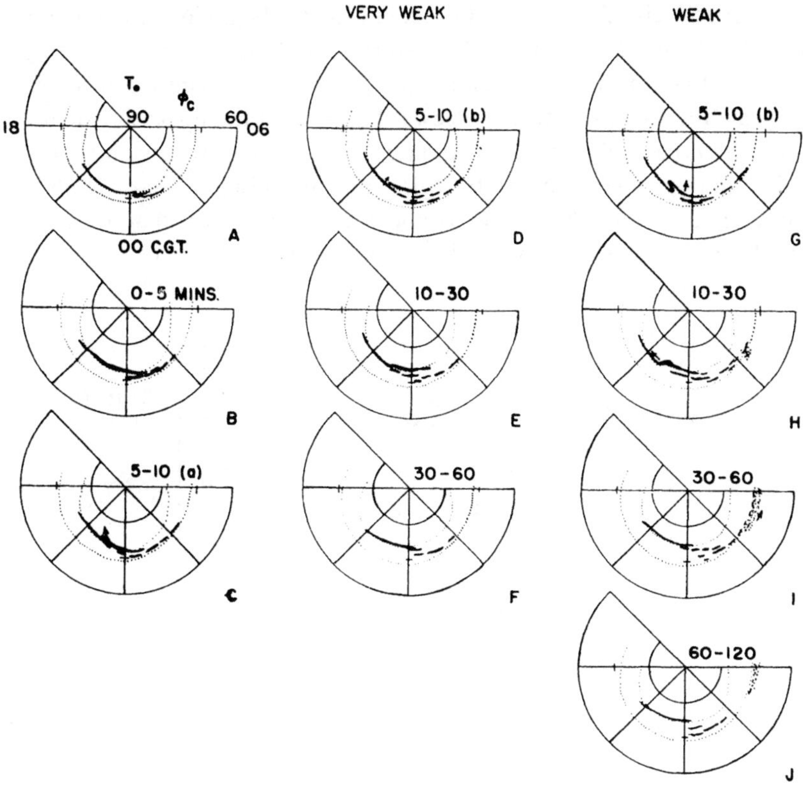

Fig. 1. The very weak (ABCDEF) and weak (ABCGHIJ) auroral and proton auroral substorms.

These coordinates were used because they provide a high order approximation of geomagnetic coordinates and time, the degree of accuracy now being demanded of such presentations. The dotted curves on the various figures represent the boundaries of the region of auroral H emissions as deduced from the spectrograms, and the heavily drawn curves and broken lines of various shapes depict auroral forms as recorded by the all-sky camera. The *a* and *b* indicate successive progressions. Where shown, arrows indicate the direction of movement of the auroral forms. The substorms have been classified according to magnitude or intensity into very weak, weak, moderate, intense and very intense. Although this classification is based mainly on the magnetic K_p index, it is laid out so that one or more significant substorm feature is added with each increasing magnitude. It must be emphasized that the diagrams do not refer to a single substorm event, but each sequence is an average representation of a given magnitude of the substorm.

C. VERY WEAK AND WEAK SUBSTORMS

The weakest magnitudes or intensities of the substorm are depicted in Figure 1. The very weak substorm is comprised of diagrams A to F, and the weak substorm of diagrams A to C and G to J. These would correspond to a K_p index of 0_0, 0_+, and

0_+, $1-$, respectively. These and all other substorms begin in the manner proposed by Akasofu (1964). At the onset of all substorms the H region in the midnight sector is more narrow than in the early evening and late morning. During the expansive phase of very weak and weak substorms poleward motion of the polar activated band (PAB) that rings the polar bulge is minor, the size of the westward traveling surge (WTS) is small, and the expansion of the H region in the midnight sector is also small. The main difference between the very weak and weak substorms is that minor patchy forms occur in the late morning during the latter, but not in the former.

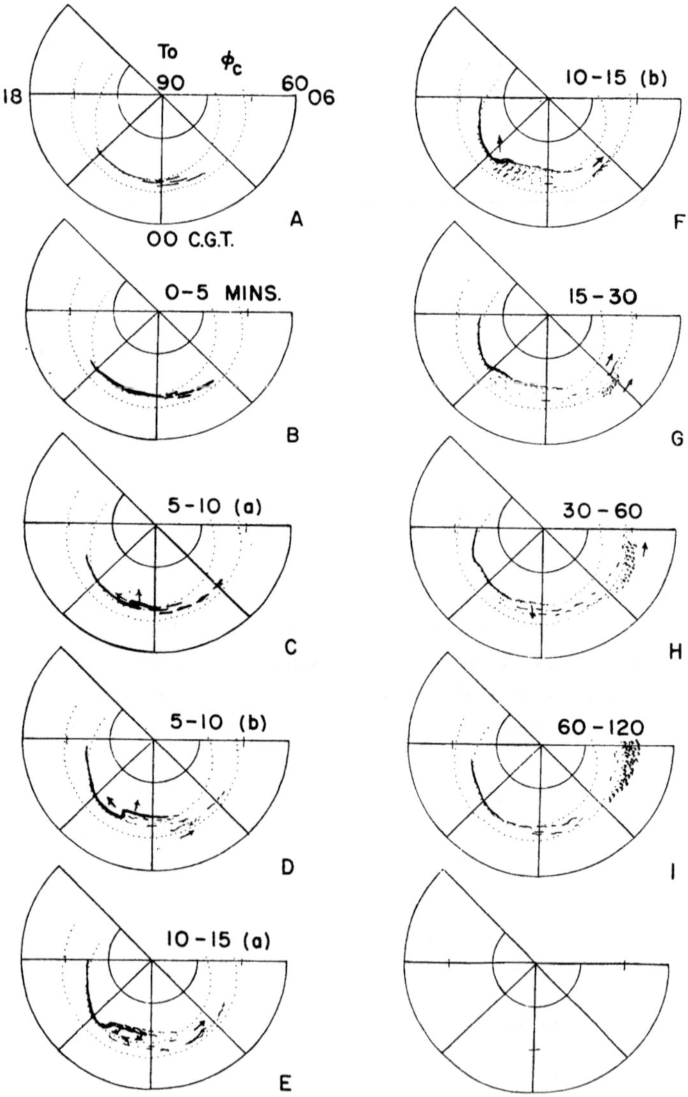

Fig. 2. The moderate auroral and proton auroral substorm.

D. MODERATE SUBSTORMS

The morphology of moderate substorms is illustrated in Figure 2. This would correspond to a $K_p = 1$, 2 or 3. The first 10 min after the onset are similar to those observed for weak substorms except that the auroral forms cover a somewhat larger area. Between 10 to 15 min after the onset a new substorm feature, which will be called N-S oriented forms, is observed. These forms are drawn in detail on Figure 3. As this bulge expands polewards, minor loops that have a clockwise rotation form just to

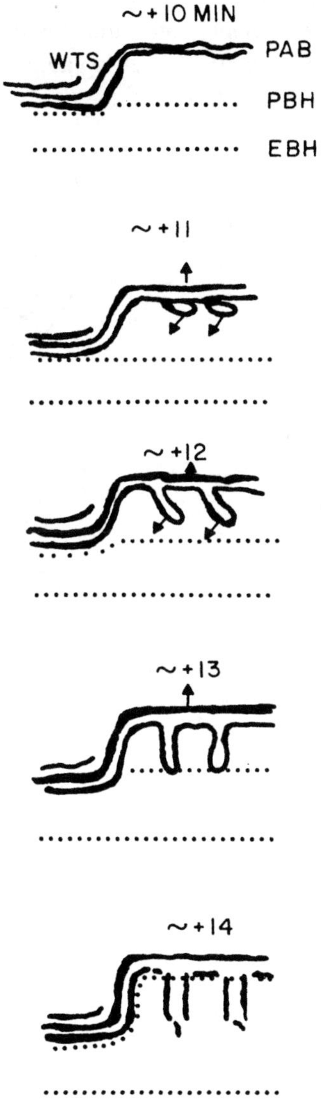

Fig. 3. The auroral and proton auroral substorm during the occurrence of N-S oriented segments. The PBH and EBH are the polar and equatorial boundaries of the H region.

the south of the PAB. These loops eventually degenerate into N-S segments. During their degeneration the polar boundary of the H region (PBH) moves rapidly poleward. In the neighborhood of the WTS the H region is essentially stationary and continues to remain equatorward of the WTS. After the sequence of events shown in Figure 3 the WTS slowly disappears, and patchy and pulsating forms begin to appear in the morning.

E. INTENSE AND VERY INTENSE SUBSTORMS

The expansive phase for intense substorms is shown in Figure 4. Substorms of this magnitude would occur for $K_p = 3$, 4, 5, or 6. Although this phase, in most respects, is similar to those of moderate intensity some additional effects are observed:

(1) Between 10 and 15 min after the onset isolated patches of hydrogenic auroral

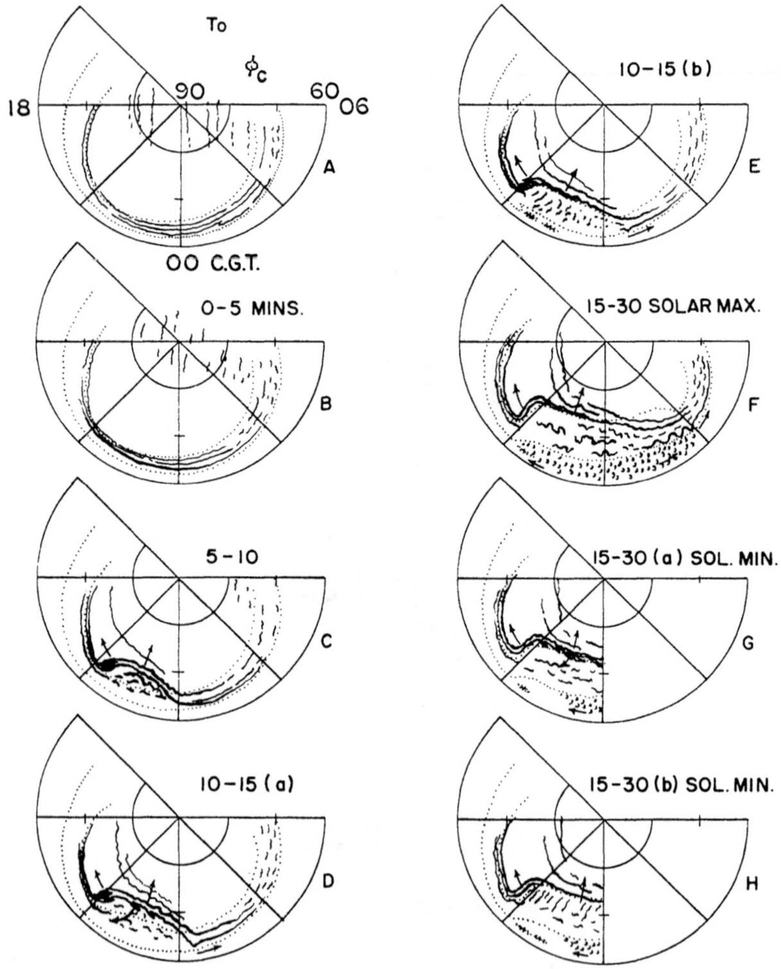

Fig. 4. The expensive phase of the intense auroral and proton auroral substorm.

forms begin to break away from the equatorial boundary of the H region near the location of the westward traveling surge.

(2) North-south oriented forms occur even though the H region is already in an expanded state. These occur during what will be called subsidiary substorms. The first occurrence of a subsidiary substorm is usually between 15 to 30 min after the substorm onset. They can reoccur a number of times during a substorm. On December 23/24 1964, no less than 11 such occurrences were observed between 2245 and 0119 hr, 90° west meridian time.

(3) Upon formation of the polar bulge a small crescent shaped form appears at its eastern edge. This form is too small to show on the diagrams but is located at the eastern edge of the sinusoidally shaped 'omega band'. It would lie on the 0300 hr line in diagram F. This small form moves rapidly eastward and with it moves the eastern edge of the expanded H region, and the eastern limit of both the omega band and the eastward drifting patchy and pulsating forms. This form seems to be the eastern counterpart of the westward traveling surge.

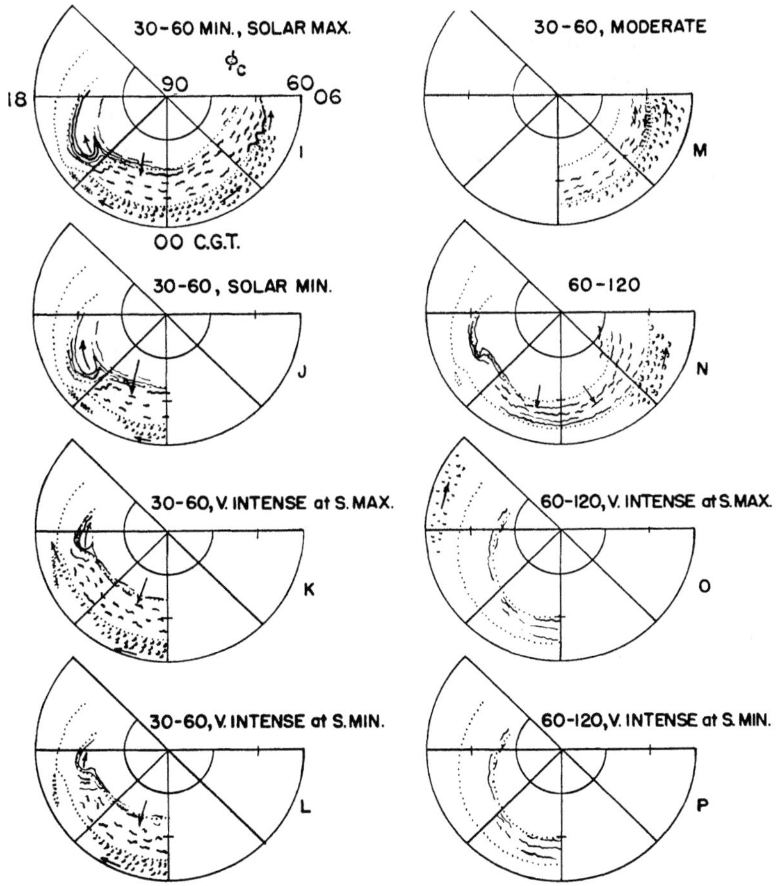

Fig. 5. The recovery phase of the intense auroral and proton auroral substorm.

The recovery phase for intense and very intense substorms is shown in Figure 5. The loop system described by Akasofu *et al.* (1965) is shown at 30–60 min after the onset. Little or no hydrogen emissions are present within this system, and the H region located equatorward of it is in a narrowed condition. The omega band is shown with the H region equatorward of its troughs. This is the usual case, but the equatorial boundary of the H region can also coincide with the sinusoidal shaped edge of the omega band. A significant feature that occurs during the recovery phase of very intense substorms, for which $K_p > 5$, is the occurrence of rather faint patches of luminosity (somewhat like a cirrus cloud) at about 60 to 120 min after the onset. This feature appears to be associated with early evening auroral absorption events at lower latitudes. There are no H emissions associated with this feature. The rest of the recovery phase is similar to those for substorms of moderate intensity.

Substorm diagrams depicting the difference between solar minimum and maximum are also included in Figures 4 and 5. The main differences are that the patchy and pulsating surfaces extend as far as 2200 CGT during the solar minimum and to 2000 CGT during solar maximum, and the patches resembling cirrus clouds occur only during solar maximum.

There has been some doubt as to whether H emission is present or absent in pulsating surfaces (Eather and Jacka, 1966; Eather, 1968; Wiens and Vallance Jones, 1969). Two types of pulsating surfaces have been identified in this study, a less common westward drifting H type, and a more common eastward drifting non-hydrogenic type. When both occur at the same time the boundary of the H region is located at their shearing interface. This is depicted in Figure 5, diagram M.

F. INTENSITY CHANGES OF AURORAL HYDROGEN

It is not possible to illustrate the intensity changes of the H emission on the substorm diagrams. However, during moderate, intense, and very intense substorms the H emission intensity in the area occupied by the polar bulge at about 10 to 15 min after the onset, increases by a factor ranging between 1 to 4 from the time of onset to 30 min after the onset when it reaches a maximum and then decreases slowly in the same area to the pre-substorm intensity.

3. Conclusion

This study of substorms has revealed several new features that should be added to Akasofu's model. By classifying substorms according to magnitude or intensity it is possible to add new features such as N-S oriented forms, subsidiary substorms and isolated segments of hydrogen aurora, as well as the boundaries of the H emission region. The proton auroral substorm can be satisfactorily described with this model.

Acknowledgments

The author is grateful to Dr Vallance Jones for assistance in this work. The material presented in this paper comes from the University of Saskatchewan Ph.D. dissertation of L. E. Montbriand, a portion of which was completed at the Communications Research Centre in Ottawa.

References

Akasofu, S.-I.: 1964, *Planetary Space Sci.* **12**, 273.
Akasofu, S.-I.: 1965, *Space Sci. Rev.* **4**, 498.
Akasofu, S.-I., Kimball, D. S., and Meng, C.-I.: 1965, *J. Atmospheric Terrest. Phys.* **27**, 189.
Eather, R. H.: 1968, *Ann. Geophys.* **24**, 525.
Eather, R. H. and Jacka, F.: 1966, *Australian J. Phys.* **19**, 215.
Hakura, Y.: 1965, *Rep. Ionos. Space Res. Japan* **19**, 121.
Montbriand, L. E.: 1969, Ph.D. Dissertation, University of Saskatchewan.
Montbriand, L. E.: 1970, *J. Geophys. Res.* **75**, 5634.
Wiens, R. H. and Vallance Jones, A.: 1969, *Can. J. Phys.* **47**, 1493.

AURORAL INFRASONIC SUBSTORMS

CHARLES R. WILSON

Geophysical Institute, University of Alaska, College, Alaska, U.S.A.

1. Introduction

Microbarographic equipment developed by the Geoacoustics Group, ESSA, (Cook and Young, 1962) has been operated to observe infrasonic waves associated with auroral activity at College, Alaska for a number of years. Typical examples of auroral infrasonic waves (AIW) are shown in Figure 1 as superposed pressure vs time traces from 4 microbarographs with the time scales shifted to give phase alignment for the coherent pressure waves. The true azimuth, ϕ, giving the direction from which the AIW has come, and the trace velocity, V, (defined as $V = c \sec \alpha$ where c is the local speed of sound and α is the angle between the wave normal and the horizontal) is also given in Figure 1 for each of the wave packets. The vertical bar represents a pressure change of 5 dyn cm^{-2} and the time scale is marked in minutes. The bandpass of the infrasonic equipment is from 10 to 100 sec. During periods of very low wind noise AIW with amplitudes of 0.5 dyn cm^{-2} have been observed. The largest AIW observed

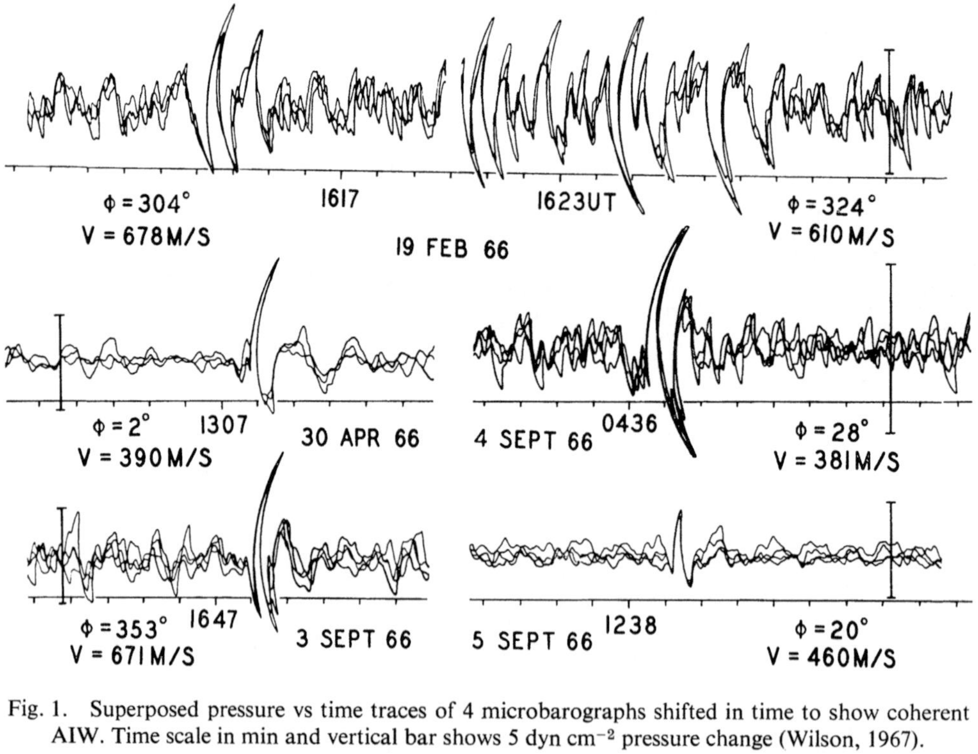

Fig. 1. Superposed pressure vs time traces of 4 microbarographs shifted in time to show coherent AIW. Time scale in min and vertical bar shows 5 dyn cm^{-2} pressure change (Wilson, 1967).

B. M. McCormac (ed.), The Radiating Atmosphere, 374–383.

thus far were 20 dyn cm^{-2}. During very large AIW substorms periods of coherent pressure waves may last for 1 hr or more. A segment of such activity is shown in Figure 1 for February 19, 1966, around 1620 to 1630 UT. However, the most frequently observed type of AIW activity is illustrated by the signal at 0436 on September 4, 1966, where the AIW wave packet rises suddenly out of the noise and lasts only for a few cycles. There is some waveform distortion introduced by the filter-amplifier system; however, a characteristic irregular dispersion can be identified in many AIW as previously observed by Campbell and Young (1963) and predicted by Maeda and Young (1966).

The average trace velocity of AIW is 513 m sec^{-1} for 1564 signals that were observed at College from December 25, 1965 to June 28, 1970. For a surface speed of sound of 335 m sec^{-1} this gives a value of $\alpha = 50°$. This is consistent with AIW being of auroral origin in the ionosphere.

2. AIW and Supersonic Auroral Motions

Wilson and Nichparenko (1967) and Wilson (1969a) found that only supersonic motions of large scale auroral arcs produce AIW. Figure 2 gives an example of an AIW received at College on January 7, 1967 at 1327 UT (0327) LT) 6 min after eastward propagating auroral arc crossed the zenith moving at 1200 m sec^{-1} from $\phi = 304°$. The highly directional AIW is shown by the arrow on the map of the projected auroral positions at College moving parallel to the aurora.

When particular infrasonic wave packets have been identified with supersonic arcs that have passed over College, strong magnetic perturbations corresponding to westward electrojets have been observed on the College magnetograms (Wilson, 1969b). In Figure 3 on the College 'storm' magnetogram negative bays in H of hundreds of γ and reversals in the Z component can be seen for each of the electrojets which produced AIW (whose arrival times are given in UT near each arrow) as they passed over College. The projected auroral electrojet arc positions are given in Figure 4 for the supersonic arc that produced both the AIW at 1635 UT (0635 LT) at College and the negative bay of 1500 γ in H at 1630 UT as shown in Figure 3. The direction of the 1635 UT AIW at College is parallel to the motion of the arc and to the total horizontal disturbance vector $\mathbf{\Delta H}$ due to the electrojet line current. The same infrasonic wave packet continued to propagate and was observed at the infrasonic array at Palmer 368 km south of College 20 min later.

When all-sky auroral data are not available the use of a line-current approximation will allow the speed, direction of motion, and zenith crossing time of auroral electrojets to be determined from their surface magnetic perturbations. Thus, particular AIW can be associated with supersonic auroral arcs even during daylight hours (Wilson, 1969b). Two examples of this technique are shown in Figure 5 for AIW observed at College. The speed of the electrojet is determined from $d/dt(B_Z/B_H)$, the variation of the ratio of the vertical (B_Z) to the horizontal (B_H) components of the electrojet surface magnetic perturbation, the zenith crossing time by the time of reversal of B_Z,

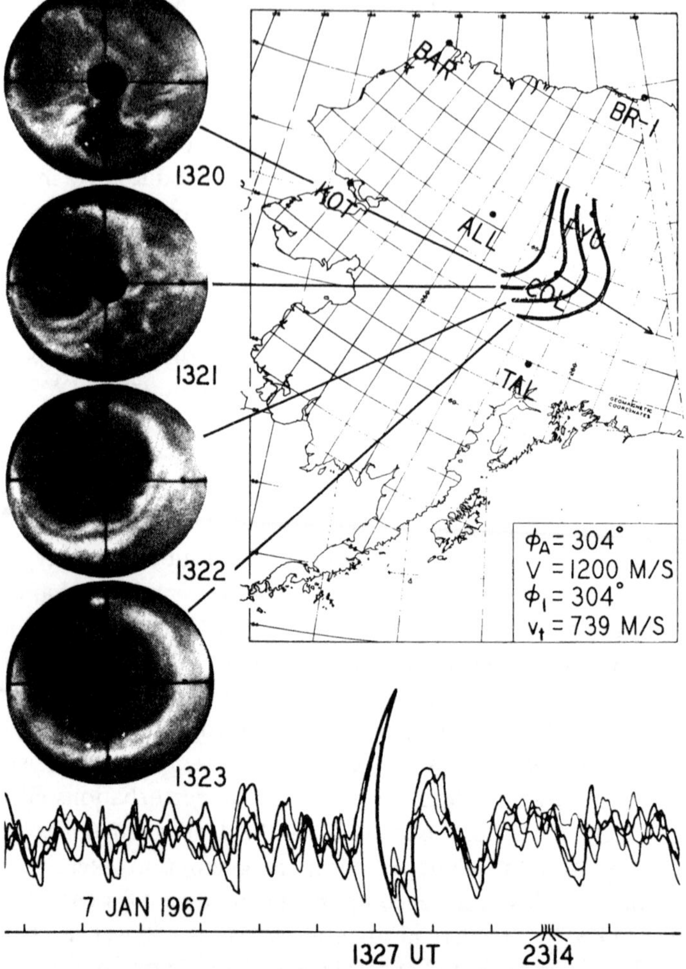

Fig. 2. All-sky auroral photographs of a supersonic auroral arc that produced AIW at College as shown at the bottom of the figure at 1327 UT January 7, 1967. Arrow on map shows direction of travel of AIW as parallel to motion of the auroral arc. ϕA and ϕ1 are azimuths of auroral motion and of AIW, respectively, while V and V_t are the arc velocity and trace velocity of the AIW (Wilson, 1969a).

and the direction of motion by the sense of reversal of B_Z at zenith crossing of the arc. Further evidence for the identification of an AIW with a supersonic electrojet is given by the parallelism of Δ**H** at the time the signal was radiated (arrival time minus 6 to 8 min) and the direction of propagation of the AIW as shown in the bottom of Figure 5. The delay times between arrival of the AIW and the zenith crossing time for the electrojets are both 7 min for the August 17, 1968 and August 11, 1966 infrasonic wave packets. This is the correct delay time for the infrasound wave to propagate down from the ionosphere to the surface (Wilson, 1969a).

As a result of the association of AIW with supersonic auroral arcs by Nichparenko

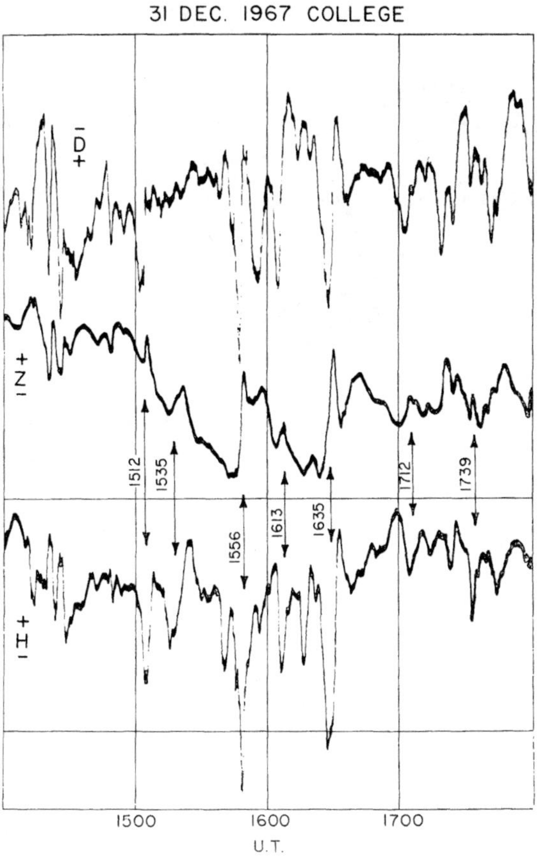

Fig. 3. College magnetograms for the infrasonic storm of December 31, 1967 showing the magnetic perturbations associated with AIW. The arrows give AIW arrival times adjacent to their associated magnetic disturbances in UT (Wilson, 1969b).

(1967) a shock wave model was developed by Wilson (1967) to explain the directional properties of AIW and the amplification of the auroral infrasound. Isotropic radiation of an acoustic pulse (of constant phase with respect to the frame of reference of the moving aurora) from an auroral arc moving with constant supersonic speed resulted in a shock pulse radiated in the forward direction because of the superposition of the infrasound emitted from each interval along the path. The delay time between zenith crossing of an auroral arc moving with a speed V_A at a height h and the arrival of the AIW shock wave at the surface was given by

$$\Delta t = h/c\sqrt{1-(c/V_A)^2}$$

where c is the speed of sound in an isothermal atmosphere. For $c = 300$ m sec^{-1} and $h = 120$ km the delay time should be about 6 min for $V_A > 2c$. The simple model predicts that the AIW trace velocity should be equal to the speed of the auroral arc.

The model was expanded (Wilson, 1969c), as illustrated in Figure 6, to show the shock front geometry both in front and to the side of a supersonic auroral form in

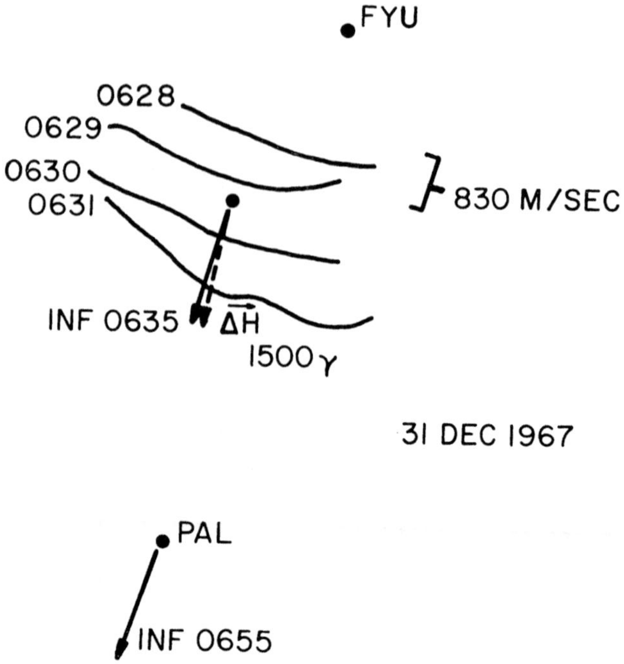

Fig. 4. Projection of supersonic auroral arc at 1 min intervals (LT) showing motion parallel to AIW received at College at 0635 LT and at Palmer at 0655 LT. Magnetic perturbation at College ΔH of 1500 γ is shown by dashed arrow parallel to direction of AIW (Wilson, 1969c).

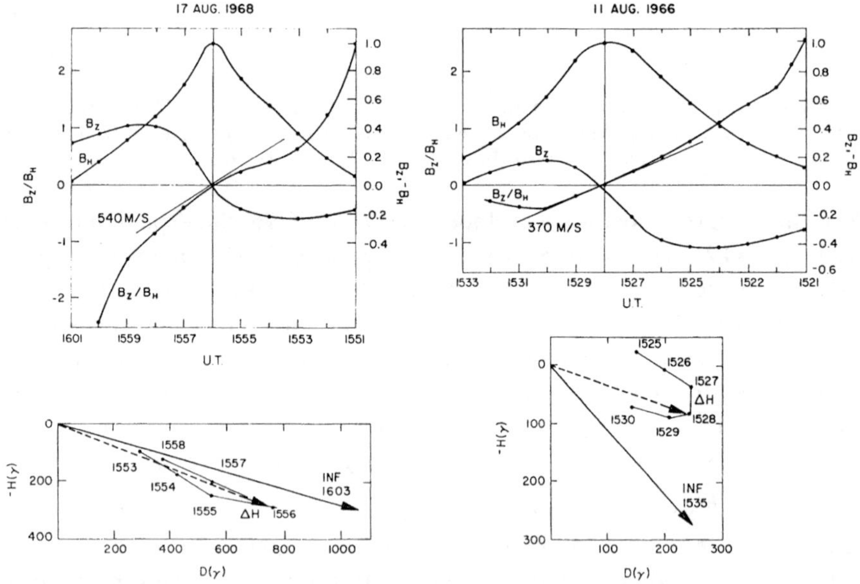

Fig. 5. Horizontal (B_H) and vertical (B_Z) components of magnetic perturbations vs time for moving auroral electrojets and their ratio B_Z/B_H that are associated with AIW observed on the ground under the electrojets. Traces for the total horizontal disturbance vector ΔH vs time and directions of associate AIW for the moving electrojets (Wilson, 1969b).

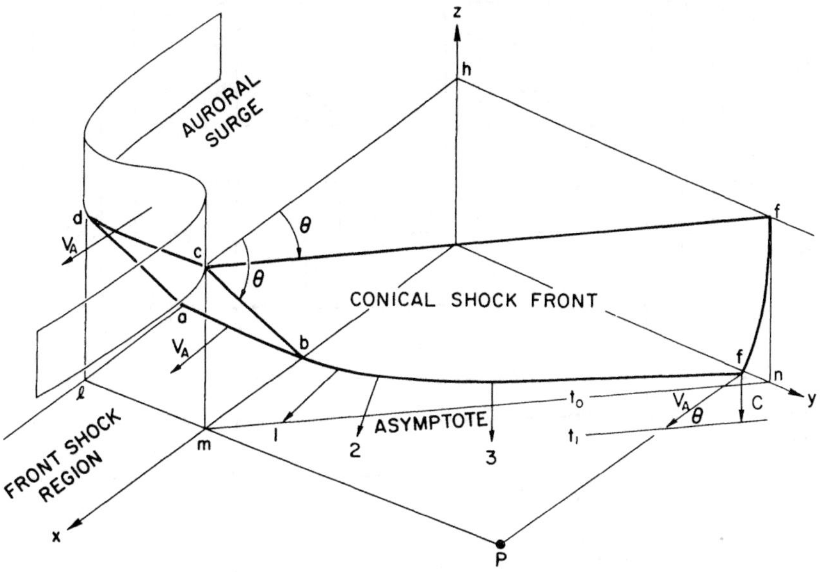

Fig. 6. Infrasonic shock surface geometry for an auroral surge at height h traveling in the positive x direction with a velocity v. The xy plane is the earth's surface. The mach angle of the shock is θ (Wilson, 1969c).

order to account for the divergence in ϕ and the difference in trace velocity for the same infrasonic wave packet observed first at College then at Palmer, Alaska 368 km to the south. When College was in the front shock region and Palmer in the side shock region of eastward traveling auroral surges in the late morning hours, the average trace velocity at College was 664 m $\sec^{-1}$ and only 515 m $\sec^{-1}$ at Palmer with an average divergence in signal azimuth of 31° for the same AIW's observed first at College then at Palmer.

Chimonas and Peltier (1970) have quantitatively investigated the problem of the generation of a bow wave by an auroral arc in supersonic translation. They find good agreement between the total arc current and the strength of the received AIW signals using College data.

3. Morphology of AIW Substorms

The general level of auroral infrasonic activity is closely related to the development of the polar electrojet. An excellent correlation can be seen in Figure 7 between the commencement of DP currents, as evidenced by the negative bays around 1400 UT on magnetograms A and B, and the arrival at College of AIW as shown by the small arrows below each magnetogram. AIW at College in the early evening hours can be associated with positive bays due to the return currents of a polar electrojet far to the north of College as shown in magnetograms C and D or with strong negative bay activity far to the east of Alaska as in the case of magnetogram E.

The variation in azimuth vs time of arrival for 1564 AIW observed at College over

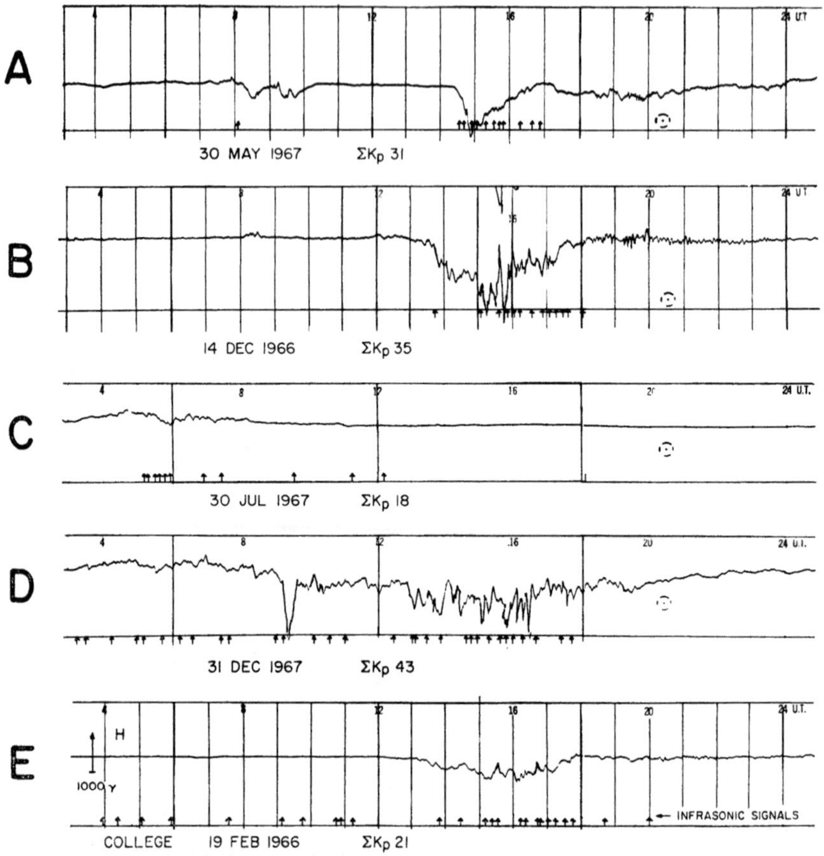

Fig. 7. Insensitive College *H* magnetograms and the occurrence of AIW indicated by vertical arrows. ΣK_P is the 3 hr magnetic index given for each storm (Wilson, 1969a).

a 54 month period is given in Figure 8. Although there is considerable scatter in the points in Figure 8 one can see that, on the average, the AIW arrive first around 2000 hr LT from about 40° and then from the magnetic meridian, $\phi = 30°$, around local midnight (1000 UT). The last signals come from west of the magnetic meridian, $\phi = 320$ to 360°, around 0800 LT. This migration of the source region of AIW from east to west was observed by Chrzanowski *et al.* (1961) in Washington, D.C. and by Goerke and Woodward (1967) in Boulder, Colo.

A more meaningful presentation of the variation in the azimuth of arrival of AIW at College is given in Figure 9. The number of AIW as a function of ϕ (the direction of arrival) is shown in Figure 9 (note scale) at the correct location of College with respect to the average auroral oval for each hour (UT) and each 20° of azimuth. It is clear from this Figure that the infrasonic signals that have been identified as AIW are coming from sources within the auroral oval. Most of the AIW at College in the evening sector (0500 to 0900 UT) were due to side shocks from supersonic westward traveling surges. In the midnight sector (0900 to 1300 UT) most of the AIW were

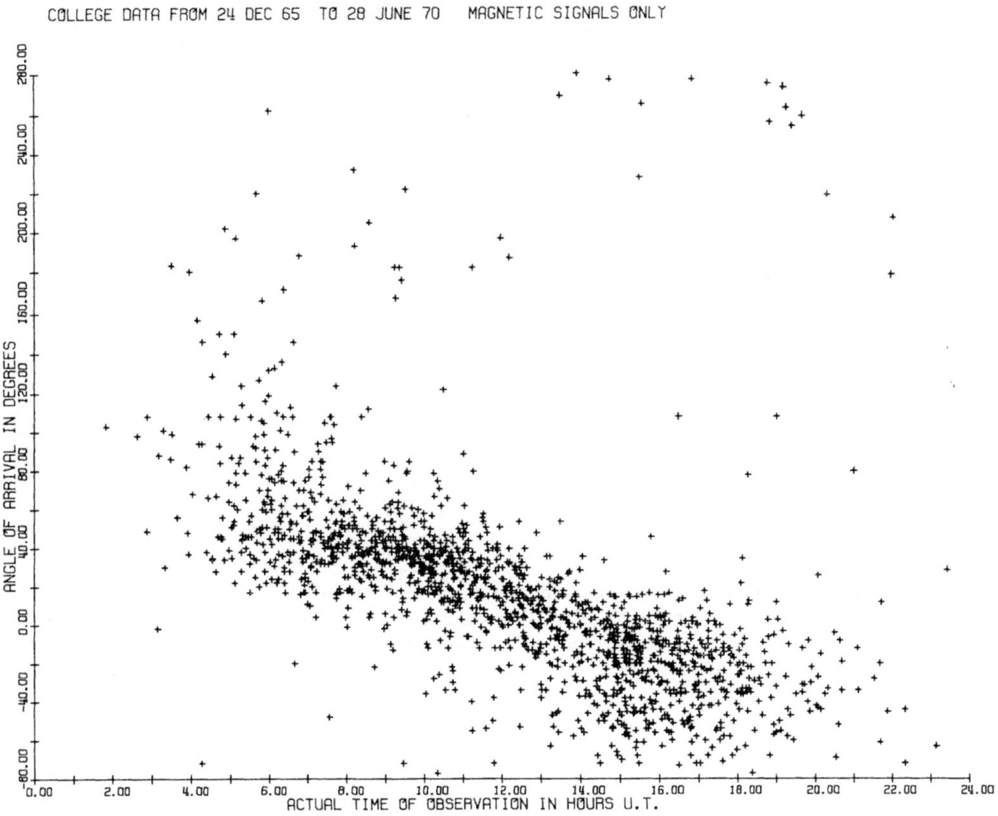

Fig. 8. Azimuth of arrival vs actual time of observation for 1564 AIW observed at College from December 24, 1965 to June 28, 1970.

direct shocks due to the southward expansion of the auroral bulge. In the morning sector (1300 to 2000 UT) the AIW were signals from eastward traveling auroral surges (Wilson, 1969c).

The morphology of AIW's follows directly from the temporal and spatial distributions of supersonic auroral motions that develop within the auroral oval during polar magnetic substorms when the directional characteristics of the infrasonic shock wave radiation are taken into account. The 'breakup' or expansive phase of auroral substorms result in the rapid auroral motions that generate the pattern of AIW's shown in Figure 9. The reversal in direction of arrival of AIW around magnetic midnight (1130 UT) at College corresponds to the well established reversal in the direction of auroral motions at that same time. Since the substorms develop first in the midnight sector and then expand into the dawn and dusk sectors it is natural that the AIW direction of arrival vectors (Figure 9) should all point toward the midnight sector in both the dawn and dusk sectors.

A small arrow head at each hour in Figure 9 gives the direction from which the AIW trace velocity is a maximum at that particular hour. A very clear pattern of direction

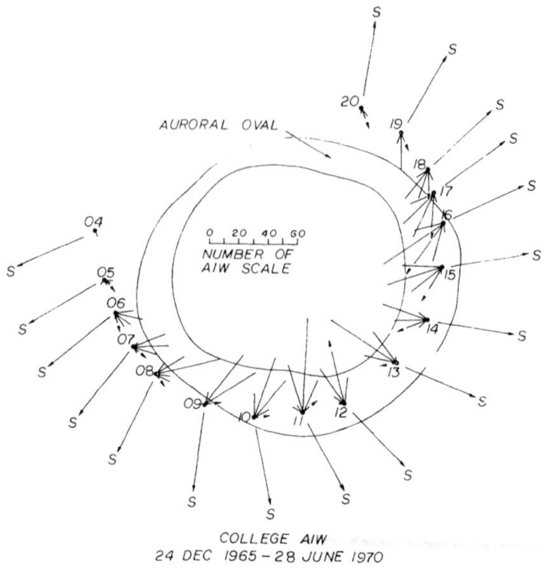

Fig. 9. Number of AIW as a function of ϕ, plotted at College with respect to the average auroral oval for each hour (UT) and each 20° of ϕ. The scale of AIW is given within the oval. Geographic south is given by the arrow S. The vectors point toward the origin in the direction from which the AIW came. Arrow head at each UT hour gives direction for which trace velocity was a maximum for that hour.

of maximum trace velocity can be seen in Figure 9. The maximum velocities are for directions parallel to the oval in the morning and evening sectors, and transverse to the oval in the midnight sector. This corresponds directly to the velocity distribution expected from the front shock and side shock effect as shown in Figure 6. Thus, say, at 0700 UT when the auroral oval expands to the extent that College would be in the front shock region of a westward traveling surge that was moving parallel to the oval a maximum trace velocity would occur in the direction parallel to the oval as shown in Figure 9.

The general pattern shown in Figure 9 has been related to the expansion of the polar substorm by Wilson (1970) in a comparison of the median global pattern for the progression of auroral absorption onsets and the distribution of AIW with respect to the auroral oval.

4. AIW Generation Mechanisms

Wilson (1967) has shown that supersonic motion of the source of infrasound within an auroral electrojet will give the directional properties and amplification of the initial acoustic pulse that are found in AIW. The actual mechanism that transfers momentum to the neutral gas is still undetermined.

When the thin veil of auroral electrons with its accompanying electrojet moves laterally through an undisturbed region of the ionosphere, it may produce a pressure pulse within a fixed volume of the ionosphere by at least four different processes: by heating from the incident electrons (Maeda and Watanabe, 1964), by joule heating by the electrojet (Cole, 1962), by Lorentz force $\mathbf{J} \times \mathbf{B}$ (Chimonas and Peltier, 1970), and by electrodynamic drift or Pedersen current (Martyn, 1953; Piddington, 1964).

Recent AIW observations at Inuvik, NWT Canada at 70° geomagnetic latitude indicate that no infrasonic shock waves are generated by poleward supersonic motions of westward auroral electrojets; however, equatorward motions of these same electrojets do produce AIW. A reversal in the direction of meteor winds at the 75 to 110 km level to an equatorward flow at the time of magnetic substorms has been found at College by Lilley (1970). These two facts indicate that Pedersen current is responsible for the transferal of momentum from the auroral electrojet to the neutral gas to produce the infrasonic waves.

Acknowledgment

This work has been supported by the Advanced Research Projects Agency, in cooperation with the U.S. Department of Commerce, ESSA Geoacoustics Group.

References

Campbell, W. H. and Young, J. M.: 1963, *J. Geophys. Res.* **68**, 5909.
Chimonas, G. and Peltier, W. R.: 1970, *Planetary Space Sci.* **18**, 599.
Chrzanowski, P., Green, G., Lemmon, K. T., and Young, J. M.: 1961, *J. Geophys. Res.* **66**, 3727.
Cole, K. D.: 1962, *Nature* **194**, 75.
Cook, R. K. and Young, J. M.: 1962, *Sound* 1, 25.
Goerke, V. H. and Woodward, M. W.: 1967, *Trans. Am. Geophys. Union* **48**, 81.
Lilley, E. R. Jr.: 1970, Masters Thesis, *Ionopheric Winds in the Auroral Zone Meteor Region During Geomagnetic Disturbances*, University of Alaska, College, Alaska.
Maeda, K. and Young, J. M.: 1966, *J. Geomag. Geoelec.* **18**, 275.
Maeda, K. and Watanabe, T.: 1964, *NASA TN D-2138*, June.
Martyn, D. F.: 1953, *Phil. Trans. Roy. Soc.* **A246**, 306.
Nichparenko, S.: 1967, Masters Thesis *Aurorally Associated Infrasonic Waves*, University of Alaska, College, Alaska.
Piddington, J. H.: 1964, *Space Sci. Rev.* **3**, 724.
Wilson, C. R.: 1967, *Nature* **214**, 1299.
Wilson, C. R.: 1969a, *J. Geophys. Res.* **74**, 1812.
Wilson, C. R.: 1969b, *Planetary Space Sci.* **17**, 1107.
Wilson, C. R.: 1969c, *Planetary Space Sci.* **17**, 1817.
Wilson, C. R.: 1970, *J. Atmospheric Terrest. Phys.* **32**, 293.
Wilson C. R. and Nichparenko, S.: 1967, *Nature* **214**, 1299.

INSTITUTE DISCUSSION

It is difficult to use amplitude data to obtain information about auroral electrojet motions.

The optimum height for observing the AIW is at the earth's surface because the noise background is lowest there.

Pulsating arcs in the morning do not seem to produce AIW; however, many higher frequency components would be outside the bandpass of his detectors.

No model for explaining the origin of AIW can adequately explain why AIW are not observed to propagate over the polar cap.

DYNAMICS OF AURORAL FORMATION

AKIRA HASEGAWA

Bell Telephone Laboratories, Inc., Murray Hill, N.J., U.S.A.

Abstract. Several theories related with the dynamics of aurora formation are reviewed. Subjects covered are the acceleration of precipitating particles that causes the auroral breakup and the formation of large and small-scale structures of an aurora. Particular emphasis is placed on the role of plasma instabilities.

1. Introduction

An auroral breakup is produced by excitations of molecules and atoms by suddenly enhanced precipitating low energy electrons and protons. Hence the formation of an aurora is a manifestation of the dynamics of those particles precipitating into the upper atmosphere.

The basic questions to be answered are therefore

(a) What is the acceleration mechanism of the precipitating particles?

(b) Why do they precipitate in such a particular location with a particular shape (thickness and width of the auroral cross section, multiplicity of the arcs).

(c) What causes the small scale ($\sim$10 km) or large scale ($\sim$100 km) perturbations in aurora?

These questions have been attacked for more than half a century but only recently reasonable answers are being given to them as a result of direct probe by satellites and rockets, and the accompanying development of plasma theory. Some guide lines will be given for the solution of these problems by reviewing several theories presented to date. In particular the role of plasma instabilities in the aurora dynamics is emphasized.

2. Acceleration Mechanism of Precipitating Particles

Simultaneous observations of enhanced intensities of low energy electrons in the auroral region and at the equator (O'Brien, 1964; Lezniak *et al.*, 1968; Parks and Winckler, 1969) and the evidence of the field aligned current associated with the auroral breakup (Zmuda *et al.*, 1966; Cummings and Dessler, 1967; Akasofu and Meng, 1969; Cloutier *et al.*, 1970) indicate that the acceleration of the precipitating particles does not take place locally in the ionosphere. This means that we have to seek for the energy source needed for the acceleration not only in the ionosphere but also in the magnetosphere, in the magnetospheric tail, or possibly even beyond. If we classify the theories of precipitating mechanism by the assumed location of the energy source, there are basically four groups of theories.

The first group considers the energy coming directly from the solar wind. Dungey (1961) considered the interplanetary field lines carried by the solar wind merge with the magnetospheric field which converts their energy to particle energies. Axford and Hines (1961) considered a viscous interaction between the solar wind particles and the

B. M. McCormac (ed.), The Radiating Atmosphere, 384–393.

magnetospheric field which produces $\mathbf{v} \times \mathbf{B}$ electric field ($\mathbf{v}$ is the velocity of the solar wind particles). They have shown that this electric field can accelerate particles to energies of tens of kilovolts. Speiser (1967) showed by calculating the trajectories of electrons and protons in the neutral sheet taking into account the transverse electric field postulated by Axford and Hines that they are accelerated toward the earth and ejected out of the neutral sheet to be precipitated into the auroral latitude. Although this group can explain the thin shape and location of the aurora, they have difficulty in explaining the multiplicity of the aurora as well as the time lag between the enhancement of the solar wind and the auroral breakup which has often been found to be random.

To explain this time lag one has to assume a storage of energy somewhere in the magnetosphere and sudden release of it. The release is most easily understood by considering some kind of plasma instability.

The second group we consider assumes the energy to be in the magnetic field in the neutral region of the magnetospheric tail. This energy can be dissipated into particle energy by field line merging. The field line merging as a cause of acceleration of particles in the tail region has been discussed by many authors (Dungey, 1953; Parker, 1957; Petschek, 1963; Dessler, 1968). Analysis of this phenomena in terms of a plasma instability called "tearing mode instability" by Coppi *et al.* (1966) presented a possibility of multiple reconnection of the field lines in the tail region which for the first time could explain the production of the multiple arcs in an aurora. Another attractive point of this theory is in considering an instability as a cause of the sudden release of the stored energy, which is consistent very well with the explosive phase of the auroral breakup.

However, a more careful analysis of this instability showed that the instability condition is quite sensitive to the anisotropy of the plasma temperature (Laval and Pellat, 1968) and the tearing mode is stabilized if

$$\frac{T_{\parallel}}{T_{\perp}} > 1 + \frac{\varrho_e}{\lambda} \sim 1 \tag{1}$$

where $T_{\perp}$, $T_{\parallel}$, ϱ_e and λ are perpendicular and parallel temperature (with respect to the magnetic field), electron Larmor radius and the sheet thickness. Although the same authors could show that under the stable condition shown by Equation (1), a different instability can be set up. This new instability can simply convert energy in $T_{\parallel}$ into that in $T_{\perp}$ and cannot directly convert the magnetic field energy into the particle kinetic energy, hence it cannot serve as a dominant acceleration mechanism.

The third group assumes the energy source in the ring current. The idea was postulated by Akasofu (1969). He considers that the ring current in which a part of the energy in the main phase is stored is suddenly disrupted in the midnight sector and flows along the field lines and connects with auroral electroject in the ionosphere. He further considers that if the field aligned current becomes unstable, the current is impeded by an anomalous resistivity produced by the instability. This will create a large electric field along the field lines which will accelerate particles in the same direction. He postulated electron-ion two stream instability after Swift (1965) as a possible candidate for the instability.

This idea is also attractive to explain the time lag between the sudden commencement and the substorm phase. Although the idea of the partial ring current as postulated by Akasofu may be debatable, the actual observations of the existence of the field aligned current stimulated thoughts of acceleration mechanisms associated with the current instability. The original idea by Swift (1965) requires a relatively large current because the drift velocity of electrons v_d associated with the current has to be larger than the electron thermal velocity v_T, i.e., (Criticism of a similar sort has also been presented by Ossakow (1968)).

$$v_d > v_T . \tag{2}$$

Realizing this fact Kindel and Kennel (1971) looked for all the possible micro-instabilities produced by a current and came to the conclusion that the electrostatic ion cyclotron instability has the smallest threshold velocity, which can be written roughly as

$$v_d > v_T/4 . \tag{3}$$

These microinstabilities, however, even if they occur may produce the anomalously resistive region only very locally along the field lines, because the typical wave length of the instability is negligibly small compared with the scale length of the current. For this reason Hasegawa (1970) looked at a macroscopic instability and showed that the observed current sheet is subject to the kink instability. The fact that the field aligned current is subject to the kink instability has also been noted independently by Forslund (1970) by assuming a current with a cylindrical cross section.

This instability has a parallel half wave length correponding to the length of the current L and perpendicular half wave length to the E–W width of the sheet current b. The condition of the instability can approximately be written as

$$b/L < B_\perp/B , \tag{4}$$

where B is the geomagnetic field and $B_\perp$ is the transverse magnetic field produced by the sheet current I, i.e.,

$$B_\perp = \frac{\mu_0 I}{2b} . \tag{5}$$

We can compare the threshold condition with that of Equation (3) by expressing I in terms of v_d as

$$I = env_d ab \tag{6}$$

where a and n are the N-S thickness of the current and the density of precipitating electrons. Equation (4) can be written as

$$\frac{v_d}{v_T} > \frac{2c}{v_T} \frac{c}{\omega_{pe} a} \frac{\omega_{ce}}{\omega_{pe}} \frac{b}{L} \tag{7}$$

where ω_{ce}, ω_{pe}, and c are the electron cyclotron frequency, the electron plasma frequency, and the speed of light. If we use as the values at the ionosphere, electron temperature of 3 eV, $n=10^{11}\ \mathrm{m}^{-3}$, $a=10^3$ m, $b=10^6$ m, and $L=1.5\times 6\ \mathrm{R_E}=6\times 10^7$ m, then $2c/v_T=6\times 10^2$, $c/\omega_{pe}a=2\times 10^{-2}$ $\omega_{ce}/\omega_{pe}=6\times 10^{-1}$, $b/L=1.5\times 10^{-2}$ and Equation (7) becomes

$$v_d > v_T/10. \tag{8}$$

Hence for the above choice of the parameters the kink instability requires the smallest threshold for the drift velocity. However, Equation (7) is quite sensitive to the choice of the aspect ratio of the sheet current b/a which is the most ambiguous parameter. If the sheet is flatter than assumed, the threshold becomes higher or vice versa. In terms of the actual value, the threshold current required for the instability becomes $\sim 10^6$ amp for the sheet width b of 10^3 km, which is quite reasonable, and becomes larger proportional to b.

One problem of this theory is the difficulty of explaining the multiplicity of the auroral arcs. However, this may be solved by combining the arc theory developed recently by Atkinson (1970), which will be discussed in the next section.

The fourth group assumes the energy in the DS current in the ionosphere. This is also reasonable because the total energy in the DS current is considered to be comparable to that in the ring current. It can also explain the time lag of the two phases discussed above. This current system produces potential variations of tens of kilovolts among different field lines which may be mapped to the magnetosphere and the tail regions. Taylor and Hones (1965) calculated the particle trajectories assuming the potential variations mapped from the ionosphere to the tail and showed that the auroral particles can be accelerated suitably by this electric field.

However, it seens rather unlikely that the equipotential surface can be mapped to a distance of 20 $\mathrm{R_E}$ because a shear flow $v_E(x)$ of a plasma produced by the nonuniform electric field $E(x)$, and magnetic field, $B(x)$, i.e., $v_E(x)=E(x)/B(x)$ (x is a coordinate parallel to the electric field) is expected to trigger a plasma instability, most likely the Kelvin-Helmholtz instability which may produce a turbulent mixture of particles from different field lines and hence on different potential surfaces. According to the theory of the Kelvin-Helmholtz instability (which will be discussed in more detail in Section 3) the instability occurs when the wave length perpendicular to the shear (in this case perpendicular to the x direction) becomes the same order as the distance over which the shear flow exists, i.e., the thickness of the shear flow. Therefore, one can reasonably expect that the Kelvin-Helmholtz instability may take place approximately at a distance of a few earth radii away from the surface of the earth. The potential difference on different field lines would disappear beyond this point.

In addition to the four sets of theories mentioned above there are other theories in which the acceleration is explained using some plasma instabilities (for example Chamberlain, 1963) or wave particle interactions (for example Hruska, 1968). However, because the total energy dissipated in an auroral breakup is comparable to that of the ring current (Akasofu, 1969), a plasma instability whose free energy source is much smaller than the ring current energy or a wave particle interaction in which the wave energy is much smaller than this scale of energy cannot be a dominant mechanism of particle acceleration.

In conclusion, explanations used by the second and the third groups of theories seem most likely at present.

3. Arc Formation

A. LARGE SCALE FORMATIONS

In this subsection *the multiple arc, the rapid northward motion* and *large scale drapery* of an aurora will be discussed.

As was shown in Section 2, the *multiple arc formation* may be explained in terms of the tearing mode instability of the neutral sheet; however, the condition of the instability is rather stringent as shown in Equation (1). Recently, Atkinson (1970) presented a unique model of multiple arc formation; the mechanism may be outlined as follows. The basic idea is a feedback oscillation in the precipitating current and voltage associated with the high conductivity region in the ionosphere. Figure 1 shows the mechanism. Figure 1a shows the cross section of the E region in the ionosphere, where the high conductivity strips are assumed to be produced by precipitating current j. Then he shows that the enhanced conductivity strips can sustain the precipitating current in turn. Figure 1b shows the conductivity variation which produces a variation of x directed electric field as shown in (c) because of the Ohms law $I_x = \sum_p E_x = \text{const}$. Such an electric field can be mapped to the magnetosphere as indicated by E_x in (d). As a flux tube directed in the z direction convects through the system with an x component in its velocity, it is subject to a polarization current

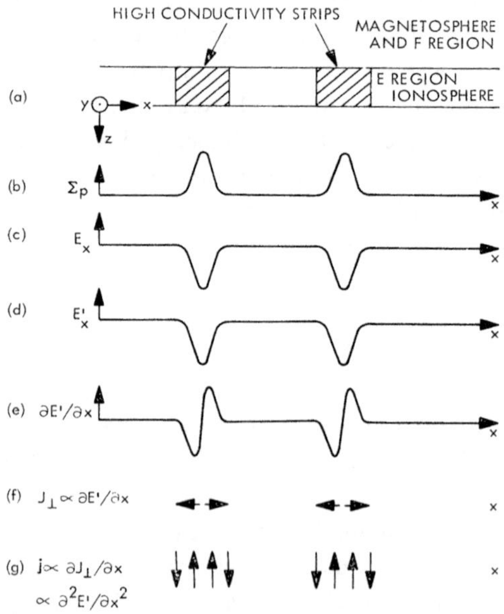

Fig. 1. Mechanism of multiple arc formation in an aurora (Atkinson, 1970).

$v_x(\partial E'_x/\partial x)$ as seen in the frame fixed to the arc. The resulting polarization current is shown in (f). This polarization current closes with the field aligned current as shown in (g). The field aligned current occurs as precipitating electrons that create the high conductivity strips initially assumed; thus, the feedback loop is closed.

The most crucial assumption in this model is the electric field mapping from the ionosphere to the magnetosphere. However, it requires the mapping to a distance of only a few earth radii, hence it might be justified.

As is discussed by Hasegawa (1970), *the rapid northward motion* of an auroral arc may be understood in terms of a flute type instability of the field aligned current. The precipitating particles carrying the field aligned current receive centrifugal force $\mathbf{F}(=mv^2/\mathbf{R})$ owing to the curvature of the field lines. This force induces drift to the particle across the field lines in the E-W direction with the drift velocity given by $\mathbf{F}\times\mathbf{B}/qB^2$, where q is the charge of the particle. Hence, electrons and protons drift in the opposite direction thus producing a current, $J_d(=nF/B)$, where n is the number density of the precipitating particles. This current sets up a polarization charge, which in tern induces a polarization current, $J_p\ [=nm_i/B^2\ (\partial E/\partial t)]$ in the opposite direction where m_i is the proton mass and E is the electric field in the E-W (or W-E in the Southern Hemisphere) direction. In the ionosphere, this E-W electric field produces a conduction current $J_c(=\sigma_\perp E)$ because the perpendicular conductivity is finite. Balancing these currents we have

$$\frac{\partial E}{\partial t}+\frac{1}{\tau_0}E=\frac{eTB}{m_iR}, \tag{9}$$

where

$$\tau_0=nm_i/B^2\sigma_\perp$$

and T is the energy of precipitating particles in eV.

For a typical parameter in the ionosphere, we may assume $\tau_0\sim10^{-2}$ sec, indicating after 10 msec or so, the conduction current becomes dominant over the polarization current. The electric field then assumes a constant value given by

$$E=\frac{eTB\tau_0}{m_iR}\sim\frac{T(V)\,B(W/m^2)}{10}\quad(\mathrm{Vm}^{-1}),$$

where we took the radius of the curvature $R=2R_E$. This electric field now induces $\mathbf{E}\times\mathbf{B}$ drift to the precipitating electrons. The drift velocity v, which is directed poleward, is given by

$$v=T(V)/10\quad(\mathrm{m\ sec}^{-1}). \tag{10}$$

If we take $T\sim5$ to 10 kV for the energy of the precipitating electrons, v in Equation (10) becomes 0.5 to 1 km sec^{-1}, in good agreement with the observation (Akasofu *et al.*, 1966).

Finally we discuss the *long wave length drapery formation.* The author believes that this is most likely produced by the kink instability of the field aligned current. As was shown in Equations (4) to (6) the instability condition is sensitive to the aspect ratio of the cross section of the current sheet b/a. One can reasonably expect that the instability will start if this ratio is smaller than 10^3 for a precipitating current of 10^6 amp. The instability is produced by the tension of the magnetic field. The bent magnetic field due to the field aligned current tries to become straight when the current is increased over the threshold given by Equation (4) and consequently the current is deformed into a drapery shape. However, because the more exact solution which is now being worked out by the author is not available yet, the actual shape of the deformation is not known yet.

B. SMALL SCALE FORMATIONS

A small scale structure of an aurora with typical wavelength of 10 km or less will be discussed. The relevant plasma instabilities are the Kelvin-Helmholtz instability and the $\mathbf{E} \times \mathbf{B}$ instability.

The classical Kelvin-Helmholtz instability (instability due to a shear flow of an incompressible fluid) is subject to occur in an auroral plasma if there exists a local excess of electron or ion charges. The nonuniform electric field generated by these charges across the direction of the geomagnetic field produces a nonuniform $\mathbf{E} \times \mathbf{B}$ drift, a shear flow of the plasma.

The growth rate γ of this instability was given originally by Rayleigh for an incompressible fluid. This was also derived later by Buneman *et al.* (1966) for an electron beam and is given by (analogy to electron beam holds only approximately, cf. A. Hasegawa, 1971),

$$\gamma = \frac{\omega_S}{2}\left[e^{-2ka} - (1 - ka)^2\right]^{1/2} \tag{11}$$

where a is the thickness of the flow with the shear, hence the thickness of the aurora, k is the wave number of the perturbation parallel to the flow, and ω_S is a measure of the shear in velocity v_d, i.e.,

$$\omega_S = \frac{dv_d}{dx} = \frac{d}{dx}\left(\frac{E}{B}\right) \tag{12}$$

and x is in the direction perpendicular to the flow. As can be seen from Equation (11), the growth rate becomes real for a wave length comparable or longer than the thickness a, i.e., when

$$ka \lesssim 1.3\,. \tag{13}$$

It was first suggested by Webster (1957) that this instability may be relevant in an aurora. This suggestion has been carefully surveyed recently by Hallinan and Davis (1970).

One attractive feature of this instability is that by studying the direction of the curl and the wave length of the perturbation, one can estimate the degree and the sign of the charge nonneutrality. Namely, if the aurora is electron (proton) rich, the electric field will be directed toward the center (edges) of the sheet from both edges (center). The

$\mathbf{E} \times \mathbf{B}$ shear flow produced by such an electric field will be directed counter clockwise (clockwise) looking up from the earth's surface in the Northern Hemisphere and clockwise (counter clockwise) in the Southern Hemisphere. The direction of the shear flow-should, of course, be the same as that of the curl observed in an aurora. Using this fact, Hallinan and Davis (1970) concludes that the aurora is predominantly electron rich.

This instability has extensively been studied for a sheet shaped electron beam by Pierce (1956) and by Cutler (1956). Although the previous analyses assumed a collisionless plasma, the result should hold even in the ionosphere provided that the surface charge accumulation produced by the instability at the edge of the aurora can be maintained; that is, if the collision frequency of protons is less than its cyclotron frequency.

The $\mathbf{E} \times \mathbf{B}$ *instability* was first found by Simon (1963). The name is not particularly appropriate because the previous instability is also produced by the '$\mathbf{E} \times \mathbf{B}$' effect. This instability is a combination of a drift wave instability produced by a density gradient and a two stream like instability produced by $\mathbf{E} \times \mathbf{B}$ drift (here $\mathbf{E}$ is considered uniform) and occurs only in a partially ionized gas. In a partially ionized gas the $\mathbf{E} \times \mathbf{B}$ drift velocity for electrons is much larger than ions because of the difference in the ratio of the collision frequency ν to the cyclotron frequency ω_c. In particular, if $\nu/\omega_c \gg 1$ for ions, they may be regarded as stationary and electrons stream through the ions. It can be shown in general that if the electron stream velocity is smaller than its thermal velocity the system is stable and becomes unstable only when, in addition to the electric field, there exists a density gradient drift parallel to the $\mathbf{E} \times \mathbf{B}$ drift, namely if ∇n is parallel to $\mathbf{E}$ (Simon, 1963; Hasegawa, 1968). Unlike the Kelvin-Helmholtz instability, this instability is a 'compressional' instability hence it will be related with the formation of striations in an aurora. There exists a wave number k in the direction of $\mathbf{E} \times \mathbf{B}$ that gives the maximum growth rate (Hasegawa, 1968) which is approximately given by

$$k \sim \left(\frac{\kappa}{\varrho_e}\right)^{1/2} \left(\frac{\lambda_m}{\varrho_e} \cdot \omega_{ce} \tau\right)^{1/2} \tag{14}$$

where κ is the measure of density gradient

$$\kappa = \left|\frac{\mathrm{d} \ln n}{\mathrm{d}x}\right| \sim \frac{1}{a} \tag{15}$$

λ_m is the mean free path of electrons

$$\lambda_m = v_E \tau = \frac{E\tau}{B} \tag{16}$$

and τ is the mean free collision time. Because the quantity in the second bracket of Equation (14) is a large number ($\sim 10^3$ to 10^4), the wave length corresponding to the maximum growth rate may be considered at least on the order of a, the thickness of the aurora. Linson and Workman (1970), Simon (1970), Tchen and Hankel (1970) and Zabusky and Perkins (1970), have applied this instability to the formation of striations of Ba clouds and showed a reasonable agreement with the observation. More recently Linson (1970) applied this instability to the auroral striations and concluded that (1) the northside of the auroral curtain is unstable, (2) the growth rate is proportional to E/a, (3) unstable wave has broad spectrum, and (4) the lower the altitude of the lower border, the faster the growth rate.

4. Conclusion

Dynamics of auroral formations are discussed by reviewing several theories on the mechanism of acceleration of precipitating particles, and large and small scale formations of auroras.

Four basic groups of theories are introduced on the acceleration mechanism depending on the location of assumed energy sources. However, none of them seems conclusive at the present stage. On the other hand, mechanisms of large and small scale formations are relatively well understood. They are shown to be produced by several different kinds of plasma instabilities such as flute, kink, Kelvin-Helmholtz, and $\mathbf{E} \times \mathbf{B}$ instabilities.

Acknowledgment

The author wishes to thank W. L. Brown, L. J. Lanzerotti and N. A. Tartaglia for their critical readings of the manuscript.

References

Akasofu, S.-I.: 1969, *Nature* **221**, 1020.
Akasofu, S.-I. and Meng, C.-I.: 1969, *J. Geophys. Res.* **74**, 293.
Akasofu, S.-I., Kimball, D. S., and Meng, C.-I.: 1966, *J. Atmospheric Terrest. Phys.* **28**, 497.
Atkinson, G.: 1970, *J. Geophys. Res.* **75**, 4746.
Axford, I. and Hines, C. O.: 1961, *Can. J. Phys.* **39**, 1433.
Buneman, O., Levy, R. H., and Linson, L.: 1966, *J. Appl. Phys.* **37**, 3203.
Chamberlain, J. W.: 1963, *J. Geophys. Res.* **68**, 5667.
Cloutier, P. A., Anderson, H. R., Park, R. J., Vondrak, R. R., Spiger, R. J., and Sandel, B. R.: 1970, *J. Geophys. Res.* **75**, 2593.
Coppi, B., Laval, G., and Pellat, R.: 1966, *Phys. Rev. Letters* **16**, 1207.
Cummings, W. D. and Dessler, A. J.: 1967, *J. Geophys. Res.* **72**, 1007.
Cutler, C. C.: 1956, *J. Appl. Phys.* **27**, 1028.
Dessler, A. J.: 1968, *J. Geophys. Res.* **73**, 9.
Dungey, J. W.: 1953, *Phil. Mag.* **44**, 725.
Dungey, J. W.: 1961, *Phys. Rev. Letters* **6**, 47.
Forslund, D. W.: 1970, *Trans. Am. Geophys. Union* **51**, 405.
Hallinan, T. J. and Davis, T. N.: 1970, *Planetary Space Sci.* **18**, 1735.
Hasegawa, A.: 1970, *Phys. Rev. Letters* **24**, 1162; 1468.
Hasegawa, A.: 1968, *Phys. Rev.* **169**, 204.
Hasegawa, A.: 1971, *Rev. Geophys.* and *Space Phys.*, to be published.
Hruska, A.: 1968, *Planetary Space Sci.* **16**, 1297.
Kindel, J. M. and Kennel, C. F.: 1971, *J. Geophys. Res.* **76**, 3055.
Laval, G. and Pellat, R.: 1968, Proc. of the ESRIN Study Group Frascati, Italy, ESRO SP-36, p. 5.
Lezniak, T. W., Arnoldy, R. L., Parks, G. K., and Winckler, J. R.: 1968, *Radio Sci.* **3**, 710.
Linson, L. M.: 1970, private communication; to be published.

Linson, L. M. and Workman, J. B.: 1970, *J. Geophys. Res.* **75**, 3211.
O'Brien, B. J.: 1964, *J. Geophys. Res.* **69**, 13.
Ossakow, S. L.: 1968, *J. Geophys. Res.* **73**, 6366.
Parker E. N.: 1957, *J. Geophys. Res.* **62**, 509.
Parks, G. K. and Winckler, J. R.: 1969, *J. Geophys. Res.* **74**, 4003.
Petschek, H. E.: 1963, Proc. of the AAS-NASA Symposium on the Physics of Solar Flares, NASA SP-50 (Ed. by W. N. Hess), p. 425.
Pierce, J. R.: 1956, *IRE Trans.* **ED-3**, 183.
Simon, A.: 1963, *Phys. Fluids* **6**, 382.
Simon, A.: 1970, private communication; to be published.
Speiser, T. W.: 1967, *J. Geophys. Res.* **72**, 3919.
Swift, D. W.: 1965, *J. Geophys. Res.* **70**, 3061.
Taylor, H. E. and Hones, E. W.: 1965, *J. Geophys. Res.* **70**, 3605.
Tchen, C. M. and Hankel, R.: 1970, *J. Geophys. Res.* **75**, 6287.
Webster, H. F.: 1957, *J. Appl. Phys.* **29**, 1388.
Zabasky, N. and Perkins, F. W.: 1970, to be published.
Zmuda, A. J., Martin, J. H., and Heuring, E. T.: 1966, *J. Geophys. Res.* **71**, 5033.

INSTITUTE DISCUSSION

I. McDiarmid asked whether auroral arcs could move northward while, by A. Hasegawa's description, a Ba cloud does not. A. Hasegawa said that due to different sites and energies involved he might expect the Ba cloud to move much more slowly.

R. Burrows pointed out that the plasma sheet in the tail may be as likely a candidate as the ring current for the energy reservoir. A. Hasegawa agreed and noted that the important thing is that the energy be inductively rather than capacitively stored.

PHYSICAL PROCESSES INVOLVING STRIATION FORMATION IN BARIUM CLOUDS

HEINRICH VÖLK and GERHARD HAERENDEL
Max-Planck-Institut für Physik und Astrophysik, Institut für Extraterrestrische Physik, Garching München, West Germany

1. Introduction

A. OBSERVATIONS

Almost all artificially injected ion clouds in the ionosphere show a characteristic pattern of striations aligned with the magnetic field. Typical experimental parameters are: height of injection about 200 to 250 km above the ground, ion densities $N \approx 10^5 \text{ cm}^{-3}$ of released Ba and a temperature $T \approx 10^3$ K. The observed time scales of the development of the striations from a smooth cloud are of the order of 5 to 60 sec, while the spatial scales range from about 200 m to several kilometers. The lower spatial limit is usually also the lower limit of photographic resolution: as slow a speed as 20 m sec^{-1} of an irregularity perpendicular to the line of sight together with a 10 sec exposure time already gives an uncertainty of 200 m in resolution. In addition, the observed thick striations could be a superposition of thinner ones along the line of sight. This leads to some uncertainty for the determination of transverse scale lengths.

The striations, which are often a prominent feature of the clouds (Figure 1), have been observed in the auroral zone (Föppl *et al.*, 1968; Haerendel *et al.*, 1969; Wescott *et al.*, 1969a), at medium latitudes (Haerendel and Lüst, 1968) as well as at the geomagnetic equator (Rieger, 1969). There is a positive correlation between their growth rate and the strength of the ambient field.

The time history of the striations in a particular experiment is, in general, not monotonic in the sense that they become a stationary feature. Quite often they appear to die out after some time and reappear later in a new pattern. Nevertheless, except in conditions of strong variability of the ambient field (Haerendel *et al.*, 1969) the time scale of this process is generally appreciably much larger than the time scale in which a distinct pattern is generated from a smooth configuration.

Where it was possible to judge with sufficient confidence, it turned out that the striations remain behind the cloud in its motion; smaller spatial scales seem to grow faster in time than the larger ones. There are two configurations where one can check this unambiguously on an optically thin cloud: When the cloud is in or very near the magnetic zenith, or if the cloud has a velocity component perpendicular to the line of sight at the dip equator (Figure 2).

B. CAUSES FOR STRIATIONS

How do these irregularities occur? Since plasma motions at cloud levels are essentially incompressible, density irregularities can originate only in the presence of a

B. M. McCormac (ed.), The Radiating Atmosphere, 394–406.

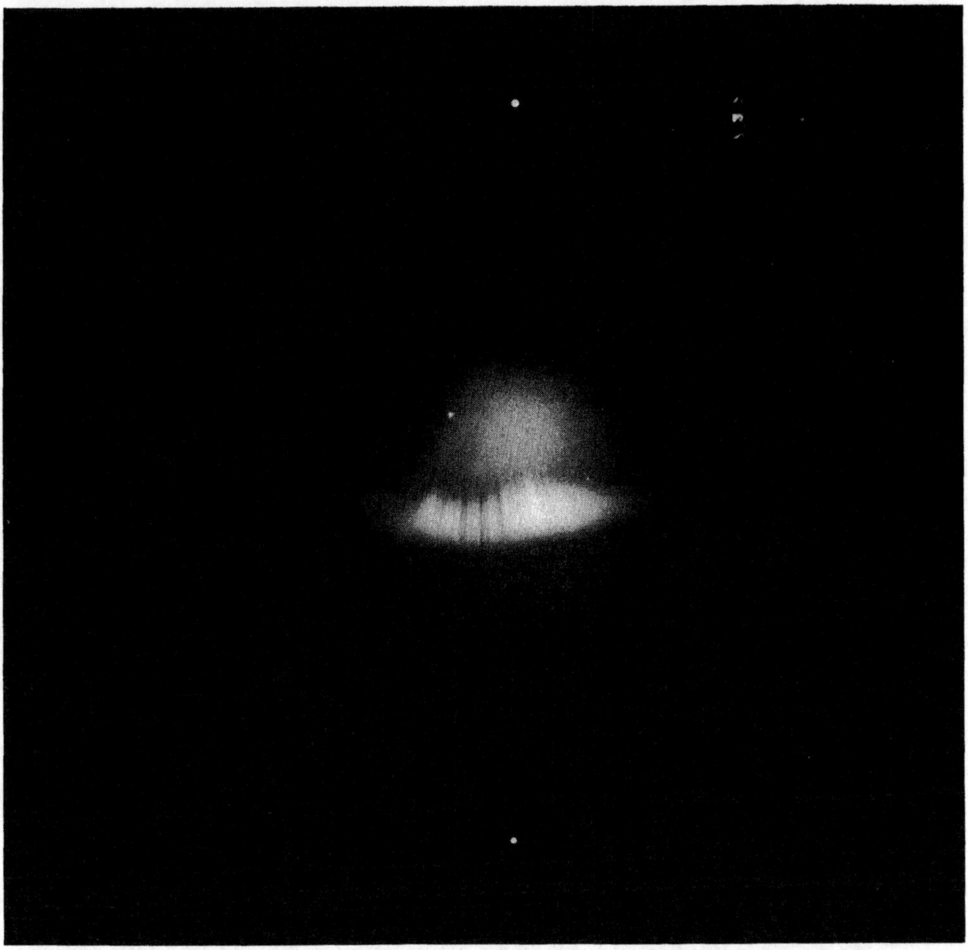

Fig. 1. A strongly striated Ba cloud released near Fort Churchill (Canada). The cloud is seen from the north and is drifting in a S-W direction. Note that the striations project out on the upper side in the direction of the neutral cloud. Thus they are nearer to the observer and don't quite follow the ion cloud.

primary density gradient transverse to the **B** field. An ion cloud artificially provides such an inhomogeneity. Electric fields with a transverse scale smaller than that of the cloud then can carry material away from it and the apparent growth of the resulting irregularity comes about just through the kinematic effect of its drifting down the density gradient. The growth rate is given by the drift speed of the irregularity (relative to the cloud) divided by the gradient length.

Consequently the whole problem reduces to the question of the origin of the necessary small scale fields. Conceivably they could be generated in the magnetosphere, in the background ionosphere, or in the cloud itself. Because of the large ratio $\sigma_{\parallel}/\sigma_P$ of the parallel conductivity, $\sigma_{\parallel}$, to the Pedersen conductivity, σ_P, a flux tube

Fig. 2a. An ion cloud released near Thumba (India) at the dip equator. Striations are seen only on the top of the cloud.

'painted' within the cloud by resonantly scattering Ba ions could be pushed upon somewhere between the magnetosphere and the ionospheric E layer and still produce the visible irregularities. This is also why the latter are elongated along **B**.

The existence of such small scale fields (typically 0.2 to 2 km wide) much above the F region in the auroral ionosphere has been demonstrated by measurements on OVI 10 and OGO 6 (Heppner *et al.*, 1968; Heppner, 1969; Maynard and Heppner, 1970). Similar small spatial scales are seen in auroral forms, although it is not clear that the apparent motion of auroral rays is connected to electric fields at F region heights. It has been observed (Wescott *et al.*, 1969b; Heppner and Wescott, 1971) that auroral forms can move quite differently than ion cloud striations. The frequent occurrence

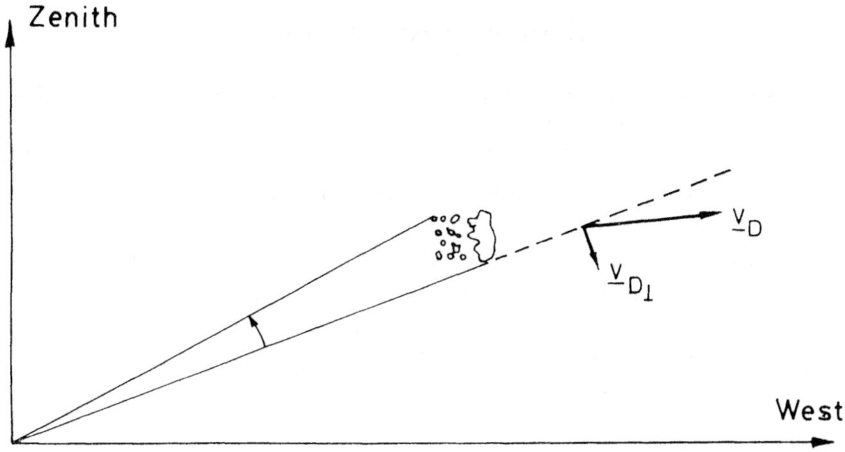

Fig. 2b. A schematic interpretation of picture 1a. The **B** field is directed out of the paper. The observer, located at the origin, sees the cloud drifting to the west and upwards. The cloud has a drift component $\mathbf{V}_{D\perp}$ perpendicular to the line of sight. If the striations, as shown here, remain in back of the cloud they appear located on its upper side.

of density irregularities in the lower ionosphere can be inferred from ionograms and star scintillation measurements. Indeed, in a theory of spread *F*, Reid (1968) has suggested that they may be responsible for ion cloud striations. There is also the possibility that ion clouds may become unstable thus producing their own striations (Haerendel *et al.*, 1969; Völk and Haerendel, 1969; Linson and Workman, 1970).

From mere optical observations of ion clouds one cannot distinguish between all these effects. Since, to our knowledge, there do not yet exist conclusive observations of irregularities in the background ionosphere on the field lines intersecting the clouds *before* and *during* ion cloud releases there is no way to decide which phenomenon caused the striations. If it was an effect impressed on the cloud from the ambient medium – and we argued that this appears quite possible – there is little more that can be said about the sources of the driving fields before more extensive overall observations are available. Ion clouds then act as tracers of these phenomena. As far as we consider ion cloud experiments entirely as a means of tracing ionospheric plasma drifts, such a striation would then be an indication of very short scale electric fields and their attendant ionospheric density modulations; whereas the motion of the cloud as a whole would indicate the large scale plasma motions.

It is quite important to see to what extent this point of view is justified. This can clearly be discerned if we investigate in which way the cloud itself can be responsible for the development of its own striations. To the point where a comparison with observations can be attempted it has not been done yet. We will, therefore, confine ourselves to a discussion of the physical processes involved. Although this is done for a specific instability these effects are of a general significance. They will always play an important role whenever ionospheric irregularity formation takes place.

2. The Polarization Instability

Figure 3 shows a model situation of this instability (Haerendel *et al.*, 1969). The cloud viewed along **B** is of infinite extent in x direction. The density profile in y direction is shown in the left part of the figure. For simplicity we assume only an E_x field to exist in the neutral wind frame; thus, the cloud drifts in negative y direction. If we now deform the boundary of the cloud in an incompressible fashion thus producing a density perturbation electric polarization $\mathbf{E}_p$ will be set up owing to the Pedersen current driven by the field **E**. The Pedersen current is carried by the ions because of their large transverse mobility as compared with the electrons. A perturbation of the back side of the cloud (with respect to [**E**, **B**]) will grow while the front side will be

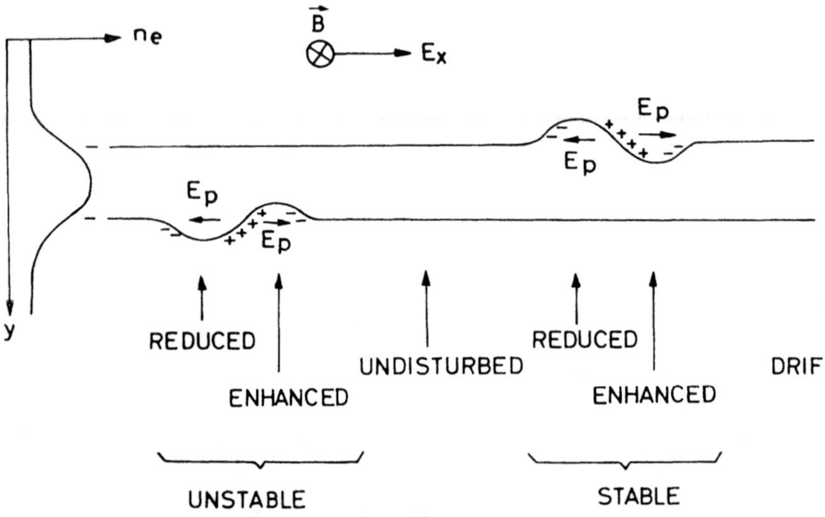

Fig. 3. Model of magnetic field aligned striations when viewed along the field lines: stable and unstable perturbations of density contours in an elongated inhomogeneity. A possible y component of E has been omitted (Haerendel *et al.*, 1969).

stable. If we were to interpret these growing perturbations as the striations we would see the latter to remain behind the cloud, i.e., to stick closer to the neutral atmosphere and the neutral part of the cloud. This is just what appears to be observed. The main question concerns the *growth times* of these perturbations. Obviously the growth time will depend on the strength of the polarization field E_p that can be built up by a density perturbation.

At this point the finite extent of the cloud along the B field becomes important, since we cannot disregard the effects simultaneously occurring in the ambient ionosphere along the field lines. Indeed, because of the large electron mobility along **B** any charge separation due to the higher mobility of the ions across **B** will be partly offset by background electrons flowing in and out of the striation along **B**. This is the same effect that occurs for the cloud as a whole (Figure 4). These electron currents will be closed by ion Pedersen currents which partly *short circuit* the polarization field. In addition, since we assume the background to be homogeneous transverse to **B**, these short circuiting currents will give rise to induced density variations in the background. We shall call them *image striations*. They, in turn, will polarize in the ambient field **E** and thus react back on the motion of the visible striations in the cloud.

Since these effects operate for any inhomogeneity of finite extent along **B** it may be asked why a striation in the cloud will polarize differently than the cloud as a whole, i.e., lead to its own growth. This is clearly a question of transverse scales. Although $\sigma_{\|}/\sigma_P \gg 1$, this ratio is not infinite and the distance $z=l$, to which a transverse field of scales applied at, say $z=0$ penetrates along **B** into the ambient medium, is $l \approx d\sqrt{(\sigma_{\|}/\sigma_P)}$ (Farley, 1959). We always take **B** parallel to the z direction. For the cloud as a whole d is rather large and the cloud can be very effectively depolarized; we will, in fact, assume that the cloud is practically nonpolarized (Haerendel *et al.*, 1967). In addition, if the cloud is not too dense and at the same time not too thin we will also neglect the

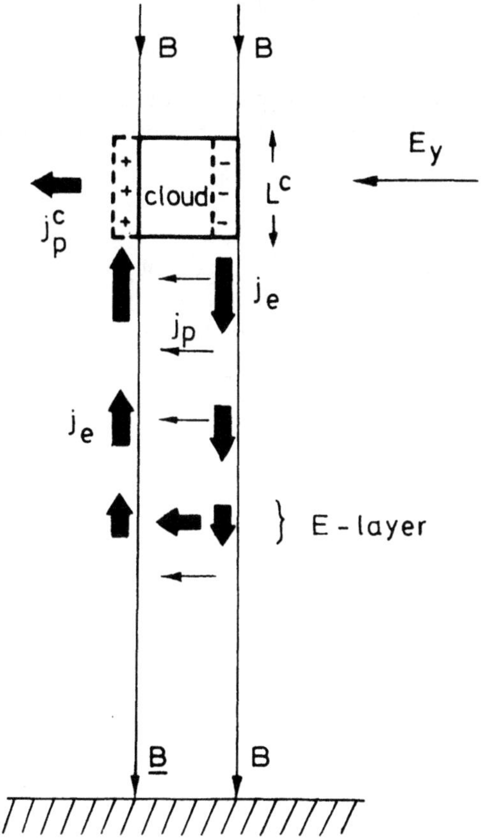

Fig. 4. The current cell due to an ion cloud. It allows an almost unimpeded Pedersen drift of the ions. The space charges developing at the front and the rear side (with respect to E_y) are neutralized through parallel electron currents and Pedersen currents flowing predominantly in the E layer where $\sigma_P{}^b$ is largest.

density variations it induces in the background (Haerendel and Papamastorakis, 1970); however, there have been experiments reported where this is not justified (Haerendel *et al.*, 1969).

For a striation which is necessarily of smaller transverse scale than the cloud, the penetration depth of the polarization field is also smaller and the Pedersen current integrated to the distance l along **B** is less effective in depolarizing the striation.

We treat this short circuiting effect in the next section, neglecting induced density variations. The latter is taken into account in Section 4.

3. The Short Circuiting Effect

Neglecting diffusion and gravity as small effects, the velocity **v** of a charged particle under the action of an electric field **E** is given by (Haerendel *et al.*, 1967):

$$\kappa_{\perp j} = \frac{c}{B} \frac{\kappa_j}{1 + \kappa_j^2} \mathbf{E}_\perp + \frac{c}{B^2} \cdot \frac{\kappa_j^2}{1 + \kappa_j^2} [\mathbf{E}, \mathbf{B}] \qquad (1)$$
$$\kappa_{\parallel j} = \kappa_j E_\parallel .$$

Here the subscripts $\perp$ and $\parallel$ mean 'perpendicular' and 'parallel' to **B**, which we take parallel to the z axis. The index j = i, e stands for ions (i) and electrons (e).

$$\kappa_j = \frac{q_j B}{m_j c} \cdot \frac{1}{v_j}$$

is the ratio of gyrofrequency to collision frequency v (with neutrals). At the cloud's level we have: $\kappa_i = 10^2$, $\kappa_e = -10^5$. In fact $\kappa_e/\kappa_i \approx 10^3$ throughout most of the ionosphere. Thus we can disregard the parallel ion motion. Because the ratio of plasma pressure to magnetic pressure is small, we can assume **E** to be quasistatic, i.e., curl free.

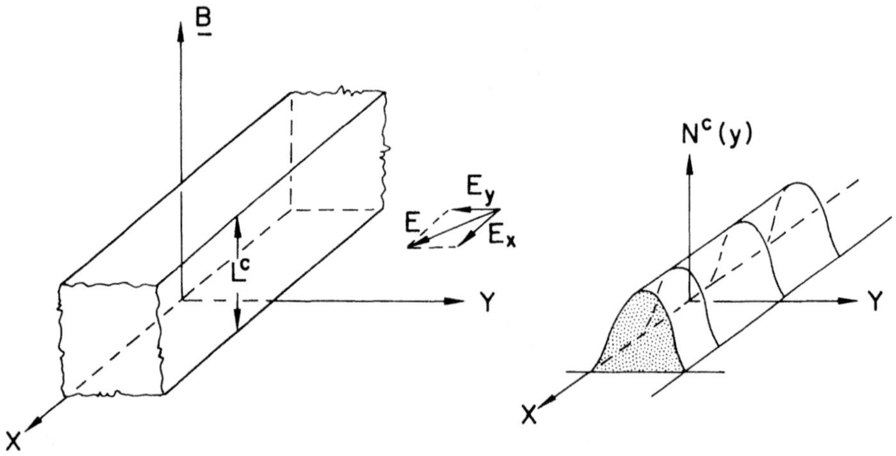

Fig. 5. The equilibrium model of the cloud. (i) It is infinitely extended in x direction, but of finite extent L^c parallel to B. (ii) A possible density profile in y direction.

We consider the model situation in Figure 5. The cloud has a finite extent L^c along **B**. For simplicity we take it of infinite extent in x direction but finite in y direction; thus $N^c = N^c(y)$. (The superscript c will denote cloud quantities.) Since the cloud is assumed to be nonpolarized, the ambient field **E** permeates the cloud and the background.

The triation of density $N^c + \delta n^c$ develops a polarization field, $\delta \mathbf{E}^c$, in the cloud. Under the action of the x component of this field it drifts relative to the cloud perpendicular to the unperturbed density contours. A small Pedersen drift parallel to $\delta \mathbf{E}^c$ is superimposed. Thus, from the ion continuity equation we obtain for the time rate of change of δn^c

$$\frac{\mathrm{d}}{\mathrm{d}t}\left(\frac{\delta n^c}{N^c}\right) \approx -\frac{1}{\kappa_i^c} \cdot \frac{c}{B} \operatorname{div}(\delta \mathbf{E}_1^c) - \frac{1}{N^c} \frac{\partial N^c c}{\partial y B^2} [\delta \mathbf{E}^c, \mathbf{B}]\, y\,. \tag{2}$$

Since the first term on the right hand side which describes the compressible part of the motion is very small ($\kappa_i^c \gg 1$) the growth of the striation is given by the second term. It describes the $\mathbf{E} \times \mathbf{B}$ drift due to $\delta \mathbf{E}^c$ across the density contours.

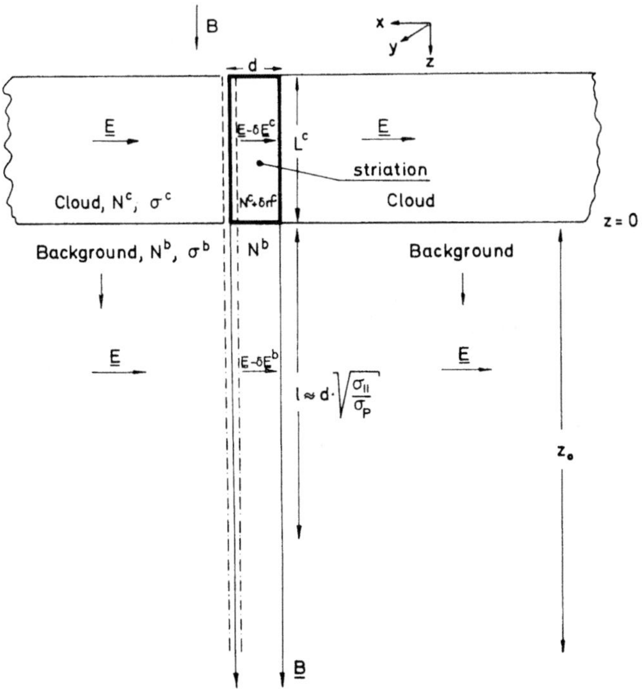

Fig. 6. Schematic drawing to illustrate the short circuiting effect. The continuity of electric current is calculated in the box made out of dashed lines (left boundary of striation) and broken dashes (in the background). Within the striation a polarization field δE^c exists, which is conducted along B into the ionosphere roughly to the distance l, giving rise to δE^b. The Pedersen conductivity in the background becomes very small at roughly

$$z_0 = \frac{\Sigma_P{}^b}{\sigma_P{}^b(z_0 = 0)}\,.$$

Now, to evaluate the time scales of growth we need to know the polarization field. For this we consider the current balance (Figure 6) in the boxes made out of dashes in the cloud and broken dashes in the background ionosphere (denoted by the superscript b). Since the layer above the cloud should not play a great role in the auroral zone, we neglect electron currents parallel to $\mathbf{B}$ at $z=-L^c$. Also, for large z (say, below the E layer) no such currents can leave the box. With these assumptions the current balance through the boxes reads:

$$L^c \cdot \{\mathbf{E} \cdot \delta\sigma_P^c + \sigma_P^c \cdot \delta\mathbf{E}^c\} + \int_0^\infty dz \sigma_P^b \cdot \delta\mathbf{E}^b = 0 . \tag{3}$$

Here $\delta\sigma_P^c$ is the Pedersen conductivity due to δn^c. The integral is to be extended over the whole lower ionosphere. Assuming δE^c to penetrate into the background to roughly the distance $l = d(\sigma_{\|}/\sigma_P^b)^{1/2}$, we have $\delta E^b = \delta E^c \cdot e^{-z/l}$.

In the case of a homogeneous background ionosphere Equation (3) is solved by:

$$\delta\mathbf{E}^c = \frac{\delta n^c}{N^c} \cdot \mathbf{E} \cdot \{1 + D\}^{-1} \tag{4}$$

where

$$D = d \frac{\sqrt{\sigma_{\|}^b \cdot \sigma_P^b}}{L^c \sigma_P^c} . \tag{5}$$

For a very long cloud ($L^c \to \infty$) or a thin striation ($d \to 0$), the 'depolarization factor' D goes to zero.

In reality the background ionosphere is not homogenous, and the Pedersen conductivity has a finite integral value, $\sum_P^b$. If the suitably generalized penetration length, l, becomes comparable to the spatial extent of the lower ionosphere, the depolarization is reduced. Formally this can be taken into account by replacing D in Equation (4) by $R \cdot D$, where the 'reduction factor' R is given by:

$$R(\chi) = \frac{1}{\chi} [\sqrt{1 + \chi^2} - 1] \tag{6}$$

and

$$\chi = \frac{1}{l(z=0)} \frac{\sum_P^b}{\sigma_P^b(z=0)}$$

D has to be taken at $z=0$. R assumes small values for very large l (thick striations). However, $R \cdot D$ approaches a finite limit, expressing the fact that the polarization field, δE^c, of the striation becomes equal to that of the whole cloud. In order to obtain the difference in polarization between cloud and striation, we, therefore, have finally to replace $\{1+D\}^{-1}$ in Equation (2) by $\{1+DR\}^{-1} - \{1+DR/_{l=\infty}\}^{-1}$.

Using δE^c from Equations (4), (5), and (6) in Equation (2), we obtain for the growth rate of the striation:

$$\gamma = \frac{\frac{\mathrm{d}}{\mathrm{d}t}\left(\frac{\delta n^{c}}{N^{c}}\right)}{\left(\frac{\delta n^{c}}{N^{c}}\right)} \approx \left(\frac{1}{N^{c}}\frac{\partial N^{c}}{\partial y}\right)\frac{c}{B}\,E\left[\{1+DR\}^{-1} - \{1+DR/_{l=\infty}\}^{-1}\right]. \qquad (7)$$

Thus, for a thick striation ($d \to \infty$), the growth rate goes to zero. A thin striation, on the other hand, is very weakly depolarized because the 'external load resistor' $1/l \cdot \sigma_{P}^{b}$ becomes very large. The growth rate then reaches its maximum value

$$\gamma_{max} = \frac{1}{N^{c}}\frac{\partial N^{c}}{\partial y} \cdot \frac{c}{B} E.$$

4. Image Striations

To see, how a density variation, δn^{b}, (image striation) is induced in the background by a striation, δn^{c}, in the cloud, we consider Figure 7. Inside the cloud the density

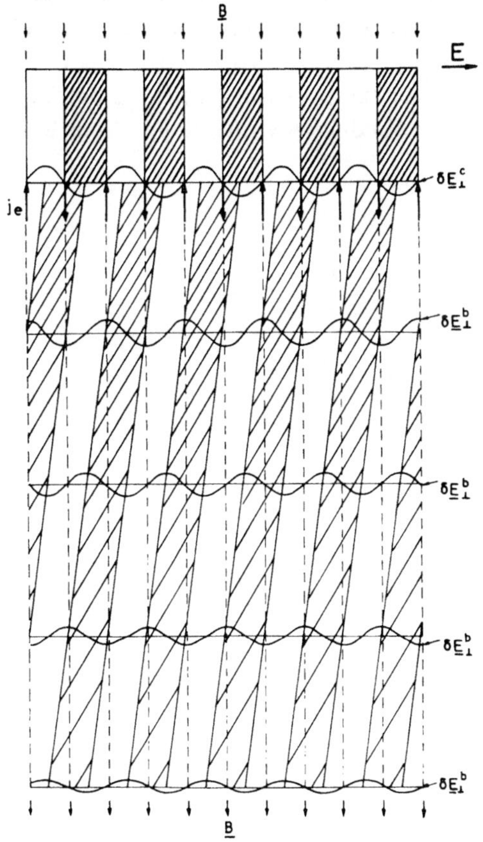

Fig. 7. The formation of image striations. They are tilted relative to B. Nevertheless, the electron currents j_e are strictly parallel to B.

perturbation δn^{c} and the attendant polarization field $dE_{\perp}^{c}$ are by 180° out of phase, since the Pedersen current must be conserved. Outside the cloud the phase relation between δn^{b} and $\delta E_{\perp}^{b}$ is different. Using equation (2) in the background:

$$\frac{\mathrm{d}}{\mathrm{d}t}\,\delta n^{b} = -\frac{N^{b}}{\kappa_{i}^{b}}\frac{c}{B}\frac{\partial}{\partial x}\,\delta E_{\perp}^{b}$$

we see that the maximum change of density occurs at the zeros of $\delta E_{\perp}^{b}$. The phase shift between δn^{b} and $\delta E_{\perp}^{b}$ is therefore only 90°. Since the electric field must be continuous in z-direction we conclude that, just underneath the cloud, δn^{b} is lagging δn^{c} by 90°. If we argue with the conservation of the Pedersen current as for the primary striations, there must be a contribution to $\delta E_{\perp}^{b}$ due to δn^{b}. This secondary field is again 180° out of phase with δn^{b} or by 90° with the primary polarization field. The progressive superposition of primary and secondary polarization fields leads to a total field, $\delta E_{\perp}^{b}$, whose phase shift relative to $\delta E_{\perp}^{c}$ grows with increasing distance from the cloud. As a result, the lines of $\delta E_{\perp}^{b}=0$ will be tilted with respect to the magnetic lines of force and so will the lines of maximum δn^{b}, which delineate the image striations. Their strength will drop off with increasing distance from the cloud. It is, however, still true that the electron currents that maintain quasineutrality, flow along **B**.

The relative amplitude of the polarization field is determined by the relative perturbation of the Pedersen conductivity integrated along **B** as far as the polarization

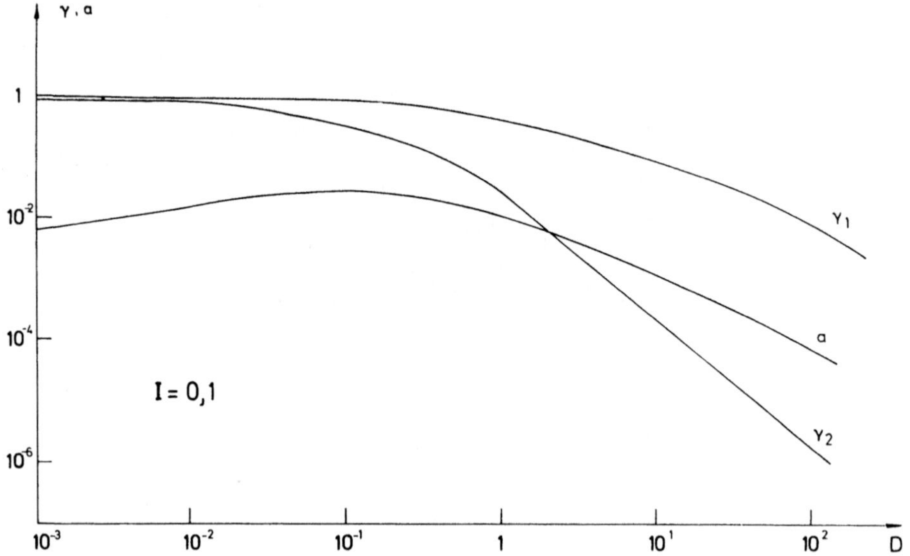

Fig. 8. A plot of the growth rate γ_1 and γ_2, normalized to the maximum growth rate

$$\gamma_{\max} = \frac{1}{N^{c}}\frac{\delta N^{c}}{\delta y}\frac{c}{B}E,$$

and of a, vs the depolarization factor D; $I = 10^{-1}$.

field extends. The existence of tilted image striations now has the consequence that the contributions to the integrated conductivity perturbation from the background are of opposite sign than those from the striations in the cloud. In certain situations then the relative perturbation of the total (cloud plus image striations) integrated Pedersen conductivity can become very small, thus leading to a strongly reduced growth rate.

Of primary interest in this context is the rate at which the image striations are built up as compared with that of the primary striations. We call the inverse of this ratio, I, the image parameter:

$$I = \frac{\dfrac{\mathrm{d}}{\mathrm{d}t}(\delta n^{\mathrm{c}} L^{\mathrm{c}})}{\dfrac{\mathrm{d}}{\mathrm{d}t}(\delta n^{\mathrm{b}} \cdot l)}.$$

I turns out to be independent of the width d of the striation. (The full expression for I will be given elsewhere.) Clearly, if the 'image parameter' I is large, the effect of the image striations will be small. Actually for $I \gtrsim 1$ it is negligible. Using the actual numbers, we find that I ranges roughly between 10^{-1} and 1; thus we expect an appreciable effect. As it turns out, the growth rate depends very sensitively on the parameter I. Figure 8 shows for $I = 10^{-1}$ a comparison of the growth rates, γ_1, normalized to the maximum growth rate for short circuiting only and γ_2, including the image effect. The growth rates are plotted as functions of D, which are proportional to d from Equation (5). Since D is observed to be of the order of 1 or larger, we see that the image effect is very large. The penetration depth l is also changed into l/a, where a, in the case of Figure 8 decreases strongly with increasing D. Thus, from these considerations it appears not too probable that ion clouds can produce their own striations, if these are not very thin. On the other hand, just because of the enormously increased penetration depth of the polarization field into the background ionosphere, a consideration of the real finite ionosphere appears necessary before we can draw reliable conclusions.

References

Farley, D. T.: 1959, *J. Geophys. Res.* **64**, 1225.

Föppl, H., Haerendel, G., Haser, L., Lüst, R., Melzner, F., Meyer, B., Neuss, H., Rabben, H., Rieger, E., Stöcker, J., and Stoffregen, W.: 1968, *J. Geophys. Res.* **73**, 21.

Haerendel, G. and Lüst, R.: 1968, in *Earth's Particles and Fields* (ed. by B. M. McCormac), Reinhold Book Corporation, New York, p. 271.

Haerendel, G. and Papamastorakis, J.: 1970, Abstract Ex 19, Frühjahrstagung der Deutschen Phys. Gesellsch., Braunschweig.

Haerendel, G., Lüst, R., and Rieger, E.: 1967, *Planetary Space Sci.* **15**, 1.

Haerendel, G., Lüst, R., Rieger, E., and Völk, H.: 1969, in *Atmospheric Emissions* (ed. by B. M. McCormac and A. Omholt), Van Nostrand Reinhold Company, New York, p. 293.

Heppner, J. P., Stolarik, J. D., and Wescott, E. M.: 1971, this volume, p. 407.

Heppner, J. P.: 1969, in *Atmospheric Emissions* (ed. by B. M. McCormac and A. Omholt), Van Nostrand Reinhold Company, New York, p. 251.

Heppner, J. P., Aggson, T L., and Maynard, N. C.: 1968, Abstract, Symposium on the Physics of the Magnetosphere, Washington D.C.

Linson, L. M. and Workman, J. B.: 1970, *J. Geophys. Res.* **75**, 3211.
Maynard, N. C. and Heppner, J. P.: 1970, in *Particles and Fields in the Magnetosphere* (ed. by B. M. McCormac), D. Reidel Publishing Company, Dordrecht, Holland, p. 247.
Reid, G. C.: 1968, *J. Geophys. Res.* **73**, 1627.
Rieger, E.: 1969, in *Low-Frequency Waves and Irregularities in the Ionosphere* (ed. by N. D'Angelo), D. Reidel Publishing Company, Dordrecht, Holland.
Völk, H. and Haerendel, G.: 1969, Abstract V-65, IAGA General Scientific Assemblies, Madrid.
Wescott, E. M., Stolarik, J. D., and Heppner, J. P.: 1969a, *J. Geophys. Res.* **74**, 3469.
Wescott, E. M., Heppner, J. P., Stolarik, J. D., Hallman, T., Davis, T N., and Romick, G. J.: 1969b, *Trans. Am. Geophys. Union* **50**, 281.

INSTITUTE DISCUSSION

It was questioned whether all Ba clouds become striated. According to E. Wescott most clouds become striated at least in the auroral zone.

D. Hunten pointed out that a geometry problem similar to that encountered in viewing rayed curves may exist, i.e., striations may appear very different if the cloud is viewed from above.

E. Wescott pointed out that in large Ba clouds a prominent feature is the steepening of the backside (nearest to neutral cloud) of the ion cloud. However, in some cases in the auroral zone it occurred on the front side. H. Völk indicated that his model cannot explain striations on the front side.

FIELD ALIGNED CONTINUITY OF HALL CURRENT ELECTROJETS AND OTHER CONSEQUENCES OF DENSITY GRADIENTS IN THE AURORAL IONOSPHERE

J. P. HEPPNER, J. D. STOLARIK and E. M. WESCOTT*

NASA Goddard Space Flight Center, Greenbelt, Md., U.S.A.

1. Introduction

Wescott *et al.* (1970), from a preliminary analysis of polar cap Ba^+ releases at 200 to 300 km, and Haerendel and Lüst (1970), from extrapolation of Ba^+ releases at 12.5 R_E, both concluded that the polar cap electric field was incompatible with the long standing assumption that the surface magnetic disturbance in the polar cap was caused solely by an overhead ionospheric sheet current. These findings created the question: what causes the polar cap magnetic variation that accompanies auroral electrojet activity? Initially, explanation was sought under the assumptions: (a) that a significant fraction of the disturbance was attributable to the polar cap ionospheric Hall currents required to complete the circuit of auroral electrojets, and (b) that the problem was to find another source of disturbance which would give the total observed disturbance when vectorially added to the disturbance from ionospheric currents. However, answers were not found under these assumptions and this prompted a more thorough examination. A variety of factors led us to conclude that practically all of the polar cap horizontal magnetic disturbance vector $\Delta \mathbf{H}$ had to be explained in terms of a non-ionospheric current. This conclusion created a second, two part, question: where and how do the Hall current auroral electrojet circuits close if not across the polar cap? A mid-latitude ionospheric circuit, also frequently assumed, was ruled out on grounds that both the nighttime electron density and magnitude of $\mathbf{E}$ were too low.

In subsequent interpretation efforts it became apparent that the Hall current continuity question had much in common with questions simultaneously being considered regarding mechanisms for generating Ba^+ striations, the decrease in $\mathbf{E}$ associated with auroral forms, etc. The framework for a new model was developed on the basis of the electric field data and the high latitude pattern of magnetic disturbances. Consideration of the effects of horizontal gradients in electron density N between 90 and 130 km is the principal factor that makes this model fundamentally different from past models. Relative to explaining observations, the effects have different consequences as a function of the dimensional scales considered as indicated by the following statements. Differential movement between adjacent flux tubes with cross section dimensions of roughly 0.1 to 10 km is believed to be associated with horizontal

* Now at the Geophysical Institute, University of Alaska, College, Alaska.

B. M. McCormac (ed.), The Radiating Atmosphere, 407–426. *All Rights Reserved.*

gradients (and/or discontinuities) in N existing over a similar small scale. This differential movement contributes to the spreading out and distortion of Ba^+ clouds that is visibly apparent in the form of striations. On a much larger scale the sign and magnitude of the ratio grad N/grad E, perpendicular to **E**, determine whether current flows 'into' or 'out of' the ionosphere along magnetic field lines and this current provides continuity for the Hall current electrojets. On the scale of the entire auroral belt-polar cap region the convective pattern and distribution of grad N/grad E combine to produce net field aligned currents 'out of' and 'into', respectively, the auroral belt magnetic time sectors 2000 to 2400 hr and 0800 to 1200 hr. These net currents, although locally irregular, will resemble field aligned sheet currents when viewed from a distance. On the low latitude side of the auroral belt (e.g., $\Lambda = 50$ to 62°) they produce smaller eastward and westward magnetic disturbance vectors, respectively, over broad regions centered near 2200 and 1000 hr in magnetic time.

In this presentation the polar cap electric field measurements that showed the need for a new interpretation of electrojet continuity are illustrated in Section 2. Section 3 notes alternative interpretations that were attempted but were found to be inadequate. The basis for a new model is given in Section 4 in three parts: (a) the continuity mechanism, (b) the field aligned current geometry dictated by surface magnetic disturbance, and (c) how the merging of continuity considerations with the convection pattern produces the required current distribution. In Section 5 we illustrate how striations in Ba^+ clouds can result from convective motion through ionospheric irregularities. For additional detail relative to Sections 2 to 4 the reader should see Heppner *et al.*, (1970). For the observations of Ba^+ cloud motions in auroral belt regions that demonstrate that auroral electrojets are primarily Hall currents, see Föppl *et al.* (1968), Wescott *et al.* (1969, 1970), Haerendel *et al.* (1969), and Haerendel and Lüst (1970).

2. Polar Cap Ba^+ Releases

Twelve Ba^+ clouds were released at 76 to 78° IN Lat from three rockets launched from the Pin-Main DEWline station, Cape Parry, N.W.T., Canada. Two flights took place during local evening twilight and the other was during local morning twilight. The three flights were on successive days and all occurred under conditions typical for disturbances with K_p near 3. Plots for the morning flight are given here to illustrate data characteristics (see Heppner *et al.*, 1970, for details).

Figure 1 shows triangulated locations of the four Ba^+ clouds vs time projected along magnetic field lines to a common reference plane at 100 km altitude. Release altitudes for the four clouds were, successively: 221, 286, 286, and 210 km. A total of ten K-46 aerial cameras and four all–sky cameras photographed the clouds from the three stations: Pin-Main, Pin-1 and Pin-2, also shown on Figure 1. Magnetometers were operated at Pin-Main and Pin-2. Relative to auroral zone releases where the clouds frequently become highly striated and breakup or spread out with increasing dimensions, the Figure 1 clouds remained compact with only several major striations per cloud.

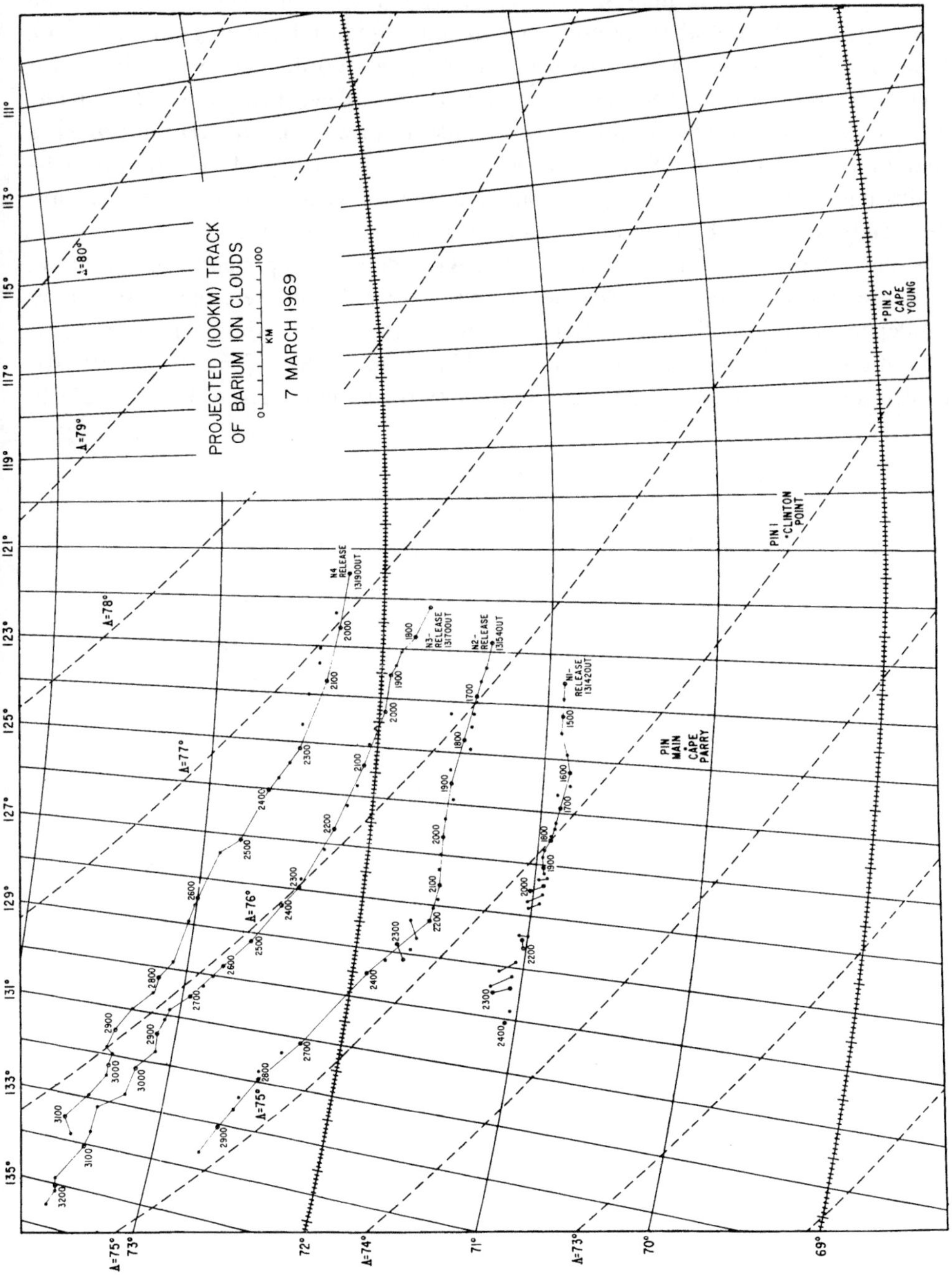

Fig. 1. Ba$^+$ cloud tracks vs time for morning launch on March 7, 1969. Locations at 20 sec intervals (dots) are given with UT, minutes and seconds, noted at 1 min intervals.

The instantaneous polar magnetic disturbance at the time of the fourth release is shown in Figure 2 together with the location of the tracks in magnetic time and geomagnetic latitude coordinates. The uniformity of the horizontal component of the polar cap disturbance when plotted in magnetic time and either geomagnetic or IN Lat coordinates, as opposed to using geographic coordinates, is a well known property of polar cap disturbances accompanying auroral electrojet activity (Heppner, 1967). In general, in winter months at $K_p = 3$ disturbance levels, the **ΔH** uniformity is apparent at >72° IN Lat on the nightside and >80° on the dayside. (Note: throughout this paper when we refer to polar cap disturbances we are not including the dayside variations which become prominent when solar elevation angles exceed 10° – particularly evident in summer months.) The **ΔZ** (vertical component) sign and magnitude are highly dependent on distance from the auroral electrojets and, as shown by an independent (unpublished) analysis, its polar pattern can readily be interpreted in terms of the distant effects of the electrojets. The distant effects of the auroral electrojets also affect, to a lesser degree, the uniformity of the polar cap **ΔH** – particularly

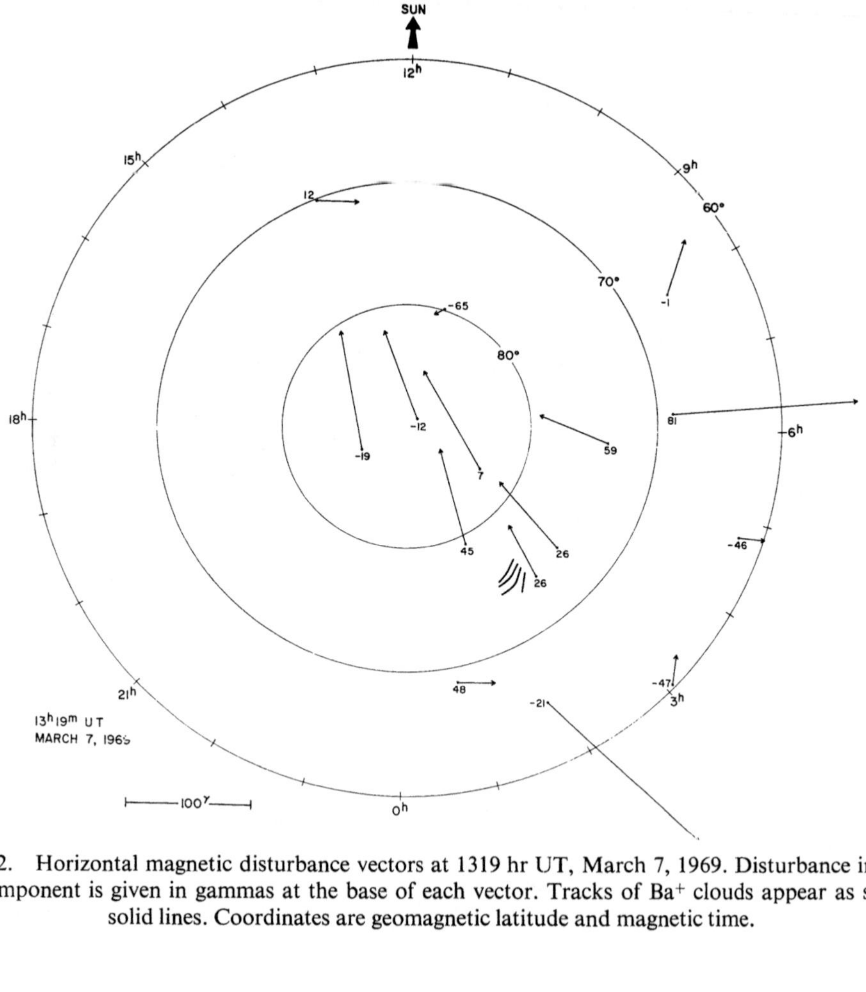

Fig. 2. Horizontal magnetic disturbance vectors at 1319 hr UT, March 7, 1969. Disturbance in the Z component is given in gammas at the base of each vector. Tracks of Ba^+ clouds appear as short solid lines. Coordinates are geomagnetic latitude and magnetic time.

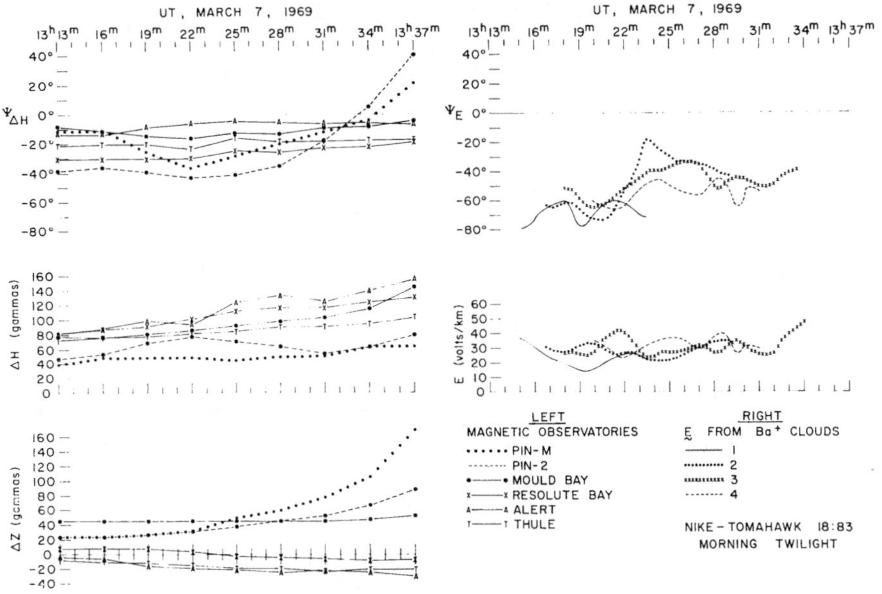

Fig. 3. Electric field measurements and polar cap magnetic disturbance vs time for morning flight on March 7, 1969. See text and Figure 4 for definition of angles ψ_E and $\psi^{\Delta}{}_H$.

for stations closest to the electrojet. For this reason at the IN Lat of the Pin stations, (near 75°) the polar cap **ΔH** is contaminated by the distant field of the auroral electrojets to the south to a greater degree than stations at higher latitudes. The best index for knowing that polar cap conditions are being measured thus comes from knowing the location of the closest aurora. In the case of this flight (Figures 1 to 3) aurora was not visible from the Pin stations during the tracking period or during pre-launch preparations. However, following the data period, about 1335 UT, aurora was observed near the southern horizon of Pin-Main at an elevation too low to appear in all-sky photographs. Magnetic records from Alaska observatories indicate that the aurora and electrojet moved northward during the flight but at the closest approach, near the end of tracking, the IN Lat separation between Ba^+ clouds and aurora was still $>6°$.

The reader should be aware that the term 'uniform' for the polar cap disturbance is relative in that it does not exclude fractional differences between observatories. These will always be present as a consequence of differences in the local earth induction. On a regional scale differences can also result from non-uniformities in the distribution of auroral belt field aligned currents as described later. For these reasons, the existence of differences does not constitute evidence for local ionospheric currents.

The polar cap magnetic disturbance throughout this morning flight is plotted in Figure 3 (left side). $\Psi_{\Delta H}$ is the horizontal angle between the **ΔH** disturbance vector and the direction of the sun as illustrated in Figure 4. The narrow ranges for $\Psi_{\Delta H}$ and ΔH are indicative of what is meant by a uniform disturbance. Values for the Pin

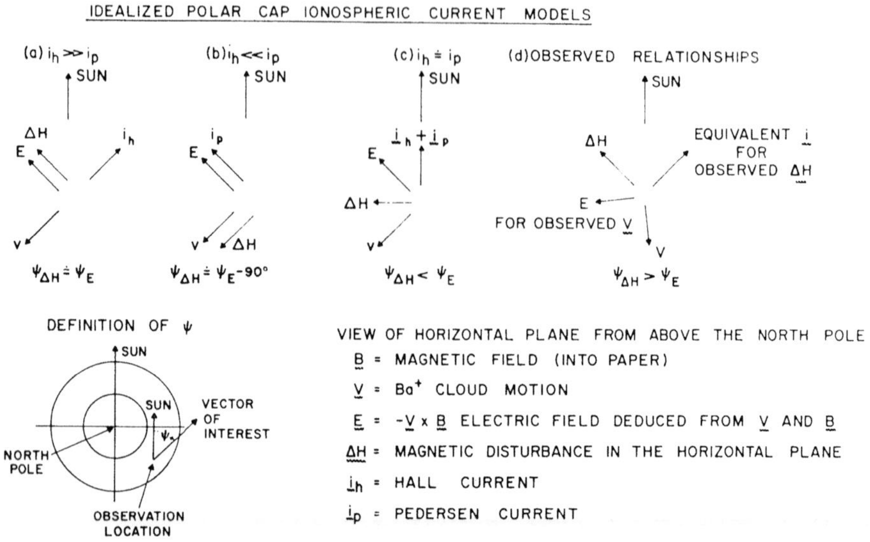

Fig. 4(a)–(c). Vector relationships for ionospheric sheet currents. (d) Typical observed relationship between **V** (or **E**) and **ΔH**.

stations are similar but it is apparent that they are contaminated by the superposition of a field from the auroral electrojet to the south. Baselines for determining the disturbance vectors are believed to be highly accurate.

The total **E** vector (Figure 3, right side), calculated from $-\mathbf{V} \times \mathbf{B}$, is described by |E| and Ψ_E, the horizontal angle between **E** and the direction of the sun, in the absence of magnetic field aligned electric fields. The assumption that **E** along **B** is negligibly small is based on the fact that fall rates for the Ba^+ clouds were normal (i.e., the same as for the neutral clouds). The Figure 3 curves for |E| and Ψ_E are smoothed curves in that each curve is the average of three curves. The original three curves were produced by determining the velocity vector **V** over 1 min intervals three times per min by using positions successively from 00 to 00 sec, 20 to 20 sec, and 40 to 40 sec. The difference between the three curves and an average curve gives an approximate measure of the combined magnitude of true irregularities in **E** and triangulation errors washed out by the smoothing process. For |E| and Ψ_E these differences were, respectively, <5 V km^{-1} and $<10°$ except for an interval near 1323 hr for cloud 2 and several points near the end of the observations. This implies that differences exceeding 5 V km^{-1} or 10° along the curve for a single cloud or between clouds in Figure 3 are real. Despite the 'bumpiness' of the average curves the measured **E** is considerably more uniform in time and space than electric fields encountered in auroral belt regions.

Relative to later discussions it is important to note: (a) that the mean |E| did not increase with time although $|\Delta H|$ did, and (b) that the shift in Ψ_E centered on the interval 1322 to 1323 hr was not accompanied by a correlated change in $\Psi_{\Delta H}$. This lack of correspondence between Ψ_E and $\Psi_{\Delta H}$, and **E** and **ΔH**, argues against attributing the polar cap $\Delta\mathbf{H}$ to overhead ionospheric currents. However, an even more important feature of the data is the displacement between Ψ_E and $\Psi_{\Delta H}$. As described in the next section these angles should coincide if $\Delta\mathbf{H}$ is caused by Hall currents in the form of an ionospheric sheet. Roughly the difference between $\Psi_{\Delta H}$ and Ψ_E in Figure 3 averages 20 to 40°. For the other two flights (evening launches) the angular discrepancy was in the same direction but greater – typically 40 to 90°.

3. Models that Fail

It is apparent that the angular difference between $\Psi_{\Delta H}$ and Ψ_E cannot be a consequence of the disturbance being either totally or partially caused by Pedersen currents $\mathbf{i}_p$ as has sometimes been assumed (e.g., Taylor and Hones, 1965; Cole, 1963) although this possibility (i.e., that the Pedersen conductivity σ_1 might be comparable to the Hall conductivity σ_2 in the polar cap) was one of the factors that initiated our polar cap experiments (Heppner, 1969). As illustrated in Figure 4b and c an overhead ionospheric Pedersen current gives an angular discrepancy of the opposite sign to that observed. Cowling currents, proportional to $\sigma_1 + \sigma_2^2/\sigma_1$, are similarly ruled out as their vector properties relative to $\Delta\mathbf{H}$ and $\mathbf{E}$ are identical to those of $\mathbf{i}_p$. They would, of course, only exist if non-uniformity was introduced in the Figure 4 idealization. The idealization also ignores questions of continuity in looking at only the effect of local currents but the more distant parts of a circuit cannot be expected to produce a larger opposite effect.

Auroral belt observations are best represented by Figure 4a (omitting the sun reference) and this also comes closest to the polar cap observed relationships illustrated by Figure 4d. We thus initially viewed the problem in terms of finding the source for a component $\Delta\mathbf{H}_e = \Delta\mathbf{H} - \Delta\mathbf{H}_h$ which vectorially added to $\Delta\mathbf{H}_h$, the field from an overhead Hall current, would give the observed $\Delta\mathbf{H}$. Continuity of auroral electrojets required that $\Delta\mathbf{H}_h$ be a significant fraction of $\Delta\mathbf{H}$. This consideration, and the $\Psi_{\Delta H} - \Psi_E$ difference, indicated that a $\Delta\mathbf{H}_e$ of comparable magnitude directed approximately toward the sun would be required.

Geometrically, two field aligned sheet currents centered in longitude, respectively, 0600 hr (current into the auroral belt ionosphere) and 1800 hr (current out of the auroral belt ionosphere) would produce a sunward $\Delta\mathbf{H}_e$. These locations and signs for field aligned currents were previously suggested by Fejer (1961) and Akasofu (1970). However, these models encounter serious difficulties. Fejer (1961) connected his dawn and dusk field aligned currents with a polar cap Pedersen current $\mathbf{i}_p$ that would produce polar cap magnetic effects contrary to the Ba^+ release results. Akasofu (1970) connected his dawn and dusk field aligned currents with an auroral electrojet band in which the ionospheric current is a Cowling current that is continuously westward from dawn to dusk. Thus he completely ignored the reversal of the electric field between morning and evening electrojets and other results from past auroral belt investigations (e.g., see references at end of 'Introduction'). However, independent of the objections to the ionospheric circuits proposed, it appeared obvious to us from various past analyses, and was confirmed by new analyses, that field aligned sheet currents centered at 0600 and 1800 hr would produce magnetic effects on the low latitude side of the auroral electrojets that were not consistent with observations. This is discussed further in Section 4B.

Another possible source for a sunward $\mathbf{\Delta H}_e$ is the polar field produced by the combined effect of cavity confinement of the magnetosphere and magnetospheric inflation. For example, to illustrate existence, the Williams and Mead (1965) model gives a sunward horizontal component of 3.2 γ at the poles. However, this model ignores internal magnetospheric plasma pressures for $R_E < 10$ which are known to be large (e.g., Coleman and McPherron, 1970; Sugiura *et al.*, 1970). We noted that inclusion of this inflation in cavity models would increase the computed sunward field at the poles in obtaining the same computed magnetopause configuration. Also, the observation (Sugiura *et al.*, 1970) that magnetospheric fields at mid-latitudes on the nightside were much stronger than computed fields was indicative that the real magnetopause at polar latitudes might be closer to the earth than computed. In total, the differences between model magnetospheres and observed field distortions are such that a stronger sunward component in polar regions is indicated. However, we were unable to justify a $\mathbf{\Delta H}_e$ magnitude of sufficient intensity by this means. The logic of relating solar wind compression this closely to electrojet activity was also questionable. It should, be recognized, however, that a time variable sunward field of this nature is always present. In subsequent modeling, (Section 4B), we regard this field as a source of error.

Explanation for the difference between $\mathbf{\Delta H}$ and $\mathbf{\Delta H}_h$ was also sought in terms of earth induction effects and ionospheric winds. Reasons for considering and rejecting these sources are outlined in Heppner *et al.* (1970).

4. A New Model

The following factors (lettered roughly in the time sequence in which they were considered) led us to conclude that the polar cap $\mathbf{\Delta H}$ had to be explained almost entirely in terms of an external source (i.e., $\mathbf{\Delta H} \doteq \mathbf{\Delta H}_e$ with $\mathbf{\Delta H}_h \doteq 0$): (a) inability to explain $\mathbf{\Delta H}$ by a vector addition $\mathbf{\Delta H}_h + \mathbf{\Delta H}_e$ (Section 3), (b) lack of correlation between time changes in $\mathbf{E}$ and $\mathbf{\Delta H}$ (one example given in Section 2), (c) lack of reasons for assuming that polar cap conductivities change proportionally with auroral electrojet changes as required to have electrojet and polar cap disturbances in phase, or alternatively, a lack of unique relationships between the magnetosphere and ionosphere that would make $\langle \sigma \rangle$ $\mathbf{E}$ proportional in polar cap and auroral regions, (d) noting from past experiments that auroral electrojet currents comparable to currents required for the polar cap $\mathbf{\Delta H}$ do not appear unless aurora is visible, and (e) lack of both theoretical and observational evidence that polar cap electron densities at 100 to 130 km altitude are sufficient to support a Hall current of the required magnitude.

The latter point is particularly significant as conductivities used in various models are often rather arbitrarily assumed. Our literature search revealed only that polar cap electron densities at 100 to 130 km must be very low. Ionosondes, which sweep upward in frequency from about 1.0 mHz, indicate that the density is $<10^4$ cm^{-3} because the *E* region reflections disappear during the dark hours (Penndorf, 1965). Model calculations by Keneshea *et al.* (1970) and Lerfald and Little (1970) indicate that values near 10^3 cm^{-3} may be realistic. The possibility that a significant wide spread *E* region ionization is created by particle influx appears remote in view of polar cap airglow measurements (Eather, 1969). A mid-latitude nighttime ionospheric return circuit for the electrojets is ruled out on the dual grounds of low electron density and very weak electric fields.

The two problems consist of (a) explaining the polar cap **ΔH** and (b) finding an alternate circuit for continuity of auroral electrojets; they were initially pursued separately in our analyses. Their merger to provide the framework for a new model was a natural, unforced, consequence of conclusions reached in the independent answers to each problem. Accordingly, below, we illustrate the two problems separately before showing how they merge to provide a model.

A. ELECTROJET HALL CURRENT CONTINUITY

Figure 5 illustrates in simplest form the consequences of a flux tube convecting across a discontinuity (or gradient) in plasma density in the lower ionosphere. In the upper ionosphere both ions and electrons are transported from position 1 to position 2 in the earth's frame. In the lower *E* region (defined approximately by $\nu_{in} > \omega_i$ where ν_{in} = ion-neutral collision frequency, ω_i = ion gyrofrequency) ions are restrained by collisions with neutrals such that their bulk velocity is small compared to the electron bulk velocity which is essentially unattenuated relative to the convective **V**. As is well known this difference in ion and electron velocities constitutes the Hall current in a uniform ionosphere (Figure 5a). However, where there is a spatial change (or localized time change) in the plasma density, either current must flow along field lines or a net charge is developed as illustrated by Figure 5b and c. This is a recognized

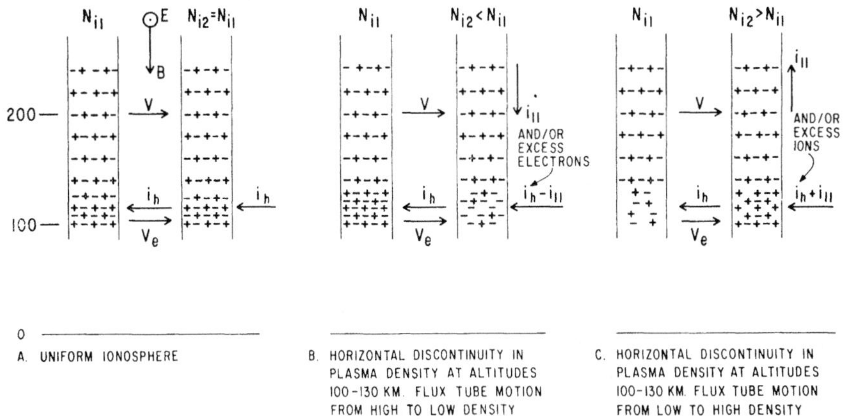

Fig. 5. Idealized effect of a plasma density discontinuity on Hall current continuity.

fact that has been ignored in past analyses except for those analyses that proceed under the assumption that charge accumulates in the lower ionosphere. Quantitative treatments of the problem posed do not exist in the literature and our analyses have not proceeded far enough to provide one here. We can, however, use observations as a guide and these indicate that the accumulation of a net charge is relatively insignificant in the continuity problem. By insignificant we mean that charge does not accumulate to the extent of producing a horizontal polarization electric field comparable to the convective **E**. As there is not a theoretical basis for assuming that a significant net charge can accumulate in the lower ionosphere this may not appear surprising. However, charge accumulation is often invoked in electrojet mechanisms and if followed, one arrives at models quite different from the one proposed here. Small scale (i.e., ray dimension) polarizations (Section 5) are known to be weak from satellite observations. Strong fields indicative of polarization at the boundaries of auroral forms are not indicated by probe measurements and are not observed when an auroral form crosses the path of a Ba^+ cloud.

In total, the above implies that we can examine the Hall current continuity in terms of the convective **E** field without introducing other fields. Similarly this means that the local effect of Pedersen currents is simply one of decreasing the magnitude of **E** in that **E** is maintained perpendicular to $\mathbf{i}_h$. This condition is considered applicable when maximum ionization exists between 100 and 130 km. Our model considerations would have little application in cases where the ionization was weak below roughly 130 km but intense at altitudes of 130 to 150 km. This is a relatively inconsequential point, however, in that various observations suggest that the existence of electrojet magnetic disturbances of appreciable intensity is dependent on the ionization being intense below 130 km. This is expected from the altitude dependencies of σ_1 and σ_2 and explains why we speak of 'plasma density at 100 to 130 km' rather than integrated conductivities.

Figure 6 illustrates the continuity problem for a geometry more applicable to electrojet currents. The condition C>1 which states that $|\mathbf{E}|$ or $|\mathbf{V}|$ is decreased within the shell of enhanced density is taken to be observational fact as it has been repeatedly verified by measurements (e.g., Aggson, 1969; Wescott *et al.*, 1970; and more recent unpublished observations). The N-S density distribution is not critical. Case (1), Figure 6, shows the consequences of restricting current to the ionosphere. Case 2(A) is included simply to indicate that the end point polarization, shown for accuracy in Case (1), is not as fundamental to the problem as the consequences of having C$\doteq$K, as opposed to the condition K>C shown as Case 2(B). Thus, the discussion of Case (1), below, also applies to Case 2(A).

$i_{\parallel}=0$, Case (1), Figure 6, requires that the electric field (or velocity) decrease in direct proportion to the increase in density (i.e., grad (NE)=0 in all horizontal directions). This means that there is not a concentration of current in the high density strip and this consequence of $i_{\parallel}=0$ is hardly acceptable in terms of various rocket and surface measurements indicating an enhanced current in space-time association with visual aurora. The proportionality $N \propto 1/V \propto 1/E$ might not appear

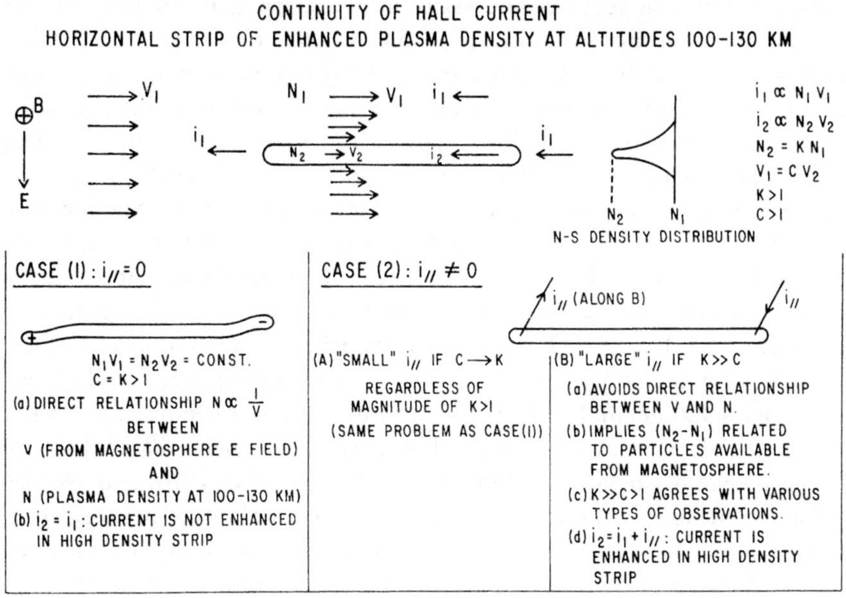

Fig. 6. Idealized illustration of continuity problem for a strip of enhanced plasma density (e.g., an auroral arc) aligned along E-W equipotentials of the convective field. View of horizontal plane between 100 and 130 km. $N = N_e = N_i$ = electrically neutral plasma density.

objectionable in view of observations that require $C > 1$, noted above. However, there are a number of reasons for believing that the proportionality cannot be exact. One is that observations indicate that $\Delta N/N$ is more variable than $\Delta E/E$ which is discussed later. Second, there does not appear to be any logical basis, or mechanisn, for assuming that N at 100 to 130 km could be this closely tied to E if one assumes that the convection is driven by the solar wind. Third, $N \propto 1/E$ leads to the observational absurdity that E becomes extremely large when and where N drops to a low value. Thus if a unique model to satisfy $N \propto 1/E$ could be found it would also require a unique threshold condition for applicability. Lack of any cause and effect relationships that would relate the existence of electrojets to increased ionization is another factor which makes one question the logic for Case 1.

Case 2(B) (Figure 6), represents a model behavior which is more realistic for the large scale properties of auroral electrojets. There is agreement with observations, the logic is not restrictive, and cause and effect is implied. **E** in region 1 (Figure 6), or in the absence of density enhancements, is representative of the driving convection unattenuated by ion drag or Pedersen current shorting; unlike Case (1), it can be weak without increasing N. A concentration of current in high density regions is a necessary consequence of having (grad N)/$N >$ (grad E)/E parallel to **E**. Field aligned currents flow in accord with the sign of grad N/grad E for (grad N)/$N >$ (grad **E**)/**E** perpendicular to **E**.

Relating Case 2(B) (Figure 6) to phenomena becomes a function of the scale considered. Small scale features are discussed in Section 5. For scales comparable to those enclosing major auroral forms (e.g., one to tens of kilometers wide and several hundred to several thousand kilometers in length) typical electron densities and values for $\Delta N/N$ can be estimated from rocket measurements and expectations from the luminosity relationship $N = k$ (brightness)$^{1/2}$. As reviewed in Heppner *et al.* (1970), with particular attention to the measurements of McNamara (1969) and Baker *et al.* (1967), the rocket measurements are in reasonable agreement with tabulations of N vs IBC (the International Brightness Coefficient) (Dalgarno *et al.*, 1965; Omholt, 1966) but indicate that the IBC values of N could be too low by a factor of 2. Using the IBC scale as a guide, conservative values are: $N = 10^6$ cm^{-3} for moderate intensity discrete forms, $N = 10^5$ cm^{-3} for glows and the fringe regions of discrete forms, and $N < 10^4$ cm^{-3} for regions remote from aurora. The latter figure could be as low as 10^3 cm^{-1} below 130 km in the absence of auroral excitation as noted previously.

Magnitudes for **E** relative to auroral activity or magnetic disturbance intensity cannot be as readily categorized. Both Haerendel and Lüst (1970) and Wescott *et al.* (1970) have concluded: (a) that there is not a simple relationship between $|E|$ and the magnitude of magnetic disturbance, and (b) that the conductivity plays a greater role than $|\mathbf{E}|$ in determining the electrojet intensity. These conclusions alone support the Case 2(B) model but we can attempt being somewhat more quantitative. In all documented cases where measurements have occurred within the magnetic shell of an auroral form $|\mathbf{E}|$ has been observed to be weak: typically $\leqslant 10$ and $\leqslant 5$ V km^{-1}. The converse, that $|\mathbf{E}|$ is always large away from active shells, does not hold in that both values > 100 V km^{-1} and < 5 V km^{-1} have been observed within < 50 km of visible aurora. However, if one neglects the extremes which can probably be attributed to variations in the driving convection, the range 10 to 50 V km^{-1} is representative of most measurements remote from visible aurora but within and adjacent to the general belt of nighttime auroral activity.

In general the above implies that $|\mathbf{E}|$ most commonly varies by a factor $\leqslant 10$ between highly conducting and poorly conducting regions. This is to be compared with factors of 10 for N between a highly conducting auroral strip and its fringe regions and factors of 100 to 1000 relative to remote regions. Thus it is locally feasible for Case 2(A) to be more representative than Case 2(B) when and where there is sufficient diffuse particle precipitation to produce an ionization $\geqslant 10^5$ cm^{-3} below 130 km surrounding discrete forms. In these cases the electrojet may more closely resemble a latitudinally confined horizontal sheet current than a line current. Rocket measurements conducted with multiple forms present (Burrows *et al.*, 1971) do in fact indicate this variability between a line and sheet configuration. In the most general case a behavior somewhere between that of Cases 2(A) and 2(B) is to be expected on the local scale of an auroral form and its immediate surroundings.

Extending the scale to several degrees of latitude and several hours in local time, to include multiple forms and longitudinal changes in form, and comparing this with adjacent regions lacking auroral ionization, the Case 2(B) behavior has to dominate. These scale considerations demonstrate that $i_\parallel$ is not necessarily confined to regions of maximum luminosity; also because it is the difference between the field aligned flux of electrons and ions that gives an $i_{\parallel}$, the local $i_{\parallel}$ is not necessarily proportional to the precipitation flux. Cause and effect is however clearly implied in that $i_{\parallel}$ does depend on the ratio grad N/grad E perpendicular to **E** and the distribution of grad N is determined by the distribution of the precipitating (i.e., ionizing) particles.

B. THE REQUIRED FIELD ALIGNED CURRENT DISTRIBUTION

An auroral belt field aligned current distribution that explains the polar cap $\mathbf{\Delta H}$ must also be compatible with the magnetic disturbance observed on the low latitude side of the electrojet region. As noted in Section 3 this test ruled out the possibility of centering field aligned sheets at 0600 and 1800 hr to explain a solar directed $\mathbf{\Delta H}_e$. Compatibility does appear however, when the field aligned sheets are shifted to explain the total polar cap $\mathbf{\Delta H}$. As illustrated in Figure 7b field aligned sheets oriented to give the polar cap $\mathbf{\Delta H}$ should produce a disturbance on the low latitude side that will appear primarily as $+\Delta Y$ (i.e., east declination) and $-\Delta Y$ (i.e., west declination) variations, respectively, in the sectors of outward and inward field aligned current. Viewed statistically, such that random non-uniformities in the sheets disappear, the ΔY magnitude should approach $\frac{1}{2}\mathbf{\Delta H}$ near the meridians of the centers of the sheets within roughly 10° of latitude on the low latitude sides.

The above expectation was tested using three independent approaches involving large quantities of data (Heppner *et al.*, 1970); however as the properties of ΔY between roughly 50 and 62° are evident from past statistical representations of equivalent current systems for bay disturbances, we will only note their verification of the ΔY shown in Figure 7(b). Taking the representations of Silsbee and Vestine (1942) and Nagata and Kokubun (1960) as examples, the reader should note the following:

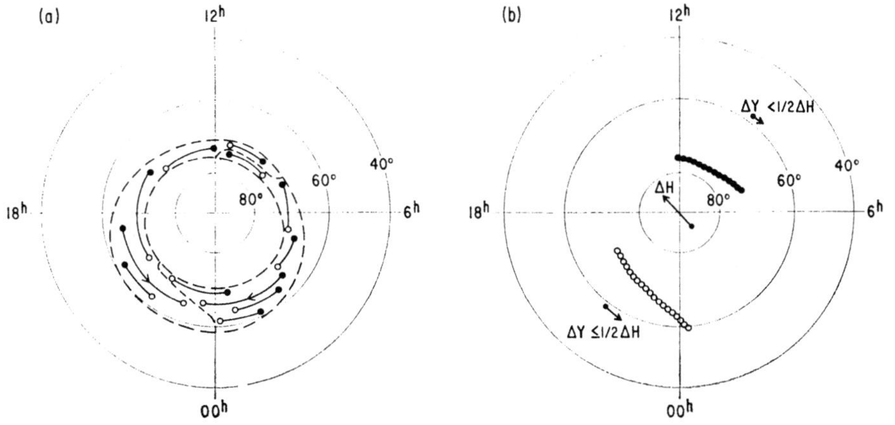

Fig. 7. Schematic illustrations of: (a) a distribution of Hall current electrojets terminated in the ionosphere by currents along magnetic field lines, and (b) the net distribution of electrojet terminations resulting in 'sheet' configurations for the net field aligned current. ΔH and ΔY vectors indicate, respectively, polar cap and lower latitude (e.g., $\Lambda = 50$–60°) magnetic disturbances attributed to the sheet currents. Open and filled circles, respectively, indicate currents 'out of' and 'into' the ionosphere.

(a) that a line drawn parallel to the polar cap currents through the center of the polar plots is also parallel, within roughly ± 1 hr, to the current drawn to lower latitudes from the regions of transition between westward and eastward auroral belt currents, (b) that the vector signs agree, and (c) that magnitudes, as illustrated by the spacing between current lines, give a mid-latitude $|\Delta Y|$ slightly less than one-half the polar cap $|\Delta H|$ in the late evening quadrant and a lesser magnitude for $|\Delta Y|$ at the same latitudes in the late morning quadrant. In total, for latitudes $>50°$ the meridianal ionosphere currents shown entering and leaving the electrojet regions in these representations produce magnetic effects similar to those of field aligned sheet currents.

The statistical line of symmetry (i.e., a common perpendicular to the polar cap $\mathbf{\Delta H}$ and the maximum lower latitude $|\Delta Y|$) varies between different analyses: for example, roughly 0800 to 2000 hr in Silsbee and Vestine (1942), and a range, from curvature, of 0800 to 1000 hr to 2000 to 2200 hr in Nagata and Kokubun (1960). Placing the maximum ΔY vector near 2130 hr in Figure 7b is based on our studies; however, the exact value is not too meaningful as these statistics show a broad region, essentially 2000 to 2400 hr MLT. Variability is to be expected both from the angular variability of the polar cap $\mathbf{\Delta H}$ and the variability of the transition regions between westward and eastward currents. As discussed by Heppner *et al.* (1970) verification of the late morning $\mathbf{\Delta Y}$ is more subject to error than the late evening $\mathbf{\Delta Y}$.

C. FIELD ALIGNED CONTINUITY RELATED TO THE PATTERN OF CONVECTION

The fact that the normal to the polar cap $\mathbf{\Delta H}$ and lower latitude $\mathbf{\Delta Y}$ vectors intercepts the MLT sectors where the auroral belt **E** fields and electrojet currents reverse is not likely to be just coincidence. The boundaries between eastward and westward Hall current belts in Figure 7a are rough averages taken from the patterns given by Heppner (1969). As suggested by Figure 7b, placement of sheet currents in the reversal regions gives an appropriate geometry.

Although the geometry indicated by Figure 7 is appropriate, a static statistical picture of this type is not directly physically applicable in terms of the preceding continuity discussion. One reason is that it ignores the timespace variability of the reversal region (Heppner 1967, 1969) but the principal deficiency is that the grad N and grad **E** relationship are obscure. For example, in Figure 7a the dashed line between evening eastward and midnight westward currents might be interpreted as representing a narrow zone of electron density and taken literally this would be misleading. The detailed convection pattern and the dynamic behavior in the reversal region has to be considered. This behavior is too complicated and varied to treat briefly but in simplest form one need only recognize that the convection, unlike the current, is continuous. That is, the E-W motions paralleling the conducting strips are a continuation of a flow from higher latitudes. The principal is illustrated by Figure 8a in terms of a uniformly conducting auroral belt with electron density N_2 and a poorly conducting polar cap with density N_1. In this case an idealized polar cap boundary becomes the ionospheric intercept of the field aligned currents. Figure 8b is a step closer to reality in that the uniform belt is replaced by conducting strips.

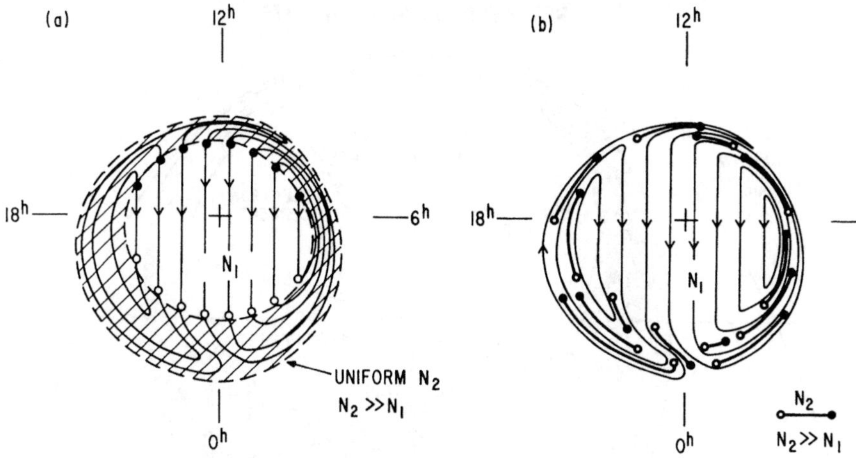

Fig. 8. Illustration of the change in distribution of ionospheric terminations of field aligned currents in going from a case, (a), where the entire auroral belt is uniformly conducting, to a case, (b) where high conductivity exists only in strips parallel to the convection. Continuous lines form the idealized convection pattern. N = plasma density at 100 to 130 km altitude. Open and filled circles, respectively, indicate current 'out of' and 'into' the ionosphere along magnetic field lines.

It is obvious that obtaining a sheet configuration like Figure 7b from the principle illustrated in Figure 8a depends primarily on the detailed configuration of the convection and the distribution of precipitating particles producing ionization below 130 km. Data regarding the convection pattern are sufficient to permit an idealization in the pre-midnight sector like that shown in Figure 8. It is consistent with both electrojet current patterns (Heppner, 1967, 1969) and Ba^+ cloud motions observed in the reversal region (Wescott *et al.*, 1970 and more recent unpublished data). Particle data are not comprehensive enough to indicate where maximum ionizing precipitation occurs. However other indicators of the ionization, such as auroral brightness, radio wave absorption, and peak values in the spatial distribution of magnetic disturbance, all suggest maximum ionizing precipitation near midnight coincident with and following time changes in the reversal region. The existence of a pre-noon maximum in magnetic disturbances between 75 and 80° IN Lat is also well known.

The combination of the Figure 8 convection pattern and a spatial association between maximum ionization and the reversal region gives some basis for believing that a true average configuration for the net field aligned current will not be greatly different from that of Figure 7b. However refinements can be expected from further analysis and one can anticipate that a need for refinement will come from the observation that the motion of aurora in the breakup (i.e., reversal region) does not, at least in some cases, parallel convective motions. This suggests precipitation by particles with sufficient energy to be greatly affected by gradient and curvature drifts in the magnetosphere and means that convective motions with a large component normal to the conducting auroral strips will need to be considered in a more comprehensive time varying model. It should also be noted that the Ba^+ motions observed in the morning

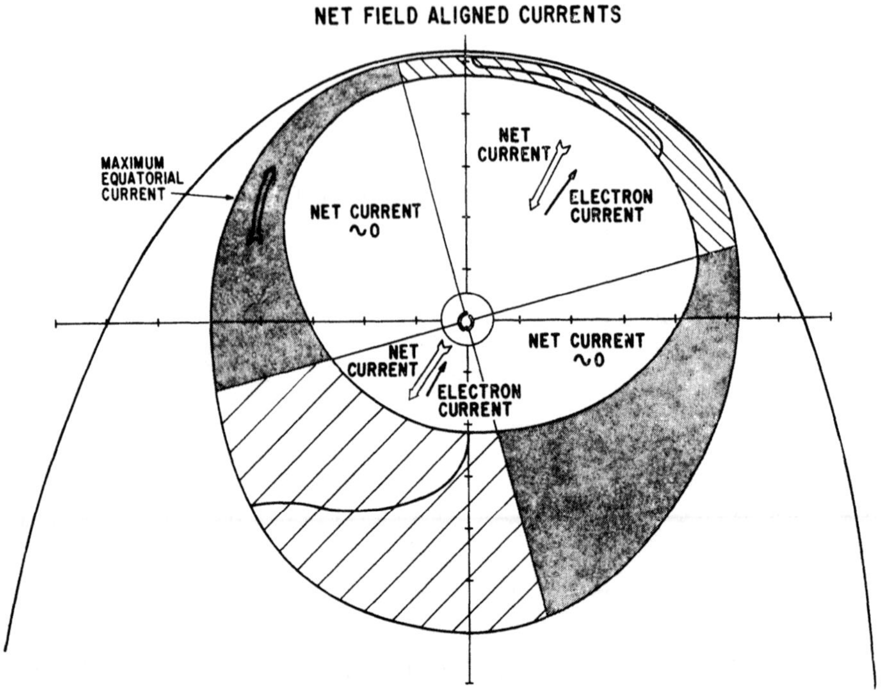

Fig. 9. Magnetospheric distribution of the net field aligned current and its continuation in the equatorial plane.

polar cap flight (Figures 3 and 4) are not parallel to the noon-midnight line. If the same angular difference relative to noon-midnight is encountered in future flights in the post-midnight polar cap, even the highly idealized Figure 8a distribution will become biased toward the Figure 7b configuration.

In presenting the concepts of Hall current-field aligned current continuity we have not attacked the problem of completing the circuit in the magnetosphere. However, the magnetospheric regions providing a net flux of electrons or ions can be grossly defined as shown in Figure 9 in terms of 90° sectors which are adequate for discussion without detailed magnetic field mapping. Continuity in the equatorial plane between these regions is drawn consistent with gradient and curvature particle drifts and the asymmetric ring current distribution described by Cummings' (1966) results. This could be taken as support for the model described but a better physical understanding of the equatorial current is needed before emphasizing this agreement. One might also pursue a circuit in which a net flux of solar wind protons enters directly on the day-side, energetic electrons from the tail precipitate on the nightside, and the electrojet circuit closure occurs on the magnetopause or in the interplanetary surroundings. This latter circuit would resemble Birkeland's models. The Birkeland type model, however, ignores ionospheric feedback. At the other extreme the role of precipitating, ionizing particles relative to electrojet currents is purely one of producing ionization and a pattern of electron density gradients; the ionizing precipitation does not represent a net current and the electrojet and field aligned currents develop within the convecting field only as a consequence of the increased ionization and the gradients in the ionization pattern. This latter extreme is much closer than the Birkeland model to the model we propose but we do not exclude the possibility that energetic particles contribute to the net current.

5. Ba^+ Cloud Striations

The formation of striations in Ba^+ clouds is currently being intensively studied by a number of investigators whose results are beginning to appear in the literature (Völk and Haerendel, 1971). Most investigations assume that the Ba^+ cloud has a role in generating the irregularities subsequently seen as striations. This may be appropriate for releases of large quantities of material and/or the low latitude releases where only a few striations appear. However, it appears highly unlikely that these approaches are applicable to the auroral belt cases where a large number of striations appear in Ba^+ clouds produced by the release of a small quantity of Ba. In these cases it appears more appropriate to assume that the structure within a Ba^+ cloud is a consequence of the existing ionospheric fine structure. The Ba^+ clouds thus do not contribute to, or reveal, the cause of the fine structure other than provide a measurement of **E** to be

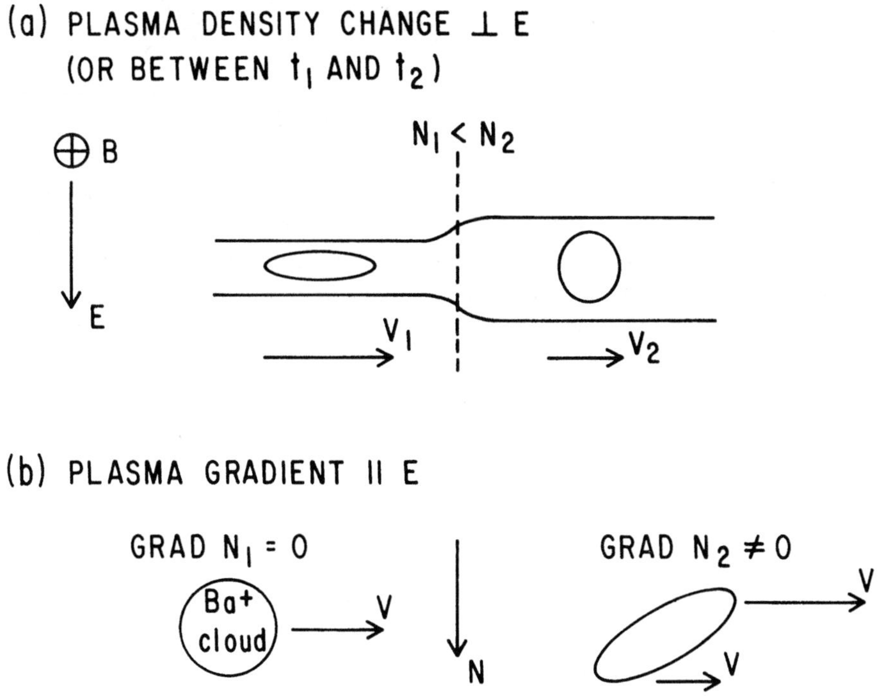

Fig. 10. Change in horizontal cross section of a Ba^+ cloud (or flux tube) above 200 km when the horizontal grad N in the lower ionosphere is (a) perpendicular or (b) parallel to **E**.

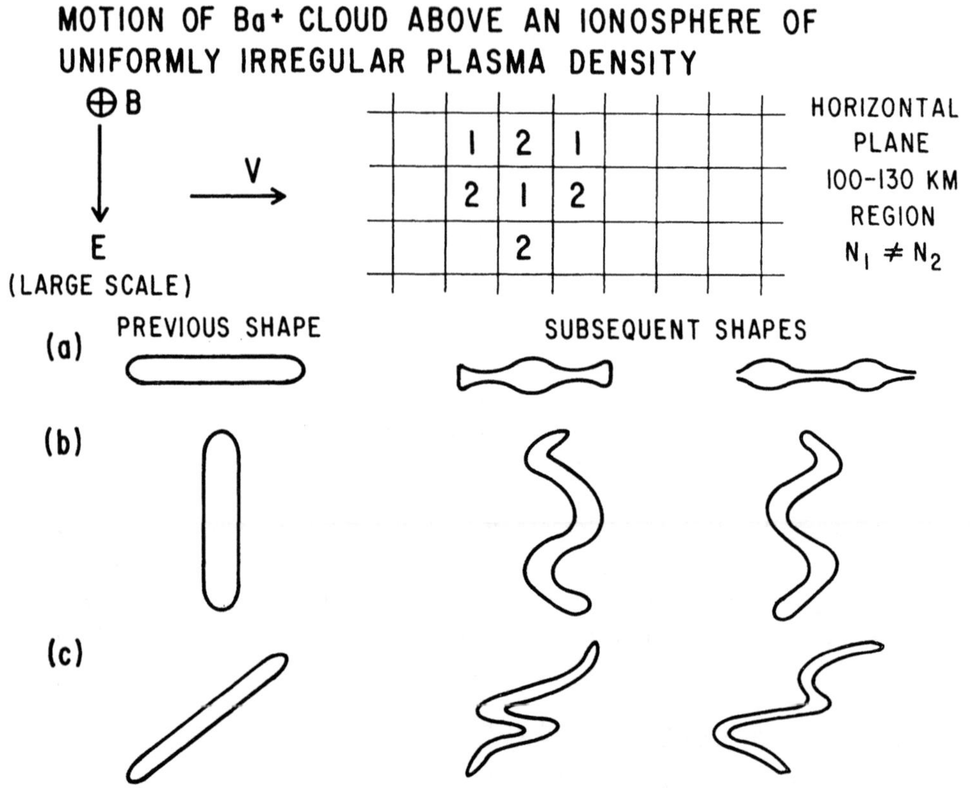

Fig. 11. Distorted horizontal cross sections in Ba+ clouds when the density in the lower ionosphere varies parallel and perpendicular to **E**.

used relative to **E** × **B** instability analyses. Here, we want to note that it is possible and reasonable to explain fine structure in Ba^+ clouds in terms of small scale plasma irregularities in the lower ionosphere by applying the relationships between N and **V** (or **E**) which are observed on a larger scale as discussed in previous sections.

Figure 10 illustrates the change in a horizontal cross section of a Ba^+ cloud (or element of a cloud, or element of a magnetic flux tube) when drifting relative to horizontal plasma gradients in the lower ionosphere using the observation that the velocity decreases when and where the plasma density increases. In Figure 11 the combined effect of parallel and perpendicular gradients in a uniformly irregular ionosphere is illustrated. It should be apparent that almost any cross sectional configuration or distribution of striations desired can be modeled by introducing: non-uniformly distributed irregularities, time variations in the irregularity structure, the time integrated effect of time-space changes in the irregularity structure, and a range of non-linear proportionalities between N and **E**. Furthermore, observations appear to justify introducing each and all of these variables. The 'apparent' configuration, of course, always depends on the angle from which the cloud is viewed.

Clearly the strength and generality of explaining auroral belt Ba^+ striations in these terms depends on the assumption that the large scale observed relationships between N and **E** remain approximately valid for cross sectional dimensions in the range 0.1 to 10 km and on the magnitude of $\Delta N/N$ over similar dimensions. A 'best guess' speculation is that for random irregularities on this scale the inverse relationship $E \propto 1/N$ may be approached analogous to Case 2(B), (Figure 6). If valid, $\Delta E/E$ would provide an approximate measure of $\Delta N/N$ which has been an elusive quantity to determine from radio measurements (see Czechowsky and Lange-Hesse, 1971; McNamara, 1971). However, in most cases in Ba^+ experiments only the integrated differential motion is measured. The spreading of a cloud relative to the path length over which the cloud has travelled becomes a rough indicator of the average $\Delta E/E$. Although this spreading is highly variable the spread relative to the path length is most often $<10\%$ and rarely reaches 50%. This is consistent with the $\Delta E/E$ values indicated by the low frequency measurements by satellite probes (Maynard and Heppner, 1970). The condition $\Delta E < E$ for scale dimensions comparable to Ba^+ striations and auroral rays lends support to our argument in Section 4A that polarization charges do not 'accumulate' to where they significantly alter the convection. It also means that attempts to explain auroral ray structures in terms of circulation about intense excess charge regions (Davis *et al.*, 1971) are either contrary to **E** measurements or only apply under very special conditions.

References

Aggson, T. L.: 1969, in *Atmospheric Emissions* (ed. by B. M. McCormac and A. Omholt), Van Nostrand Reinhold Company, New York, p. 305.

Akasofu, S.-I.: 1970, in *Particles and Fields in the Magnetosphere* (ed. by B. M. McCormac), D. Reidel Publishing Company, Dordrecht, Holland, p. 34.

Baker, K. D., Pfister, W., and Ulwick, J. C.: 1967, in *Space Res.* **VII**, North-Holland Publishing Company, Amsterdam, Holland, p. 665.

Burrows, K., Stolarik, J. D., and Heppner, J. P.: 1971, submitted to *Planetary Space Sci.*

Cole, K. D.: 1963, *Planetary Space Sci.* **10**, 129.

Coleman, P. J. and McPherron, R. L.: 1970, in *Particles and Fields in the Magnetosphere* (ed. by B. M. McCormac), D. Reidel Publishing Company, Dordrecht, Holland, p. 171.

Cummings, W. D.: 1966, *J. Geophys. Res.* **71**, 4495.

Czechowsky, P. and Lange-Hesse, G.: 1971, this volume, p. 314.

Dalgarno, A., Latimer, I. D., and McConkey, J. W.: 1965, *Planetary Space Sci.* **13**, 1008.

Davis, T. N., Hallinan, T. J., and Stenbaek-Nielsen, H. C.: 1971, this volume, p. 160.

Eather, R. H.: 1969, *J. Geophys. Res.* **74**, 153.

Fejer, J. A.: 1961, *Can. J. Phys.* **39**, 1409.

Föppl, H., Haerendel, G., Haser, L., Lüst, R., Melzner, F., Meyer, B., Neuss, N., Rabben, H. H., Rieger, E., Stöcker, J., and Stoffregen, W.: 1968, *J. Geophys. Res.* **73**, 21.

Haerendel, G. and Lüst, R.: 1970, in *Particles and Fields in the Magnetosphere* (ed. by B. M. McCormac), D. Reidel Publishing Company, Dordrecht, Holland, p. 213.

Haerendel, G., Lüst, R., Rieger, E., and Völk, H.: 1969, in *Atmospheric Emissions* (ed. by B. M. McCormac and A. Omholt), Van Nostrand Reinhold Company, New York, p. 293.

Heppner, J. P.: 1967, in *Aurora and Airglow* (ed. by B. M. McCormac), Reinhold Publishing Corporation, New York, p. 75.

Heppner, J. P.: 1969, in *Atmospheric Emissions* (ed. by B. M. McCormac and A. Omholt), Van Nostrand Reinhold Company, New York, p. 251.

Heppner, J. P., Stolarik, J. D., and Wescott, E. M.: 1970, (GSFC preprint X-654-71-50) submitted to *J. Geophys. Res.*

Keneshea, T. J., Narcisi, R. S., and Snider, W.: 1970, *J. Geophys. Res.* **75**, 845.
Lerfald, G. M. and Little, C. G.: 1970, *Radio Sci.* **5**, 1017.
Maynard, N. C. and Heppner, J. P.: 1970, in *Particles and Fields in the Magnetosphere* (ed. by B. M. McCormac), D. Reidel Publishing Company, Dordrecht, Holland, p. 247.
McNamara, A. G.: 1969, *Can. J. Phys.* **47**, 1913.
McNamara, A. G.: 1971, this volume, p. 301.
Nagata, T. and Kokubun, S.: 1960, *Rept. Ionosph. Res. Japan* **14**, 273.
Omholt, A.: 1966, *Ann. Geophys.* **22**, 217.
Penndorf, R.: 1965, *Geomag. Aeron.*, Antarctic Research Series 4, Am. Geophys. Union, p. 1.
Silsbee, H. C. and Vestine, E. H.: 1942, *Terr. Magn. Atm. Elec.* **47**, 195.
Sugiura, M., Skillman, T. L., Ledley, B. C., and Heppner, J. P.: 1970, in *Particles and Fields in the Magnetosphere* (ed. by B. M. McCormac), D. Reidel Publishing Company, Dordrecht, Holland, p. 165.
Taylor, H. E. and Hones, E. W.: 1965, *J. Geophys. Res.* **70**, 3605.
Völk, H. and Haerendel, G.: 1971, this volume, p. 394.
Wescott, E. M., Stolarik, J. D., and Heppner, J. P.: 1969, *J. Geophys. Res.* **74**, 3469.
Wescott, E. M., Stolarik, J. D., and Heppner, J. P.: 1970, in *Particles and Fields in the Magnetosphere* (ed. by B. M. McCormac), D. Reidel Publishing Company, Dordrecht, Holland, p. 229.
Williams, D. J. and Mead, G. D.: 1965, *J. Geophys. Res.* **70**, 3017.

PART VIII

SUMMARY AND CONCLUSIONS

INSTITUTE REVIEW

A. VALLANCE JONES
Astrophysics Branch, National Research Council of Canada, Ottawa, Canada

As at previous Institutes a panel was assembled to present a final review of the most significant points emerging during the two weeks of lectures and discussions. The aurora results from the precipitation of particles via the magnetosphere and the particle observations both in the magnetosphere and immediately above the aurora are discussed by B. Whalen. An increasingly complete description of the morphology of the magnetosphere is being built up together with a better theoretical description of the processes by which particles are trapped and accelerated. These points are discussed both by B. Whalen and T. Hartz who dealt also with radio noise from aurora and auroral CNA. The resulting visual auroral forms, their distribution, and temporal behavior are summarized by A. Egeland. Progress in the understanding of processes in aurora and in the airglow go hand in hand with the understanding of the behavior of the ionized component of the atmosphere. This interrelationship was discussed by L. Thomas. A closely related but technically separate phenomenon is that of radio aurora of which the most recent development and theories are summarized by R. Unwin. Finally, the large topic of chemical and physical processes in the upper atmosphere is discussed by D. Hunten. This includes both processes caused by particle precipitation and airglow processes driven ultimately by solar radiation. A few interesting contributions to the Institute, such as C. Wilson's account of infrasonic emissions from aurora escaped discussion in any of the above categories. It is clear from the summaries that more powerful and general theories of processes in the magnetosphere-atmosphere are being formulated and that increasingly powerful observational techniques are being employed to test them. As is usual, some of the observations over the past few years have yielded results in conflict with the theories but such is the manner in which progress is achieved.

Particle Observations*

I will attempt to summarize the reports presented during the Institute which are directly concerned with auroral particle precipitation. Early in the conference I. McDiarmid presented a magnetospheric model which he concluded was more or less in agreement with satellite observations of the energetic particle populations in the magnetosphere. Two regions which were defined in this model, the magnetosheath and the plasma sheet, attracted the most attention during the conference with regard to particle precipitation.

Some details of the interactions at the boundary between the solar wind and

* Presented by B. A. Whalen.

B. M. McCormac (ed.), The Radiating Atmosphere, 429–441. *All Rights Reserved.*

the magnetosphere were presented by D. Winningham, R. Sharp and myself. D. Winningham presented a fairly convincing picture which indicated that the solar wind plasma penetrates directly into the magnetosphere on the dayside in the regions near the neutral point. D. Winningham's data strongly supported the original contention of the Lockheed group that the soft electron precipitation observed on the dayside was from the neutral point. R. Eather presented a limited number of observations which he indicated were in good agreement with the particles observed by D. Winningham.

I presented some ion measurements which indicated that particles precipitating in the nightside auroral oval also originated in the solar wind.

The suggestion (which I believe was originally due to V. Vasyliunas) that the auroral oval is the low altitude extension of the plasma sheet was given some support during the Institute by both R. Sharp and I. McDiarmid. In detail, they introduced data to show that the minimum radial extensions of the plasma sheet mapped down to the low latitude boundary of the auroral oval and the high latitude boundary of the oval apparently mapped out to the high latitude boundary of the plasma sheet. The softening in the electron participation (which is now fairly well documented) as one moves from low to higher latitudes in the auroral zone was related to the equivalent softening in the electron population observed in the plasma sheet as one moves from the central plane to higher latitudes. In fact R. Sharp presented data which showed that the softening observed at the low latitude termination of the plasma sheet (which was reported by V. Vasyliunas and referred to as a cooling of the electron plasma) was seen in the Lockheed data at the low latitude boundary of the auroral oval.

Continuing on with this picture one might hope that all that is needed to explain nightside auroral precipitation is to introduce some mechanism which would precipitate particles trapped in the plasma sheet. Some support for this contention was given by D. Bryant. He observed particle precipitation in a sounding rocket during the breakup phase of a substorm at a time when ground base measurements indicated that auroral pulsations were in progress. His particle measurements showed a dispersion in arrival time which he related to an equatorial source which operated simultaneously on all particles to precipitate them into the auroral ionosphere. His pitch angle distribution analysis indicated that pitch angle diffusion was responsible for the precipitation. Interestingly enough, earlier in the Institute, G. Shepherd showed some ground based optical measurements in a similar event. Dispersion in electron arrival time was evident in his measurements which he also interpreted as being due to an equatorial electron source.

The concept that all auroral precipitation can be explained in terms of interactions at the equatorial plane is somewhat shattered when one observes the details in auroral precipitation as shown in some of N. Davis' auroral movies. When faced with the fine structure and rapid motions present in the auroral arcs, one feels intuitively that interactions at large distances cannot be solely responsible. It seems reasonable to introduce relatively local interactions, such as the instabilities described by N. Davis and his associates and A. Hasegawa, in their proposals to explain the small and large scale auroral forms.

D. Evans discussed the spatial and temporal structure of auroral precipitation and summarizing ground based measurements and using particle observations from sounding rockets he concluded that the small scale structures observed in auroral emissions were not present in the electron precipitation above the aurora. He also concluded that the local acceleration of electron precipitation was probably necessary to account for some of the energetic electron precipitation measurements.

In summary our understanding of the interactions of the solar wind with the magnetosphere and the interrelation of the magnetospheric population of particles and auroral precipitation have improved somewhat since the last Institute. There are problems which require intensive study, and in particular I feel that the following are the most important. First, what processes occur in the magnetosphere during an auroral substorm? The solar wind probably supplies the energy but how is it transferred into the magnetosphere and stored? Is there a triggering mechanism which starts the substorm and what are the main acceleration mechanisms which occur during the storm? Second, what are the sources of magnetospheric electric field and what roles do they play in particle acceleration and precipitation? Third, what effect does the ionosphere have on the magnetospheric particle population and does it directly influence auroral precipitation?

Radio Observations and Particle Precipitation*

The radio noise emissions are looked on as a tool for studying the aurora and some of the related phenomena. An excellent summary of the status of the noise observations, both LF and VLF, was presented by T. Jørgensen. He indicated that most of the observations fit into two main categories: the chorus and ELF hiss in the one, and auroral hiss in the other. In general, chorus appears to be mainly a daytime phenomenon, appears predominantly in the geomagnetic latitude range of 60 to 70°, seems to show a sequential relationship with substorms, and can be identified with our zone of "drizzle" precipitation. The source apparently lies at great distances in the magnetosphere, either in the equatorial plane or not too far removed from it. J. Vigneron noted that chorus seldom exhibited conjugacy. The generation mechanism is apparently the electron cyclotron process involving the so-called transverse resonance instability.

Auroral hiss, on the other hand, is associated with the auroral oval, with a distinct preference for the high latitude side of the oval. Two diurnal maxima stand out in the results presented by A. Hughes and myself, one near noon or early afternoon and the other just before geomagnetic midnight. The latter maximum appears substorm related while the former does not, though this conclusion is much more obvious in the satellite data than in observations from the ground. In the case of the ground based measurements *D* region absorption may affect the results. The auroral hiss is very

* Presented by T. R. Hartz.

broadband, apparently extending from a few kHz up to a few hundred kHz with indications of a decreasing intensity with increasing frequency. Spacecraft measurements have localized the source of the emissions to the general vicinity of, though principally above, the satellites. The LF measurements at 200 kHz on ISIS 1 show a close correlation with the measured electron flux at about 100 eV. The source mechanism seems to be the Cerenkov process.

The radio noise observations seem to fit very well into the two-zoned pattern identified by Hartz and Brice, with chorus and ELF hiss primarily in the one zone and auroral hiss in the other. The daytime VLF and LF results seem to be a special case of low energy electron influx that is associated with the magnetospheric neutral points. This influx has characteristics different from those of the auroral oval at night, and I have suggested that a clear distinction should be made in the general auroral-particle precipitation pattern to indicate this. Thus I have proposed a modification to the midday portion of the Hartz-Brice diagram that should make that pattern more meaningful in terms of magnetospheric processes.

Several other presentations fit into this general framework: a brief one by D. Jelly and a more lengthy one by H. Trefall, both with regard to the sequential dependence between the nighttime substorm and phenomena in the outer or drizzle zone. The nighttime riometer or magnetometer events with a sudden onset, marking the start of a substorm, appear to be followed by phenomena (absorption and X-ray events) in the outer zone that are most prominent in the morning sector. The question is, does this represent an eastward drift of particles from the night sector, or a related local acceleration? H. Trefall seems to favor an acceleration over the whole 0 to 6 hr sector, rather than just near midnight, and an association between that extended acceleration mechanism and his X-ray observations in the drizzle zone. The alternative picture of an acceleration mechanism operative near midnight, followed by a build up in the outer zone, was favored by D. Jelly. Clearly, this is a subject that requires further study and I believe D. Jelly indicated that work was now in progress to map riometer data for particular events in all time sectors just to settle this point.

The other paper that I might mention was that of R. Eather and S. Mende. They reported on auroral observations from aircraft which identified regions of soft electron flux. In both the day and night sectors these appeared at the high latitude edge of the oval. The electron energies were estimated from the intensity ratios of the 6300 Å and 4278 Å auroral lines. For the daytime region they obtained electron energies of about 150 eV, while at night the indicated energies were about 600 eV. Unfortunately, their observations did not cover the morning and afternoon sectors sufficiently to give a complete diurnal picture. Nevertheless, the results are most interesting in the light of our general precipitation patterns.

Auroral Morphology *

(1) Based on optical data it may be concluded that there is only one complete electron

* Presented by A. Egeland.

auroral oval for both day and night. The exact form is not agreed upon. The oval may widen, move, and it also shows a considerable amount of structure.

(2) The shape and position of the oval is related to magnetic activity. The daytime oval moves much more rapidly equatorward with K_p than the nighttime oval. As K_p increases from 0 to 1 the nighttime oval moves equatorward and reaches a plateau. It does not move further until K_p has increased to >3. During disturbed conditions the oval is more or less circular.

(3) If the satellite data are grouped according to prevailing K_p and Dst values, the effects of DP and DR currents can be found. As Dst varies from 25 to $-10\ \gamma$, the oval first moves poleward and then it moves considerably equatorward. Generally, the DP currents are related to a broadening and brightening of the auroral emission region. DR currents seem to control the extent of this broadening by narrowing the zone and moving it equatorward. It is suggested that more direct measurements of DP and the neutral sheet current be introduced.

(4) For $K_p=0$ to 1, and $<67°$ IN Lat, the aurora is conjugate and has the same intensity in both the north and south. Therefore, it is produced on closed field lines in the GSFC field model. For more disturbed conditions ($K_p=2$ to 4) auroras equatorward of 64° are still conjugate. However, at latitudes $>64°$, the Southern Hemisphere auroras are generally 100 to 350 km eastward from corresponding aurora in the northern hemisphere. The average intensity is much higher in the north. The field lines $>64°$ seem to be highly distorted and may even be open. Magnetic field models are needed which take into account the tilt of the field with respect to the spin pole and the effects of the various storm time currents.

(5) Electron auroral substorms seem to fit S. Akasofu's model very well, but there are a few deficiences. There seems to be a marked tendency for an equatorward movement of the oval before the onset of the substorm as seen on the magnetogram. More information is needed on substorm effects on the dayside.

(6) L. Montbriand introduced certain new substorm features which will necessitate modification of Akasofu's model in order to include the morphology of H emissions. The most significant addition is the documentation of characteristic N-S oriented segments developing from the omega bands in the polar bulge at substorm time of 10 to 15 min. The H region at onset of the substorm is narrower in the midnight sector than in the morning and evening. It expands in the bulge after occurrence of the N-S segments. At this time isolated segments of H aurora break off from the equatorward edge (H region) in the vicinity of the westward traveling surge. Subsidiary substorms usually occur between 15 and 30 min but the H region is already in an expanded state when they occur.

Two types of morning patchy forms were identified, namely a less common westward drifting H associated one and an eastward drifting non-H type. During the morning phase the H region gradually assumes pre-substorm conditions

(7) During quiet times ($K_p<2$) the H and electron aurora overlap. During disturbed conditions both move equatorward, but the electron aurora moves faster on the morning side.

(8) The quiet H arcs in the evening are conjugate and include electron precipitation. During disturbances both precipitation regions move equatorward, but the H region moves faster.

(9) Equatorward of the electron oval there is a zone of high pitch angle protons associated with Hβ intensity up to 30 R for $K_p<4$. Because a significant amount of luminosity may be associated with anisotropic pitch angle distribution, this may explain previous difficulties encountered in matching H emission wavelength profile measurements assuming isotropic precipitation pattern. Profile observations in the magnetic zenith just equatorward of the main emitting region are needed to check this.

(10) The conversion efficiencies for luminosity are not well known. A value of 200 R of 4278 Å for 1 erg cm^{-2} sec^{-1} seems to be reasonable. The conversion efficiency of H emission seems to be too small, perhaps because of insufficient corrections for the contribution from energetic neutrals.

(11) There is need for direct height measurements of the polar glow together with higher energy proton measurements.

The Ionosphere in Relation to Aurora and Airglow*

The information of ionospheric interest presented at this Institute has chiefly related to the *F* and *D* regions. It will be convenient to consider these two regions in turn and to distinguish between the results for different latitudes.

The 6300 Å emission from O observed at *F* region heights at low latitudes is one of the best understood features of the nightglow; excitation of $O(^1D)$ atoms is believed to occur during the dissociative recombination of O_2^+ molecules, and the close relationship between the emission rate and the loss of ionization has been clearly demonstrated. At the present meeting J. Blamont showed some very interesting observations of the 6300 Å emission made onboard the OGO series of satellites. A marked asymmetry was observed in the latitude variation of intensity at low latitudes, the northern tropical arc being more permanent and intense than that to the south. Since this asymmetry was present during January to June, it is difficult to relate it to the effects of meridional winds as proposed to explain differences between the northern and southern peaks in the daytime *F* region equatorial anomaly. In an analysis of 6300 Å and *F* region data for Hawaii, W. Brown found that the recombination of every O_2^+ ion produced an $O(^1D)$ atom and this result was confirmed by laboratory measurements reported by Zipf.

Observations of predawn increases of electron temperature at *F* region heights and of 6300 Å emission intensity at mid-latitudes in winter, and their interpretation in terms of photoelectrons arriving from the sunlit conjugate point, have attracted considerable attention in recent years. The reference made by L. Broadfoot to a similar evening enhancement in the 1N band of N_2^+ at 3914 Å would imply a contribution to *F* and *E* region electron production by photoelectrons from the conjugate points at

* Presented by L. Thomas.

twilight periods in winter. It would be expected that a similar twilight effect should be observed in the 10830 Å emission of He, believed to be due to the fluorescence of metastable He(2^3S) atoms produced by collisions with electrons of energy about 25 eV.

Attempts to interpret 6300 Å intensities in terms of *F* region parameters, either by the use of the Barbier formula or with more exact equations, have usually indicated a residual component. It has been suggested that this might result from the production of O(^1D) atoms from reactions of $N(^2D)$ and O(^3P) atoms provided the rate of this reaction is faster than that of quenching of $N(^2D)$ atoms by electrons or O_2. The detailed analysis of data by J. Noxon showed no such residual component except in winter. J. Noxon also made the interesting suggestion that the energy gained by an electron during quenching of an $N(^2D)$ would be shared with other electrons and this effect could account for the observation that the electron temperature at *F* region heights at night exceeds that of the neutral gas.

Observations of the atmospheric drag on earth satellites and ground based measurements of the Doppler broadening of the 6300 Å line have shown that the atmospheric temperature at *F* region heights increase during magnetic storms. Satellite observations of the 6300 Å line width in the dayglow by J. Blamont showed the latitude dependence of this atmospheric heating, the temperatures increasing from 1000 K to about 1300 to 1500 K at geomagnetic latitudes above 35 to 40° in both hemispheres, the temperatures at lower latitudes remaining unchanged. Measurements associated with *M* arcs (which are also related to *F* region disturbances at the time of magnetic storms) were also described. Chappell presented OGO 5 observations at the time of *M* arcs which supported the suggestion that these arcs, and the associated disturbance in the *F* region, are associated with changes near the plasmapause. In connection with *M* arcs, D. Winningham reported ISIS satellite observations which indicated a substantial flux of precipitating protons in the vicinity of an arc. The emissions associated with such protons, together with a report by R. Schaeffer of 5577 Å emissions from an arc, would seem to throw some doubt on the spectroscopic purity of the emissions from *M* arcs.

The maintenance of *F* region ionization over the polar cap during winter has been the subject of some considerable study in the past. The substantial fluxes of electrons of energy about 150 eV observed by D. Winningham near 80° IN Lat shortly before geomagnetic noon, and confirmed by aircraft observations of emissions by R. Eather, would represent an important source of ionization over a limited area.

Perhaps the results of most direct interest in studies of the ionospheric *D* region at mid and low latitudes were the NO concentrations deduced from rocket borne observations of the dayglow reported by J. Pearce. His revised values and those obtained from recent measurements by Meira showed concentrations rather similar to the previous results of Barth. On the basis of these results it is evident that the photoionization of NO by Ly-α radiation represents the main source of ionization in the middle *D* region.

The height distribution of $O_2(^1\Delta_g)$ (of interest in the production of ionization and the negative ion chemistry of the *D* region) was considered by E. Llewellyn and D. Evans, and H. Rundle. They drew attention to the secondary layer of this metastable molecule located above 80 km. It appears that this layer is enhanced in the winter.

Laboratory measurements concerned with the positive and negative ion chemistry of the *D* region have emphasized the importance of the theoretical studies of an O-OH atmosphere already undertaken for investigations of OH and other airglow emissions. The study described by R. Gattinger drew attention to the importance of taking proper account of absorption of the Schumann-Runge bands by O_2. His analysis, which made use of the measurements of Hudson *et al.* (1969) and therefore included the effect of predissociation, showed important changes in the photodissociation rates of water in the lower mesosphere from those given by less complete treatments. The effect of eddy diffusion was included in the model, and R. Gattinger suggested that changes at nighttime in the intensity of the IR Atmospheric O_2 bands, which he attributed to the recombination of O, and of the OH emission could be due to changes in the eddy diffusion coefficient.

Considerable evidence was provided of enhanced concentrations of minor constituents at *D* and *E* region heights in the auroral zone; K. Baker and J. Noxon described observations of 1.27 μ emission on rockets and aircraft, respectively, which showed increases of $O_2\,(^1\Delta_g)$ concentrations not correlated with changes in 3914 Å emission. A. Stair mentioned that whenever aurora was observed aircraft measurements showed enhancements in the OH band near 1.655 μ. E. Zipf reported mass spectrometer observations of enhanced NO concentrations in the auroral zone. The mode of excitation of the $O_2\,(^1\Delta_g)$ molecules and OH radicals, and the production of NO during aurora are of considerable interest in themselves. Furthermore, the possible transport of long lived constituents to mid-latitudes could have important consequences for the mid-latitude *D* region. In this connection, D. Hunten pointed out that the photochemical lifetime of NO is determined by the N concentration. The atomic concentration at *E* region heights might be reduced in the absence of sunlight during winter and the lifetime of NO correspondingly increased. In any case, NO molecules transferred by mixing processes to mesospheric heights would have very long lifetimes and horizontal transport to mid-latitudes would be possible.

Radio Aurora and Plasma Instabilities*

The radio aurora has been discussed and is still under investigation. The morphology and general behavior are now fairly well known. A. McNamara showed overwhelming evidence that much radio aurora is basically due to a plasma instability, probably the two stream instability of Farley. The $\mathbf{E} \times \mathbf{B}$ or drift gradient instability was discussed. Other plasma instabilities should be investigated. The theorists need to incorporate amplitudes of the irregularities into their models. The full treatment is urgently required but will be very difficult.

* Presented by R. Unwin.

Experiments show that there is more than one type of radio aurora which needs to be better identified in time and location to compare to theory. The CW technique is not good for identifying the type. The best instrument is the pulse Doppler back-scatter radar with good height and range discrimination.

For fundamental studies of instabilities it would be best to use the equatorial electrojet. Plasma instabilities occur there, certainly the two stream and possibly the drift gradient instability. The situation is stable compared to the auroral region where conditions change very rapidly.

Direct observations of irregularities giving rise to radio aurora and their power spectra, along with electric field measurements in the auroral region, are needed. A. McNamara has made a good start in such studies.

The stage is set for settling the age old argument on the relation between radio and optical aurora. G. Lange-Hesse has shown the general association between areas of optical and radio aurora at the outer edge of the auroral oval, considerably better than what has been done before; but his range and azimuth discrimination must be better. J. Heppner and E. Wescott showed that *E* fields are depressed near and within optical auroral forms and this suggests that there will be either a negative or positive association in space depending on the type of plasma instability involved, which gives a reason for the past confusion on the topic. Much more precise measurements are needed. However, the two phenomena are essentially different – one involves plasma instabilities and the other a consequence of particle precipitations. Radio *aurora* is really a misnomer; if it had been called, say 'high latitude *E* region irregularities' much confusion may have been prevented. The observations of radio aurora have been too isolated in the past. They should be related more with other phenomena, substorms, electric fields, and currents in the ionosphere.

I strongly support A. McNamara's suggestion that observation of the radio aurora can be a useful tool in investigating ionospheric currents and electric fields. The electric field could be monitored when it exceeds the threshold for plasma instabilities with 3 or 4 pairs of pulsed Doppler radars with their beams intersecting at 63 to 65° IN Lat to measure the direction and amplitude of the *E* field.

Atmospheric Processes*

First of all, I have a few additional comments on NO. The stability of NO is important, particularly because it depends very much on the environment; in the polar winter NO is unusually stable. Nitric oxide is destroyed by reactions with N atoms and very likely with $N(^2D)$ metastables. Any other reaction of NO, say with O or O_8, will produce NO_2, which reacts with another O atom and replaces the original NO. The only true loss of NO is by combining two N atoms. In the dark there are practically no N atoms. Therefore, NO is almost completely stable, and of course there is a lot of darkness in the polar cap during winter. This is a totally different situation

* Presented by D. M. Hunten.

from the lower latitudes, and the difference should be reflected in the NO chemistry.

The second way for NO to be unusually stable is to have a substantial quantity of it. It will consume any N atom as soon as it appears; unless a large amount of N atoms are produced, the large amount of NO cannot be destroyed. We have seen from the measurements that E. Zipf briefly reported that this is the case in at least some auroras (see Zipf *et al.*, 1970). We do not know how widespread this phenomenon is, nor its time dependence during the polar winter, but we know it can be important.

Now it seems to me very likely that a great deal of this excess NO which has actually only been observed in the *E* region will be mixed downward into the *D* region during the polar winter. Once it gets into the *D* region it is quite stable, even under normal conditions. The time constant is more like weeks than days below 95 km, even at low latitudes. On such a time scale, we may expect transport of this NO to lower latitudes, to the region where people live and take ionograms and observe such things as anomalous winter absorption. To some extent, this explanation was foreseen by Sechrist (1967); but at that time almost nobody was aware of the factors involved in the production and stability of NO. We now know that the NO production and loss processes are largely independent, and this gives us a tremendous amount of scope for variability, particularly at high latitudes. This should be a fruitful field of investigation in the next few years.

Now let us consider mesospheric aeronomy and particularly the work on OH emission which has been discussed by several people. First of all, a little bit about R. Gattinger's very interesting paper. My criticisms reflect the fact that he has tackled an extremely difficult problem and that there still are some important factors which may cause his results to be considerably modified when they are included. I have exactly the same criticisms of the published work of Hesstvedt (1968) and Shimazaki and Laird (1970).

The problem has to do with the upward transport of H_2O and the eventual Jeans escape of H atoms. Dissociation products of H_2O are important directly in OH production, and indirectly in catalyzing the destruction of O and O_3. The equation of eddy diffusion is not difficult to treat, but the correct application of boundary conditions is extremely subtle. Hesstvedt (1968), for example, treated a slab extending from 60 to 100 km, and applied local photochemical equilibrium at these boundaries. This condition must be justified individually for each species, because transport through the boundary may represent a much larger term than any of the chemical processes. The more recent work uses different boundaries and conditions, but none of them are really satisfactory.

The escape flux of 10^7 to 10^8 H atoms cm^{-2} sec^{-1} must be balanced by an upward flux of H_2O, half as great, at all heights below 70 or 80 km. Once this flux has been put into the equations, we face another difficulty – we don't know enough about vertical transport rates, usually represented by an eddy diffusion coefficient. There is a real possibility that a bottleneck to the H_2O flow is present, somewhere around 50 km, and that H densities at all greater heights are controlled by it. This situation is treated by Urey (1959), but we still don't know enough about the atmosphere to make any progress, and it may be years before we do. I should mention that the transport may well be due to large scale circulations, rather than eddy diffusion in the usual sense.

R. Gattinger's work includes the predissociation of O_2 by the Schumann-Runge bands. The corresponding continuum is mostly absorbed above 90 km; the bands penetrate considerably deeper, and the weaker absorption is more than made up by the larger flux of radiation. These statements apply even more to the Herzberg continuum, which has been the only source of O_2 dissociation included in previous treatments of the lower regions. Though the existence of the bands has been well known, the usual practice has been to avoid the complications and use an average absorption coefficient. The demonstration by Hudson *et al.* (1969) that most of the bands predissociate has stimulated considerable interest in a better treatment. Vibrational levels above $v'=2$ can rapidly transfer to a different electronic state in which the atoms fly apart. The treatment of the absorption of sunlight is complicated, because the wavelengths that predissociate are most strongly absorbed. For a foreign gas, like H_2O, the effective absorption coefficient is different.

In my opening paper, I said that we had a rather serious lack of knowledge of the total intensity of the OH vibration-rotation bands; during the last two weeks this lack of knowledge has been largely remedied. H. Gush showed us his balloon results in which the missing fundamentals were actually observed and found to be not as intense as had been expected. An outstanding piece of work by Murphy was reported by A. Stair on the transition probabilities among the vibration-rotation bands. These will give us, when published, quite definite information from which we can at last estimate just how much energy is being radiated in the Meinel vibration-rotation bands. In my paper, I compared the Chamberlain estimate of the total OH intensity (4.5 MR) with the effective photon flux in the Schumann-Runge continuum (0.5 MR). It may be that the band absorption raises the input somewhat. But more to the point are the results on the OH intensity, which is beginning to look like about 1 MR. With a photon yield of 2 or 3 per recombination the rates seem to balance nicely. The comparison does tell us in a direct manner that a very substantial fraction of the odd O association is going by the OH route.

Another old problem is the competition between two mechanisms for the excitation of OH. The Bates-Nicolet-Herzberg mechanism is

$$H + O_3 \rightarrow OH^{\ddagger} + O_2 \tag{1}$$

$$OH + O \rightarrow O_2 + H \tag{2}$$

where the double dagger represents vibrational excitation which can be radiated by OH. The Krassovsky mechanism is

$$O + O + M \rightarrow O_2^{\ddagger} + M \tag{3}$$

$$O_2^{\ddagger} + H \rightarrow O + OH^{\ddagger}. \tag{4}$$

Since Equation (4) populates levels well above $v=9$, it is assumed that the high levels are removed by the reverse of Equation (1). It is clear that the Krassovsky mechanism is now in deep trouble because of the observations reported here by several people (also Moreels *et al.*, 1970) that the OH intensity tends to decrease at sunrise and then recovers later on in the day. There is no reason, that I can see, why the Krassovsky mechanism should behave that way and it can probably be reduced to the status of a minor contribution at most. The dawn decrease is natural for the $H-O_3$ mechanism because of the rapid destruction of O_3 by sunlight; the subsequent slow rise of the H concentration eventually restores the original intensity. The effect is naturally absent in the 'slow twilights' examined by Shefov (1969) at high latitude; a dropout is only to be expected when sunrise is a rapid event compared with the 1 or 2 hr required for recovery. J. Pearce's results on the UV dayglow reminds us again that the photo-electron-excited components look very much like an aurora of intensity *I*. The exception one must remember is that the effective height is somewhat greater. The dayglow looks more like a type A red aurora than a normal green one.

A. Stair presented some remarkable results of a spectral feature at 9.6 μ, measured by a rocket borne spectrometer. The natural interpretation looks to O_3 emission, but as A. Stair pointed out the observed intensity gradient is much too small to be reasonable. He mentioned the possibility of thermal emission by silicate particles, which have an absorption band at the same wavelength. Though he clearly did not favor this explanation, I think it deserves serious consideration. Clouds of dust, and single dust particles, have been observed frequently by dayglow experiments, and in fact constitute a serious hazard in this work (cf., Wallace and Hunten, 1968). A radiating dust cloud would bear little or no relation to the ambient atmosphere, and would explain the observed gradient much better than O_3.

A. Stair also showed us some observations of a possible auroral continuum in the 4 to 5 μ region. If verified, this radiation could be important to the energy balance in aurora. The intense emission from $O_2(^1\Delta)$ in some auroras aroused considerable interest. My impression is that there is a zone of $O_2(^1\Delta)$ uncorrelated with the individual auroral forms. It seems almost certain that the energy being emitted by $O_2(^1\Delta)$ cannot be coming from precipitating particles whose energy would have to convert 100% into 1.27 μ emission.

According to J. Walker, the principal aeronomical effect of an *E* field is to heat the positive ions to a very high temperature. This raises many important possibilities which none of us have had time to think about. In the standard work of Megill and Carleton, essentially no attention was paid to the ion temperature, which J. Walker says will probably be a lot higher than the electron temperature. He pointed out that ions with mean energies up to $\sim$0.5 eV and a Maxwellian tail of several eV may well be involved in all sorts of hot-ion chemistry. Perhaps this is a clue as to what is going on with NO in auroras; like the $O_2(^1\Delta)$, it may be related to the electrojet, not to particle precipitation. The anomalous NO concentrations reported by E. Zipf, 2 orders more than normally found in the *E* region, could perhaps be produced by J. Walker's mechanism.

The Ba ray excitation reported by *E.* Wescott may have some relation to the accepted explanation of Na e mission in aurora. N_2 is vibrationally excited by auroral electrons; though we do not know the loss rate for $N_2^{\ddagger}$ it should have a large concentration. Excitation of the metal then occurs by energy transfer

$$N_2^{\ddagger} + Ba^+ \rightarrow N_2 + Ba^+ . \tag{5}$$

The high vibrational excitation of N_2 is nonequilibrium, though v' may be as high as 10, a high vibrational temperature is not implied for N_2 as a whole. Energy transfer to Na has been verified by Fite and his students, although there is evidence that a little kinetic energy may be required.

References

Hesstvedt, E.: 1968, *Geophys. Norweg.* **27**, No. 4.
Hudson, R. D., Carter, V. L., and Breig, E. L.: 1969, *J. Geophys. Res.* **74**, 4079.
Moreels, G., Evans, W. F. J., Blamont, J. E., and Vallance Jones, A.: 1970, *Planetary Space Sci.* **18**, 637.
Sechrist, C. F., Jr.: 1967, *J. Atmospheric Terrest. Phys.* **29**, 113.
Shefov, N. N.: 1969, *Planetary Space Sci.* **17**, 1629.
Shimazaki, T. and Laird, A. R.: 1970, *J. Geophys. Res.* **75**, 3221.
Urey, H. C.: 1959, *Handbuch der Physik* **52**, 363.
Wallace, L. and Hunten, D. M.: 1968, *J. Geophys. Res.* **73**, 4813.
Zipf, E. C., Borst, W. L., and Donahue, T. M.: 1970, *J. Geophys. Res.* **75**, 6371.

CONCLUSIONS

BILLY M. McCORMAC, J. N. BRADBURY and JOHN E. EVANS
Lockheed Palo Alto Research Laboratory, Palo Alto, Calif., U.S.A.

The purpose of the Institute was to ensure that the experimental and theoretical results were presented and adequately discussed and no attempt was made to reach agreements or determine popular opinion. Each individual must evaluate the importance and credibility of this information for himself. These personal conclusions about the Institute are presented below from the point of view of future research requirements. The Institute did much to identify important areas of agreement and disagreement and focused attention on multidisciplinary research and key problems.

Ultraviolet, visible, and IR emissions from the earth's atmosphere are aurora and airglow. There are no generally accepted definitions of aurora and airglow as there are significant exceptions to all possible definitions. It has been suggested that if the excitation is generated by energy deposited by energetic charged particles in the vicinity of the auroral oval then the emission is aurora, otherwise, except for Rayleigh and other nonselective scattering and thermal emission, the emission is airglow. Since it is difficult to define aurora and airglow, it is important to ensure that the meaning for each use is understood. Aurora and airglow were investigated from two points of view: the first was for studying atmospheric processes, while the second was to help analyze magnetospheric phenomena.

It was concluded that in future Institutes emphasis should be placed on physical and chemical processes and phenomena in the upper atmosphere rather than on specific observational techniques. Auroral morphology and magnetospheric dynamics will not be included unless slanted toward the upper atmosphere. It was decided to conduct a 1972 Institute, Physics and Chemistry of the Upper Atmosphere, in northern Europe. The Program Committee members are Drs. Rolf Boström, Alv Egeland, Donald M. Hunten, Günther Lange-Hesse, M. Nicolet, Anders Omholt, H. I. Schiff, Lance Thomas, Alister Vallance Jones, and Gilbert Weill.

The persons mentioned below attended the Institute and either have chapters in this book or made their contributions during the discussion.

1. Atmospheric Airglow Emissions

The primary energy source is solar radiation. Primary emphasis is now being placed on physical and chemical processes and phenomena, wherein many types of airglow and other observations are considered.

The chemical and physical processes of the atmosphere are far from being resolved, especially in the *D* region. The height profiles of many constituents are not known at the higher altitudes. Atmospheric mixing processes, such as eddy diffusion, are are not well understood but affect details. Vertical motion of H will have an impor-

B. M. McCormac (ed.), The Radiating Atmosphere, 442–449.

tant effect on OH processes. There are important transport considerations. Significant quantities of NO may be produced in the polar winter ionosphere, transported to the *D* region where it has a lifetime of weeks, and then transported in the *D* region to low latitudes. Magnetic storms may result in the production at high latitudes of $O > 90$ km which has a lifetime > 1 week and can be transported to lower latitudes. More attention needs to be given to simultaneously measuring several different facets of the same phenomenon.

A. OXYGEN PROCESSES

Oxygen processes are very important. Much energy is stored in dissociated O_2. Important reactants are O, O_3, H, and NO wherein H and NO act as catalysts (Hunten). The net effect of these reactions is the association of O. Photodissociation of O_2, O_3, and H_2O is also important but is difficult to calculate. Airglow emissions may result from deexcitation of the end products.

In the nighttime *F* region, airglow is produced by various reactions which remove the dominant positive ion O^+ and by low energy electron excitation.

A number of excited states are produced, in particular three important metastable species: $O(^1D)$ 6300 Å, $O(^1S)$ 5577 Å, and $N(^2D)$ 5200 Å. It is useful to observe these metastable species to try to understand upper atmospheric processes. Observational results do not support the height profile predictions for $O(^1D)$, $O(^1S)$, and $N(^2D)$. Additional observations and processes seem to be required.

Use of the Barbier formula to interpret 6300 Å observations seems likely to lead to more confusion. It is just a four parameter fit to the observations and does not seem to be based on the correct physics.

B. ALKALI METALS

The role of the alkali metals in atmospheric chemistry still has not been determined. The neutral atoms of Na, Li, and K and the ions of Ca have been observed at twilight while only Na has been observed in dayglow.

The observations were reviewed (Llewellyn; Rundle). The relative quantities of $Na:K:Li \sim 1000:5:1$, but there is much variation in the data. The seasonal and diurnal variations of Na have been well documented for 20 yr. The peaks of dayglow and twilight Na layers both seem to be at 90 km, as is the twilight layer for Li. The K layer altitude is uncertain but may be a little lower. The layer scale heights are 3 to 4 km, much smaller than that for the atmosphere.

Three models have been considered to explain the alkali metal observations (Hunten; Rundle): meteor ablation, photochemical, and aerosol. These models all fail to adequately explain the concentrations, and the seasonal and diurnal variations. For example, the observed meteor behavior is not reflected in the observations. The photochemical model fails to provide the correct scale height, and the aerosol model needs a suitable evaporation and sink mechanism.

A fundamental difficulty lies in not knowing the total atomic and molecular inventory of the alkali metals. It is not known if the observed neutral atoms are a very small or a very large fraction of the inventory. There may be a significant lack of knowledge of the reaction chemistry of the various alkali metal oxides.

C. CONJUGACY

Investigation of the magnetic conjugacy of various airglow phenomena often invokes useful constraints on interpretations.

In general, as shown by the 6300 Å intensity data obtained from OGO 4 and OGO 6 there is no conjugacy between the southern and northern tropical red arcs (Blamont). There is a large variation in the arcs as a function of UT and LT. Sometimes there is no arc in the Southern Hemisphere when there is a very strong arc in the north. At other times the southern arc is weaker than the northern arc. Sometimes there are strong arcs in both the north and the south but no conjugacy. The OGO satellites also observed $O(^3S)$ radiation at 1304 Å. There is no correlation of $O(^3S)$ and $O(^1D)$ between $\pm 40°$ latitude. These data seem to demolish current theories for explaining the tropical red arcs, as these theories invoke a magnetic field symmetry.

D. ATMOSPHERIC TEMPERATURES

Better temperature data are required for the electrons, ions, and neutral gas, particularly in the *F* region.

Doppler temperature profile observations of 6300 Å at mid and low latitudes show that (a) the temperatures agree with exospheric temperatures inferred from satellite drag, (b) there are rapid changes in temperature, and (c) horizontal winds in the *F* region may be detected through distortions in the spectroscopic line shape.

Some preliminary interpretations of the 6300 Å intensity and line shape temperature data obtained from OGO 4 and OGO 6 were presented (Blamont). At high latitudes the temperature is 1300 to 1500 K at 275 km altitude compared to ~1000 K for the mid-latitudes and equatorial region.

E. DISTURBED CONDITIONS

Magnetic storm associated changes occur in the *F*2 and *D* regions while there are essentially none in the *E* region (Thomas).

The *D* region ionization effects may persist for 10 days with the largest effects occurring 2 to 4 days after the storm. There is no satisfactory model to explain the *D* region effects. Observations for changes in [NO] and $[O_2(^1\Delta_g)]$ should be made. The negative ion chemistry may be modified. Enhancements observed in OH indicate that there is an increase in [H] or $[O_3]$ which could be caused by heating.

There is a marked decrease in *NmF2* with little change in peak altitude. The *F* region storm effects show increases in OH and $O_2(^1\Delta_g)$ emissions but their lifetimes are too short (10^{-2} sec and 60 min, respectively) to allow significant transport; however, O and NO can be transported from the auroral region to the mid-latitudes. The *F* region changes may be due to (a) reduced electron production rate [this may be reflected in a change in $O^+(^2D-^2P)$ at 7319 to 7300 Å], (b) increased electron loss rate [an increase in 6300 Å from an increase in O^+ reactions due to temperature and an N_2 vibrational excitation increase may be observed], (c) *E* field induced ionospheric motion, and (d) thermospheric winds. The wind velocities need to be measured.

The trough (the latitudinal variations of *F*2 electron densities) moves equatorward with an increase in magnetic activity. The field line through the trough seems to coincide with the plasmapause (Thomas). The mid-latitude red arc exhibits similar behavior. It is believed that the enhanced 6300 Å is produced through collisional excitation of O by hot electrons. There is heat conduction from the plasmasphere which heats the *F* region electrons. Satellite observations have shown a $T_e > 4000$ K at 1000 km altitude over a red arc. ISIS observations show a precipitating flux of protons (~ 1 erg cm^{-2} sec^{-1}) in the vicinity of a red arc (Winningham). Correlations between the plasmapause and red arcs were also described (Chappell). Red arcs are observed to occur during the rapid recovery phase of a large magnetic storm. During the storm, ring current protons are injected into the magnetosphere and drift into the plasmasphere. In the 10 to 20 hr following injection, the ring current particles are lost, probably through wave-particle interactions in the plasmasphere. The red arc is located on the field lines through the plasmapause. Ring current particle energy thus is transferred to electrons and into the *F* region.

It may be that the polar region nightglow and *F* region ionization also depend on heat conduction.

More observations to investigate heat conduction are required.

2. Auroras

Auroras occur in a narrow band in the shape of an oval which is offset a few degrees from the magnetic pole. The oval is continuous, widens and shifts equatorward with increased K_p, and also has much structure (Egeland). Although the auroral oval may appear to be continuous on the basis of widely separated observations, there may be several separate processes involved. Better observations are required to resolve this point. The relation of the polar cap aurora to the auroral oval is not known. Most of the so-called polar cap auroras, in fact, may be part of the auroral oval.

The instantaneous shape of the auroral oval depends on UT (secular variation of magnetic storms) and LT. Data plots without regard to UT and LT lead to erroneous results. There is no completely satisfactory UT parameter; for example, mixing data taken under different K_p lead to erroneous conclusions. Mixing data taken with very much spread in LT combine different auroral region phenomena. This is especially true in the vicinity of magnetic midnight where data taken at the same LT may include pre-breakup, breakup, and post-breakup phenomena. There is no utility anymore in statistical auroral plots where the several parameters are not limited to small ranges. Such statistical plots may conceal any physics and do not indicate any correlations or exclusions. Much of the scatter in morphology data would be removed if data were referred to the instantaneous latitude of the auroral oval and a careful choice of LT was exercised.

A. AURORAL CONJUGACY

Aircraft observations of auroral conjugacy have been made in both hemispheres (Davis). In quiet times at lower latitudes the auroras are conjugate and show similar intensity variations and geometry. At disturbed times, the Southern Hemisphere electron auroras at high latitudes are displaced eastward and equatorward of the Northern Hemisphere auroras by 100 to 350 km. These displacements result from distortion of the magnetosphere. The intensity variation is also not similar. At the same time that these electron auroras are not conjugate the equatorward diffuse *H* auroras are still conjugate and no displacement is found.

There are a number of types of periodic behavior in visual auroras. Flickering aurora has a frequency of 10 ± 3 Hz intensity variation and occurs just prior to all or almost all auroral breakups. Pulsating aurora has a period of 0.1 to 10 sec and occurs during post-breakup. Pulsations are observed to show conjugacy. More details on the conjugacy of auroral phenomena are still needed.

B. INFRARED EMISSION

The auroral IR spectra are difficult to obtain because of the infrequency of suitable aurora (strong and stationary) and observing conditions (Vallance Jones). There are detailed spectra to about 2 μ with gaps at 1.3 μ, 1.38 μ, and 1.88 μ because of H_2O absorption bands. Above 2 μ thermal emission in the atmosphere is important and measurement must be made from a high altitude.

Recent observations have shown that the IR bands of $O_2(^1\Delta_g)$ are quite bright. The noon time intensity of the 1.27 μ emission is about 25 to 50 MR. This emission is excited through the photolysis of O_3. In the vicinity of a daytime aurora, the 1.27 μ emission is enhanced by $\sim$5 MR.

The nighttime 1.27 μ emission is $\sim$100 kR; however, in the vicinity of an aurora the emission is increased by $\sim$200 kR. Observations show that the $O_2(^1\Delta_g)$ emission does not correlate with the auroral form. A zenith emission of $\sim$5 MR has been reported on the basis of a horizontal observation from a rocket; however, it has been argued that there was a failure to properly convert the horizontal observation to a zenith value and that in this case it would be 100 kR.

Simultaneous observation of OH at $\sim$1.66 μ and 1.5 μ with 5577 Å and 4278 Å photometry shows that the auroral oval emission in OH is not coordinated with particle precipitation and that the OH emission is considerably more intense than at mid-latitudes (Stair). The OH emission is not enhanced in the polar cap even during a PCA event.

Estimates have generally placed the vibrational temperature of N_2 considerably above the gas kinetic temperature, but it is difficult to predict a temperature because of the uncertain efficiency of $O(^1D)$ quenching. This temperature is important for predicting the energy transfer to CO_2 and its consequent radiation. Rocket observations using an electron gun gave a temperature of 1200 K at 135 km altitude during an aurora which was 20 kR at launch and 2 kR at the end of the observations (Stair).

Significant improvements are being made in IR detectors. It seems reasonable to predict that there is a very important future for investigating the upper atmosphere in the IR.

C. PARTICLE PRECIPITATION

Although there have been many improvements in the picture of particle precipitation, many crucial facets are still unexplained. We still do not understand auroral substorms and their energy sources. Although the solar wind can provide the energy, the method by which it is transferred and stored throughout the magnetosphere is still unknown. What is the triggering mechanism, if any, and what is the acceleration mechanism for a substorm? The required efficiency is only a few percent. What are the sources of the magnetospheric electric fields and what important roles do they play in particle acceleration and precipitation? It still remains to be resolved if the ionosphere affects the magnetospheric particle population and auroral precipitation. Is there an important polar wind?

The precipitation patterns have improved considerably. There are much excellent low energy data which are being analyzed with other observations at specific times and locations. It now appears clear that the electron and proton precipitation regions do not cross over (Eather and Mende; Hartz) as was originally suggested on the basis of crude statistical analyses.

There is no evidence of a diffuse mantle aurora. An upper limit of 2×10^{-3} erg cm^{-2} sec^{-1} for diffuse energetic particle precipitation across the polar cap was obtained.

An oval coordinate system was presented (Eather and Mende). Its use has led to the important result that the energy of the precipitating soft electrons decreases monotonically with increasing latitude on the nightside. Their analysis shows how erroneous results can be obtained from mixing data of different K_p. The average ratio of 6300 to 4278 Å not separated to limited K_p ranges shows a peak whereas the ratio should increase monotonically.

The spatial and temporal structure of auroral particle precipitation was discussed in detail (Evans). It was suggested that the small scale structure observed in visual observations of an aurora is not present in the precipitating electrons above the aurora and local acceleration of precipitating electrons is suggested to account for some of the energetic electron precipitation observations. However, simultaneous observations at 4278 Å and 5577 Å are interpreted to show that the incoming electrons are structured and that there is no need for atmospheric processes to produce such structure.

Correlation of X-ray observation details with visual auroras is difficult. Since visual auroras are produced by low energy electrons and X-rays by high energy electrons, there may be a different acceleration mechanism involved and different locations. Spatial resolution of the observations is also quite different. Arrays of balloons to measure the temporal and spatial behavior of X-ray events have made a significant improvement in the X-ray event morphology description (Trefall).

Pulsations do show correlation. If there are X-ray pulsations then visual pulsations are always seen. The modulation mechanism thus seems to affect both high and low energy electrons. Pulsations are widespread as they are always correlated between all balloons of an array.

D. OBSERVATIONS

Radio wave modification and generation are among the many observables associated with particle precipitation, ionization, currents, and other auroral and magnetospheric phenomena. Interpretation of radio observations requires the use of correlated visual, particle, and ionospheric data. Radio scatter observations are sometimes referred to as radio aurora; however, the term aurora signifies an upper atmosphere emission feature. By no stretch of the imagination is radio aurora a proper description and its use seems to adversely bias the interpretation of the observations. Instead of referring to the scattering region as radio aurora it should be called 'high latitude *E* region irregularities' (Unwin).

Radio scatter data at high latitudes indicate that the scattering is not from overdense regions and that the microstructure is produced by one or more unidentified plasma instability mechanisms. Although the two stream instability type mechanism is popular not all data support it. The microstructure has been measured with rockets and is capable of reflecting VHF and UHF signals (McNamara). The characteristics of the backscatter are very sensitive to frequency and care must be taken in comparing results.

Observations with VHF CW bistatic systems show that the radio scatter is from the PEJ. Back scatter only occurs when the PEJ current exceeds a threshold value (magnetic field change of 100 γ) and the geometry satisfies a rough perpendicularity requirement. The backscatter does not occur from ionization associated with the visual auroras (Lange-Hesse). Backscatter may occur from both the westward PEJ at night and the eastward PEJ in the daytime.

Observations at 30 to 3000 MHz with pulsed radar backscatter show that the altitude of the scatterers is from 90 to 130 km. The off-perpendicular angular dependence seems to vary widely and signals are often received at large angular deviations from perpendicular. Both positive and negative Doppler shifts are observed and indicate that the scatterer velocity varies from a few 100 m sec^{-1} to 2 km sec^{-1} (McNamara). The radar returns may be more sensitive to the spatial extent of inhomogeneities than to their magnitude. The microstructure in the auroral region has been measured with a rocket and found to exist between 96 and 130 km. The minimum size observed is the order of 10 cm.

E. AURORAL MORPHOLOGY

An investigation of auroral morphology requires an understanding of the particle population and precipitation behavior in the magnetosphere. The real current system is unknown but must include auroral electrojets in the ionosphere and Birkeland currents at each end of the electrojet connecting to source regions in the outer magnetosphere. The change in the current system was discussed in detail (Boström). Several observations indicate that changes occur in the state of the magnetosphere during some hours before the substorm onset. It appears that the magnetospheric current system is built up before the substorm onset and at onset is discharged.

F. AURORAL E FIELDS

Observations show highly variable $E_\perp$ fields of the order of 100 mV m^{-1} in the auroral region. Above the *F* region, satellite observations from OV1 10 and OGO 6 show that the structure in the *E* field is 0.2 to 2 km. Because of the structure in $E_\perp$ ion drift will cause field aligned currents resulting in joule heating.

Inside an auroral arc joule heating effects are much less than particle energy deposition and can be ignored; however, outside the auroral arc joule heating dominates the energy budget. The heat goes into the ions and neutrals and it is estimated that $T_{ion} - T_{gas} \simeq 1100$ K. Such a high ion temperature could significantly affect the ion chemistry of the ionosphere.

There is no evidence that field aligned currents directly excite $O_2(^1\Delta_g)$; however, an ion temperature of 1000 K above the neutral gas temperature could produce such an excitation. Emission would be associated with regions near auroral arcs but not in the arcs themselves.

Almost all artificially injected ion clouds show striations aligned along magnetic field lines. The time of development from initiation of a striation is 5 to 50 cm. The two dimension superposition of striations onto a photograph shows a ripple in the intensity of 200 m to several km. The true size distribution could be larger or smaller (Völk). There seems to be a positive correlation between striation growth rate and the magnitude of the ambient *E* field.

Observations clearly show that auroral forms move differently than Ba clouds; the velocity of the latter being considerably smaller. The importance of the ion cloud's size and density for its own striations has not been resolved. There are no observations of the background *F* region ambient irregularities to determine if ion clouds are tracers of *F* region phenomena, i.e., irregularities, plasma drift, and small scale *E* fields.

GLOSSARY

Most of the abbreviations and terms utilized in this book are obvious. Several whose meaning may not be apparent are listed below.

Auroral Absorption (AA). Measure of cosmic noise absorption through the ionosphere in the auroral oval.

Auroral Electrojet (AEJ). Current system in the ionosphere of the auroral region.

Auroral Oval. Locus of auroras in latitude as a function of time, which has an oval shape.

Bow Shock. Collisionless shock set up by the interaction of the solar wind and the earth's magnetosphere.

Closed Magnetic Field Line. Earth's magnetic field line which is continuous through space from one hemisphere to the other.

Conjugate. Used herein for magnetic conjugacy and refers to the opposite ends of the same closed magnetic field line.

Cosmic Noise Absorption (CNA). Surface measurement of the absorption of cosmic noise, usually about 30 MHz, passing through the ionosphere.

Equatorial Electrojet. A current system in the ionosphere, flowing generally along the earth's equator.

f_0F2. Maximum electron density in the F region of the ionosphere.

FP. Fabry-Pérot.

Geomagnetic Micropulsations. Magnetic field fluctuations in the period range of 0.2 sec to 10 min.

Hall Currents. Current flow perpendicular to both the electric and magnetic fields.

$h'F$. Virtual height of the maximum electron density in the F region of the ionosphere.

IN Lat. Invariant latitude, Φ.

IN LT. Invariant local time.

IN Pole. Invariant pole, where $\Phi = 90°$.

Invariant Coordinate System. McIlwain's B, L space magnetic coordinates.

IR. Infrared radiation covering from about 7800 Å to 1000 μ.

K_p. Quasi-logarithmic scale, from 0 to 9, measuring the range of activity of the most active component of the magnetic field within a 3 hr interval.

L. McIlwain's invariant shell parameter, whose units are expressed in R_E at the magnetic equator.

LBH. Lyman-Birge-Hopfield band system.

LT. Local time.

Magnetic Bay. Positive or negative deviations from the normal magnetograms, having a characteristic shape of the shore line of a bay.

Magnetopause. Boundary of the earth's magnetosphere.

B. M. McCormac (ed.), The Radiating Atmosphere, 450–451.

Magnetosheath. Region between the magnetopause and the bow shock.
Magnetosphere. Region inside the magnetopause.
Magnetotail. Region of the magnetosphere extending in the antisolar direction beyond the trapping region.
MLT. Magnetic local time.
M Substorm. Magnetospheric substorm.
Neutral Sheet. Narrow region about 1000 km thick in the middle of the tail of the plasma sheet where the magnetic field falls to a very low value.
NmF2. Density of electrons at the maximum in the *F* region of the ionosphere.
Open Magnetic Field Line. An earth's magnetic field line which is connected to the interplanetary magnetic field.
Pedersen Current. Current flow along electric field which is perpendicular to the magnetic field.
PEJ. Polar electrojet.
Pitch Angle. Angle between the instantaneous velocity vector of a charged particle and the direction of the magnetic field.
Plasmapause. Boundary at about L of 3.5 to 4 inside of which the plasma density is much higher.
Plasma Sheet. Thick slab of hot plasma in the magnetosphere.
Polar Cap. Region inside the auroral oval.
Polar Cap Absorption Event (PCA Event). High energy proton precipitation in the polar cap producing high cosmic noise absorption.
Polar Electrojet (PEJ). See auroral electrojet.
Pre-dawn Enhancement. Enhanced optical emission produced before normal sunrise behavior as a result of charge particles from the sunlit conjugate region.
R_E. Earth radius.
Ring Current. Current of trapped low energy protons at $L = 3$ to 6.
Solar Wind. Electron, proton, α-particle and other charged particle emission from the sun.
Trapping Region. Region of closed field lines wherein charged particles can bounce from one hemisphere to the other and can drift all of the way around the earth.
UT. Universal time.
UV. Ultraviolet radiation and extends from 100 to 3800 Å.
VK. Vegard-Kaplan band system.
VLF. Very low frequency and is from 3 to 30 kHz.
VLF Chorus. Radiation consisting of a multitude of overlapping rising tones, usually in the band of 2 to 4 kHz and it peaks in the morning hours, sounding like those of a distant bird colony.
Whistler. Radio signals in the audio-frequency that 'whistle'.
1N. First negative band system.
1P. First positive band system.
2P. Second positive band system.

INDEX OF SUBJECTS

ASTROPHYSICS AND SPACE SCIENCE LIBRARY

Edited by

J. E. Blamont, R. L. F. Boyd, L. Goldberg, C. de Jager, Z. Kopal, G. H. Ludwig, R. Lüst, B. M. McCormac, H. E. Newell, L. I. Sedov, Z. Švestka, and W. de Graaff

1. C. de Jager (ed.), *The Solar Spectrum. Proceedings of the Symposium held at the University of Utrecht, 26–31 August, 1963.* 1965, XIV + 417 pp.

2. J. Ortner and H. Maseland (eds.), *Introduction to Solar Terrestrial Relations. Proceedings of the Summer School in Space Physics held in Alpbach, Austria, July 15–August 10, 1963 and Organized by the European Preparatory Commission for Space Research*, 1965, IX + 506 pp.

3. C. C. Chang and S. S. Huang (eds.), *Proceedings of the Plasma Space Science Symposium, Held at the Catholic University of America, Washington, D.C., June 11–14, 1963.* 1965, IX + 377 pp.

4. Zdeněk Kopal, *An Introduction to the Study of the Moon.* 1966, XII + 464 pp.

5. Billy M. McCormac (ed.), *Radiation Trapped in the Earth's Magnetic Field. Proceedings of the Advanced Study Institute, Held at the Chr. Michelsen Institute, Bergen, Norway, August 16–September 3, 1965.* 1966, XII + 901 pp.

6. A. B. Underhill, *The Early Type Stars.* 1966, XIII + 282 pp.

7. Jean Kovalevsky, *Introduction to Celestial Mechanics.* 1967, VIII + 427 pp.

8. Zdeněk Kopal and Constantine L. Goudas (eds.), *Measure of the Moon. Proceedings of the Second International Conference on Selenodesy and Lunar Topography held in the University of Manchester, England, May 30–June 4, 1966.* 1967, XVIII + 479 pp.

9. J. G. Emming (ed.), *Electromagnetic Radiation in Space. Proceedings of the Third ESRO Summer School in Space Physics, held in Alpbach, Austria, from 19 July to 13 August, 1965.* 1968, VIII + 307 pp.

10. R. L. Carovillano, John F. McClay, and Henry R. Radoski (eds.), *Physics of the Magnetosphere. Based upon the Proceedings of the Conference held at Boston College, June 19–28, 1967.* 1968 X + 686 pp.

11. Syun-Ichi Akasofu, *Polar and Magnetospheric Substorms.* 1968, XVIII + 280 pp.

12. Peter M. Millman (ed.), *Meteorite Research. Proceedings of a Symposium on Meteorite Research held in Vienna, Austria, 7–13 August, 1968.* 1969, XV + 941 pp.

13. Margherita Hack (ed.), *Mass Loss from Stars. Proceedings of the Second Trieste Colloquium on Astrophysics, 12–17 September, 1968.* 1969, XII + 345 pp.

14. N. D'Angelo (ed.), *Low-Frequency Waves and Irregularities in the Ionosphere. Proceedings of the 2nd ESRIN-ESLAB Symposium, held in Frascati, Italy, 23–27 September,* 1968. 1969, VII + 218 pp.

15. G. A. Partel (ed.), *Space Engineering. Proceedings of the Second International Conference on Space Engineering, held at the Fondazione Giorgio Cini, Isola di San Giorgio, Venice, Italy, May 7–10, 1969.* 1970, XI + 728 pp.

16. S. Fred Singer (ed.), *Manned Laboratories in Space. Second International Orbital Laboratory Symposium.* 1969, XIII + 133 pp.

17. B. M. McCormac (ed.), *Particles and Fields in the Magnetosphere. Symposium Organized by the Summer Advanced Study Institute, held at the University of California, Santa Barbara, Calif. August 4–15, 1969.* 1970, XI + 450 pp.

18. Jean-Claude Pecker, *Experimental Astronomy.* 1970, X + 105 pp.

19. V. Manno and D. E. Page (eds.), *Intercorrelated Satellite Observations related to Solar Events. Proceedings of the Third ESLAB/ESRIN Symposium held in Noordwijk, The Netherlands, September 16–19, 1969.* 1970, XVI + 627 pp.

20. L. Mansinha, D. E. Smylie and A. E. Beck, *Earthquake Displacement Fields and the Rotation of the Earth. A NATO Advanced Study Institute. Conference Organized by the Department of University of Western Ontario, London, Canada, 22 June – 28 June, 1969.* 1970, XI + 308 pp.

21. Jean-Claude Pecker, *Space Observatories.* 1970, XI + 120 pp.

22. L. N. Mavridis (ed.), *Structure and Evolution of the Galaxy. Proceedings of the NATO Advanced Study Institute, held in Athens, September 8–19, 1969.* 1971, VII + 312 pp.

23. A Muller (ed.), *The Magellanic Clouds. A European Southern Observatory presentation: principal prospects, current observational and theoretical approaches, and prospects for future research. Based on the Symposium on the Magellanic Clouds held in Santiago de Chile, March 1969, on the occasion of the Dedication of the European Southern Observatory.* 1971, XII + 189 pp.

In preparation:

25. G. Fiocco (ed.), *Mesopheric Models and Related Experiments. Proceedings of the 4th ESRIN-ESLAB Symposium, held at Frascati, Italy, July 6–10, 1970.*

26. I. Atanasijević, *Selected Exercises in Galactic Astronomy.*

27. C. J. Macris (ed.), *Physics of the Solar Corona. Proceedings of NATO Advanced Study Institute on Physics of the Solar Corona, held at Cavouri-Vouliagmeni, Athens, Greece, 6–17 September 1970.*

SOLE DISTRIBUTORS FOR U.S.A. AND CANADA:

SPRINGER-VERLAG NEW YORK, INC., 175 Fifth Ave., New York, N.Y. 10011

GPSR Compliance

The European Union's (EU) General Product Safety Regulation (GPSR) is a set of rules that requires consumer products to be safe and our obligations to ensure this.

If you have any concerns about our products, you can contact us on ProductSafety@springernature.com

In case Publisher is established outside the EU, the EU authorized representative is:

Springer Nature Customer Service Center GmbH
Europaplatz 3
69115 Heidelberg, Germany

Batch number: 10815974

Printed by Printforce, the Netherlands